My Dear Sara

CIVIL WAR LETTERS 1861-1865

7TH CONNECTICUT INFANTRY REGIMENT VOLUNTEERS

Edited by
Robert Adams

Copyright © 2021 by Robert Adams

First Edition – 2021

All rights reserved.

No part of this publication may be reproduced in any form, electronic or mechanical, without the permission in writing from the author, with the exception of excerpts from Stephen Walkley's book, or any descendant of a soldier that served with the 7th Connecticut Infantry Regiment.

Printed in the United States of America

Cover Design by Robert Adams with help from Book Marketeers
Book interior design by Asya Blue Design.

ISBN 978-1-513-687223 (eBook)
ISBN 978-0-578-65489-8 (Paperback)
ISBN 978-0-578-94652-8 (Hardcover)

1. Civil War, Military, Biography & Autobiography

ABOUT THE COVER
The background handwriting is Edwin Barden's. The standing picture is also of Edwin —- but colorized.

ACKNOWLEDGMENTS
Thank you to Mr. Sean Spoonts for his editorial suggestions and discovery of the 1905 book by the 7th Connecticut veteran and unit historian Stephen Walkley.

Thank you to Abby Weitkamp whose amazing typing skills made the final transcription almost effortless, and with appreciation for the editorial skills of Cheryl Johnson.

TABLE OF CONTENTS

INTRODUCTION 1
Chapter 1 How it Began 5

SECTION 1 – 1861 7
Chapter 2 Tybee Island and the Capture of Fort Pulaski 40

SECTION 2 – 1862 47
Chapter 3 James Island 116
Chapter 4 Secessionville 126
Chapter 5 Pocotaligo South Carolina 192

SECTION 3 – 1863 219
Chapter 6 St. Augustine Florida 284
Chapter 7 The Second Charge on Wagner 311
Chapter 8 Siege of Wagner 323
Chapter 9 Fort Sumter 332
Chapter 10 Fort Sumter 344
Chapter 11 The Dingie Plan 363

SECTION 4 – 1864 395
Chapter 12 At Bermuda Hundred 433
Chapter 13 The Richmond Campaign 456
Chapter 14 Bermuda Hundred on the James River 461
Chapter 15 Richmond Campaign 466
Chapter 16 Ned's Enlistment Ends 472
Chapter 17 On Return to Duty 473
Chapter 18 Expedition to New York 493
Chapter 19 Laurel Hill, Virginia 513

SECTION 5 – 1865 515
Chapter 20 Back to City Point 518
Chapter 21 Capture of Fort Fisher 528
Chapter 22 Fort Anderson Capture 540
Chapter 23 Wilmington 547
Chapter 24 Lincoln Assassination 561
Chapter 25 President Jeff Davis Captured 570
Chapter 26 Home Sweet Home 619

EPILOG – THE FAMILY (Linked to 'Ned' Barden) 624
INDEX 627
ABOUT THE EDITORS 633

INTRODUCTION

This book was over 160 years in the making. It is a Civil War history, adventure, and love story written during a difficult time for our nation. A young country—divided.

Edwin Janes Barden (Ned) enlisted in the Union army at 26 years old, with the locally formed 7th Connecticut Infantry Regiment Volunteers, and was assigned to Company G, from his hometown of Canaan, Connecticut.

The U.S. Army Regiment was formed in New Haven Connecticut on 13 September 1861 and consisted of 1,018 officers and men.

The next day they began movement to Washington, DC where it was brigaded with the Sixth Connecticut and Third New Hampshire and Seventh New Hampshire regiments, under command of Yale Law School graduate BGEN Alfred Terry.

Edwin Barden signed on for a three-year enlistment, as did the bulk of the men. Many were men that had previously signed on for three months. "The three months men" had signed on before the government realized the enormity of the rapidly building rebellion.

He had a love interest at the time. Sarah Maria Jones (written to as 'Sara') was his dominant interest and saving grace during his four years of war. His older brother Jesse Barden enlisted later on 20 December 1863 with Company H, 2nd Connecticut Heavy Artillery Regiment and served until 18 August 1865.

References to Lizzy (Jones) suggest she was Sara's sister.

Libbie and Sara B. are mentioned often. They are Edwin Barden's sisters.

The entire 7th Connecticut Infantry Regiment, after three weeks of constant drill, then moved to Annapolis, Md., arriving October 8th, where drill was continued until October 19th. It then embarked on a steamer for Fortress Monroe, Va., attached to MGEN Horatio Wright's 3rd Brigade with GEN Sherman's Department of The South, and prepared to sail under sealed orders.

Leaving Fortress Monroe October 29th, they encountered a series of heavy gales which wrecked some vessels and scattered the fleet. The regiment arrived off Port Royal, S. C., November 4th.

At the bombardment of Port Royal and the capture of Forts Walker and Beauregard on the 7th of November 1861, the Seventh was the first regiment ashore and into the rebel fortifications, and its colors were the first to float over the soil of South Carolina since her secession. This historic fact was noted by Governor William Buckingham in a congratulatory order which was read before every Connecticut regiment then in the field.

17 September 1861 to 20 July 1865

Officers Killed or Mortally Wounded: 11
Officers Died of Disease or Accident: 4
Enlisted Killed or Mortally Wounded: 157
Enlisted Died of Disease or Accident: 192

The casualty summary above was recorded by Captain William H. Pierpont, Co. D, 7th C.V.)

The by-company listings below is copied from the 1905 book by the 7th Connecticut veteran and unit historian Stephen Walkley:

CASUALTIES BY COMPANIES.

	Killed.	Died of Wounds or in Prison.	Wounded.	Captured.	Total Casualties.	Total Mustered.
Field and Staff		1		2	3	10
Non-Commissioned Staff					1	6
Company A	22	14	54	47	137	233
" B	15	17	45	38	115	196
" C	8	10	37	33	88	182
" D	13	14	48	27	102	200
" E	10	8	32	9	59	163
" F	7	9	33	14	63	220
" G	6	3	48	13	70	200
" H	7	8	58	23	96	172
" I	10	10	34	30	84	204
" K	9	7	45	21	82	186
Unassigned recruits						25
Totals	107	100	436	257	900	1997

CASUALTIES BY BATTLES.

	Killed.	Died of Wounds or in Prison.	Wounded.	Captured.	Total Casualties.
Fort Pulaski, April 10, 11, 1862			2		2
James Island, June 16, 1862	14	7	64	5	90
Pocotaligo, October 22, 1862	1	4	30		35
Morris Island, July 10, 1863			5		5
Assault on Wagner, July 11, 1863	16	9	43	54	122
Siege of Wagner, July 12 to October 16, 1863	3		11	2	16
Olustee, February 24, 1864	8	13	43	21	85
Chester Station, Virginia, May 10, 1864		1	9	1	11
Drewry's Bluff, May 16, 17, 1864	27	21	77	64	189
Bermuda Hundred, June 2, 1864	5	22	29	58	114
Bermuda Hundred, June 17, 1864	7	4	16	25	52
Deep Bottom, August 14, 15, 1864	1	1	11	2	15
Deep Run, August 16, 18, 1864	9	5	26	5	45
Chapin's Farm, September 29, 1864			3		3
Near Richmond, October 1, 1864	1	1	3	7	12
Newmarket Road, October 7, 1864	1	2	8	2	13
Darbytown Road, October 13, 1864	4		7		11
Charles City Road, October 27, 1864	1		1		2
Fort Fisher, January 15, 19, 1865	1		9		10
Skirmishes and Picket Duty, Nov. 8, 1861 to June 4 1865	8	10	39	11	68
Totals	107	100	436	257	900

It will be noted that the above list of casualties is largely in excess of that given in the report published by the Adjutant General's Department of Connecticut. This is accounted for by the fact that the Adjutant General's report seems to have been compiled from the official reports of battles given immediately after they occurred and before all the facts were known, while the above are compiled from the margins of the roll printed on pages i to lxvii. For instance, a man might be reported as captured and afterward found to have received a wound of which he afterward died. In the Adjutant General's Office report he would be reported as captured. In this report as captured, wounded and died of wounds. The Adjutant General's report tells how many men suffered casualties. This tells how many casualties they suffered.

The letters that follow were discovered by my mother Claudia Adams-Estes in a suitcase following the death of her father VADM William R. Smedberg, III in 1994. They are written by Corporal Edwin Janes Barden, her great-grandfather, to his girlfriend, and later his wife, Sarah Jones (he always wrote to her as "Sara") while assigned as a Union clerk in General Ulysses Grant's headquarters.

My mother began a years-long effort to transcribe each letter. It became a passion to pass to her family and ensure transcription accuracy.

Each letter was written in pencil or pen and in a beautiful script, and Sara kept and numbered each letter received. Mom preserved his words exactly as written with misspellings and punctuations as he wrote them.

You will note that spelling and punctuation improve over time. Before her death, Mom proudly gave me a 433-page notebook with the typed pages of her transcription. I immediately saw the need to obtain these documents in electronic form and we made that happen.

It has taken me months to create this book from those electronic files many years later, as the saved files were incomplete and only covering the first three years. My blessing was the original 433-page hard copy original – and amazingly – the four huge notebooks containing the original letters that she had carefully placed in clear sheets in chronologic order. I found all those original letters at the time of her death October 2016.

Another amazing find was a <u>1905 book by Stephen Walkley</u> the 7th Connecticut Infantry Regiment historian. I have quoted his brilliant summaries at the beginning of most chapters to orient the reader to wartime activities surrounding the comments in Ned's letters.

As the editor and second transcriber I have added e-links and an index to names, places, weapons, ships, and strange words that have not well survived history.

I hope you will enjoy this Civil War journal and love story.

Editors Note: There are underlined words, names of people, and phrases in this print book that correspond to e-links provided in the e-version of this book. They are all also listed in this book's index. For readers that wish to learn more about these historically significant underlined items, you can go to the book's website where the e-links are active in the website's Index. www.dearsaraletters.com

CHAPTER 1

HOW IT BEGAN

From his 1905 book by Stephen Walkley the 7th Regiment Connecticut Volunteers:

Disastrous as was the rout of the Union forces at Bull Run, on Sunday, July 21, 1861, it was doubtless worth more to the Union cause than would have been a victory. It taught the North how great was the task before it; but more than that, it sent a sting of shame throughout the country which made thousands eager to enlist and wipe out the disgrace.

Among those who keenly felt this sting, were Colonel Alfred H. Terry of the Second Connecticut Regiment and Joseph R. Hawley, captain of rifle company A of the First.

It was not shame for themselves; they had brought off their commands in good order and in official dispatches were mentioned with honor. Stedman of the New York World wrote of the three Connecticut regiments brigaded under General Tyler: "The Connecticut brigade was the last to leave the field of Bull Run, and by hard fighting had to defend itself and protect our scattered thousands for several miles of the retreat." Colonel Terry and Captain Hawley were ardent patriots, and during their short term of service had become sincere friends.

Before parting, they pledged themselves to each other to go home and begin recruiting a regiment for three years or the war. A little more than two weeks later, their three months having expired, they were mustered out and went home; Colonel Terry to New Haven, and Captain Hawley to Hartford.

Governor Buckingham, on August 15th, issued general orders directing that volunteers be accepted for the Sixth, Seventh, Eighth and Ninth, three years' regiments. Colonels Chatfield and Terry were appointed colonels respectively of the Sixth and Seventh; and those regiments were ordered to rendezvous at New Haven. Captain Hawley at once commenced recruiting a company, with the hope of joining one of them.

While the right flank company was forming, a similar spirit was moving throughout the Commonwealth. The announcement that Terry was to be colonel of the Seventh brought to his standard squads and companies from all over Connecticut.

When camp was established at Oyster Point (now City Point), New Haven, the regiment rapidly filled; it represented every county, and one hundred and thirty-four out of the one hundred and sixty-eight towns of the state.

Following this history through the war, and knowing the survivors, one cannot but feel that they were fair samples of what Connecticut homes, schools and churches have done to produce good citizens. Not all of native stock, all had caught the true American idea. Of all religions, a few claiming to be of no religion, there were many who would have felt it no honor to be known as "a Bible class company;" yet all honored right, truth and goodness, and were ever ready to stand boldly in their defense.

Whether because sacrifice ennobles men, or whether only the noble will offer themselves for sacrifice, I am sure that anyone who knew the inner and outer life of those men would acknowledge that they were "Nature's Noblemen."

ALFRED HOWE TERRY.
First Colonel Seventh Connecticut Volunteers.
Afterward Major General, U. S. A.

JOSEPH ROSWELL HAWLEY.
Second Colonel Seventh Connecticut Volunteers.
Afterward Brevet Major General, U. S. V.
Twenty-four Years in U. S. Senate.

LETTERS FROM EDWIN JANES BARDEN TO HIS FIANCE
AND LATER HIS WIFE, SARAH MARIA JONES,
WRITTEN DURING THE CIVIL WAR

SECTION 1 – 1861

The following note was on the wrapping paper (in Sarah's handwriting) around one batch of letters and states "Letters rec'd from E.J.B. while in the army. 1863"

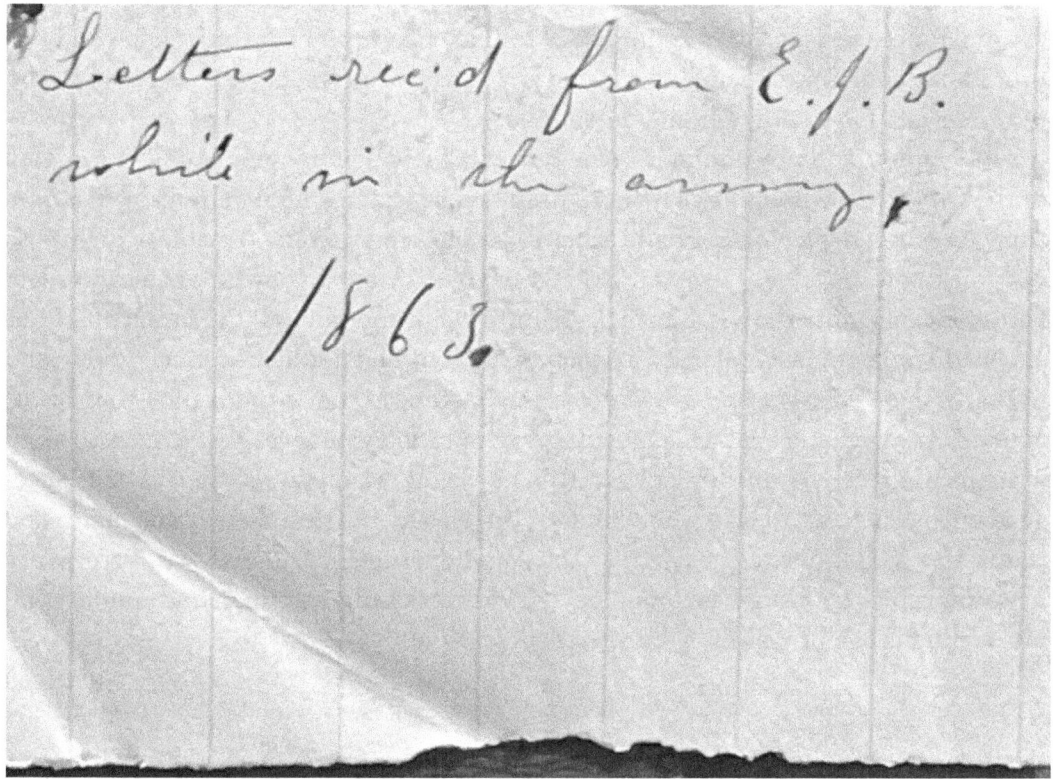

(Note: Sara numbered and dated each of Ned's letters as she received them. The first 73 letters are written in pencil before ink is used, and most are very well preserved)

No. 1 Rec'd Sept 12th 1861

New Haven Sept 11th 1861

My Darling Sara,
    Words cannot express the misery I experienced yesterday as the hours passed away and it seemed as though I must part from you perhaps forever without the opportunity of again seeing you and bidding you good bye. Over and over again I cursed my stupidity in not writing to you last week, but I thought each day I should surely come home the next and see you before you could get the letter, and when we finely found that we wer to stay over the Sabbath it was to late (to) write. Thank God

I was permitted to see you once more although it was but for a moment. Thank you dearest for the ring it shall never leave my finger untill you require it. You cannot imagine how I prize it for long years I have waited for it now it has come and under circumstances that make it doubly dear to me, cheer up my darling "hope on hope ever" keep your courage up like a brave girl as you are remember that I look to you to remember me to pray for and defend me.

    I cannot stop to write more now but will write again immediately. I came up to the City to get some things which I need before leaving. and am writing this on a show case in a drug store. I had rather a serious time coming down last night was very sick before we got here and obliged to take a carrage down to the camp. am better this morning though I have been quite weak and faint all the forenoon, it has stormed tremendiously all the forenoon. Our tent has not leaked much and would not have done so at all had it been properly fastened. many thanks for your letter which I received and read this morning, write to me whenever you can and I shall write as often as possible, must close this now as it is time to return to camp. be of good cheer darling. for He who "holdeth the waters in the hollow of his hand, and counteth the Nations of the earth as the small dust in the balance" will surely care for and protect me wherever I am, may His best blessing rest upon and comfort you my darling is the prayer of "your own dear

Ned"

(Editor's note: Following is a note written at the bottom of Ned's letter in Sara's handwriting.):

Isn"t it too bad. I can't sit by you? for I'm going home this noon. I looked for Ned last night (untill) very late, because James Deane told father George intended to come if he could get a pass. & I supposed Ned would come if he did. (Now) must I give up seeing him. It will kill me. I have two letters for you to read & his answer for you to look at. this you can read now if you like. Oh dear! what shall I do.

No. 2 missing

No. 3 Rec'd Sept 16th 1861
New Haven (written in pencil)

Camp English
<u>New Haven Sept 15th 1861</u>

My darling Sara

This is Sunday morn but I hardly think it will be very wicked under the circumstances to spend a portion of it in writing letters. We have been to breakfast and Mr. Dexter and myself have taken our portfolios and found a good nice place in the school away a little distance from the tents and have seated ourselves to write. Our camp ground is beautifully situated about a mile and a half from the city in a large meadow which slopes gradualy down to the harbour on one side while the road runs aland on the other and a nice road it is to. for a long distance it runs perfectly straight is very level and shaded on each side by poplar trees. Our company are nearly at the extreme end as you approach from the city. In the furtherest corner near the road a daguerreotypist has pitched his tent and it is a leaning my back against this while sitting on the ground that I am writing this. It is the most convenient place I have found since I have been here. We have to write in all positions sometimes lying flat on the ground, or on the floor of our tent any way to hold the paper so that we can get the pencil on it. I have got me a new portfolio smaller than the one I gave you that was so large that I could not carry it. The key to yours I lost when I went from the Store into Bank. I tried at first to find another but did not succeed and finding that I had but little use for it any way did not look much. I think you may be able to find one somewhere without much trouble, and it will be more convenient. I sent you a half written letter yeaterday and ment to have written another the same day. but after drill Landon, Olin, Hawthorne, Moore, two Brintons, and myself got a pass and went up to the city. we did not return until about six and then wer so tired that we did not feel much like writing. We got our army caps first at night and this afternoon we are to go to church in full dress. I had some photographs taken yesterday but cannot get them untill some time tomorrow. I had a small size full length taken. I have your two last letters in my pocket but the boys are on both sides of me and I cannot get them out to read and answer in detail as I wish to. So I will write just what I can think of and whatever I miss you must ask me about again. One thing. I get my letters just as it happens. sometimes at night sometimes not untill morning but usually the same day they are mailed. I was called away at the close of the last sentence to go to church we all expected to go this afternoon but the Capt kindly gave us leave such of us who wished to go this forenoon also, all of our Salisbury boys went except Morris and Dexter Reid and Sweet. One of our own Canaan boys is very sick. and Dexter staid partly to care for him. When we returned (which was about a half an hour ago) we found the camp all in commotion. the report is that a telegraphic dispatch has been received this forenoon ordering the immediate movement of the two regiments. we expect to go tomorrow night. It has thrown us all into confusion as we are no ways ready yet. the first effect it has had has been to set us all to writing as if for Dear life. Moore was buisy with the rest of us when the Colonel sent to our Capt to send him a good penman and he has sent Moore. The officers are all a flying around as buisy as can be. going to church this afternoon has been summeraly disposed of. telegraphic dispatches I hear are being sent to soldiers absent in the country and altogether things look as though we wer to move very soon. We have not yet received our underclothing but as we Salisbury and Canaan boys are all well supplied it makes no difference in that respect with us. We have not received our arms knapsacks or any accouterments yet. Neither have we been paid off yet. all these things are to be done before we can go I have just been called off from my writing to receive some underclothing. won't get a chance to write a dozen consecutive lines any way. have a dozen letters that I ought to write before I go, but they will have to go unwritten. I should like dearly to have a needl book of your own making, but I am afraid I shall not be able to get it. however go on and make it just as though you wer agoing to give it to me. I will have it sent to me if possible if not keep it and look upon it as mine. I am worried about my photographs we wer to have them tomorrow noon but even if we succeeded in getting them ourselves I fear we shall not be able to send them away. Oh dear I can't keep my thoughts together anyway. we have just been called together again for another dose of clothing. Friday night we wer serenaded by about fifty or sixty young ladies and gentlemen. the singing was not anything extra but the good feeling that promoted it and of which it was the exhibition go far to cheer our hearts and strengthen our hands. the singers wer in two large and one small Omnibus Capt Hitchcock thanked them in a neat little speech which was replied to by Mr. North who seemed to be the leading spirit among them. he made a short but excellent speech. he said he heard one lady say that she wished she could kiss every soldier. the opinion of the boys was that she would get blistered lips before she got through. my own private opinion is that she would get awful sick of her bargain before she got very far around. After giving our company a serenade they passed on back through the regiment singing as they went. I have said before I think that we are encamped clear at the lower end of the ground. but it is generaly conceeded that we have the most company of any. whether it is because our officers are all young and unmarried and the handsomest men on the ground or whether it is because we have the likeliest looking body of men on the ground I cannot tell prehaps both have some thing to do with it. last night we had another serenade by a lot of little girls they sang very sweetly and were loudly cheered by the men this afternoon we shall probably be overrun with ladies and gentlemen they commence coming now. We had a fine flag presented us a short time ago by Mr. Townsend our patron. again on Friday we had a famous mess of splendid green corn enough for the whole company given us many of our men have friends in the city and they visit them in camp bringing them cakes & pies all sorts of "goodies" I should like dearly to see you and some friends here. but it will be impossible now. if we realy are to go tomorrow we shall not go either to Staten Island or Hempsted I think but go directly on towards Washington. if we do it will probably be a long long time before I can see you again. all hope of coming home must now be given up it is sad sad indeed to be parted thus but we must submit. I heard an excellent sermon this forenoon we all (about fourteen of us) attended Dr. Eustice" church the text was in the sixth chap. of Ephesans I do not recollect the verse and have not my Bible with me out under the shade but it is the one about girding on the whole armour and standing fast we went in and sat together filling three seats in the gallery the minister alluded to us and prayed for us very earnestly. I am very glad I went now for it may be some time before I can go again. you must act your own pleasure my dear with regard to making our engagement known to your father. I do not know what is best in relation to it prehaps it would be best to wait a little longer and prehaps not. I should be perfectly willing that your father alone should be made acquainted with our relation but am not quite so sure about letting Harriett know it It would suit her most to well. George had a request in one of his letters that he would hand the inclosed "something" to me. it did not state what it was and I conclude you must have forgotten to put it in. Please try again. maybe we'll have better luck. I will inclose the desired lock of hair if I can find George to cut it off. I think I will direct my letters to you at Lime Rock so that they can reach you as soon as possible. I shall not be able to write to Lizzi just now but will do so as soon as I can. I must stop writing to you now my dear. perhaps I will add more before I mail it if not good by.

Monday morn
Dear Sara. I have heard nothing more since I stopt writing last night. Several wagon loads of knapsacks and other things came into camp last night. our Capt is very

particular that we should all be within hearing distance during the day. our sick boy is better this morning now my dear I must bid you good by for this time. I think I shall get a letter from you today. direct as heretofore untill I tell you otherwise. good by darling accept many kisses from your own

Ned

No. 4 missing

No. 5 Rec'd Sept 24th 1861

Camp English
(partially written in ink) <u>New Haven Sept 18th 1861</u>

My darling Sara

I received you letter this morning and will set about answering it immediately knowing that at best it will be difficult to finish it today.

I meant to have written yesterday but our company was on guard and I had no time to write

I suppose of course that you don't understand how the guard business is managed and as it may interest you I will tell how it is done in camp here. One company does guard duty every twenty four hours. the company is divided into what is called three releifs. the first releif commences duty at half past six in the evening and stand guard two hours then the second relief take their place and the first turn in and go to sleep when the second have been on two hours the third releif is called and goes on duty and the second turn in and go to sleep. after the third releif has been on duty two hours the first releif is called and commences duty again and the third releif turn in to sleep. So you see we get two hours on guard and four hours off. (Starts writing in pencil) this continues through the twenty four hours when another company takes the place of the one on guard. this is the regular routine, but the other night when we went on guard it was necessary to greatly increase the guard so the whole company was obliged to turn out and stand untill ------ when we wer releived by another company for two hours and then we commenced our regular routine, again we wer obliged to break off again during the day and it threw us into confusion so that finely it happened that I got about eleven hours off guard duty instead of eight which was my share. When it rained the hardest I was paceing back and forth with a gun on my shoulder. I would not have cared any thing about it though if I could have seen any good coming from it. But I felt very well satisfied when I looked up and saw the Sixth regiment preparing to leave. during all that heavy rain in the afternoon the men were taking down and packing up their tents and preparing to march. they finely got away about five o"clock. they marched up to the City and wer put aboard a Steamboat and in the evening we saw them going out of the harbor and could hear them cheer as they passed our camp. the cheers wer loudly responded to by our men and soon they wer out of sight. bound for Washington as I learn from the morning paper. whether we expect to follow them today prehaps tomorrow at the furtherest. so my darling it will be impossible to see you again. prehaps ever this side of the grave but I have but little fear but what I shall see you again and before long to. I hav just seen a squad of men sent to the city to pick up and send into camp men absent over their time. everything looks like moveing and I think we shall get away today.

(The letter continues on the next page-new day, but same letter)

Camp of the 7th Reg. near Washington D.C.
Sept 21st, 1861

My Dear Sara

Well here I am at last in the Capital of our country a place which one month ago I never expected to see. Just as I was finishing the last sentence I was called away to get some of my equipments and all the rest of the day I was buisy getting ready to move at length about half past two we struck our tents and wer soon after formed in line of march. we marched from the camp ground to the steamboat landing about three miles and it was a tedious march it was, we had on our coats and overcoats with our knapsacks guns and all the rest of our equipments on us and about us it was very warm. we marched the whole distance without once stopping to rest, the mud in the streets of the city was ankle deep and when we got aboard the boat I was wet through with sweat, the streets the entire distance were lined with men women and children shouting and waving flags and handkercheifs many of our men are from N. Haven and it was one continual farewell the entire distance, on board the boat we had to manage to rest the best way we could some of the men wer in the berths some in chairs and some stretched out on the floor. I wandered around the boat awhile in search of a place and at last found a corner unoccupied on the upper deck near the end of the boat. unstraping my knapsack and laying it down for a pillow then laying my gun down so as to lie partialy on it and gathering my other equipments in my hands as well as I could I lay down and tried to sleep but it was impossible. my position was just where the crowd was passing incessantly and the singing whistleing swearing and shouting continued. a little doze was the best I could get. At about twelve o"clock we reached Jersey City where we wer transferred to the cars. when I got up from my poor attempt to sleep I found that I was sea sick. we had to form in the lower deck the air was very close and I thought I should surely have to give out. poor Olin was paying tribute to old ocean over the side of the boat at a great rate. I had no inclination to smile at him for I feared each moment that I should have to join him. luckily we soon got out in to the open air. and I felt better. I should then have got along very well but about half of the Sixty in our car commenced smokeing some miserable penny Jersey cigars and another quarter wer smoking pipes you can imagine what a nice atmosphere it was for a sea sick man, in order to help it along and make me enjoy it one ugly truth which I harbor in my mouth commenced acheing furiously. as the day advanced I felt better. we reached Philadelphia about ten o"clock. here we had a fine dinner furnished gratuitously by the Union loveing people of the city. we wer detained there several hours and finely started about three o"clock. while waiting we wer each supplied with ten rounds of cartridges to be used in case of emergency as we wer to go through Baltimore and wer not quite sure about our reception. of our ride here I cannot give any description hardley for the reason that I had no means of knowing the names of half the places we passed through. a look at one of the little war maps that are so thick about the country will give you a better idea of our route than I can. I will mention the names of the places we passed through so far as I know them. they are Jersey City New Brunswick Trenton Boardentown Philadelphia Newcastle Willmington Haver De Gras and Baltimore. those I believe are the principle places. I got but little sleep aboard the cars but would once in a while catch a short nap. we reached Baltimore at about two o"clock in the morning of course every thing was still. it was a beautifull moonlight as bright as day just as we wer comeing to a stop we passed a fine band of music which greeted us with the "Star Spangled Banner" and displayed a beautifull flag. but few persons were in the streets all was quiet but as we passed along (we had to march two miles through the city) many heads wer visable at the windows and frequently small flags and H"chfs wer waved at us, which we responded to by nods and waveing our hats. occasionaly some one would enquire where we wer from how many of us there wer & once or twice we wer cheered a little. at Baltimore we wer put into miserable baggage cars with holes cut into the sides for windows we had to ride forty miles in them (the distance from Baltimore to this place) we had a train of thirty five cars and onley one engine we did not make very good time and as we wer an extra train we had to wait for other trains altogether it was an uncomfortable ride and I was glad enough when it was ended. At different places between Baltimore and here there are small companies of men posted to gaurd the road at one place where the train waited to meet another the boys went out to wash in a small brook and some of them picked up severel balls near the place the road had been torn up and there had been a skirmish between the union & rebels troops. this was some time ago in the earlier part of our troubles, at present all is quiet along the road. About eight we reached Washington. here we wer quartered in a building put up for the purpose untill about noon when we wer formed in line marched to our present quarters a distance of three miles. It was intensely hot the sun was shining brightly and as we marched along clouds of dust wer raised altogether it was a very sever trial to new beginners and some old soldiers who wer along said that if we could stand it we could stand anything that we would be called upon for. a few of our men in the whole regiment gave out and wer brought up in the army wagons. some of the men hired little boys to bring up their knapsacks but as it was against the orders to do so I concluded to carry my whole "rigging" as long as I could and then go down all together. however I stood it very well and today I am as well as any man of the ground. we are all very tired and wer excused from drill this forenoon. this afternoon we have been drilling with our rifles just at night we all went down to a brook that empties into the Potomac about two miles from here to bathe the water was to muddy for my purpose and so I did not bathe. we wer to have had a regimental drill at sun down tonight but it rained to hard for it. so we escaped that and I am right glad of it for I am to tired to stand up. our camp is not near as pleasant here as it was at N Haven. it is up on a range of hills but as they extend a long distance in every direction and are covered with trees we have no view at all. the camp of the <u>6th Reg Lt Cav</u> is close beside ours. all around us in every direction are the camps of different regiments extending for miles around. the drum and fife are playing incessantly and in every direction companies and regiments are drilling. We can hear in the distance the heavy roar of artillery. but it is only for practice that they are fireing altogether it begins to look decidedly war like. as for us we shall not be called on for any service unless in case of great emergency for six weeks, or more as it will take that time for us to drill and get accustomed to the use of our arms. Now my dear I must close this poor disjointed letter I have done the best I could under the circumstances. I have been severel different times writing it scratching a line whenever I could get a chance. Your last kind letter I have not replied to I know but will try to as soon as I can. but if I am put through as fast hereafter as I have been the last week you will have to take such as I can scratch off when and how I can. One thing I wish to say. I thank you a thousand times for your engagement ring and for giveing yourself to me believe me dearest it makes me very happy and if I have failed to express my gratitude as I fear I have I beg you to forgive me I have been obliged to write when I was not in the mood for it because it was necessary as it was then or not at all. and in writing when my mind was not on it I have not said what I wished nor what I have felt. Believe me my darling I am yours and yours only untill you decree it otherwise. which I pray may never be. We have been seperated suddenly and under very distressing circumstances but cheer up my dear one it may not be as long as we fear and we will hope and pray that there are days of happiness yet in store for us when our faith and love shall have been sufficently tried. again I ask you to keep up good courage be as cheerfull as possible and try not to worry. from my slight acquaintance with Mrs. Barnum I think she is a very fine women and would be true to whatever trust was committed to her and a safe adviser as for Martha you know I could never have much faith in her although I like her well enough and she has always seemed to like me. you may remember me to her if you like. most of my other intimate friends you know espesialy all who are acquainted with our matters. I think I will direct my letters hereafter to you through Julia. I don't like either of our postmasters on Leakville on Lime Rock although I don't suppose either of them would be hardley mean enough to open a letter comeing from me to you under the circumstances, I have sent you a couple of small photographs I don't like them and am sorry I did not have any good ambrotype instead but I really was so hurried up and so little time that I hardly knew what I did want. Prehaps I shall have an opportunity to have something of the kind taken by and by. but do not much expect it. Please tell Sara B. that I am a going to write to her just as soon as I can. Pitt and Mrs. Moore both have asked me if I was writing to any one they knew and when I told them yes they requested me to give thier love to you. of course I did not tell them who I was writing to but Pitt knew well enough I fear I shall have to trouble you a little further the ring you gave me is rather large for my finger and I am afraid I shall lose it if you could go over to the Falls and find me a ring a little smaller than this one so that I could wear it before? and then keep it on I would be greatly obliged to you I think Salmson the jeweller has some plain cheap ones that would be just right I suppose you tried this one on your finger of course and so could judge about the size of another one. I would not trouble you again but I do not know when I shall be able to go where I can get one myself and am greatly afraid I shall lose this one. you can send it to me in a letter wrap a small piece of paper neatly around it putting the ends of the paper inside of the ring so as to fill it up as much as possible then stick it in one corner of the letter so that it will not move about. George is agoing to have a box of things sent to him and if you have the needle book done you can put it in some of his things and send it to me. I am in hopes you will be able to send me the ring by mail sooner or I would have you put that in also prehaps it will be more convenient for you to do so if it is of course you will do it, I am finishing up this letter on Sunday afternoon haveing written some of it in the morning. we wer obliged to have our arms knapsacks and all our accouterments inspected this morning by the major of the regiment it took up over two hours of our time we shall not probably be interupted much more today though there is not much telling for they order us around when we least expect it we shall not have any service today as I believe our chaplin is not on the ground one of our Canaan boys is quite sick today he is a tough hearty robust fellow and about the last one that I thought would be sick I hope he will be better tomorrow For my part I don't want to die here I had much rather fall on the battlefield I saw a man being carried to his camp home today It was at a little distance off but one of my comrads told me that they wer agoing to bury a man I should think there wer eight or ten men marching ahead of a wagon in which wer to men seated while the dead lay on the bottom of the wagon. it looked so loneley and mournful amidst the many thousands within sound of a voice to see onley a dozen men carrying a fellow man to his last resting place that I could not help shuddering but it makes little difference where or how the body is disposed of if the soul is onley at rest with its Creator. but after all one can hardley overcome their repugnance to being thrust so carelessly aside.

    now my darling I must bid you good by as the man is just going to the city with letters and I am anxious to have you get this soon as possible. Please excuse all mistakes as I am not agoing to have time to look it over and correct it please write as soon as possible and tell me all the news

direct your next letter as follows
    Edwin J. Barden Washington D.C.
    care of <u>Capt. E.S. Hitchcock</u> 7th Reg C V
    Pardon the haste with which I am closing this good by darling. Yours ever
                                                      E.J.Barden

No. 6 missing

No. 7 Rec'd Oct 4th 1861
    Washington

<u>Meridian Hill</u>
Washington D.C. Sept 27th/61

My Darling Sara

    I have just received and read your good long letter or rather letters and I can hardly express my gratification I have been here a week today and have received only one letter and that a short one from Frank. I knew it was no fault of yours for I knew you would have written if you only knew where to direct. but I had other letters owing to me before I left N. Haven and had they been sent there they would have been forwarded and reached me in a day or two after we arrived. I have sent one a day since being here and if half are answered I shall have some letters by and by. I am very gratefull to you dearest for all you have written in that good long letter received today. why do ask me if I like long letters as if I could weary of hearing from or of you. no my dear if you onley knew how much happiness a letter affords us you would never think of wearying us with thier length. I wish you could see our company when the Captain comes out in the street with his hands full of letters and papers and sings out "<u>c</u>ompany G this way" the eager anxious faces which quickly surround him the avidity with which those who have letters seize them and hasten to break them open and the looks of disappointment and the expressions of chagrin and vexation with which the unfortunate ones turn away would be quite a study for a painter. So my dear don't never fear of writing all you can find time to. I don't know hardley how to go to work to answer all your letter and as I have to write a little at a time when I can it may be that some of it may pop unanswered. You are right in saying that with all the sacrifices we may make in leaving pleasant homes and jepradizeing our lives in the army we do not suffer so much as those loved ones do whom we leave behind. It is one continual buz of excitement in camp from early dawn untill late in the evening. and there is no time to spend in thinking of and mourning for departed or desertd friends I do not mean to say that we do not think of and regret the situation from loved ones who mourn our absence, but that we do not have the opportunity to indulge in long and profitless (it would be so to us) greif. It is onley when sick and obliged to be alone in the tent whilst our comrades are drilling that there is time to think of and long for the pleasures of home. I think there is more than a probability that we shall have an opportunity to get a meaner view of "Dixie" than we have at present, There are large bodies of troops being moved over the river and every preparation seems to be makeing for a "big" battle. if it is delayed a week or two longer I think it is more than likely we shall be invited over to look on at least. Our company officers wer all among the three months men and at the Bull Run battle and they are on tip toe to try again. We did not have much of a fast day, we did our usual amount of drilling and between drills has a short service. our chaplain has arrived he is rather a preposessing looking man and I am favorablely impressed with him but have not yet made his acquaintence. We have got settled down again into one regular routine of camp life and as I believe I have not written what it is here and as it may be interesting I will describe it.

The drums beat the revielle at a quater past six at which time the roll is called and all who are able and are not on guard must be present and answer to thier names at a quater to seven the morning drill commences and lasts half an hour. breakfast at half past seven guard mounting for the day commences at half past eight. directly after breakfast those who are sick are reported to the Surgeon at nine we commence and drill untill eleven dinner at twelve at three drill again untill five, dress parade at sunset, supper immediately afterwards   at nine the drums beat the colors and the roll is again called. at half past nine all lights must be put out and everything still for the night. the guard duty is done more systematically here than it was in N Haven. four or five men or more if necessary are detailed each morning from every company in the regiment for the duty, they are on guard duty for the next twenty four hours. In the afternoon the ten companies are drilled altogether in a regiment drill Sundays we do not drill but there is an inspection of arms which occupies an hour or so and that dress parade at sunset. There has a large number of troops left the camp around us within the last two days for the other side of the Potomac and large bodies of them have passed us from other camps. And now by dear I will try to answer some of your letter. I am not sorry that you have made confidants of Mr & Mrs Barnum, but really I don't care to have it genraly known. I dislike to have everybody know so much more about my affairs than I do myself. If we wer to be married soon I should not mind it so much. you did perfectly right in informing your father though we wer exposed to greater publicity by so doing, he has a right to know of it. But you will see that I could not well say anything to him about the matter for you would never place me in position to do so. I am heartily glad the matter is settled at last I feel better about it I feel as though I could make some calculation for the future if my life is spared. Still you have taken a step which you may one day regret do you consider that it is quite probable that three years may pass before I shall even have an opportunity of seeing you and at the end of that time if I am released alive and well I shall be right where I am now penniless and without any buisiness. I do not mention the last two things because I fear them for I have been well acquainted with the situation heretofore but because it will take time to remidy both. And in the length of time which must necessarily intervene before I shall be able to offer you a comfortable home you may and probably will have opportunities of doing a great deal better than to wait for me. Do not think my dear that I write this just for the sake of haveing you again and again disclaim all care for any thing of the kind and avow your willingness to wait for me at all hazards. I know very well that ladies never barter thier pure love for wealth or situations, but I wish at least to show you that I think of these things and appreciate your self sacrifice in so willingly engageing yourself to me under so unfavorable circumstances when you have again and again refused to do so when I at least thought the prospects brighter. Do not think my dear because you have so willingly devoted yourself to me that I wish to deprive you of all other gentlemens society or friendship. I can see nothing improper in your giveing the gentlemen refered to your photograph if you had promised him one neither do I wish you to reserve copies of your letters for my perusal I can never bring myself to think so meanly of you as to wish to know everything you may be called to do and say in my absence. We do not expect that every gentleman of your acquaintance is agoing to know positively of our engagement and of cours you will receive some attentions from them you might not if they did Your own good sense will always guide you under these circumstances and my darling never fear that I am afraid to trust you. I do not wish you to become a nun on my account untill I return, I wish you to enjoy yourself to the best of your ability and to the extent of your opportunities. I think this course both reasonable and right, With regard to the slippers you spoke about I took the precaution to provide myself with a good coarse pair before leaveing but shall be happy to find a pair worked for me on my return. Do not worry about the ring. I shall wear it any way you may send me the other one when you can get it and I will wear it to keep this on. I have been a long time writing this letter and I know that you are anxiously waiting for a letter so I will close this and send it along though I have not written half I intended to. write as often as you can and send them along and when directed to the care of Capt Hitchcock they will follow us.

    Good by darling much love from

Ned

No. 8 missing

No. 9 Rec'd Oct. 1861

<u>Annapolis Oct 13th 1861</u>

My Dear Sara

    Its a cold uncomfortable Sabbath morning and I am to chilly to read and so I am agoing to try to write. We came out into camp yesterday afternoon as I expected we had to leave our snug quaters in the city and take to our canvas houses again. they seem cold and uncomfortable after living in brick and morter a few days. but I think the sun will come out warm by and by and dry up the ground a little more and then we shall begin to thaw out. My acheing tooth has troubled me long enough and as it is acheing finely this morning the prospect and probability is that we shall dissolve partnership before night. We are camped about a mile and a half from the city. I think the ground is better than that we had in Washington but we are stuck down among woods and swamps and can't see any distance. The remainder of Wrights' brigade to which we are attached and which consists of four regiments is all encamped close by us. <u>Shermans battery</u> which belongs to our division came into town yesterday. There is a large body of troops here now and they continue to come each day. Lieutenant Gale who married Miss Brinton the day after we came away called on us yesterday his regiment forms a part of the same division we are in and if we are sent off on an expedition of any importance as we expect to he will go with us. he came here a few days before we did and his regiment is camped near the city on the college grounds, he made but a short stay and promised to come out this evening and see us again. We came through the city yesterday on our way here. and so I had a partial view of it. Its a little one horse concern with narrow crooked streets and old fashioned houses and presented about as busy an appearance as Salisbury Center. The State House is the most important looking building and standing on a little rise of ground can be seen from all directions.

    <u>Afternoon</u>, We do not have to drill on Sundays, but instead have an inspection of arms and all our accouterments which is worse than a drill a great deal as we are obliged to stand in the ranks with knapsacks and every thing on for about two hours while every thing is being examined by one of the staff officers. George was on guard yesterday and last night as our removal disturbed the usual routine of guard duty he had a rough time of it. It seems as though we wer getting further and further from civilization at every move we make and know less and less what is agoing on about us. we get N.Y. papers occasionally but they are always several days old. thank you

for the papers you send. They are very acceptable. Martha sent me a paper the other day please thank her for me and say to her that I am glad to be remembered by her and should not be angry at the repetition of the offence Tuesday afternoon) I left off writing on Sunday to attend service. after service we met to form a Sabbath School but had so little time that we could only form a few classes. directly after that came dress parade and then it was night and not feeling very much like trying to write concluded to defer it till morning to have time to finish my letter and mail it before drill. but it came my turn to go on guard and so I had to bid good by to writing for yesterday. I had a very comfortable time on guard the weather was fine I had a good lot of men on my relief and though the night was rather cool yet we built up a rousing fire in front of the guard tents and when off guard spread my rubber blanket on the ground and lying down with feet to the fire and covering up with blanket and overcoat slept like a top. today I do not have to drill but do not feel much like doing anything but a good nights sleep will set me all right. we expect to leave this place in a few days. you will see by the paper a large naval expedition is being fitted out which according to account is to rendezvous here. at any rate I was told that six large steamers wer in the harbor yesterday and more comeing in. I think we shall be on ship board in less than a week. I dread it for I think our situation there must of necessity be much more unpleasant than when in camp. you need not get me the ring as it might be lost in trying to send it to me. I have placed the ring on the little finger of my right hand as it fits there closely I am in no danger of losing it. I am sorry to have to make the change but its the best I can do. the smoking cap I should like very much. and if you have sent it by Sam Wolcott I shall be glad. but I think it will be useless to attempt sending it by mail if we can succeed in getting letters occasionally we shall do well. I enclose a few envelopes for you to use. they will be more apt to reach us than any others. Remember me to Mr and Mrs Barnum. I don't know about your letting her (Mrs B) read my letters you should remember that they are written for you and you will be likely to be somewhat partial and not criticize them so closely as an uninterested reader. I fear they will not stand much of a test as literary productions however much you may prize them. I shall have to close this now as I am anxious to send it along. I will commence another directly. and as I shall not return your two last letters in this will try to answer them more satisfactorily. keep up good courage my dear. do not think of me to much and above all do not worry about me. ever remember that the same God who cared for me while in the Store and Bank still has charge of me and if my whole trust is in him he will never leave or forsake me. I intend to write to your father as soon as possible. I have so many letters to write now that after I have finished a letter to you and then think how many more I ought to write I get tired and throw down the pencil. good by darling I will write again and send as soon as possible. Ever your own

Ned

No. 10 & 11 missing

No. 12 Rec'd Oct 26th 1861

Steamship Illinois
At Sea October 22nd 1861

My Darling Sara

    I am lying in my berth. have been trying to sleep. but although drowsey enough cannot quite make it out and as I think I shall probably be able to get a tolerable nights rest shall give up trying to sleep now and attempt to write a few lines to you darling. I am the more anxious to write now for fear I shall be sea sick and unable to write when I wish to. I want to send you a letter the first opportunity on our arrival at Fort Monroe. We started from Annapolis about ten o"clock Monday morning soon after we started I went on guard (for we keep guard here just the same as on shore) I did not care much about it as my position was on the bow of the boat or on the hurricane deck most of the time and so had a good chance to see what was to be seen the sail during the day was very pleasant but as our boat kept about the middle of the bay we could see but little except the general outline of the shores on either side. At dark we came to anchor and the engines blew off steam and we lay quiet through the night I got along very plesantly until about twelve when it commence to rain a little and blow considerable both together makeing it very uncomfortable we changed the quaterers of the guard from the bow of the boat to the upper or what is called the hurricane deck (we wer not allowed to go below to stay) of the ship and huddling around the huge smoke pipes endeavored while off duty to make ourselves as comfortable as possible I managed to sleep about an hour and a half during the twenty four and so am excusible I think for not feeling very bright today, the weather is rainy misty and very unpleasant today and I have been on deck but few minutes since I came off guard, We commenced our journey again this morning soon after daylight and have been traveling slowly all day takeing soundings as the sailors call it or in plain English measureing the depth of the water frequently and stopping entirely often. It is now four o"clock and the vessel has again come to a full stop and cast both anchors the weather is misty and cloudy and it is evident that the officers are unacquainted with the channel. one of our roommates has just come in and says that we will not start from here again in a fortnight unless it clears off. so if this weather holds I shall have plenty of time to write. It is getting dark and I shall have to stop soon for today anyway. Some of the men are getting sea sick. I expect my time will come soon. Wednesday morn. We got under way again early this morning and are now (eight o"clock) approaching land again and I presume the Fort. I may have to close this rather abruptly for although we may lay off the Fort some days I am anxious to send you a line the first opportunity. I beg you to excuse the appearance of this sheet and have no doubt you will when I tell you that I have written it lying on my back in my berth holding the paper as best I could before me while the constant motion of the ship makes it difficult to keep ones pencil on the line even if the light was sufficient to enable me to see them. My quarters although bad enough are king to those occupied by some of the men in the centre and lower hold of the vessel. my berth is in what they call a state room. said room being about fifteen feet long by six wide and seven high. in it are six berths each berth occupied by two persons. the berths are one above another in three teir. my berth is one of the upper ones and so if the ship gets to pitching about much I am calculating how will be the best way to tumble out so as to come down easiest there are two small windows in our room and by keeping them open all the time we obtain aplenty of fresh air. and it is this more than anything else that makes our room preferable to those about us. Pitt has come in here this morning to write I saw George with his portfolio out on deck a short time ago it is so cold out there I could not stand it to write there

(pencil changes to very light)

I have just been out on deck we are near the Fort I see the chaplin of our regiment on deck and as he takes charge of our letters for us I am agoing to hand him this when I get a chance good by darling I will again soon ever your own loveing

Ned

No. 13 missing

No. 14 Rec'd Oct 29th 1861

Steamship Illinois
Off Fort Monroe Oct 26 1861

My Darling Sara

I have at last been able to set myself about writing to you again. Not that it is a hard and unpleasant task but because circumstances have prevented. I closed and handed to our chaplain (who takes charge of our mail) a letter for you just as we wer comeing to anchor at this place. which I hope you have received. George has written and mailed letters since I have written but one and that to Mr Randalls' people. George received a letter from you yesterday. I have looked anxiously for a letter the last three days but have been disappointed. not having received one from anybody. I think I can appreciate your feelings somewhat when you are disappointed in not getting a letter from me. I should have written yesterday but we had orders to go on shore at nine o"clock and we all prepared to go at that hour and then waited expecting to go every moment and did not finely leave the ship until after twelve. when we returned it was to dark to write as we have no light sufficient to write by after dark. I don't know why we wer sent on shore unless it was to accustom us to disembarkeing and see how quick we could do it in case of necessity. Three regiments belonging to our brigade wer landed. the fourth. the sixth is divided part on one ship and part on another and wer not landed. I was rather unwell and went to the surgeon and got excused from going but afterwards feeling some better and anxious to go got ready and went with the rest. A part of the men wer taken directly from the ships in small boats and the remainder including our company wer put on board a ferry boat and taken as near the shore as the depth of the water would allow the boat to approach when she was anchored and we landed from her in small boats. It was rather exciteing and I enjoyed it. when the boats reached the shore the rowers would run them up as far as they could on the beach and we jumped out the best way we could. the surf was beating up on the beach considerable and some would get nicely wet by landing just as a wave broke on the shore. I watched my opportunity and landed dry shod not even takeing the polish of my shoes. which I had "shined" up nicely in the morning. There did not seem to be anything particular to do. we marched a little way on the beach then formed in line of battle went through a part of the manual of arms. then stack our arms and rested. we staid perhaps an hour strolling about on the beach picking up shells and enjoying ourselves as well as we could, then we shouldered our arms and started for the fort. we wer in hopes we should be marched through it and be permitted to see how it looked inside. but wer not. we onley marched close to it and by the big gun(s) "Union" of which we have read so much and which is mounted outside the fort between it and the beach. then we went to the wharf and on board a ferry boat again which took us directly aboard the ship. on the whole it was quite a pleasant trip and many who started with cursing the officers for making us unnecessary trouble wer quite thankfull for it at the end. I enclose you a little shell. It is neither rare or beautifull but being picked up and presented by a friend from a distance may give it a value it would not otherwise possess. A sad incident occured on board our ship this morning just before I sat down to write. a man from company A droped a basin overboard and being an excellent swimmer and not wishing to lose his basin stripped off his coat and jumped into the water after it. the tide was going out very rapidly at the time and he soon found himself unable to stem it. as he had jumped over and it was known he was a good swimmer his critical situation was not known untill it was to late to save him. the boats wer promptly lowered when it was found he was in danger but it was to late. he sank the last time just before it reached him. It is a sad case. It is the first death in the regiment. I understand he leaves a wife and two children. Such an incident at home would cast a gloom over the entire neighborhood. here a person would not think anything had happened unusual. the men are strolling about carelessly makeing comments on his foolishness smokeing playing cards swearing is just the same as ever. A single human life is thought but little of here. and the utter indifferance with which men seem to regard life and things pertaining to both time and eternity is sickening. I did not witness the scene myself but Norton saw the whole of it. The view from our ship is decidedly warlike. The harbor is full of large and small vessels of all kinds. I should like to count them but think I could not very well. Large ocean steamers like our own loaded with men. smaller sail vessels loaded with coal and provisions. with here and there a large line of battleship with her guns looking grimly out of her port holes. smaller steamboats are almost constantly plying between the larger ones and the shore. while last but not least is the fort with its three hundred mounted guns looking quietly out upon the scene. The fort is a very qu1te nice looking place with its grass covered walls and green trees and easey looking dwelling houses the tops of which are seen over the walls. altogether wer it not for the cannon it could look quite like a gentlemans pleasure grounds. the Rip Raps of which we read are in plain sight. they look like any large irregular pile of stone with a few buildings on them and the masts of some ships to be seen over the tops of them. beyond in the distance is <u>Sewells point</u> where the rebels have erected batteries and where by the aid of a glass a rebel flag can be seen flying. Of our own destination we are in blissfull ignorance and probably will be kept so untill the end. I hope for the health and lives of the men that we may not be kept on board the ship very long. if we are in our present close quaters and should be taken into a hot climate nothing but Infinite Mercy can preserve us from disease and death. I do not fear so much for myself as I told you in my other letter. my quaters are both tolerably light and airy but the men are some of them very badly situated as regards both. Lieutenant Gale has just called on us again. he is agoing on

the same expedition. he tells me we may sail today. the chaplain is just conducting a releigious service up here on the hurricane deck where I am writting. I ought to stop and attend but I am in a hurry to finish this. that it may be sure to be mailed from this place. Lieut Gale tells me that letters had better be directed to us at this place so if you please you may direct here in future. I find it impossible to write as I wish to or what I feel. When alone at night or what is the same when my companions are asleep. I can think of much that I would like to say to you and much that I intend to write when it comes light. but its impossible to find a quiet corner where one can sit down and collect thier thoughts sufficiently to put them on paper in readable shape. I shall continue to write when I can and have letters ready to send whenever there is opportunity. Remember my dear that whereever I am I always think of you and look forward to time when I shall be premitted to return and claim you for my own. I shall hope to hear from you again before I leave here. but if I do not good by.
    much love my darling from

Ned

No. 15 missing

Editors note: These next two letter are referring to the Battle of Fort Royal. It was the first amphibious battle of the war with Navy ships and Army forces combined to capture Fort Walker and Fort Beauregard.

No. 16 Rec'd Nov 14th 1861

Steamship Illinois
Off Port Royal Nov 5th 1861

My Own Dear Sara

I must write you a few lines this morning as we are expecting to go on shore. but I have learned by past experience to expect to be disappointed several times before finealy getting started on any expedition and therefore although we shall probably go on shore during the day I may yet have a chance to write you a good long letter on board the ship. We started from Fort Monroe just a week ago today. I was on duty that day and had no chance to write. the next day I was taken with sea sickness and have not scarcely been able to hold my head up since. much less to write. I am sorry it has been so for I wanted to have written several letters and to have written you an account of our journey which I shall not now be able to do at present. I presume George has written a good account of it (to) Julia as he has not been as sick as I was.

We arrived at this place yesterday morning and found a large portion of the fleet awaiting us. one of our vessels I learned was lost during a gale which lasted us over twenty four hours extending through Friday and Friday night and part of Saturday. I hear differant accounts with regard to the number lost on board of her. but as near as I can learn it is about twenty. Our gun boats engaged the enemy works on shore yesterday afternoon and I have seen my first war. but at such a distance that onley the smoke and sound off the fireing was perceptible. This morning the fireing has recommenced. our ship has got up steam and is standing in shore. My darling I cannot write to you this morning amid all this confusion and excitement. I am onley agoing to direct these few lines to you that you may know why I have not written more. If my life is spared I will write again the first opportunity. I have no fears in particular for this days work at least but I am sick weak and faint and if I wer at home should be abed. Forgive me my dear for not writing more. good by God bless you. much love from

Ned

No. 17 Rec'd Nov 22nd 1861

<u>Hilton Head S.C.</u> Nov 14th/61

My Own Dear Sara

    I have a little time this morning and am in hopes to be able to finish a letter to you before being called to do any duty. The weather is very warm here and while you are probably warming by a good fire I have my coat off and am looking around for a cool place to write. I hear that another mail arrived on the island yesterday and if so I shall hope to receive a letter from you. I sent you a letter two days ago which I finished hastely as I was going on guard. I got through the duty very well although it was more severe than any I have done before. I slept out doors part of the night and a part I lay in the guard tent. We do not drill much of any now as half of our regiment are at the other end of the isleand and it takes us that are left about all the time to do the guard and fatigue duty. We are settled down in our tents again and all together things begin to look quite homelike. Our camp is just outside the captured fort and extends along the sea side on ground that was last used as a cotten field. the soil is a fine sand. no turf at all. and it is just like pitching a tent in the sand betwen the Bank and the Store there at the Falls. there is not a bit of shade and on the whole we have a dirty disagreeable camp. It is uncertain yet whether we remain here or go to the other end of the island where the rest of the regiment is. I think I should rather remain here for they say that the musquetoes are as large agan there and that in addition to them the gnats are very thick. The musquetoes and other insects are very thick and troublesome. we have to lie down in the sand and take our chances with them and they have decidedly the best of it. A great many troops are being landed here. a large storehouse is being put up. the fort is being repaired and strengthened and it looks as though the government intended to make this a strong and permenant post. We found twenty seven guns and a large amount of ammunition at this fort. at the other fort across the channel I understand they captured twenty three guns. also at the other end of this island we found two fine brass field pieces. the rebels had spiked one of them but so imperfectly that it was easily removed and the cannon is a good as new. A Lieutenant belonging on board the Wabash was here yesterday to see an acquaintance of his belonging to this company. from him we learned some additional information with regard to the fight. That vessel received some forty shots in all mostly through her rigging. only one man was killed during the fight. he was captain of one of the guns and had his leg shot off by a ball. he coolly turned to a comrad and asked for a knife to cut away a little flesh by which it still hung. on being carried below to the ships hospital he asked the surgeon to assist him in turning over his leg. the surgeon not understanding his request asked him which leg when he impatiently replied "which leg why you damned fool I haven't got but one" in less than five minutes he was dead. We hear a great many rumors as to what has been done by our army in other places and the success that has attended our arms. a good many of the men are in high spirits and think they will soon be on their journey home. but as all the reports lack confirmation and I am not as sanguine as a good many are of a speedy termination of our national troubals I do not think there is any immediate danger of our being turned over to the cold charities of the world right away. I have had two calls from Lieutenant Gale since we landed. he has been here today. his regiment is quartered within a mile or so of us and if it is possible I mean to get a pass and go and see him next Sabbeth as he is a minister I think I shall be able to spend the day more satisfactorily than in our own camp. We hear the music from the bands of other regiments around us all during the day and in the evening some one of them is playing most of the time. it makes it pleasant and is very enlivening. our own regiment has no band. the evenings now are very fine just about like August moonlight evenings at home. We lived very sumptusly for a few days after our first arriving here on sweet potatoes fresh beef and so forth but we have had to come down again to our old bill of fare. at present we have enough to eat and after the wretched cookery we had on ship board our own darkey cook makes things very edible again. I suppose it will soon be thanksgiving up in old Connecticut. I haven't seen or heard of the Govenor of this states proclimation for thanksgiving if we continue to succeede as well as we have heretofore I think he will feel more like calling for a fast. I suppose you will have to "sing" my favorite piece for your friends and I shall have to take what comfort I can in thinking I hear you and in hopeing to stand by your side and listen to them agan at some future time. When you write to Charlie remember me to him. tell him I do not forget him though I have apparantly done so and failed to write to him as I agreed to. give my kind regards to your cousin Salina and say to her that I still feel disappointed when I think of the very short visit I was able to make with her during her stay at L. Tell your father I had commenced a letter to him and should have sent it to him directly after landing here had it not been for the job of sea sickness which I had while on the voyage and which was all I could attend to. I learn from some of our boys that Mr. Randall is sick and that J. Bostwick is in the Bank. if you know any thing about it whether he is much sick and what ails him I would like to have you write me. I very seldom received a letter from him when well. and as it is now some time since I have written to him I hardley expect to hear from him direct very soon. I think you told me you thought of calling there some day. I should be glad to have you. Mrs. Randall likes you and I think your call would be a pleasant one and I know Mrs R. will be glad to see you. The young ladies of Salisbury need have no fear of being forgotten by us (if they do care to be remembered by us) they often come in as a subject of conversation while sitting in our tent during the evenings which are now getting somewhat long. The rememberance of the friends and society we have left behind us and which is God wills we hope to again join will I have no doubt do much to restrain and keep us from the doing of anything unmanly or that would give pain to those we love. I have never sent any word to Julia because George is here and is a much more punctual correspondent than I am and gives her good descriptions of our travels, still I would like to have you tell her that I often think of her and her kindness to me. I shall probably never be able to repay her and George for all they have done for me. George is very good to me now and I hardly know how I should get along without him. not but what all the boys are kind and good but we do not all of us think as far as George does. I did not finish my letter yesterday finely. today (the 16th) is quite windy and in the morning was quite cool but as the sun comes out it grows warmer. I have got to go on guard at four o"clock at the generals headquaters. at that post they are very particular. Every thing about the men and thier accouterments must be in apple pie order. white gloves on etc. the rest of our company goes out on picket guard so that we shall all be on duty tomorrow. Our mail has again arrived this morning. I have received another of your dear letters (Nov 3d) and the two papers. many thanks for them. we can get no papers now onley what are sent from home and any papers of late dates that our friends send us are thankfully received. I have not time to answer your last letter now. I may have time to answer it tomorrow. believe me my dearest I long as earnestly as you do for the time to come when I can return to you my loved one. I believe that as the distance betwen us increases my thoughts return more and more to home and friends and many an air castle is built in moments of leisure when indisposed to read or write which are suddenly dispelled by the call to "fall in" for some new duty. good by my darling I will write again soon remember me to Mrs. Barnum and all enquireing friends

ever your own dear Ned

No. 18 missing

No. 19 Rec'd Dec 3rd 1861

_Hilton Head
Near Ft. Walker Nov 21st/61

My Own Dear Sara

I am out on a sort of picket guard our rather outpost of our brigade the regiments of our brigade take turns in furnishing a guard for this post and the companies composing the regiments releive each other every twenty four hours. It came our turn again yesterday. Our companie did the guard duty here last Sunday but I was on guard at the camp that day and so did not come with the rest. We left our camp yesterday afternoon about five o"clock the distance is about two and half miles we got here just about dark and if we have good luck we shall be releived about the same time tonight. We have a fire and after posting our guards (who are only a few rods off) we sit around the fire lie down and sleep or any way to pass the time untill it is time to releive the guard again. We take a days provisions in our haversack consisting this time of a large piece of raw bacon a plenty of hard bread and tea and sugar to each man. In the place where we are posted there are a plenty of oysters that can be got at low tide. we are close by. in fact on the beach. We cut off a slice of bacon stick it on a long sharp stick and roast it over the coals. letting the fat which runs out drip on our hard bread for butter. we take our tin cups with us and heat our water and make our tea all in the same cup. Its not St Nicholas style exactly but it does first rate. especialy as this kind of buising is hungry work and hungry you know is the best sauce. I have eaten very heartily as I did not sleep much and as I sat by the fire talking with the Lieutenant who had us in charge. I kept cooking and eating. I have had a nice meal of roasted oysters alone this morning. A party of our men has gone out on a forageing expedition and I presume they will bring in some fresh meat of some kind. I have a little leisure time and so thought I would try to write you a letter, but it will not (be an) answer to keep in a ladies pocket or bear inspection. My hands are all begrimed with dirt and grease and I cannot clean them in salt water and there is no other to be had to wash in. my paper is some scraps which I have carried in my pocket untill it is very dirty so you see ya'll not get a highly perfumed letter this time. The prospect is that we are agoing to have a rain in fact it has sprinkled a little already, but I hope it will hold off untill we get back. Our tents are getting to be very poor and I don't believe they would be of much use against one of the heavy southern rains which prevail at this time of the year. The half of our regiment still remains at the other end of the <u>island at Braddock Point</u> and guard drill and fatigue duty still follow each other so closely that it is difficult to find spare time. The men scold a good deal but its no use we are in for it. and if we onley succeed in escapeing in the end with a whole skin I'll try not to grumble much. Day before yesterday a man died in the hospital who belonged to our regiment. he was a sergant in Co. E. he has been sick sometime was in the hospital when we landed. I hear that he was a married man and leaves a child or two. yesterday he was burried. his companie is at Braddock Point and the funeral escort was furnished by another Co. There is no parade made at the burial of a private. a plain unpainted pine coffin is prepared. the escort consists of I think about twelve men. besides the bearers. they are preceded by a drum and fife <u>playing the Dead march</u>". the chaplain brings up the rear. I have not attended at any burial yet and don't know exactly what the service is at the grave but it can't be much for it is very short as soon as the grave is filled up the men shoulder thier arms the drum and fife strike up Yankee Doodle or some other quick air and the dead is left to repose in his narrow home forgotten by all. but the dear ones he may have left at home who all unconscious of his suffering and death are doubtless looking forward to the time when he will be again premitted to return and gladen thier hearts. Our forageingers have returned. they made thier appearance on the opposite side of a little creek which runs just in front of our camp fire at low tide it is but two or three feet deep. but now the tide was up and the water over a mans head. they had a fine sheep which they had killed and dressed hung on a gun and borne on the shoulders of two of the men while the third one carried the rifles of the other two. the question now was how to get across. we could not think of waiting three hours for the tide to fall and there was no boat to be had after various ways had been proposed they laid the mutton down and started back away up the beach out of sight to find some drift wood to make a raft of. as soon as they wer fairly out of sight a man a man on our side of the stream striped off his cloths and swam across and takeing the sheep in one hand returned in safey. now came the fun. the men soon returned loaded with drift wood. but no sheep was to be seen and you may imagine that they wer three as sheepish looking men as you ever saw. they could not think at first where it had gone but as it was impossible to retain our own laughter long they wer soon satisfied on that point. they finely swam across and we cut up and divided the mutton. then came the cooking. George had brought with him a small basin and after he and G. Olin had cooked and eaten thier dinner. I borrowed the basin cut up my meat in small peices and cutting up some of my fat bacon and putting in with it proceeded to cook and eat the best dinner I have tasted since leaveing home. About noon it cleared off and the sun came out bright and warm and the rest of and in fact the whole day has passed very pleasantly indeed. we have been releived just about the time we came on duty. Nov. 21st. Sunday. I meant to have finished my letter in the evening after my return to camp but was to tired and sleepless and have not had a chance to write a line since untill now. Thanks to <u>Lieutenant Colonel Hawley</u> who has returned from Braddock Point and taken command of this part of our regiment, we are not obliged to drill any today or do any fatigue duty we had an inspection of arms and a dress parade this morning which occupied about an hour. the rest of the day untill about three o"clock we have to ourselves. At three o"clock is guard mounting and it takes about thirty men out of our company for that duty today most of our boys will have to go. I was on yesterday and night before last. and shall not have to go on again today. its getting to be very unpleasant buisiness. the weather is much cooler and cold winds are blowing most of the time. clouds of dust are continualy blown in our faces filling our eyes ears and nose and every place where dust can be blown. There has been another death in our regiment. A private from Co. A. Lieutenant Gale has called on us again this afternoon his regiment is oniey a short distance from ours and I have been trying to get a chance to go over and see him but have not and shall not be able to do so. he thinks their regiment will leave the island this week on an expedition which is now fitting out it is supposed for some place in Georga. We have not had any mail now for over a week and it seems as though it was months. we are all very anxious not oniey to hear from home but also to get some N.Y. papers so that we may be able to know what is going on. We have had no religious service since we came on the island except a short prayer at dress parade there is no convenient place to hold it even if we had time to attend it there is not a bit of shade anywhere very near. and to sit down in the sand either with the sun shining hot or the wind blowing clouds of dust in our faces would be out of the question. There has not enough rain fallen since we landed to lay the dust. but I suppose we shall get enough of it by and by as they say that when it once commences it rains for a fortnight right along. Our patron after whome the company was named Mr. James M Townsend has sent out a small library to us. some of the books are in our tent now. I have onley had time to look at them which is about as much as I shall probably be able to do for I can't find time to read my bible half the time. They are a good selection I should think and if we can onley find time to read them will be a valuable acquisition. George says he is agoing to send home for some things he needs. I am agoing to send for a few small articles. George does not wish it known he is agoing to send as he thinks a small box will be more apt to come directly than a large one. and so we do not let the boys here know that we are sending and you need not mention it up there as every one of our friends would wish to send something and so they would soon have a car load instead of a little box. I know that this may look selfish but I can't help it. the other boys have the same chance to send that we do. and besides any little goodies we may get will of course be divided and we always use any articles of comfort or necessity in common. The things we are to send for are things we wish to get as soon as possible and it will take a much longer time to fill up and get off a large box. Tom Norton had a large can of butter sent him in the box which we receved at Washington. he did not open the can there but put it into the orderly sergants box of things among the cooking utensils and the other day the box was brought on shore and opened. the can was found safe and on opening it the butter was found to be sweet and good as when first put up. It proves quite a treat I tell you and makes our hard bread go down with a good relish. My health since I came on shore and recovered from the effects of my sea sickness and the march and exposure of the first two or three days is really very good. I am standing it quite as well as any of the boys. I have a hard cold which is of course very disagreeable but there but few but what are troubled in the same way. Last night the 25th we had a big white frost such as we have in old Connecticut and one of the boys brought in quite a large piece of ice a quater of an inch thick. its the first frost I have seen since leaveing home. I have been a long time writing this letter and as we never know exactly when there is agoing to be a chance to send letters I will close this and put it in the Chaplains hands. I may write two or three more before this goes and you may get them all together. And now my darling what shall I say to you. Shall I tell you the old story over again that I love you better than ever and am looking forward with impatience to the time when in the kind Providence of our Heavenly Father I may be agan premitted to return to my home and the arms of you my dear one who have promised to link your fortunes with mine and share with me the trials of life. May God in his infinite mercy grant to preserve us both from all the danger and adversities that beset our paths. and grant us yet many years to live and enjoy each others society and at last receive us both to the mansions of eternal rest which are prepared for those who love and serve our Lord is the prayer of your own ever loving

Ned

No. 20, 21, 22 & 23 missing (though there are two #24's)

No. 24 Dec 21st 1861

Camp of the 7th near Fort Wells
Hilton Head S.C. Dec 6th/61

Dear Sara

The mail closes in the morning I am told, and as I shall have no time to write in the morning and am anxious to send you a line if nothing more by each mail. I am agoing to scratch a few lines tonight although it is most time for roll call and I shall have to hurry. I beleive I have not written a word this week. I have been very buisy a good deal of the time and when not buisy very tired. We are getting along very well now and are fully well settled down into camp life. The weather is very moderate yet but we find it plenty cold enough standing out on guard nights. We had a heavy thunder shower last monday night. just such a shower as we have at home in June or July. it rained a considerable during the night. and the next day it was very cold and windy. Monday night there was an alarm. the long roll was beaten and all the regiments turned out. we made very good time for raw troops. that is we got into line of battle with all our accouterments on ready for action in five minutes after the drum beat. It was about three o"clock. very cold. we had first tumbled out of our warm nests and fallen in without our overcoats and it soon became anything but pleasant standing there waiting for. we did not know what. I expected we should have to stay there untill morning but finely after about an hour we wer dismissed to our quaters. it was nothing but a false alarm and I think likely done on purpose by the officers to see how quick we could get out in case of an emergency.

Monday Dec 9th 1861

I did not have time to finish my letter the other evening. and so you will look in vain for a letter from me by that steamer. I don't have time lately to think. however.

Our roll call now is just before light about 1/2 past six breakfast is at seven. inspection of arms dress parade and drill from eight to ten. we are obliged to black our shoes. bruch our cloths. clean our guns etc which takes all our time before breakfast and inspection. at eleven there is a non-commissioned officers drill untill twelve. and just for being eighth corporal in Company G I come in for that hours extra drill which just spoils my spare time in the afternoon. we have dinner at one and at three commence drilling agan and continue untill five. then comes supper by which time it is dark and we are generaly so tired that I for one don't feel much like trying to write. I am sorry my dear that I have not written to you often. but I do not promise that I will do better in future. for if we don't have more leisure time which is not very probable I certainly cannot but i'll do the best I can. Yesterday was the Sabbeth and we had service agan for the first time since we have been on the island. it was very warm. the sun shone clear and bright. real July weather at home. the regiment was drawn up on the parade ground and formed in a hollow square. the chaplain. choir. colonel. lieutenant colonel occupying the center. in this way the whole regiment is brought into easy hearing distance of the chaplain. we had a very interesting service. our singing is mostly the tunes used in the Methodist revivals and our Sunday Schools. the text was "Jesus Christ the same yesterday today and forever" Hebr 13"8". in the afternoon I attended communion. it was held in the tent occupied by the colonel for a dining room. there wer but twelve or fifteen of us in all. but we had a very pleasant meeting. I think the most of those present wer methodist by thier talk. Our first Lieutenant was present and assisted in the ceremony. I think he is an Episcopalian. In the evening I attended prayer meeting in the chaplains tent. about a dozen present. it is the first time I have attended one since I came from home. there is to be one every evening this week. I mean to attend when I can for I feel the need of something to enable me to withstand the temptations which surround me. and to I wish if possible to spend a few moments occasionly out of the hearing of the profanity and Evill talk which is universal in the street and throughout the camp.

 Tuesday morning Dec 10th

 Didn't succeed in finishing my letter last night after all. This afternoon we have got to go out on picket and I am agoing to mail this before I go as there may a mail leave here for home before I get back which will be tomorrow night. The weather is excessively warm and I don't fancy a five mile march this afternoon very much. the trees down here are still green. the leaves on them haveing turned but little as yet. today and every day this weather the men go in batheing and wer it not for the Almanac we should hardly know but that it was June instead of <u>Dec</u> I enclose you a copy of the Camp Kettle. the first newspaper ever published on Hilton head. it is as you will preceive quite an institution and purports to be sold for the moderate sum of three cents. but as the scarcity of the articles affects the price. they cost anywhere from ten to twenty five cents. and are scarce at that. The regiment in which they wer printed has left the island and gone I suppose up to Beaufort but I am not certain. four of our company went up to Beaufort yesterday and returned this morning. they went up on the steamer Winfield Scott but they don't know for what as they had nothing to do. Barnes was one of the lucky ones. it was quite a pleasure trip and I should like to have gone but the trouble is one does not know when he is detailed on any such fatigue duty. where he is going and nine times out of ten it turns out something unpleasant. I hope ya'll forgive me for not writing oftener and not punish me by doing as you are done by, hopeing to hear from you agan soon. I remain as ever your loving

Ned

CHAPTER 2

TYBEE ISLAND AND THE CAPTURE OF FORT PULASKI

From the book:

Captain Gillmore was instructed to report whether it was practicable to reduce or capture the fort, and if so how. Fort Pulaski was a brick work of five sides or faces including the gorge, casemated on all sides, walls 7V2 feet thick and 25 feet high above high water, and mounted forty-eight guns. A full armament would have been 140 guns.

Captain Gillmore reported that he thought it practicable to breach the fort from Tybee Island and recommended ten ten-inch mortars, ten thirteen-inch mortars, eight heavy rifled guns of the best kind and eight columbiads.

This was a bold scheme. It was contrary to the military science of that day. A standard military work reads as follows:

"An exposed wall may be breached with certainty at distances from 500 to 700 yards, even when elevated 100 feet above the breaching battery; and it is believed that in case of extreme necessity it would be justifiable to attempt to batter down an exposed wall from any distance not exceeding 1,000 yards, but then the quantity of artillery must be considerable, and it will require from four to seven days firing according to the number of guns in battery and the period of daylight, to render a breach practicable."

On the 21st of February the first vessel with ordnance and ordnance stores for the siege arrived in Tybee Roads. From that time until April 9th the Seventh

Connecticut was constantly engaged in landing and transporting ordnance, ordnance stores and battery materials, making fascines and roads, constructing gun and mortar batteries, service and depot magazines, splinters and bomb proof shelters for the relief of cannoneers off duty, and drilling at mortars.

The Seventh Connecticut was detailed to serve the five mortar batteries, mounting seven ten-inch and eight thirteen-inch mortars."

Major General Hunter, commanding the Department of the South and Brigadier General Benham, commanding the northern district, both arrived with their staffs April 8th.

Just after sunrise on the morning of the 10th General Hunter dispatched Lieut. J. H. Wilson of the Topographical Engineers to Fort Pulaski, bearing a flag of truce and a summons to surrender. Colonel Olmstead in command of the fort replied, "I am here to defend the fort and not to surrender it."

On receipt of this reply the order was given to open fire commencing with the mortar batteries, agreeably to instructions previously given. The first mortar to be fired was a thirteen-inch from Battery Halleck. Captain Sanford had written on the shell, "A nutmeg from Connecticut; can you furnish a grater?"

On the morning of the 11th a little after sunrise the batteries again opened upon the fort. After three hours an entire casemate had been shot away and by twelve o'clock the one next to it was in the same condition. This opened the way to the magazine and the fire was directed upon it. To resist longer would be foolhardy and not brave, for a shell penetrating the magazine would cause an explosion which would destroy both the fort and garrison. Colonel Olmstead was brave but not foolhardy and at two o'clock raised a white flag in token of surrender, and the batteries ceased firing. Acting Brig. Gen. Q. A. Gillmore was dispatched to the fort to arrange terms of surrender which are given in the appendix.

By this capture there fell into our hands forty-seven guns, a great supply of fixed ammunition, 40,000 pounds of powder and large quantities of commissary stores; also 360 prisoners.

No 24. Rec'd Jan 1st 1962 (2nd letter numbered 24)

<u>Tybee Island Georgia</u>
Friday Dec 20th 1861

My Own Dear Sara

    We have made another move you see. and came to another stop. I mailed a letter to you day before yesterday just after we received marching orders. we packed up immediately and about noon struck our tents and about three in the afternoon took up our line of march for the landing place a distance of about a mile. I was glad it was not more for as we did not expect to march far we had loaded ourselves down with a good many things we should never attempted to carry far but which are very convenient when in camp. We immediately commenced embarking on board the <u>steamship Marion</u> but as we had to be taken aboard in small boats the last load (of which your humble servant constituted a small part) did not get on board until after dark. we had to wade out to the boats as they could not come up so that we could step into them dry shod. but we just pulled off shoes and stockings. and put them on dry when we got aboard. so that we dodged the pleasure of lying with wet cold feet all night. I will just say here that our landing here was effected in the same way and I am happy to tell you that I did not catch cold in both operations. There was a number of horses and a quantity of provisions to be taken aboard which occupied nearly all night. so that we did not start untill early in the morning. We wer given quaters for the night below. I had a passable berth. but did not sleep much on account of the noise on deck. The sun rose clear and bright and we had a pleasant sail of about four hours reaching here about ten o"clock. The first thing in the morning on comeing on deck the mail which had arrived in port and had been brought aboard during the night was distributed. It came just in the nick of time. and hearts wer made brave and hands strong for whatever was before them by hearing again from home. My own heart was gladend by the receipt of two of your dear letters for which many thanks. We did not get all of the mail as there was quite a number of bags which they did not have time to distribute before we left. We landed immediately on our arrival here. The <u>76th N. Y. Reg composed of Germans</u> wer the only troops on the island and they were glad enough to see us I tell you. They have had to work hard since they have been here which I think is nearly a fortnight. and have been obliged to keep out a large guard which is by no means the easiest part of soldiering. <u>Fort Pulaski</u> is in plain sight as we came up the channell and between two and three miles distant I should think. It is within canon shot of where we landed. The rebels practice a little every afternoon and thier shot strike close to the fort which our troops are building. they fired several times the afternoon we landed and one shot struck in the water so close to the Marion that she weighed anchor and hauled off to a greater distance. The works which at this place are more particularly designed for defence. but we are agoing to throw up batteries at another place on the island and make preparation for an attack.

(Saturday Dec 21st) One of our corporals went last night with the engineers to stake out a part of the works. and forty men from our regiment under command of Lieut Mills of our Co. worked all night on them. it is oneley a mile and a half from fort Pulaski and so we have to work nights as in the day time they would probably fire upon us. This island is realy of considerable importance to us I should think for if we can erect strong batteries there why we can't shell them out of Pulaski unless indeed they are a great deal stronger than we know of. The Tybee light is made here within a quater of a mile of our camp. I have been nearly to it and might have gone close to it if I had chose to. When our gun boats came down and drove the rebels off the island they failed to keep a guard on the island over night. and the rebels returned and set fire to the lighthouse burning out the inside entirely and rendering it useless untill we can rebuild the staircase landing to the top. the walls being of brick or stone are not seriously injured, There is a tower near the beach built of oyster and other shells and lime or some kind of cement. it is circular. the walls ten or twelve feet thick and about sixty feet high I should think. I do not know by whom it was built. some say by the Spaniardes but I don't beleive it. prehaps you may have read some description of it in some paper and know more about it than I do. but it has been built a long time. I went up on the top of it yesterday and took a look off. fort Pulaski of course is in plain sight and near enough for us to see the flag on the ramparts. with a glass the men can be plainly seen practiceing with the guns. (Seceesh) steamers can be seen flying up and down the river. they can be traced for a long distance by the dense black smoke which they make arising from the use of pine wood in the furnaces. I enclose you a little shell which I picked out of the wall. the earth works we are throwing up include this tower about in the centre of them. yesterday a body of sailors wer at work hoisting some large timbers into the top of it we do not know for certain but suppose our folks are agoing to mount a gun in the top if the old thing is firm enough and would stand the shock it would be a capitol place as we could throw shells right into Pulaski every time. I realy beleive that I am beginning to be of some use in the world or at any rate Uncle Sam means to make me earn my bread by the sweat of my brow. I have shoveled dirt in the trenches and rolled logs and hauled and lifted heavy gun carriges untill my hands are as stiff and hard as a stone masons. Yesterday I worked all day helping to mount guns. we got one twenty four pounder (a rifled peice) in position and got the carrage in position for a larger <u>eight inch Columbiad</u>. today thirty six of our company are out agan on fatigue and the rest of us are or wer to be employed about camp I am takeing time to finish this letter. We have onley about half of our regiment here. three companies are left at Hilton Head they being on the other end of the island when we came away. they have not been with us at all since the day after we landed. all the sick were left behind so that our whole number is onley about five hundred. Five of the Salisbury and two of the Canaan boys wer left behind , Norton Woolcott and Sweet are just getting over the <u>measels</u> Dexter has been unwell for several days. but was better when we left. McNeil had a hard cold but I should have thought might have come well enough but the doctor refused permission. The two other boys have been sick some time and one of them although able to be about I am afraid will not be able to stand soldering. There has been a large body of rebels on this island. our present camp occupies the site of one of thiers. while at a little distance from here in the woods they had another and a larger one which I believe they burned on leaving. I visited it yesterday morning it has all been burned over whether by them or not. We go from bad to worse every time we move with regard to camp ground. the sand fleas gnats and mosquitoes are as bad agan here as at Hilton Head while great black spiders lizards centipedes and other kindred vermin are in abundance. But I suppose we shall not have time to think of such "light afflictions" long as we shall probably encounter greater dangers. Yesterday two of our gun boats shelled a rebel camp on little <u>Tybee island</u>. a small island lying near. I hear that the rebels left the place. we could see the boats and the fireing but as to the effect. we onley have camp rumor. A large raft of timber came floating down to us in the afternoon our boats made for it and succeeded in saveing the whole of it, it is very acceptable being just what we want. it probably escaped from its fastening in rebeldom and floated down with the tide. Now my dear I have written all that I can think of about ourselves that would be interesting. I wish I could think of forty things that I want to say to you in answer to your letters. you will have to ask some of your questions over the second time I fear as they will escape my notice when I get a chance to write. I will write to Mr. R an enquire about the magazines. I presume they are safe but that he has forgotten to send them. as I requested him. Lizzie need not think that I forget her because I do not mention her oftener I think of her often and would like to write but must forego the pleasure untill I have more time. however if she has taken to reading my letters to you I think all things considered she ought to be satisfied. If she will presist in doing so she must read them hastly and not criticize as they are written hurriedly and intended for no eyes except those that I know will bear with thier imperfections. The paper and stamps which your kindness and forethought have furnished me are very acceptable indeed. There has been and I presume will soon be again an opportunity to procure paper here. but just at present it can't be bought but some of the boys have a supply all the time and what one has in our tent is common property untill the borrower can make it good agan. the postage stamps can't be bought here for love nor money and your kindness has kept me supplied while others have been sorely in need. I will send you some money shortly to repay those you have sent and pay for some more. I don't see how either old Mr. Barnum or Mr. Barnums people can think so favorably of me unless indeed it is because they know so little of me. I am very gratefull to them all for thier kindness and good opinion of me. and I will strife hard to merit it. for I am painfully conscious of the fact that I do not at present. remember me kindly to them all if they enquire for me. I am in hopes to see them all agan and enjoy thier hospitality and your own dear company in thier pleasant sitting room. I assure you I often think of the good fire and and easy arm chair which they used to vacate for our accommodation. I am sure I should not know how to act if I wer to attempt to sit in a chair now. not haveing sat in one since leaveing home. If Mr. Brinton enquires about his son you can tell him that he is well and hearty. he is a first rate fellow and seemingly enjoys himself first rate. our boxes have not come yet. we shall not be able to get our mail or any thing else as quick I know for it will all go to Hilton Head first. Do not worry my dear because you could not send me something. I assure you that I can think of nothing that you can send me more than you have already.

<u>Dec. 22nd.</u> I did not have time to finish my letter yesterday as I had to leave and go on guard unexpectdly. I had a cold windy night for it and what was worse the sand fleas in such clouds as to resemble a northern snow storm. I have a little speck or two in my eye yet which I am unable to get out and it is very troublesome. It was so cold that I did not as usual write any while on guard and so am finishing my letter this evening. I am all alone in the tent the other boys all being out on fatigue employed mounting guns on the masked batteries which have been constructed to <u>command Pulaski</u>. This is Sunday evening my darling. I cannot spend it with you. I wish I could. I fancy you will think of me this evening and wish me near. May the time soon come when I can agan hold you in my arms and restore you to your own <u>favorite</u> place. I shall hope to hear from you agan soon. and will write whenever it is possible good night darling   pleasant dreams

Ned

LETTERS FROM EDWIN JANES BARDEN TO HIS FIANCE
AND LATER HIS WIFE, SARAH MARIA JONES,
WRITTEN DURING THE CIVIL WAR

SECTION 2 – 1862

No. 25 missing

(Letterhead - Miss Liberty "Then conquer we must For our cause it is just, And this be our motto, In God is our trust.")

No. 26 Rec'd Jan 13th 1862

Tybee Island Ga Camp of 7th C.V.
Jany 3rd 1862

My Dear Sara

It is nearly a week since I have written to you and then my letter was anything but satisfactory. to me at least. I am not in a writing mood today but as I have a few moments leisure I am determined to try as I am conscious that I have taxed your patience by not writing oftener. to much already. It is a damp misty unpleasant day and I have a hard cold which makes me very uncomfortable. now don't get alarmed because I have told you the truth once and owned that I am not entirely well. I have no doubt but that before this reaches you I shall be entirely well agan. There is no news since I wrote you last. we are jogging along doing fatigue duty as usual. A forty of our boys that we left at Hilton Head came down last night and of course we had a lively time for a little while. one would have thought that we had been separated months instead of a fortnight. Some of the boys are still at the hospital at H.Hd but are all doing well. Sweet will probably not come down here at all as his discharge is made out and he will probably go home in the next steamer. I should have liked to have seen him before he left for home very much as I could have sent home some trifles which I shall not be able to either keep or send any other way. There has been fighting going on for two days up in the vicinity of Bluffton we could heare the fireing from here various reports are currant as to the result but nothing certain is known here yet, nearly all the troops are under marching orders that are at H.Hd. Norton who was one of the boys that came down last night reports that they met a steamer having both mail and express abord. just entering Port Royal harbor so we shall hope to receive news from home agan in a few days. and prehaps some of our long looked for boxes have come to. I received a kind letter from Mr. Randall the other day. he sayes that he has sent ma box of "goodies" and prepaid the Express charges clear through. just like him. any thing that he undertakes he will surely finish up entirely. The act of oppression of which the soldiers are now complaining is opening of thier boxes and the abstraction therefore of every little bottle of liquer or wine which they may contain. It certainly is a great outrage and one of which I think the soldiers have a right to complain. If they onley confiscated where they found whole boxes of liquer I would not complain but to take every little bottle of choice wine or liquer which friends at home send us to be used as medicine is I think the height of meanness. Me Randall also informed me that Mr. Barnum had gone to N.Y. to purchase rubber overcoats for us. we are all highly pleased at the prospect of having them and are very gratefull to our friends for thier kindness and generosity in furnishing us with them. They will be a valuable acquisition to our wardrobes and if we are to have such a rainy season as has been represented to us. they will be invaluable. One of the German regiment was killed by the rebels the other day. he with a number of his comrads wer out at a short distance from camp when they wer fired on from fort Pulaski and he was instantly killed. the shot cutting him nearly in two. he was buried the day after within a few rods of our camp. New Years day passed off very quietly with us. The Germans had a holiday. the night before they had a quart of whiskey apiece dealt out to them and of course they wer very jollie. I was on guard that night and the next day and as our quaters wer near thier camp we got the full benefit of thier noise. as might be expected thier holiday ended in a fight in which clubs wer freely used and one man nearly killed. New Years day was a most beautiful one. clear and warm. our regiment did not go out on fatigue duty that day but had a much harder days work at drilling and inspection. we wer also mustered in preparetory to geting paid off. I have just learned that the mail is expected to leave this afternoon some time. some time here means just when it happens. prehaps in a few moments and prehaps not till night our onley way to be sure and be (season)? so I think I will close this and send it along. It is difficult to realize that it is the first of Jany. the weather (with the exception of today) is very fine and even the nights are not as cold as they have been. I hope I shall be premitted to see you before another new year comes around. If this war is not ended and the troops discharged before that time I shall have no hope of getting home before the expiration of the three years. with much love my darling I remain ever and truely your own

Ned

No. 27   Rec'd Jan 18th 1862

<u>Camp of the 7th Reg C.V.</u>
Tybee Island. HQ Jany 10th/62

Dear Sara

I have just been writing a sort of a letter to Sara B. Pitt is sick with the measels and as he is anxious to have his folks know of all his enjoyments I wrote this morning to inform them of his good luck. he is getting along finely and will soon be about. has not been very sick but very miserable as is usual with that disease. I am better of my cold than when I wrote you last but have not entirely recovered yet. shall return to duty in a day or two. yesterday we received some papers from home. I received three from you. but no letters came for some unaccountable reason. we wer all very much disappointed as the papers though gladly received. comeing under such circumstances wer nothing but an aggravation. Some of the boxes came however and for a short time partialy healed the wound. George's two barrels came and Mr. Landon's box. my own box from F.V. and some others that are expected. did not come. but its better not to have them all come at once. The things all came in good condition except the pies which wer entirely spoiled. My things wer all in good order much obliged to you my dear for the gloves. they will be sure to remind me of you and will keep me all the warmer. the paper and envelopes and pencils wer most acceptable. It is rumored that we shall move from here shortly if we do I hope my box will not come untill we get settled agan. The rebels have been mounting a new gun on Pulaski and a day or two ago they tried its efficiency on us. thier shells reached further than any they have thrown before and exploded within a few rods of where our men wer at work. no harm was done however. last Sunday was a rainy day. the first stormy Sunday we have had since landing on H. Head and it was the first day that we have nothing to do of any kind since that time. for that day of rest we have not to thank God fearing officers. but a kind and mercifull Creator. who has more pity towards his sinfull, ungratefull children than those children have for each other. We all wished we could step in and hear Mr Reid's N. year's sermon. but as we could not. we have to charge it to the profit and loss account of our soldiers life. but we felt that we should be remembered if not in the sermon by many hearts that had been made sad by parting with us and that wer looking anxiously for our return. God grant that ere another N. Y. come around we may all be returned in safety to our homes and the fond hearts awaiting us. I earnestly hope my darling that ere that time has elapsed I shall be premitted to clasp you in my arms and hear from your own lips those words of endearment an affection which now have to travel so many long miles by sea and land to reach me. I think you girls would be quite a show down here and would create a great excitement in camp. such a thing as a real live lady away down here among an army of soldiers may be dreamed of by some dareing lover. but the reality would be out of the question. they would require a strong guard of soldiers to prevent thier being sufficated by the egerness of the crowd to agan catch sight of what makes homes happy and life pleasant, no my dear you will have to wait untill the time comes when I can come to you unless indeed we should be ordered north. which is not at all likely. The weather is very fine now although it has been cold and disagreeable for a few days back. still we do not suffer from cold as our brethren further north do. God grant that before the greatest danger to which we are likely to be exposed approaches. that is the hot weather. this war may be ended and our services no longer needed here. I have but little to write today. I can oneley tell over the old story of love and devotedness. and as to me it is ever fresh and new hope you will not worry of it. I expect some letters from you when our mail does come. I will write agan soon. much love from Ned

No. 28 Rec'd Jan 24th 1862

<u>Camp of the 7th Reg C.V.</u>
Tybee Island Ga Jany 16/62

My Dear Sara

Our long expected and earnestly looked for mail has at last arrived. It came yesterday afternoon and you may be assured that we wer a much more light hearted and better matured set of fellows at night than we were in the morning. I was the fortunate recipient of six letters. five from your own dear self and one from Lizzie. please thank her for me and tell her I will write to her as soon as possible. many thanks dearest for your own dear letters. and all you have written therein. A large quantity of express matter came for the regiment on the same steamer but none for us boys. Two more of our boys who we left behind at H. Hq on the sick list when we came away also came down and joined us. Woolcott is not yet able to join his company. McNeil has been quite unwell for several days but I think there is nothing serious in his case. one of the Canaan boys young Brinton (not the one whose father enquires of you) is still at H.Hd unable to join us. all the rest of our crew are in good health and spirits. There are many things in your letters that I wish to answer and will try to dispose of some of them. but if I don't do it in very good style and make a rather disjointed letter of it. please excuse me for I am continualy interrupted in my writing today. you see it is a rainy day and we are all in the tent. some reading and some writing and of course have some talking to do. You need not trouble yourself to send me any more paper now as I have an abundant supply. thanks to your kindness and shall look out better in future and not allow myself to get short again. I shall be obliged to you however for a stamp occasionaly as they cannot be obtained here. You ask why I did not ask Martha to write to me. I should like very much to have her if I had flattered myself she wished it. but I dare not solicit any more correspondance while I am so remis and unfaithfull with those I now have. but prehaps I was to cool with Martha and if so I am sorry. and if you think so I will write to her again. I am glad Mr French has at last made you another call. I think he ought to be neighbourly at least. but I think it is something more than friendship that promts his so faithfull attendance on Miss Green. I have no doubt Frank will make an excellent husband but I have always been a little suspicious of him since he broke off his engagement with Josie. it is true I do not know all about that affair as you do and prehaps I do wrong in attaching any blame to Frank. I am not intimately acquainted with Miss Green but think she is a very fine girl. I have not yet heard from Frank since you asked him to write. I am sorry you could not go Lenox as your friends wished. I think you would have enjoyed it very much and would have been justified in dismissing school for a week and takeing a pleasure trip. I saw your friend Mr Sabine's name among the list of graduates at the medical college in Pittsfield a short time since. We are very sorry to hear of the death of Elisha Cleveland. so far as I know he was a very likely young man and had his life been spared would doubtless have made a good and efficient soldier and done his country good service. though his is the first death among the Salisbury volunteers. we have no reason to think that it will be the last. It is hardly supposible that of the whole number who have come forth at our country's call from Salisbury none will be required to make thier last resting place in a distant and hostile soil. I am glad that young Cleveland is to be brought home for a burial. I know that it makes no difference where the body lies so long as the soul is at peace with God. but I have always thought that if I was called away in early life or before I wer married (if such happiness is in store for me) I should wish to be laid in our own quiet church yard by the side of my mother. true I have never known what it is to have a mother but if ever poor mortal yearned to know a mother's love and experience her care and tenderness that mortal is myself. but for some wise reason I have been denied the blessing. still it has ever been a pleasant thought to me that if I am so happy as to go to heaven I should meet her there and that we shall know each other. You must not think that we are working ourselves to death for we are not. we do not work hard at all. still I do not like it so well as I do drilling. I do not see any thing that looks like going back to H Hq very soon and think it likely we may be kept here for some time yet. I do not apprehend any danger from the enemy more than the damage that a stray shot or shell strikeing among an unsuspecting working party may do. there is no probability of an attack from them. but little of our attacking them at present at least. Our boys do not fear a war with England. the general feeling seems to be. let it come. we are ready for it. every one seems to owe her a grudge and willing to pay what they owe. they say we have whipped her twice and can do it agan and effectively(?). I do not think myself we shall have a war with her or that we should have had if Mason and Slidell had not been given up. I for one am sorry that they have been given up. I think that so good a chance to hang two such leading rebels who have been among the foremost in causeing all this trouble suffering and bloodshed ought not to have been suffered to be lost. I think you must have quite a family at home now with boarders and all. I

how presume it is not a first rate place for retirement and meditation. does your father get along now: has he recovered from his lameness: I suppose you will answer me and very properly to by asking why I don't write to him and find out. I can oney make the old excuse. that I am agoing to soon as possible. Have you heard from Josie lately. I have not yet received a letter from her. true I don't know as I have a reason to a expect one. but since you wrote me that you had given her my address I must confess that I have caught myself looking for a letter occasionly. I rather mistrust that she has had another offer lately. do you know: Then Mr Reid thinks you expect a ring from Beaufort does he: I hope he'll not be disappointed. he certainly will not if I live. I shall not be able to fee him so munificently as he had been that evening. but I can assure his reverance that it will be done as <u>willingly</u> and with as much pleasure to the parties concerned. How was it that such straight out Episcopalians as the Burralls should call on Mr Reid to preform the marriage ceremony. should suppose they would hardly consider it binding upon the parties unless done up in the orthodox style by a minister of thier own persuasion. There seems to be a general time of marrying in town this winter. I wonder of there will be any chance for the soldier boys when they get home if they are so fortunate: I should like very much to have one of those pictures you had taken last. why didn't you and Lizzie have your pictures taken together and send it to me. you can easily do it by mail. As you say I don't like Mr Myres although I have nothing personal aganst him I don't like him because I know him to be an utterly unprincipled man. he appears well enough but I never wish him to be intimate with a friend of mine. I am sorry the shell broke to pieces. I have ·others but will not attempt to send another at present. I enclose a little flower that I picked in the woods yesterday. Lieut Gale has gone home recruiting. he was anxious to get the chance and was trying hard for it before we came down here. he is home before this probably. I have no doubt that Mr Edgar would be glad to see you. I'll give you a line of introduction if you'll call on him. Pitt sees me writing and says "give my love to her" I shall have to stop writing now for I cannot keep my thoughts together amid the confusion. Please remember that I still love you and am looking forward to the time when I can return and enjoy your love and affection. I need not ask you to write. for you do not fail me ever. good by darling. love from

Ned

No. 29 Rec'd Feb. 14th 1862

Camp of the 7th Reg C.V.
Tybee Island Gg Jny 26th/62

My Dear Sara

I have nothing particular to write this afternoon but as I have a few leisure moments I shall be happy to devote them to you. We are haveing more leisure this Sabbeth than usual but Terry was unwell. Col Hawley is gone so that we did not have our usual regimental inspection and the company inspection did not occupy over a half an hour. We had our usual service. the chaplin preached from <u>Psalms 20th ch. & 7th verse</u>. a very interesting sermon. This afternoon he invites the singers to meet and practice in a sunny place near his tent. and also proposed to read Beachers' Thanksgiveing sermon to any who wish to hear it. The weather though clear and bright is quite cold with a stif breeze. and rather uncomfortable to be sitting around without exercise. so I think I shall not attend. there is to be a prayer meeting this evening and if it is not to unpleasant I shall try to be there. There has never been much of an attendance at the prayer meetings. oweing is a great measure I think to the want of a proper place to hold them. The chaplain's tent is a common officer's tent and will not seat of over a dozen persons. Singing and a sabbeth school have also suffered for the same reason. But it is now proposed to have a large chapel tent which will obviate the difficulty. The tent is expected to cost one hundred dollars. fifty dollars have been already subscribed by the officers of the regiment who would willingly raise the whole sum did they not think that the privates would feel more interested in the matter and more at home in the tent if they wer allowed to pay a share towards it themselves. So the opportunity is given them to subscribe to the amount of ten cents apiece. A large amount you may think. but when you remember that probably five hundred men in the regiment will be likely to give the ten cents apiece it will make you see the desired sum. It has been lately decided at headquaters that the quatermasters should transport such tents whereever regiments are supplied with them so that there will now be no difficulty in takeing one with us whereever we go. Yesterday I was out on fatigue and had as usual charge of a party of men though a much larger one than common. ordinarily a corporals squad does not consist of but ten or a dozen men but it happened yesterday that I had over thirty and it took me the most of the time to keep track of them. But I slipped away in the afternoon and went up to the top of the lighthouse. A thing I have often wanted to do. but heretofore there has been a guard at the door. and none but commissioned officers could go up. now the guard is removed. and I took the first opportunity to gratify my curiosity. I think I have told you before that the rebels on leaveing burned out the inside destroying the staircase and the lantern. access is now had to the top by means of ladder and platform. from story to story. The view from the top is very fine though the atmosphere was hazy. a kind of an Indian summer sort of a day and I could not see a great distance. This afternoon we have a new excitement. A fleet of fifteen or sixteen vessels consisting of gun boats. transports. and schooners have passed the island within a few miles of us. We conjecture that they are agoing to attack Savannah. and that they expect to reach the place by the means of one of the round about channels without going by F't Pulaski. By the aid of a glass we could see that the transports wer loaded with troops. Success to them. whatever be thier destination. Our own works on the island are progressing slowly and are arriveing at completion. We some expect to leave when we get the works done but there is no telling. George and his companions have not returned yet. we are very impatient to get hold of those boxes. but the sea has been rough and as they are probably in good quaters with nothing to do. they do not care to run any risk in crossing. **Jny 27th**

George and his party returned last evening. they have been haveing a good time as I expected. they visited Dawfuskie island where the remainder of our regiment is stationed. did not bring any express or mail as they found Col Hawley had got it loaded on board of a schooner to come another way. the schooner has come in this morning but is not unloaded yet. Well my dear I can think of nothing more to write that would be likely to interest you. I don't know what has got into me lately. I cannot command my thoughts sufficently to get them in readable shape on paper. I hope my dear you will excuse me for not writing more. I will try to write often if I do not succeed in writing much. we wer expecting to hear fireing this morning from the fleet that passed us yesterday but all is quiet yet. may that. and every other effort that is put forth to crush out rebellion and restore peace and prosperity to our beloved country be successfull. and the time speedily come when the 7th Reg C.V. can be disbandied & I can return to the arms of "the girl I left behind me" is the prayer of your ever faithfull

Ned

No. 30 Rec'd Feb. 14th 1862

Camp of the 7th Reg C.V.
Tybee Island Ga Jany 22nd/62

My Dear Sara

    I am on guard agan today. We go on now at two o"clock in the afternoon and are releived at the same tome the next day. when we went on yesterday it was pleasant but looked some like rain. in the evening we had a hard thunder storm and it rained nearly all night and now we have the prospect of a rainy day. We had another mail yesterday and I received your letter of the 8th. A party started for <u>Braddocks Point on Hilton Head</u> yesterday noon to bring over a lot of express matter which is there. and among which we hope to find our boxes. they rowed over in a small boat. George went with the party and I think he will be sorry he enlisted in the enterprize for rowing is hard work for one unaccustomed to it. It is quite cold and windy and I don't think they will attempt to return today. I have no news to write for all is quiet with us as usual. we are still at work on the fort. and batteries and the "?" occasionally send us thier compliments in the shape of a few shells. The boys are all usually well. I wrote a letter to Lizzie day before yesterday. I dare not say much to her because I feared that the old folks would get thier eyes on it. and expect the privilige of reading it. does she hear from Charlie often: has she met him since I came away. if she is disposed to complain because when (she) does not see him often enough you can just remind her that she is as well off as you are and that it is as fair for one as it is for the other. I wrote to Charlie a short time ago but have not had hardly time to hear from him since. 23rd Yesterday turned out to be a cold damp day. to unpleasant to write and so I did not finish my letter. Today is cold windy and stormy and we have not got to go out to work. but it is hardly comfortable in the tent. George has not yet returned and we did not expect him yesterday. today is so stormy that he cannot possibly come. and the sea may be so rough that they cannot come in a day or two. I am keeping your letters untill I can get a chance to send them home some way. as you say I have quite a lot of them. I should have sent them by John Sweet if I could have seen him before he started. This is a real homesick day for me at least. I beleive we think more and more of home the longer we stay away. I can think of but little else when not buisy but home and what I shall do on my return. I have often to check myself by the reflection that it is not at all sure I shall be premitted to return and if I wer inclined to be superstitious I should almost fear that it was a bad omen to be thinking about it so much. Putting them all together we spend many hours talking of our homes, our friends and the girls. particularly the latter. well you must not blame us. we might spend our time much less wisely. more foolishly and wickedly. recollections of a pleasant home and kind friends will never arrouse evil thoughts or incite to many wrong actions. I pity from the bottom of my heart the man who is never home sick. For the want of something more interesting to write I'll (tell) you how we pass the time down here on Tybee. revielle beats at a quater before seven. when we have to turn out at roll call. breakfast at seven. fatigue call is at eight when we turn out for work. surgeons call is at nine when all the sick go up to the doctor's tent for advice. recall from fatigue is at twelve. dinner directly after. return to work agan at two. work untill five when the recall sounds agan. retreat is beat at sunset when we have another roll call. supper directly after. tattoo is at eight. taps half after. when the lights are put out and everything quiet for the night. When we drill instead of work. which is not often. we have I think two hours in the forenoon and two in the afternoon. Sundays if it is pleasant we have a regimental inspection commencing at ten o"clock. but we have to commence directly after breakfast to get ready for it and have to work faithfully to. to get our guns and accouterments in proper trim it usuly lasts untill near twelve. then comes the church call. and all have to be in the ranks and march out to the service to the music of the drum and fife. Service usuly is about an hour long. After dinner we generally have to strike our tents in order to air them and all our things. At half past four comes our dress parade so you see that the Sabbeth is anything but a day of rest with us. I can usuly get as much leisure time on a week day. This I think is entirely wrong. it is certainly unnecessary and an uncalled for violation of one of the strongest of God's commands. I think if we ever get home we shall be able to appreciate our Sabbeths and Sabbeth day privaliges better than ever before. I certainly hope my dear that dreams do not go by "contraries" always at least for I earnestly hope that some of yours may come to pass. You know that you have had several that you have told me of. but would not tell me what they wer. I mean to remember them or about them and some day I may have an opportunity to ask you in person to tell them to me. which I shall not fail to improve. As for me I find it difficult to get to sleep long enough to dream for we are so tormented with fleas that in warm nights it is impossible to get anything like a decent nights rest. I do not see how it is hardley possible for men to stay here in hot weather for I should think the insects would render it intolerable. But I can assure you my dear that I have day dreams a plenty and I doubt whether your night dreams are more pleasant and I might add more wild. It has been one great fault of my live thus far that I have <u>dreamed</u> to much and <u>acted</u> to little. It will not do. this life of ours is a battle. and every one who would succeed must act well thier part. it will not do to depend upon others or to get them to act for us. nothing but active. ernest and perservereing effort will ever make the man or give success in life. had I but acted up to this truth years ago. I (would) not have to regret so much ill spent time and unimproved advantages. I well remember Mr Reid's donation and how well I enjoyed myself. it is of no use wishing that I could attend this year. but I hope you will go and enjoy yourself. you must take a sleigh ride for me if you can get a chance for I don't expect to see a flake of snow this winter. But I think the rainy season which we have been looking for has come at last. and very unpleasant it is to. I hope you enjoyed your contemplated visits around amongst your scholars. Lizzie says that you are looking better. I hope it is so and that you are feeling well to. I am sorry that I am not able to make this letter more interesting. will try and do better in my next. good by darling. much love from

Ned

No. 31 Rec'd 24th Feb. 1862

<u>Camp of the 7th Reg C.V.</u>
Tybee Island Ga Feby 10th 1862

My Dear Sara

Before you receive this you will have looked long and in vain for letters from me. I say long. for a fortnight I know seems a long time for lovers to wait. I have been sick for the last ten days and unable to write. Today I am trying to write lying on my back and sitting and lying around any way to do it easy. I have a sort of a slow malarious fever have no appetite at all and have not had since I was taken sick. I do not know how much longer I have got to remain on the sick list. but shall not be able to return to duty right away. I am able to be about and in fine weather feel better when out of doors stirring around, but for two days now it has been wet and stormy and of course I have been confined to the tent. Wells and Hawthorn have been sick all the time I have and are still under the weather. All the other boys are usually well. Our Boxes all came in a heap the last day of Jany. the day I was taken sick so that they have done me but little good as I have not had the least appetite for the "goodies" and could not help eat them. Hawthorn and Wells are in the same boat with me. and unable to partake very freely. Mr Randall sent me a box of nice things. and they all came in excellent order. Our Hospital has been moved down here since I wrote you last. and we now have our tent full agan. all the boys (I mean those in our tent) are now in and have slept together for the first night since leaveing Annapolis. I think now that we are likely to remain on this island for some time. Our new tents have have come and we are agoing to pitch them here. Besides it is currantly reported and beleived that we have been removed from Wrights brigade and if it is so and we are a sort of detached regiment we shall be as likely to remain here as to go away. Three of our companies are still on detached service over on Dawfuskie island and are like to remain there for the present.

Feby 12th I did not succeed in finishing my letter the other day. Yesterday was cold and unpleasant and I could not write very well. Today I think I am some better but I get along dreadfull slow. The day is fine but rather chilly this afternoon. our regiment have been employed for the last two days in prepareing the ground for our new camp. I see that they have pitched some of the tents this afternoon. presume we shall have to move tomorrow. I dred the change as the tents are little small narrow contracted affairs onley seven feet wide by nine in length. slanting directly up from the bottom to the ridge pole which is onley a little higher than my head. Into this small compartment five men have got to stow themselves with all thier arms equipments and clothing. and here they have got to live. if liveing it can be called. There is not near the room for five men in one of them that there is for ten in the old ones. There will be a few tents that will not have to have but four men in them but I do not expect to be able to inhabit one of them. I do not know as there is any war news to write you. I have forgotten if anything worthy of note has transpired. We have heard no news from the expedition that I mentioned as haveing passed our island one Sabbeth. our gun boats are lying in a little sound over back of <u>Dawfuskie island</u>. we can see them from our fort. it is said that they can enter the Savannah river from where they are by comeing within three quaters of a mile of fort Pulaski. I presume if it is so. that they will enter the river some fine day when they get good and ready. Today a detachment of men from the gun boats rowed down quite near Pulaski. they appeared to be takeing soundings. the rebels opened fire on them from the fort. but they kept about thier buisiness very cooly and when through rowed off. They have not fired on us I believe since I wrote you last. We have three heavy masked and bomb proof batteries now completed. each one mounts a heavy eight inch columbiad and they are ready for use any day. We think they will do good execution aganst the fort. and now darling it seems as though I ought to make some more excuses for not writing sooner but I know you will believe me when I tell you that although I have been able to be about more or less for several days yet I have not felt able to write sooner. my head has troubled me a good deal and you know that it is hard work to write when ones head is out of order. Still you often speak of writing to me when your head aches. you women are always more self sacrificeing than we men are. O dear Sara if God spares our lives and premits them to be united I fear you will some day be tempted to regret the sacrifice you have made for me. I am glad you are enjoying the sliding it is fine sport. I enjoyed it very much last winter. and a good deal of it to. I wish I could be with you prehaps we might have a sleigh ride occasionly I have forgotten that I had expressed any dislike to haveing you go with Mr Crowell. I know that I have expressed my opinion of him pretty freely and not very favorably of him. He is able to act the part of a gentleman well and I know he would not dare do otherwise in the company he was in that evening. prehaps I judge him to harshly. At any rate I am glad you went that evening and had such a pleasant time. I stand a poor chance to have any sleigh rides this winter. I did make one of the party from Falls Village last year. I remember it well. a bitter cold day. an exceedeingly lively time at Pecks on our return and a severe cold. which abut used me up for several days. I have heard no account of the one they had this winter. I Miss Martha K. likes to make a considerable ado about small matters some times when it suits her convenience. prehaps she did not hear so much said about me as she makes believe. I am real sorry you did not attend that musical convention. you would have enjoyed it so much. I am sorry to learn that you have been so unwell and hope ere this you have recovered your health. The evenings are very fine now. bright moonlight. There is a violin in camp and for two evenings past we have had music and dancing untill roll call. the negroes of which there are any quantity hanging around the camp. have done the principle part of the dancing. the large box which was sent to us turned bottom side up serveing for a floor. We have had several heavy rains of late and in consequence the grass is springing up quite green and the wild flowers blossoming freely. I send you one or two samples. they are not well preserved. but I do not understand the art. when fresh they wer quite pretty and very fragent. On Dawfuskie island they have peas high enough to bush. potatoes up. and strawberries. and peach trees in blossom. Now my dearest I must bid you good by for this time. I am agoing to send a lot of letters home by Sweet who has finely had to come down here. I think he will take them to you all right. I also send a few envelopes for you to use. I am sorry I could not get more of them. but I knew they wer nearly gone. Do not get low spirited to fear I am forgetting you. you are dearer to me than ever and I am still your own loveing

Ned

Feby 13th. Dear Sara I am agoing to enclose my letter and put it in the chaplains mail bag this morning but it may not go untill I have a chance to write agan. Besides the wild flowers which I mentioned yesterday I enclose a rose. and a geranium leaf which came from Stoddards plantation on Dawfuskie island. Also some leaves of a camelia which came from the widow Mungens on the same island. They wer brought over by one of our boys who went over as one of a boats crew. He brought back some splendid boquets of flowers. all of which wer growing in the open air. I am feeling some better this morning and am now in hopes to be about agan in a few days. give my love to Lizzie and Julia. Does Lizzie hear from Charlie lately: I wrote to him some time ago. but have received no answer. I wonder if he received the letter. please remember me kindly to my enquireing friends. especially to old Mrs Barnum who enquires so kindly about me. I do not know Sara dear when if ever I am agoing to see you again. I do not see anything that looks like a speedy termination of the war. but still the end may be nearer than we think. But we must both keep up good courage and hope for the best. If my health is onley spared me I shall be most happy. Mrs Randall wrote me a long good letter which came in my box. It did me more good than the contents of the box. for I could enjoy the letter even though I was unwell. I must answer it the first opportunity. I have got clear behind hand with my correspondance since I have been unwell. Good by darling remember that I am as ever your own loving

Ned

CIVIL WAR LETTERS 1861-1865

No. 32 Rec'd March 6th 1862

camp of the 7th Reg C.V.
Tybee Island Ga Feby. 25/62

My Darling Sara

I have this day received three more of your dear good letters. accept many thanks for them. You are indeede a faithfull correspondant. and faithfull to when fear many others would be disappointed at my tardiness and be at least disposed to complain. I wish I might be more punctual. and be able to feel sure that if you do not receive one letter at least by every mail that goes north that its not my fault. I mailed you one letter about ten days ago which is the onley letter I have been able to write since the last day of Jny. at which time I was taken sick. When I wrote the letter to you was recovering fast and hoped to have been able to return to duty long ere this. But I had a sort of relaps or a fresh attack of the disease which brought me down much lower than I had been any time before. I am now getting well agan. But am very weak and should find it very difficult to sit up all day. I am compelled to get about more or less as I have to wait on myself. But prehaps it is better for me than if I had some one to wait on me and nothing to stir me up to action. I believe we wer just agreeing to more into our new camp and new tents when I wrote my other letter. I do not like them as well as the old kind as there is not near the room. we have five great broadshouldered men in our tent and it is utterly impossible for us all to lie on our backs at once. Last night one of our boys had a large blister seven inches square on him drawing all night. Of course we could not touch him. It made very cramped quarters for us. And I spent a miserable night. But poor Dick had the worst of it. He was the onley one in our tent that was worth a rush to wait on me and now he is down I cannot do much for him. Three or four large schooners are anchored here waiting to be unloaded. They have on board large morters with all the necessary carrages and things to mount them with. They are being unloaded and put in position as fast as possible but the wind blows a good deal lately and the surf runs very high so that there is but little time that the men can work at unloading. Of course they are being placed in position to bear on Pulaski. and I really should not wonder if there was some music here by and by that will stir up the boys a little. The Dutchmen took three <u>secessh</u> over near the fort somewhere a day or two ago. they wer full blooded and talked big. one of our men told one of them we wer agoing to have Pulaski soon. he was very incredulous. And replied that he "diden't see it". I hope we shall. There has been a deal of heavy firing for a few days past. It is off somewhere in the direction of Savannah. what it has amounted we do not know. But it is said that the communication of the fort with Savanah is nearly cut off. I received a letter from Frank French along with yours. I shall answer it soon as I can. He said he supposed he could not write any news as well as a great deal of <u>love</u>. Cool isn"t he. I have just read your letters over again. It makes me sad when I think how long you have to wait to hear from me when a letter from me is so highly prized by you. I am sorry my dear but I know you will excuse me when you receive my last letter and know that it is because that I am unwell. that I have not written. The only letters I have written since some time in Jany have been to you. I am sorry that I did not try to write sooner than I did. I could not have made out much of a letter but prehaps a few lines would have been better than nothing. I am very glad you have been able to enjoy yourself so well and pass the time so pleasantly. I wish you could have had your sleighride to N. Canaan but one gets disappointed about a sleighride occasionly. I suppose you will be somewhat more lonely now that Martha and her friends are gone. She is a queer creature. I can oneley explain her singularity on the ground that she is a "Richardson" and they are a peculiar people. I cannot blame Mr. R. for releasing her from her engagement under the circumstances. I think you mentioned to me once before that after showing you one of his letters she tore it up and threw it in the fire. I do not think that it is either ladylike or generous to treat a man in that way. If the lady thinks she has been hasty in forming an engagement and as time passes on she finds her love for the gentleman on the wane untill she is satisfied that she cannot be happy with him better come at once to the point and tell the gentleman frankly that she wished to break off her engagement. at the same time giveing him her reasons for such a course. I think the ladys conscience would then be clear. while the gentleman would still respect her. which I should think it would be very hard for Mr Rosette to do in the present instance. I am writing very plainly but I am writing to one with whom I have no secrets and to whom I am apt to open my heart. If you do not agree with me please say so. I have not of course written the above for Martha's eyes or ears. as I do not care to incur her displeasure. I am glad Lizzie has been able to see Charlie agan. I'll warrent there was a happy meeting. does she get any chance to correspond with him now: I wish he would answer that letter I wrote him some time ago. I remember Mr Root I saw him the evening I spent at his house. I am sorry to hear of his death. I always remember the family with kindness and gratitude and sincerely regret this great affliction that has fallen on them. If you write Mrs Root agan I should be glad to be remembered to her and express my sympathy with her in her bereavement. You see my darling that those who stay at home are called away. as well as those who go to the war. In reply to your question whether you shall send Dr Sabine your photograph. I should say yes of course I can see no impropriety as he has urged you so hard for it. The situation he has applied for is a very desireable one indeede but it ought never to be held by a young practicioner just from his lectures. the lives of men are to precious to be intrusted to the care of such inexperienced men. we have abundance of proof of that in our own regiment. If I ever get home I can tell you a story of medical practice in the army that I think will astonish you. Don't think my dear that I am at all envious of your friend or wish him anything but the utmost success in all his plans but haveing as I think suffered from the effects of ineficent doctering I have sympathy (for) those who do or are like to suffer from the same cause. I have no doubt <u>Dr. Sabine will be a good M.D.</u> when he has had practice. the want of which is all I have aganst his getting the desired situation. I hope my dear when you receive my letters you will alter your mind with regard to the "out of sight out of mind" buisness. I do assure you my dear that I think of you much and often. especially has it been so during the long tiresome days of my sickness. I love you as dearly as ever and long earnestly to fold you again in my arms and tell you that I do love you and that with me at least "abscence has not conquered love" nor will it ever. I have no fear that my home. should I ever be so happy as to have one. will not be the most attractive place I can find and I hope my wife will not have to spend many lonely evenings on my account. I am sorry to hear that your Father appears troubled and is unwell. I have never written to him yet. its an inexcusible neglect on my part. but I will certainly try to write to him as soon as I get able. Its raining this afternoon tremendiously the overcoats would come in play now if they wer here. its a heavy thunder shower such as we have in July at home. I did not receive a letter from Mr Barnum. presume he did not have the opportunity to write when he expected to. we have learned to wait a long time for things sent us from home and so are not disappointed at not haveing received them yet. If shipped when you say they wer they should have been here before now but there is a great carelessness and delay with soldiers boxes which is as annoying as it is unjust. We shall all feel under the greatest obligations to Mr Barnum not onley for his kindness and liberality but for the interest he has taken in seeing the matter attended to in proper shape. Another poor fellow has gone to his long home out of our regiment he died in the hospital. was buried sunday afternoon. of course I could not attend but I lay and heard them pass. the drums and fifes playing the dead march. then after the short service was over and the coffin lowered I could hear the three volleys fired over the grave then the procession reformed and the drums and bugles struck up a lively air and one not knowing what had been done would have thought they had been at some gay scene. We have had no releigous meetings of any kind for some time. o how I miss the Christian privalages which are so plenty at home. It is so hard to keep near to God. no one in the buisy throng around seems to think there is such a being but his name is used so freely it does not seem that man can be so dareing in the abuse of his creator. but it (is) at so. young men who now swear every word almost. Have told me that at home they seldom swore. On young man was pointed out to me who seldom swore when at home and was a prominent member of the Sabbeth School. now he is as bad a man to swear as there is in the regiment. The means of grace we soldiers enjoy are very limited it is true but we are so apt to neglect those we have and almost before we are aware of it we have slidden away. Soldiers are a careless and thoughtless kind of men and although I have no doubt but that the loud mouthed boaster and profane swearer would tremble and be almost if not quite ready {when any sudden danger overtook him} to cry for mercy. yet soon as the danger is over he would be the same as before. I do not wish to fall away and if life is spared go home a worse man than when I came away. I wish to grow better as it is every Christians duty to do. I must close my letter now with an apoligy for such poor writing. my sickness has unstrung my nerves and that with weakness has made it almost impossible to write so that I could read it myself. I am finishing my letter the 26th as I did but little more than commence it yesterday. I hope to be able to write agan soon. Please give my love to Lizzie and say to her that I am glad she succeeded so well with Charlie. Please remember me to my friends at Mr Barnums who are all kindly remembered by me. I wish I could write some news that would interest old Mrs B. but we soldiers know nothing of what is going on untill the news has been to NY and back. if I could write anything that would interest her I should be glad. The shower is over with now and the sun has come out bright. They tell me that a garden the Dutchmen have made up near the light house is looking splendidly. the vegetables being up and as far advanced as ours at home are in June, Peach and Plum trees in full blossom. so you see it does not look much like sleighing down here. Good by my darling I am as ever your own

Ned

No. 33 Rec'd March. 1862.

camp of the 7th Reg C.V.
Tybee Island Ga Mch 3rd 1862

Dear Sara

    I have nothing in particular to write to you today. but I thought I would try to make out a letter as I have time and feel able to write. I am much better than when I wrote you before. but still very weak and unable to do duty. I never had any disease handle me before as this does. I have not the least appetite. and have not had since I was first taken sick. always before when I have been sick and began to get well I have had a craveing appetite. I buy apples and oranges and these taste good and do not hurt me. I eat them freely. apples at present are quite plenty and cheap. We wer paid off the first day of March. that is we received the two months pay due us the first of Jany. so that we have still two months pay due us. We have the promise of receiveing that before the close of this month. but quite likely we-shall not see it untill another two months have passed. Uncle Sam does just as he has a mind to with us now. There has nothing new transpired since I wrote last. The men are still at work at the morters and on the road over which they are to be drawn to thier destination. There has been three sloops of war here about ever since we have. two of them wer steam. the Pawnee and Wyandotte and the third the Vandalia a sailing vessel. The last day of Feby a fleet of gun boats passed our island and the two steam sloops of war went out and joined them. I understand that they wer destined for Fernandina. a place of some importance in Florida. Of course you will hear of thier success long before we shall. Another poor fellow went to his long home from the hospital yesterday. he was taken sick since we landed at Hilton Head and his disease ran into the quick consumption and although his discharge was all made out yet for some reason his departure was delayed untill death overtook him far from home and friends and he is takeing his last sleep beneath the oaks and palmettoes in the enemy's country. I have at last succeeded in writing to your father. I finished and mailed the letter this morning after being three days nearly in writing less than two sheets. I did not succeed in making out a very interesting letter I fear. I did not mention our affairs. or you. I did not know whether you wished me to or not and as I do not see as it can be of any use at present I should chance to let the matter rest untill   shall be nearer the consumation of our desires may that happy and long wished for time speedily come. I beleive that I wrote to you that we wer to have a large chapel tent. it has come and they pitched it yesterday morning and used it in the afternoon for service. I did not attend as it was quite chilly and I did not think it prudent. the tent is a large nice one. will hold a hundred I should think comfortably. it will be capitol for prayer meetings and singing schools. I had to think of home yesterday. it was just six months since the last time I attended church at Salisbury and that to was communion day. ·It brought it back to my mind all the incidents of that. to me exciteing day. I had no buisness to be excited but you know I was for I could not conceal it. God alone knows where I shall be at the end of another six months. many of the boys are confident of being home then. but I have but little hopes of the wars closeing so soon. We are reported to have gained great victories out west and I hope all will prove true but those victories must be followed up by still greater ones ere we can hope for an end to the war. As I gave you warning on the start that I could not write much of a letter you'll not be disappointed to find me closeing off without accomplishing much. but I am determined to write oftener than I have heretofore if some of my letters do fail to be interesting. they are singing out that the mail goes right away and I think I'll try and get this in. I shall hope to receive some more of your dear letters by a mail we are expecting to receive in a day or two. Good by darling remember that I am ever and truely your own dear   Ned

No. 34 Rec'd March 20th 1862

camp of the 7th Reg C.V.
Tybee Island Ga Mch 8th 1862

Dear Sara

As I hoped when I wrote you my last letter we received a mail a day or so after I wrote and I received five letters. three from your own dear self. and from Lizzie. and one from Mr Barnum which was written and mailed the same day that one of your was that I received the mail before. So you see oμr mail matter does not always come and go strait. Lizzie's letter was written at North East and mailed at Millerton. I have been trying to catch up in my correspondance for a few days I back and have partially succeeded I but have still a number I wish to write. One reason why I have so many to write just now is that there wer a number of persons to whom I have owed letters ever since I enlisted haveing never written to them at all. I am trying to write all I can now for I shall be able to go on duty in a day or two now. and then good by to leisure time. I have now a good many little hinderances so that after all I cannot make out to write much more than one letter a day. I read a long letter the other day from H. Bushnell to Lee Wells. he (Hezzy) is in Michigan City in a store. and well suited. there was nothing special in the letter which covered three sheets. of course it was a good letter. If I onley had the Bushnell faculty of writing letters I could be more punctual and give my correspondants less cause for complaint. We have had real March weather since this month came in. windy and cold. night before last was the coldest night we have had since we landed in Dixie. water froze in the basins over half an inch thick. today it is warmer and quite pleasant. but I have managed to catch a beautifull cold during some of the chilly nights it does not make me sick but is real uncomfortable I as all colds in the head are apt to be. It seems as though I were to have a death to chronicle each time I write to you lately. The last death is that of our Adjutant who died on Thursday. he had been sick but a short time. and had the best care that can be taken of a sick person in camp. his disease was typhoid fever. This is the first death among the officers of the regiment. Several others of them are quite sick and among the number our Lieu Colonel. but I am glad to say that he is getting better. I think the men would rather lose any other officer than him. The Adjutant's body is to be sent home to New Haven where he has a mother liveing. Am not sure of but think he was an onley son. The rank of an adjutant is that of first Lieutenant but they have the pay of a Captain and are allowed a horse besides. He is the lowest staff officer. I think our own first Lieutenant stands a good chance for the vacant office if he wishers it as he has several times taken the adjutants place when absent or sick. I shall be very sorry to have him leave the company. As usual I have no war news to write. our regiment are hard at work as they will have to be for two months to come. The news received by the late papers raised the spirits of the boys very much and some of them begin to think seriously that they will be home before the summer is over. I am not as sanguine as they but shall be ready if alive to go as soon as they do. I think there has still to be hard fighting done and a good deal of it before the Rebels are affectively subdued. and when it is done it will be some time before all the troops are discharged. I have answered Frank French's letter. Lizzie wants me to write to Charlie. I will if I can get a chance. but I have written to him since he has to me but prehaps he expects me to write first all the time. One of my tent mates has fallen in love with my cap and wants one like it real bad he asked me yesterday if I would not write to the lady who knit mine and ask her to knit him one and send it to him by mail and he would pay all expenses. of course I redaly complied with his request. most of the Salisbury boys have been supplied with them by thier friends. but the Canaan ladies do not seem to know how to knit them. at any rate they have not sent any of them down to us. If you can find time to knit my friend a cap and send it to him or me I will take it as a special favor to myself. The gentleman's name is Richard Hawthorn. He has been sick a great deal and although better now I fear is not agoing to be very tough while in the army. A cap the size of mine will fit him. And now my darling I must close. you have indeed been very carefull how you have written lately much more so than I really wish you had for indeed your letters have been almost cool along back. I do not wish you to restrain your feelings when writing to me. write just as you think and feel. I am so glad you had been able to enjoy yourself so well this winter. I feared that the winter night prove a long and dreary one. bad to get about and difficult for anyone to find much enjoyment. But instead from all accounts I should think you have had a very pleasant winter. Never mind dear we hope there is a good time comeing when he whom you love will be agan able to ask you out to ride and spend not onely one day but many many days months and years in your society and then when our journey here below is ended. we hope to join hands in that "better country" and together rejoice in our Saviour's presence through all comeing ages. I shall hope to hear when next you write that all your little patients have recovered from thier sickness and that none of them have been called to go to thier long home. My own heart has been made sad by hearing of the death of my friend Manley. I think you have heard me speak of him. he was a distant relative. but we wer brought up close together and intimacy and strong friendship made him seem almost like a brother. his sickness has long been lingering. But he had long before made the Savior his friend and death had no terrors for him. Still it was a hard trial for him to leave his wife and child. he has often told me that was all he cared for in this world. still he took a lively interest in the affairs of our country and as he took my hand and bade me good by at parting he said "if I wer only well I should be with you." We all have our cares and sorrows either fancied or real and "every heart knoweth its own sorrow" But it is a comfort to know that we are in the hands of one who "doeth all things well" and that all the events of his Providence though they seem dark and mysterious are all ordered in love and mercy. I should have loved most dearly to have joined in worship with you on the Sabbeth you mentioned. I believe I shall know better than ever before how to prize the privaliges and blessings of out quiet home Sabbeth if I ever return. The first Sabbeth in March I longed to be with you. And now darling

I must bid you good by agan. As I mentioned in a previous letter I am in hopes to write to you oftener than I have done though I may not succeed in writing very long or interesting letters. I need not ask you to write often for you are always faithfull. Good by darling much love from

Ned

No. 35 Rec'd April 2nd 1862

Camp of the 7th Reg C.V.
Tybee Island Ga Mch 20th 1861*
(*date should have been 1862)

My Dear Sara

I received your letter of the 7th day before yesterday. today is the first chance I have had to answer it and I hasten to improve it. I have been out at work every day but one since I returned to duty untill today and I thought it was time for a rest. It has rained tremendiously during the morning and is raining by spells still. real April shower. The fatigue party started out once and returned untill the heavy rain was over when they wer agan called out. They put us right through now day and night. Since I wrote you last there has been thriteen large cannon landed on the beach. These have to be drawn in the same way the morters wer and the same distance. Among the canon are five large ten inch Columbiads weighing over fifteen thousand pounds apiece. long Parrot guns which are rifled and carry a long distance. and I am told some of <u>the James rifled canon</u> of which we hear so much. Large quantities of powder. shot. and shell together with the gun carriges are also being unloaded, and at times line the beach. All these things have to be brought on shore in large boats called lighters from the schooners which bring them, and which are anchored at some little distance from the shore, but as near as they can safely carry, It makes the beach present quite a lively appearrance during the working hours of the day. A number of teams consisting of horse carts and four horse wagons are continually passing to and froe conveying the ammunition and other ordinance stores to the various magazines prepared for them. The reason we have to work nights is that the place we are to work at is in sight. and very near Pulaski. so near that our men can hear the rebels on the fort. besides for quite a distance we have to draw the guns right in plain sight of the fort and it is hardley likely that the rebels would let so fine an opportunity to drop a shell among us with the fair prospect of killing a score or so. pass without improveing it. I have not yet been up to the place where our batteries nearest to Pulaski are being constructed. but think it quite likely I shall haave a chance tonight unless it rains very hard. and prehaps even then I think I am quite well agan now and able to endure as much as ever. our overcoats came last night. just in the nick of time for this rainy spell. and wer gladly welcomed by us. They are very nice indeede and will be of great service to us. I shall write to Mr. Barnum acknowledgeing the receipt of them as soon as possible. in the mean time you need not tell him they have come unless they enquire. I have not written to you as often as I meant. And promised myself to. it does not seem but two or three days since I wrote you last. but it is over a week. How fast time flies when one is employed. I shall make no more promises about writing but will do the best I can. I shall not attempt to keep up with George who writes page after page and letter after letter. what he finds to write about so much I am sure I do not know unless he writes down at night all his thoughts during the day. for surely this monotonus life of ours cannot furnish materials for any very extensive literary work. Since I wrote you last there has been another death in our regiment caused by that which causes so much misery and wo in so many families all over our land. liquer. The man was an habitual drinker. but had been sick some time and had become quite weak. by some means he procured the liquer of which he drank very freely. drinking untill he was insensable. fromm which condition all the efforts of the surgeon to restore him wer unavaling. and in that state he died. Another warning to us all. but how little it will be heeded by those whome we think most need it. An accident occured to one of our fatigue party yesterday. by which I fear a young man has been made a cripple for life. although not a bad one. or at least bad compared with what it would be to lose an arm or a leg. The man had hold of the ropes close by the wheels betwen which the heavy canon was hung. it happened that at one place on our road some men wer cutting small brush to be used somewhere about our works. and had piled them on both sides of the road very close so that there was barely room for the wheels to pass. the man got entangled among the brush his foot slipped under the wheel and before we could stop it had passed over it. crushing nearly one half of his foot in a shocking manner. I saw it after the shoe and stocking wer removed and it looked bad enough. but he bore it like a "soldier". Soldiers you know are not expected to have any feelings or to care for any wound less than the loss of a head. or some such important member. I was not at any time during my sickness in the hospital. I was not sick enough to render it necessary for me to go there. On many accounts if I wer much sick I should prefer to be there. each patient is provided with a rude bedsted wide enough to hold him comfortably and he has more room than he possibly could have in his own tent. I have been in our hospital a considerable and think that the patients are as well cared for as could possibly be expected under the circumstances. They have the best medical service in the regiment. and if the weather is cold the hospital is warmed and blankets furnished to those who have not enough of thier own. besides the hospital patients fare better than we well folks and better than one could who was sick in thier tent. I do not consider myself particularly fortunate in relation to my tent mates. not but what they are all good fellows in thier way. but I do not exactley like thier ways. but do not wish to complain to any one but you. They are F. J. and Henry Brinton, A. E. Barnes and R. J. Hawthorn. the last mentioned I get along the best with. he has been sick a good deal since he enlisted though always tough and strong at home hardly knowing what sickness was. and I think it has made him more impatient than it is natural for him to be. Our boys are all well now except one of the Brintons (F.J.) who is troubled with the rheumatism and has not done any duty since soon after we landed at H. Head. I am glad to hear that the children have all escaped so easily and hope that Lillie will soon be well agan. If pity for (her) will induce you to write me. I hope you will still continue to pity me often though as I often said I have no need to reproach you for not writing often. And now my darling I must close. I managed to finish a letter to Lizzie this morning after having been a week in writing it. I shall not be able to write as much now as I did while getting well. I can assure you I am glad to hear that our cause is prospering and that the prospect of a speedy termination of our difficulties is flattering. I am (as I have ever been) anxious to return to you my darling but do not want to leave as long as my poor services are needed. I sometimes flatter myself that I have a few friends in Conn. but it is a great comfort as well as pleasure to know that there is one there who beyond all others loves me most dearly and who watches. waits. and prays for my return. The Bible tells us that the prayers of the righteous avail much and also gives the assurance that the prayer of faith shall be answered. and I am comforted. encouraged and my courage strengthened when I remember that I am not forgotten in your petitions to "Our Father in Heaven". Praying that your life, health and happiness may still and ever be precious in the sight of Him in whome we hope and trust. and that his kind Providence may grant us a speedy, and while life shall last, a never to be broken reunion. I remain as ever your own

Ned

No. 36 Rec'd Apr 3rd 1862

Tybee Island G Mch 12th 1862
Camp of the 7th Reg C.V.

My Dear Sara

    We received another mail three days ago and I received your two letters of the 25th & 28th for which please accept many thanks. I also received the havelock in good condition. it is a nice one and I am more than obliged to you for your thoughtfullness of my comfort and your kindness in sending it to me. I know you will make a good wife for you seem to be always studying my comfort and happiness. You will be glad to know that I am better so much so that I have been on duty this forenoon. I am agoing to proceede carefully for a few days and then I hope to be able to take my regular turn with the rest of the men. Prehaps you would like to know what I have been doing. We have been hauling on of the heavy morters up to its place. They are loaded upon four large wheels made on purpos larger than any wagon wheels you ever saw. then long ropes are attached and about two hundred men take hold and draw them to thier destination. It is hard work and they get over the ground very slowly. The morters are not all to be put in one place. but in several differant batteries. some of them have got to be placed very near the fort and those have to be placed in the night. Tonight the men have got to draw one over to that place I am excused from night work by the Doctor for a few nights but shall probably get enough of it before the week is all done. The shells are being unloaded and carried to the magazines as fast as possible. We have received a small addition to our force on the island consisting of two or three companies of sappers and miners. island consisting of two or three companies of sappers and miners. and two companies of the third Rhode Island regiment who are artillerists and who are to man the batteries during the attack if we ever get redy to make one. We have heard that there is a mail and express matter at Hilton Head and that we shall probably receive it in a day or two. We are also told that no mail will go from Hilton Head to the North untill after the expected attack on Pulaski. If such is the case. and I am more than half inclined to beleive it. you will probably receive this about the first of May. But I shall continue to write and send them from here whenever I have an opportunity. Tell Lizzie when you see her that I have written to Charlie. and am going to answer her letter as soon as I can get a chance which I think will be before long. If you get one or two of my last letters you will see that I have at least made out to write to your father. I did not think far enough to send any particular message to your mother for which I am very sorry. I should certainly be most happy to have her write to me. and have no doubt

George would be glad to receive her favors also. I presume you will have a chance to read my letter to your father. and have no doubt you will be disappointed in it. for. for some cause or other I could not write well at all when I attempted that one. I had the pleasure of reading a good long letter of his to George received by our last mail. With regard to the letters you speak of I beg pardon for my neglect in not acknowledgeing the receipt of them as I ought. The way in which so many have passed apparently unnoticed is this. I would receive some times four or five at a time. when I came to write which has freguently not been untill two or three days after thier reception. I would prehaps be in so much of a hurry that I had no time to read them over. and so I have answered them in a sort of a general way. which when I have reread your letters afterwards has been entirely unsatisfactory to myself but your kindness has kept me----so well supplied that I have not been able often to revert back to them and answer them in detail. Feby 13th. I did not finish my letter yesterday partly because I was tired and partly because I did not feel like writing. I have reason to be sorry for it today as the mail has gone from here this morning. I have been out at work agan today at the same work we did yesterday. and worked much harder than I meant to and am very tired. I never can bear to see things drag along or rather not drag at all. as was the case a part of the time this forenoon and so I worked harder than I meant to but it will not hurt me any. The officers in thier hurry to get a good deal of work done hurried the men off so early that many of them had no time to eat thier breakfasts. of course the men wer angry and resented it as much as possible. they would not pull onley just when they wer a mind to, although they seemingly obeyed orders and it seemed as though they wer doing thier utmost when in reality they would not draw a pound. While we wer on the beach this morning loading the morter, the little ferry boat Mayflower came over from Dawfuskie bringing the remainder of our regiment. the three companies that have been separated from us ever since a couple of days after we landed at Hilton Head. They have been under the command of our Major who is very unpopular with the men. and they seem very much pleased to rejoin the regiment and we are certainly very glad to have them for it makes us a full regiment agan. and just at this time we need thier assistance very much. The officers of these companies are considered by the boys to be the best lot taking them through of any in the regiment. Today the weather is very fine and warm. I noticed the butterflies fluttering about this forenoon. I will send you more flowers whenever I can find them for which I have had no opportunity lately. there are no flower gardens on the island and wild flowers are not very plenty just around here. I am glad those I sent wer in good shape. I was afraid there would be no color in them by the time they reached you. I should have been glad enough to have gone with George I can assure you but they could have no drones along. and I do not know anything about rowing which requires both good muscle and strong lungs. so I could not go. Two of the companies who have just joined us are agoing to pitch thier tents right on behind ours and as it requires a good deal of clearing and leveling before the ground is fit they are at work at it this afternoon. It made it so noisy in our tent that it was difficult to write and so I have taken my writing materials to the chapel tent where I am finishing up your letter. There is a long table fixed up in the center of the tent and a plenty of seats and it has already become quite a place for writing and reading. there is a plenty of papers magazines and tracts which affoard quite a quantity of reading material. While I am quietly writing a heavy canonnadeing is going on up the river in the vicinity of fort Pulaski. I do not know what is going on but presume the seccesh boats are trying to come down to Pulaski and our batteries are fireing at them.

    I am sorry to have you low spirited on my account but it gives me pleasure to know that you feel my abscense and that you are hopeing and longing to have me return. I do not think much on the future except so far as it relates to my return home. I confess to you that when I do think further I cannot make the way as smooth as I could wish it. I know it is wrong as by so doing I may be loseing faith in that all wise Providence that I beleive is continually watching over us for good, but you know that it is one of my failings that I am apt to look on the dark side of things. I do hope when next I hear from you that all the little ones will be well. I hope I may soon be permitted to hear you sing agan my favorite songs. music is getting to be very scarce here in the army. the men have so much to do that they do not feel much like singing and I must say that they have exchanged singing for grumbling. I had part of a dream the other night. I do not dream much now but by mistake I commenced and got so far as to think that I hear Carrie sing one verse and them I woke up. There are not many of the contraband negroes on this island. they come over occasionly in boats from Hilton Head where they are very plenty and from Dawfuskie. I have had no opportunity of converseing with them much. And now my dear I must close. I fear it will be a long time before you will receive this or any other letters from the south. But hope I am mistaken. I hope to hear from you soon may God bless and protect you is the prayer of your own

                                                                                      Ned

No. 37
Rec'd April 8th 1862

                                                          Camp of the 7th Reg C.V.
                                                          Tybee Island April 1st 1862

My Dear Sara,

    I have the prospect of a leisure day or part of one at the least. and concluded I would write you a letter to begin with. A Steamer came in yesterday afternoon from Hilton Head. She reports that a large mail had just arrived, but she could not wait to have it assorted. So we shall hope to hear from our friends at home once more in a few days. I have no news to write. The works are being rapidly completed, and the ammunition and ordinance stores are being unloaded and placed in the magazines with all possible dispatch. Everything is being got ready, and I shall be mistaken if before this reaches you we have not commenced the attack on the fort. The captain has been pleased to give me a small chance in with the rest. I am to be one of a party to fill catriges and shell. Our company and company C are to man one battery of two morters. The captain is pleased with the one assigned to us and thinks it a good position. Some of the batteries contain three. Some two and some but a single morter. Where there are two or more morters the magazine is placed as nearly as may be in the center. the powder of course is all within and <u>supposed</u> to be safe. the shells are piled up outside as near as possible to the entrance. the filling is done as the shells and catrages are needed during the action. Six men are required at a time for the operation three of whom are inside filling catriges and three outside the shells. all equally exposed I should think with those who work the morters. We went up yesterday and <u>took lessons</u>. we did not have to drill any on it for it is all simple and easily done, but the day before the fireing is to commence we are to go and fill a few first to begin with and to get our "hand in". George is to be one of the party. It is now thought that all will be in rediness within a week. The report is that a company of our troops wer taken prisoners on a small island near Savannah last Sunday. At all events there was fighting going on in that direction for we could hear the fireing of musquetry down here. George was not very well yesterday but is better today all the rest of us are well. not all either Wolcott has been down several days with a hard cold I believe. have not heard from him in a day or two. George's barrell and box wer received last Saturday, and have been living on his bounty most of the time since. Mrs Jones sent me a nice box of cookies which came in good order and for which I am very much obliged. Mr John Cleaveland sent "the boys" a nice donation of sugar which was most acceptable. I beleieve all the rest was from our usual benefactors. I know Liz sent George a box of splendid cookies to which I have paid particular attention. I think I"ll mention in my next letter to the North Pole that she can make good cookies. I have not much to write this morning and cannot think of any thing but that mail which is at H. Hilton it seems as though if I oneley had my share of that I should know what to write. I forgot to mention while writing about the batteries, that each one is named. Ours is battery Halleck among the names of the others are Stanton, Lyons, Grant, Lincoln, Burnside and one or two that I do not recollect. Today (the 1st) will be a buisy day at home for a great deal of buisiness is always done. every body is praying every one else that he owes and if he fails in so doing he will be apt to be reminded of his obligation before night. But it is like all other days with us, as buisy and no more. I hope if I live another April will not find me on Tybee Island or in Uncle Sams' service anywhere.

    I cannot write this morning darling excuse this poor apoligy for a letter. I will try to do better next time. Will write as soon as we receive that mail. Let me assure you once more of my love, and earnest desire to be with you and remember me as ever your own

Ned

No. 38 rec'd April 8th

Camp of the 7th Reg C.V.
Tybee Island Ga Mch 23rd 1962

My Dear Sara
    I have just received your letter of the 4th. A small mail came in today and I received yours and one from Frank French. I had been thinking all the forenoon that

I would write to you today. but I hardley think I should have got at it after all if I had not received your letter. Today is Sunday and there are but a few men at work and the rest of the regiment are haveing a day of rest. I do not know but we shall all have to go out and work tonight and do not care but little so long as we have not been obliged to work during the day. We have had a short service this forenoon. It was held in the open air. the weather is quite chilly today but the sun shone brightly most of the time and with my overcoat on I was not uncomfortable. The text was Genesis 4th. 9 & 10th. I fear I was not in a right frame of mind for I could not interest myself in the discourse as I ought. This afternoon there is to be a bible class commenced in the chapel tent I ought and should like to attend but I want to write to you. and if possible answer Frank's letter. At home where the mail goes every day it would make but little differance whether I wrote today or not. but when we know nothing about when the mail is to do. and our onley way is to write whenever we can get a chance and so be ready for it. prehaps I may be excusible if I stay in my tent and write. I have nothing new to write you for we are at work as usual. Yesterday I was on duty. our buisness was drawing up a heavy cannon from the beach to the place we always leave them. when we wish to take them up to Goats Point. which is the nearest place to Pulaski on which we are placeing batteries. It is about half way there that we leave them when we work in the day time. The rest of the way is nearly all in sight of the fort. and of course it would not do to be seen drawing cannon or at work there. The road is nearly in a direct line between the lighthouse on this island which commands a fine view of Pulaski and the surrounding country. and the fort itselfe. while we wer on the road Pulaski opened fire on the lighthouse. one shell striking within a few yards of it. and just back of the Dutch hospital. providentially it did not explode. if it had it could scarcely have failed to have killed a number. they fired two more both of which exploded before reaching the lighthouse and very near to our party. the last one in the air a little to the right of and directly opposite of us. We hear the whistle of the shells directly after the report of the gun. and can hear it comeing nearer and nearer seemingly directly towards us although it may be many rods away either side of us. The music is not the most agreeable in the world and I don't care to have the shells come much nearer to me than they did yesterday. But George had the most of a benefit after all. he was at work up at Goats Point on the batteries. the rebels opened fire on them about nine o"clock in the morning and during the day fired a dozen or more shots some of them striking and bursting very near them. much nearer than they did to us. The rebels evidently begin to have an inkling of what we are up to. and mean (to) trouble us all they can. fortunately we have got our works well along. and with a little care. I think they can be finished with but little if any loss of life. Then we shall be ready for them. and I hope be able to bring them speedily to terms. It does not seem possible that they will be able to long withstand the terriffic storm of shot and shell which we are prepareing for them. It is true they are snugly ensconsed behind a strong brick fort. and that they have casemate guns which they can work in comparative safty while we are onley hid behind sand heaps. but when they come to see the preparation we have made and that we are determined to subdue them at all hazards. I flatter myself that they will make an early surrender. Thier escape is entirely cut off and they cannot leave if they wish to. But I am inclined to think that they calculate to hold the fort and beat us off. I am inclined to think that they will be mistaken. though they have doubtless a strong force and some good artillerists within the walls. In the afternoon yesterday we wer employed in filling up shells at two of the morter batteries. It seems as though nothing could be able to withstand such a shower of these iron monsters as we expect to send over to the fort. They are bad things to handle and in spite of the caution given if the men get thier fingers badley smashed. I do not know as all these little incidents which are all that at present releieve the monotony of our lives. and the descriptions of what I have been doing from day today will be interesting to you. but I have little else to write and am desireous of makeing my letters interesting to you if I can. Please thank Mrs Barnum for her kind rememberance of me. and tell her that I am now quite well agan and able to do my duty the same as ever. I am rapidly regaining my lost flesh and if I keep on a short time as I am now shall soon be as "bad" as ever I do not much think you will leave Mr Barnums very quick. I have no doubt you are oftimes lonely and long for some society. but it is such a good place and the chances of getting one better or even as good in all respects are so extremely small. that I should be willing to put up with a good deal rather than leave them. Talk of your not knowing how to appear in society. what do you think of me. and who do you suppose will be seen in my company if in the good providance of God I am premitted to return home haveing forgotten what little I ever knew of the usuges of society. who will be willing to be seen in company with the rude uncouth soldier. You will see that I am answering your letter of Mch 4th. haveing received and answered yours of the 7th some days ago. It shows how good our mail arrangements are. Franks' letter received today with yours is dated the 9th. nice work somewhere. And now darling I must bid you good by again. I am sorry that I cannot be with you to try at least to help you while away some of your lonely hours. but this at presant may not be. Hoping that our heavenly Father will protect and preserve us both and grant that we may be happy both here and here and hereafter I remain as ever your own affectionate

Ned

P.S. our boys are all well now as usual and in good spirits hopeing to be home soon

Ned

No. 39 Rec'd Apr 10rd 1862

camp of the 7th Reg C.V.
Tybee Island Ga
Mch 26th 1862

My Dear Sara.

  Night before last we receved another mail, and I was happy to receive my usual allowance of three letters from you. It happens so that it generally gets two or three at a time sometimes more and if we happen to get two or three mails near together I sometimes get but one. Before I received your last letters I had managed to catch up with my correspondence so that I owed no one a letter, but you can rest assured that I am right glad to be laid under obligation again. Your letters are dated the 9th, 14, 16th. I shall not be able to write you three in reply to them today, in fact shall think myself fortunate if I succeeded in finishing this for I am sadley out of time for writing this morning. I have at last answered Mr Barnums letter. I dreaded it and put it off as long as possible, for I knew that I could not write one that would do the matter in question justice. I could not throw the responsibility of writing, upon either of the other boys for the business had been done with me, and there was no way for me onley to do the best I could. Yesterday I worked at unloading shells, on board of a schooner, and got nicely cough to. I had no appetite in the morning for my breakfast of fried pork, and so did not eat it, and went off without any breakfast. I expected to work on the beach, and thought if I got hungry I could easily buy something there, but come to find out the sutter had moved his tent away and I went aboard the vessel, and to work breakfastless. Noon came and with it no dinner, neither could I get on shore to get any so I worked all day with nothing to eat save a miserable hard cracker that one of the men gave me. I began to feel rather faint long in the afternoon and was glad enough when the last lighter was loaded, and we could come on shore. Going without my dinner is nothing, but come to put breakfast along with it, and it is not so pleasant. George was out all day at work and was obliged to go up to Goats Point at night and expected to have to work, but fortunately for him the shovelling had not been done that was necessary and he did not have to work. I am having a leisure day today. There is not near as many of the men employed now as there has been.

All that are taken from our regiment now are a few men from each company during the day to unload the vessels and a few more at night to work on the batteries. Our regiment has got to man eight of the morters during the bombardment. A number of men from each company are detailed for the work our company furnished forty. Instead of detailing the men as is done for fatigue duty and in fact all other purposes, the Captain called for volunteers, and I think nearly the required number came forward. I tried hard to get a chance in but the Captain refused to let me go because I have been sick a little and he said I could not stand it. So I am to be left out and obliged to stand and look on or to be set about some dirty work while the rest of the boys can have a chance to do something that will count. The Captain also refused some of the rest of our boys. I do not know exactly who are to go as I cannot see the list but, no doubt all the strong <u>good looking men</u> will be taken. I do not hardly think George will go if he wants to for he is buisy now and I think he will be wanted to work at carpentering all the time. Still there is not telling. General Sherman came down here yesterday and took a look all around, inspecting all the works, &c. Three ladies, one of them wife of an officer here, and I presume the other two equally well off, came down on the same Steamer and wer regarded by the soldiers as quite a <u>curiosity</u>. They do not belong in our regiment and I have oneley seen them at a distance. I assure you my dear that you are altogether to much concerned about my health. I am at any rate as well as I was at home and probably better than I should have been by this time had I remained at home about my businiess. It is nothing for the men to be sick occasionaly as I was, and much worse sickness if I wer at home or at least sick spells as bad. As for getting a furlough, it is entirely out of the question even if I desired it, which I most certainly do not. They would not give a man a furlough to go home and recruit himself up if they waw he needed it ever so bad. When a man is so far gone with disease that they <u>know</u> he cannot live, they will sometimes give him a discharge so that he may go home and die, but to go home to get well does not seem to be down in their programs. But I assure you my dear and beg you to believe me that at present I do not need either thier furlough or discharge. I am able to do my duty now and my hope is that I may continue to be able to do all that is required of me during the war. I should like very much to be in the tent with George, but it cannot be arranged so. George is very kind to me and does all for me that he can. I do see the N.Y. papers every mail. Some of the boys always have plenty of them but I do not get much time to read. I like to get hold of our county papers occasionly, as there is most always something that interests me transpiring because it is near home. I should like to drop in some evening <u>before</u> long and hear those new songs. One of them is an especial favorite of mine although I never herd it but once. "What shall be my angel name," it is beautifull both words and music. Our boys in camp are singing some very good verses. I do not know theier title but they are about "Ellsworth," they are sung to the tune of "Annie Lisle" a very pretty song which I heard a few times before I came away. I have forgotten whether I got it for you or not. I think I tried once at least, at any rate. You may know both the song and the verses to which I refer. If not I can easily get a copy of the verses and send you. I gave Frank a pretty good benefit in my last letter and when I write agan shall not fail to follow him up, untill he will be glad to cry "quits" I suppose that his and Josie's affairs are all settled. Some day my dear if we live I shall hope for an explanation from you of thier estrangement. buy the way Josie has never written to me although I have expected or at least had reason to hope she would. If I had her address I do not know but that I should venture to write her at the risk of being thought impertinent. I do not know how much Josie is in your confidance and do not much care you know all the circumstances of our acquaintances and that you Lizzie, Josie and myself wer quite intimate long before Sara B. was admitted. I have ever felt a strong friendship for her, but do not think and have no reason to expect that she values me as more than a passing acquaintance and, on the whole do not know as she would care to hear from the absent "soldier boy." I have a few of those envelopes done up together with quite a number of your letters in a small package intending to send it to you by John Sweet who at the time was expecting to go home in a short time. I think I will open the package and send you the envelopes as I did the other ones. I am sorry there is not more of them but I got the last the captain had. Lieut Gale is indeed having an easy time of it, but under the present circumstances I doubt whether even the charms of a young, and loved wife would keep me so long away from my duty. I may some day say some things to you with regard to him but not at present he may yet prove himself more of a patriot than I am at present inclined to give him credit for, but do not say a word of this to anyone. I see that I have answered your refrences to my comeing home and I think I need say no more about it. I onley do not wish you to think that I am satisfied where I am and have no wish to come home, and no wish to see you and the rest of my friends. I think you understand me and will not misconstrue my meaning. Tell Miss Lizz that prehaps you do write nonsense, but not to me, and that she ought not to judge your letters by those she writes to a fellow up somewhere towards the north pole. Please remember me to Auntie when you see her and say to her that it would give me great pleasure to "empty" a half dozen glasses of her beer. She knows my capacity for that buisness and so would not fail to make up a good suply. Don't let Lizzie use the measels all up. have her be carefull of them and tell her I should like to have her likeness taken when she is well broken out with them. You will have a delightfull time of it at home if they all have the measels while you are there. I should prefer to have school keep right along.

Mch 28th I did not finish my letter the day it was commenced and so lost the chance of sending it by the mail that went from here last night. Yesterday I was at work unloading shells from the vessels. It was very windy in the morning and the water was so rough that we could not work on the vessels. but in the afternoon it was more calm and we went on board them. it was quite rough then and we wer obliged to make two loads of what would have been taken at one had the water been smooth. As it was I managed to be awfull seasick and paid my tribute to old Neptune in a more liberal than to me pleasnat manner. I do not beleieve that I was ever made for a sailor. but I get vexed once in a while and think that if I ever live to get discharged I will go to sea untill I can stay there and not be sick. Two steamers came in in the afternoon each towing a loaded vessel. there is now three vessels in the harbour that we have not touched besides the ones we are at work on. so that there is no danger of our getting out of work at present. I think if those "sleepy" girls had to take up thier quaters down on Tybee in the same way that we do. thier slumbers would be less sound. what with gnats fleas and mosquetoes they would be troubled to get the amount of sleep necessary for thier health. Tell Liz I suppose we did not have much supper to get and that it onley made her exercise enough to get her well waked up. I think it would be pleasant to dream of home sometimes. it would be better than nothing. but I cannot do as much as that. in fact I have hardly been sound asleep enough to dream for the last two weeks. We still have encouring news from the war all around us and the more sanguine of the men are jubilent over the prospect of an early return home. And now darling a few words more and I must close. do not never. never fear that your letters are uninteresting and do not fear to write just as you think and feel. You cannot write to often or to much onley do not praise me I beg of you. I am not good. I know myself and my own heart better than you do and I have and. am thankfull for it. sufficient judgement to show me that I am weak vain and selfish. No one thinks a fractional part as well of me as you do. and you will alter your mind if you ever get acquainted with me. I was in hopes to overcome some of the evil of my nature while in the army but instead I find it difficult to hold my own. If there is a place on earth that will call out. all the basest passions of the human soul and expose the evil that is therein I believe it is the army. It has come to be a saying with the men that if you want to get acquainted with and know a man you must live awhile with him in the army. Now my dear good by for this time. remember me to all friends and beleieve me as ever your own loving

Ned

No. 40 Rec'd April 21st/62

Camp of the 7th Reg C.V.
Tybee Island Ga April 5th 1862

Dear Sara

The expected mail arrived two days ago and with it your letter of the 20th. It is very warm today but if the gnats will let me alone I'll try to get along with the heat and answer the letter. In the first place I am very well indeede. and so are the rest of our boys except Wolcott who does not get along very fast. There is no news. its the old story with us work day and night. but we are getting along with it. and the batteries are being rapidly completed. Still it will take nearly a fortnight yet I guess to finish. The trouble is that when one job is done and we begin to think that we are most through. they set us right about an entirely new one. We wer reinforced yesterday by five companies of the eighth Maine Reg. We have not a very large force on the island now. it will not exceede three thousand. There are rebels enough within ten miles of to eat us up if they could get on the island. I think I mentioned in my last letter that I was to assist in filling the shells. no different arrangement has been made and as the men are still drilling upon the morters I think we shall have to man those eight yet. George is still employed every other day up at Goats Point. He huries to get up at three o"clock and start for his work and does not get back untill late in the evening. so he does not have to work onley every other day. I shall have to own up to haveing the blues occasionly. it is wrong I know. and more particularly at the present time. it is so uncertain whether I shall ever live to return. but they do not last long. in the present buisy time with us it is difficult to retain any one impression more than over night. I received a good long letter from Mrs Randall. a short one from her sister. and one from Carrie by the last mail. o dear I do wish this buiness was finished up and we on our way home. It may be fun for the officers and others in authority to be back at thier ease. and be waited on by thier servants. but it is anything but fun for us. Prehaps I missjudge. maybe everything is being done as speedily as possible. but it does not look so to us. If Mr. Barnum makes any comments on my letter (which he will not be very apt to do before you) I wish you would please report. I was surprised to hear of the sudden death of Henry Dodge as you say it is a severe blow to his father. But he who doeth all things well has done what he knoweth to be best and he can comfort the bereaved ones. My friend Hawthorn is the son of a farmer in Canaan. is about twenty one years old about my highth but a considerable hevier not bad looking nor remarkably handsome has been sick a good deal since he enlisted but I think he will come out tough yet. I like him the best of any of my tent mates but we are liable to have a falling out most any day. although I told him a spell ago that I would not get mad at him any way. You see we are both quick tempered as we can well be. and so if one or the other does not look out we should be likely to get together by the ears.

Apl 6th I left off writing yesterday afternoon to go down to the beach and bathe. the weather is so hot now that we go into the water most every day. when I came back I could not get to writing agan and so my letter was left unfinished. Today is the Sabbeth. we have had an inspection. and our usual service. The Text was Acts 20 & 24. I did not get very much interested in the sermon in fact we all have to pleade guilty to getting asleepe once. But I hardly think you would blame me for it if you knew how difficult it is for us to sleep nights. it is almost impossible since the hot weather has come on. the fleas have become perfectly ravenous, it is impossible to get rid of them. because the sand is full and they can hop on us at any time. George had to give in last night. he has not been troubled by them much before but last night they wer to much for him. and he was up with the rest of us about the middle of the night. It seems to me that you have rather a short vacation. I thought you had a month usuly. If you spend the time at home I doubt your getting much rest. and as the time is so short I suppose you do not intend going away. Well darling my mind refuses to furnish me with any thing that I consider worth writing today and am not going to try to write more now. But for your consolation before I close I'll just say that I love you dearly and hope I shall live to prove it. There has a small steamer just come in and I hope she has brought mail. but it is not likely. but she will probably take away what there is to go from here. I do not mean that a mail shall go North without a letter for you. Good by my dear one do not forget to pray for me that I am able to do my whole duty and act well my part in any place of danger I may be.

Ned

Editors note: The following letter refers to the bombardment and <u>capture of Fort Pulaski 10-11 April 1862</u>.

No. 41 Rec'd April 21st 1862

Fort Pulaski Ga Apl 12th 1862

My Dear Sara

We are Sara at last in possession of the fort. We have gained a glorious victory. and with but the loss of one man. two wer slightly wounded. but hardly enough worth mentioning. We commenced the attack about seven o"clock Thursday morning and betwen two and three yesterday afternoon the white flag was raised on the fort. I should almost thought you might have heard our cheers up in Connecticut if the wind had been in the right direction. Our regiment had the honor of being the first to enter and take possession. We immediately commenced makeing preparations for the transfer from the island to the fort. but it was nearly two o"clock at night before we wer in here. Our battery fired the first shot and for a short time drew the enemys fire. but soon directed their guns in another direction and we wer comparatively free from danger. I received three letters from you the day before the attack and will answer as soon as possible. Have no time to write any particulars now for I am afraid I shall be to late for the mail. Three men wer wounded in the fort one of which will probably die. All our boys are well and in excellent spirits. With much love to you darling I am as ever your own

Ned

No. 42 Rec'd 23rd April.62

Fort Pulaski

<u>Fort Pulaski Ga</u> Apl 13th 1862

My Dear Sara

I have made a good beginning this time makeing a mistake the first thing. but my head is not very clear this morning for what with the labor and excitement added to a hard cold. I am not very clear headed this morning. We are quatered for the present in a small building outside the fort. It is a very pleasant Sunday morning with a fine cool breeze. I hope we shall not have anything to do today for I do not feel like work and wish to write a letter to you. I suppose of course you will see a detailed account of the bombardment in the papers. and much better ones than I can give but as prehaps you will like to hear what I have to say about it and at the risk of being tedious I will tell as nearly as possible my <u>experience</u> for the last few days. Last

Monday those of our regiment who wer to work the morters drilled on one which had been mounted down on the beach for that purpose. We fired it several times and managed to get quite tired. Tuesday forenoon we rested but in the afternoon wer obliged to get to work at dismounting and removeing the morter. and agan mounting it in the position it was to occupy in the battery. We did not get through and back to camp untill between two and three oclock. I managed to get just two and a half hours sleep that night. Wednesday forenoon we rested and at noon received a mail I had just time to read my letters when we wer ordered to fall in and go up to the battery (about two miles) and drill. came clear back to supper and at nine o"clock agan fell in with our rifles and accouterments all on. blankets rolled and slung across our shoulders. one day rations in our haversacks and all in light "marching order" as we call it. we marched up to the battery we wer to work and stacking our arms and throwing off our traps took shovels and went to work at completeing the battery. This we did not get done untill between twelve and one. Then throwing ourselves down in the best sheltered place we could find (for the wind blew like the mischeif) we rested as well as we could untill morning. We wer up with the earliest streak of daybreak before the stars are gone. Everything was got in rediness and we wer soon impatiently waiting the order to fire. Soon the generals aid went galoping by towards Goat Point and shortly after a boat with a flag of truce went over from there to the fort demanding its surrender. which of course was refused. the commandant replying that "he was placed there to defend the fort not to surrender it". Soon the boat returned. the aid went hurrying back past to head quaters to report and in a few moments agan rode up this time to our battery. with the message to the captain that "the general presented his compliments to the captain and wished him to open the ball immediately". The men quickly sprung to thier places and in a few moments the first shot was screaming through the air towards the fort. The batteries on our right immediately opened fire. those on our left at Goats Point being ordered to withold thier fire untill after the enemy had replied to us. In a short time the fort returned our compliment and soon the fireing became general all along our lines and from the enemy the roar of the guns became almost incessant. It was splendid. A more beautiful day never dawned. The air was perfectly clear and we could trace our shells through the air almost to the fort and seldom failed to see them strike. sometimes in the fort. sometimes on the parapet makeing the walls crumble and the dust to fly and agan outside in the water or on the land sending the water and dirt high into the air. then a cloud of smoke would rise from the fort and directly a shot or shell would come whistling through the air. sometimes over our heads then falling short. and agan bursting in the air almost over our heads. I tell you it was exciteing. At first thier fire was directed on our battery and some shots struck very near us but as the fireing became general from the other batteries the enemy directed thier fire more on them and left us comparatively unmolested. I was not on the first party at work and with the remainder of our releif lay in a place prepared for the purpos called a splinter proof. For four hours we. lay-there. and then took our turn at the work. At the expiration of our four hours the first party took hold agan and continued fireing untill dark when the fireing ceased for a while and those of us who had been at work all day wer releieved by a third party. From our battery and one other the fireing was continued through the night at intervals of twenty minutes. We returned to camp got some supper and slept untill about three in the morning when we wer aroused up. got a cup of hot coffee. crammed our days provisions into our haversacks and agan set out for the scene of operations. This morning we commenced fireing about sunrise as the day advanced the wind came up and blew furiously filling our faces with the fire sand besides seriously interferring with our fireing. I forgot to mention that the first day. about noon we shot away the rebel flag. you can hardly imagine how the boys cheered and danced about untill a shot from the fort sent us ducking under the cover of our batteries. In a short time it was hoisted agan but on a short staff and in another part of the fort where it remained untill it was hauled down and a white flag raised. We had had luck with our large columbiads oweing to some defeat in thier mounting. five of them dismounted at the first fire. two of them wer remounted during the day but the others being at Goats Point. where the fire of the enemy was hottest. could not be remounted. untill night two of them wer brought into action agan in the morning and did good execution. The fire from the fort was more brisk the second day and at one o"clock when our releif left the morters I thought it quite likely that we should have two or three days work before we should bring them to terms. I had just got my dinner eaten and nicely stretched out for a good nap (for we wer so tired that some of us could sleep perfectly sound within two rods of the morters and amil the almost incessant roar of the guns around) when some one sung out there goes up a white flag. I could not believe it and thought the men wer trying to fool us but as they continued to insist upon it I ran out and was just in time to see the rebel flag hauled down. and there sure enough was the white flag. this was about three o"clock. I think the wildest Hurrahing that wer proceeded from human throats was to be heard on the shore of what Whittier calls "lone Tybee" for the next fifteen minutes. Soon the generals aid went galloping by waveing his hat in response to our cheers. Shortly after two boats put off from Goats Point for the fort and orders came for the seventh to hold themselves in readiness to go into the fort immediately. But immediately in the army means any time when the officers get ready. and so we did not get started untill sun down. Those of us who wer at the batteries did not return to camp after our things but came away just as we wer. those who wer in camp brought away their things. We marched up to Goats Point where there is a little wharf and a small steamer was waiting to take us across the channel. Oweing to the narrowness of the channel and the difficulty of getting about in it. we did not get across and into the fort untill about one o" clock. We did not go into the main part of the fort. at least not our company. The fort is built of brick. five sided. the side least exposed to an enemy is used as quaters for the officers and for kitchens in this-side there are no guns. but on the out side there is another fort built of brick and earth. but onley built for one tier of guns. this outer fort is for defence aganst a land attack. and was calculated to mount a large number of guns. but there wer none. except three small brass pieces called howitzers. here we bivuacked for the remainder of the night. sleeping what we could. The fort is surrounded by a wide ditch filled with water and soft mud to the depth of twelve feet. it is crossed by a draw bridge and cannon are placed in the wall so as to rake the entrance completly. The scene inside needs to be seen to be realized. I cannot give anything like an adequate description of it. We succeeded in nearly breeching the wall the first day. and the next morning got the hole through. after it was once fairly started the breech rapidly increased in size. every shot knocking away the brick and raining clouds of dust. The shot and shell soon began to go clean through the breach and to tear through the officers quaters and kitchens which are a perfect wreck. But they soon began to tell in another and far more dangerous quater. the magazine is an angle of the wall nearly opposite to that through which our shot had torne thier way and into the walls of this they soon began to make sad havock. The rebels could stand no longer. at any moment a shot might go through the magazine and blow the whole garrison into eternity. they had made a good though ineffectual resistance and it was worse than useless to attempt to hold out longer. They claim not to have not had any killed. three are wounded and in the hospital. But we do not hardly beleieve that there wer more killed. some of the wounded I beleieve have acknowledged that there wer ten killed the first day. I do not suppose the truth will ever be known. But it seems incredable that three hundred and eighty three men could remain in the fort during such a perfect hail as there must have been of shot. fragments of shell. brick and splinters for thirty eight hours and onley three be wounded. Our loss was one killed and three slightly wounded. We remained in and about the fort all day Saturday looking about and talking with the prisoners and keeping as comfortable as possible for it was a cold windy day with a fair prospect of rain. Outside and a few rods distance from the fort are a number of houses some of which are used as hospitals in times of peace and the others probably occupied by attendants. just at night we left the fort. seven companies of us and took up our quaters in these houses. the remaining three companies and one company of the Rhode Island regiment. remained in the fort. At our quaters in one of these houses I commenced this letter and have been doing what I could since to finish it. Monday it rained furiously all the morning untill near noon makeing it miserable getting about on account of the mud which is the most sticky stuff I ever saw. you can hardly scrape it off. About noon we took our things and marched back to the fort in which seven companies of us are quatered in the casemates. the other three companies are provided for outside. Those of us who wer at work on the batteries came away without bringing any of our things with us and they have not yet come. a party have set out this morning to go over to Tybee after them. I do not --expect to get half of my things. but will hope for the best. they wer scattered all over the tent. Today. Tuesday I am in my quaters trying to finish this letter. It is a pleasant day. the men are variously engaged. some employed on fatigue clearing away the myself rubbish. some fixing up bunks for themselves and few like writing or reading. The prisoners wer removed on Sunday. They wer put on board the steamers McClellan and Ben DeFord and I suppose taken to Hilton Head to await further as to thier destination. but I do not know they may know where orders to go with them and take them I directly do not there. know I talked a considerable with them. Some of them wer <u>rabid secessh</u> and pretty firey but most of them that I talked with wer quite moderate and willing to talk. I think the Irish of which there wer a good many wer the most rabid as well as the most ignorant. There was one company composed mostly of the natives. poor whites. they seemed to be a kind of harmless creatures if left to themselves. most of them had enlisted because they had nothing to do and wer told that if they did not they would be drafted and would not get any pay. and so of the two evils choose the least. they wer enlisted for twelve months. They wer all well clothed in thier kind of cloth which was mostly coarse gray. One company however wer better dressed. wearing broadcloth pants with plenty of red trimmings. They wer all anxious to leave immediately and wer impatient of the delay. I was at the wharf when the most of them went on board the steamer. they wer marched down under a guard. they had thier luggage with them. they wer not allowed to take thier knapsacks but packed thier things in trunks. or bags or made them up in bundles. They seemed to be in good spirits considering thier situation. We exchanged "good byes" and "take care of yourselves" with them as they passed along. I do not think they had any reason to complain of our treatment of them while in the fort. The sick and wounded are still in the fort and are to be sent to Savannah. They had plenty of provisions in the fort for six months but wer on half allowances of coffee. soap. candles. salt. and one other article that I have forgotten. One very intelligent man. an ordely sergent. told me that coffee was not to be had in Savannah and that tea was four dollars a pound there. Yesterday a most terrible and distressing accident occured on the parapet of the fort. I was sitting where I am now at work on this letter when I heard the report of what I supposed to be a cannon but an instant after a terrible groaning told me at once that one of the numerous unexploded shells that wer lying about the fort had burst and some one was injured. Running out. paper and pencil in hand. I saw where the accident had occured and running up on the parapet. saw a sight that I shall not soon forget. five men torn and bleeding from gastly wounds lay within a space of ten feet square. two wer dead. killed instantly. and three others torn and bleeding. legs shattered. arms broken. hands blown off and faces terribly mangled. Two of them died during the night. they wer so badly injured that they wer not taken down but wat could be done for them was done where they lay. one poor fellow. a boy sixteen years old. had one arm broken. the other one blown off at the wrist. his right shoulder all blown away. the lower part of his face torn away and his whole face blown so full of powder that you could push your finger right through the flesh and yet he lived until late at night and could and did talk considerable. The fifth one lost his leg. he stands a small chance to recover. It seems they wer around cleaning up around the guns and picking up the loose shells thier captain had removed the cap from the shell and taken it down to show to the Colonel. the men had empted the powder out and wer knocking the shell against a piece of granite to jar out a little that remained when the shell exploded. one piece of the shell flew clear across the fort strikeing aganst a brick furnace for heating shot and near which in William Reid back and knocking others wer standing. bounding back it struck Reid the breath out of him and leaveing its imprint in black and blue on his skin. he saw the piece comeing and dodged it. had it hit him before strikeing the wall it could not have failed of seriously injureing him at least. several others had narrow escapes from pieces of the same shell. one man loseing the skirts of his coat. The men belonged to the Rhode Island company. The shell was one of the most destructive kind. called <u>the James projectile</u>. the solid shot of the same kind made terrible havock in the brick walls of the fort, some of them comeing through the breach would go through solid timber eighteen inches thick. then through a twelve inch brick wall tearing to pieces all that opposed it or came in its way. The rebels say that it was those that (did the) work. they are thrown from riffled guns. No one in our regiment received even a scratch. All are well and in good spirits. Our regiment was the first to enter the fort. We do not know how long we shall remain here. but I think not a great while. It is hard for you my dear to be so much occupied at the time when you should be at rest. I earnestly hope you will not be sick yourself. We are just hearing rumors of a great battle out west. I cannot beleive the reports that come to us. but hope that if there has been a fight we have been success full. I think we are gaining ground slowly but surely. I suppose. our next move will be on to Savannah. the rebels told us we would have hard work to take the city. they gave the number of men defending it all along from thirty to sixty thousand. but they are woefully ignorant of what has been done for the last two months. knew nothing of our victories or of Beauregards retreat from the Potomac and it is not likely that they know much better about the defenses of Savannah. I asked one of them if they did not get thier papers. he said not often, when the mail came the officers took the papers themselves and they did not see them much. They have had communication with Savannah all the time by means of small boats. and received a mail the morning of the fight. The officers told them that there had been a great battle in Tennasee. that they had killed and taken prisoners thirty thousand of our troops. that general Buell was killed on our side (Editor's note: Buell was not killed in this very bloody battle at Shiloh) and Johnson on thiers. There are fifty four guns mounted on the fort & besides the small ones I mentioned outside of these thirteen are mounted on the parapet and the remainder in the casemates. Five ten inch morters. three outside near the water and two on the parapet. Five of the guns are dismounted. I enclose a piece of a rebel flag found in the fort. it was not I suppose the flag that was flying during the action which I beleieve could not be found. it is supposed it was burned together with others company flags and the like. There was doubtless a large amount of stuff burned before our boats reached the fort. I also send you a couple of specimens of southern currency. I hardly think they will be bankable with you. but some of the Irish prisoners stoutly insisted that they would be as good some day as our U.S. Treasurey notes. Poor as the money I make, it is what the soldiers take as pay for thier services or part pay at least. some of them did have southern Bank Notes but one is as good as the other at the North. The men wer very glad to exchange thier shin plasters for uncle Sam's coin and a great many wer bought by our boys to send home. Well my dear I am afraid I shall already have wearied you with my letter. You will find it difficult to read and full of errors which I beg you to excuse for I have been several differant times writing it and most of the time in a hurry. I have had a poor pencil and been obliged to write on a board and I fear that by the time it reaches you. you will not be able to read it. I am very sorry darling that you had to look in vain for letters. but I assure you it was not my fault for up to last week I have not failed in sometime to write you either one or two letters a week. Still I see no reason why Georges letters should go and mine not. Please remember me to grandma Barnum and any others of the family that enquire for me. I shall try to write to your father agan soon and to all my other friends if I have time. Good by Darling accept much love and constant rememberance from your own

Ned

No. 43 Rec'd April 28th/62

Pulaski
Fort Pulaski Ga Apl 19th 1862

My Dear Sara

    Three days ago I received your letter of the 6th and last night I received three more from you dated the 3rd 8th & 11th. How the odd one got strayed away from the others is the question. Many thanks darling for your continued kindness in writing so frequently. I have come to watch the delivery of the mail with jealous eyes and if I do not get one or two certain I am inclined to overhaul things and see if there is not some mistake. We have not received a mail in a long time but what there has been from one to three letters in it for me from you. The first mail that came this week George did not get a letter. the first time he has failed in getting from one to seven since we enlisted. last night he was more successfull and received five. today he has gone on guard. for the first time in a long while for he has had a steady job of work and has not had to do guard duty. I expected to have had to have gone on today myself but escaped some way. I do not want to go on first yet for they have a very large guard now and it makes very hard work for all concerned. I presume in a few days the guard will be lessened. Some of the companies are drilling today for the first time in over two months. so many of our company are detailed on differant kinds of work that there was not enough left to drill and so I have this opportunity which I am trying to improve of writing to you. I do not see why Mr Barnum should send that letter up to you. he must think that you are very much interested in me. I think myself that the boys would do well to make a public acknowledgment of the favors we have received from friends at home. and we may do it yet. But we are so divided up now that it is not easy to get any concentrated action. by and by if we remain here and get settled down into the routine of garrison duty we may have more time to think and act upon the matter. At present Mr Moore is the Adjutants clerk Mr Landon and Sam Wolcott are sick in the hospital and the rest of the boys not on guard are on fatigue. So you see that we are all pretty much engaged. besides we are not allowed any light at night except in lanterns. and as there are but few of them. we onley get about light enough to make darkness visable. so that for the presant at least but little reading or writing can be done in the evening. We are not together now as we wer in the tents. but are distributed in the casemates as our names come on the roll book in alphabetical order. except the sergants and corporals who are distributed as they stand in rank. which brings me away down in the last one. Wells and Reid are with me and I have a good fellow to share my bunk with. Sergant Peck. The casemates are not divided off. but one continual arch the whole length of the fort. The embrasure and the space occupied by one gun is called a casemate. our company occupies five of them. It is a noisy place when the men are all in and there is no such thing as quiet or privacy. Still we have more room to stir about than in our tents. The weather for a few days past has been splendid and the mud around the fort is getting dried up. So much for our folks now for other matters. Tell Sizzie that I am not agoing to write to that young highflyer of hers agan very soon. I have written him two letters and have not heard a word in reply. I do not beleive that he is so buisy with his studies that he cannot find time to just write me a line at least. Whatever letters I send you are for safe keeping and of course they are for your perusal without any referance to their authors. Sara B. of course writes a good letter for a sister as good as one could wish but you need not fear of my prefering hers to yours. You are to modest by half. I wonder you do not get sick of my disjointed. misspelled letters. they are always written in a hurry and as I never was a good speller. haste does not improve my orthography. Talking of pancakes and maple syrup. you ought to see the pancakes we have occasionly. flour and water with a little salt mixed up together. heavy as lead sometimes half burned. sometimes not half cooked. nearly always about cold before we get them. they would not prove a very tempting dish at home. but with molasses and sometimes sugar on them I usuly manage to demolish my allowance. O' we have some tall cooking you better believe. By this time your fears for my safety are doubtless allayed. It turned out that we wer not in much danger and to try ones courage after all. Such a battle as the one near Corinth for instance. I went and assisted the Adjutant yesterday afternoon do some copying. I should like to get some of the many clerkships that are in the army for it would be a benefit to me in keeping my hand in at writing. but I do not expect to for there are any number of men who stand ahead of me and I have long ago given up all thoughts of any promotion while I am in the army. I had rather be a soldier all the time if I could could escape the fatigue work than to do anything else. I am sorry that you should think my dear that my wife's probable expenses are troubleing me and clouding my thoughts of the future. I assure you it is no such thing. but I do not want my wife to have to work to support me. I presume I can find employment at some thing. I have no doubt of that if my health is restored. It is more on account of the uncertainty of my return home than anything else that I dislike to lay plans for the future. I love you dearly and long to call you my wife and wish that I could surround you with all that makes life pleasant and attractive and cannot bear to think that prehaps by marrying you I may remove you from a pleasant home and the comforts if not some of the luxuries of life to the discomforts of shareing a poor man's lot. please forgive me darling for what I have said. you wish me to confess these thoughts are not agreeable to me. it would be much more to my taste if I could see oneley the bright side and believe that of course all would be well. I believe. however. That it is best to look these matters square in the face. You seem to have done so and your proposition to assist me by giveing music lessons seems to be the result of your deliberations. many thanks for your kind offer. I think Mr Rosette must be in trouble. if I wer oneley acquainted with him I might be able to advise him in the matter. at least tell him what I should do wer the case my own. you seem (to) be fortunate in the possession of love secrets. I imagine you will soon be fully competant to advise in all matters of the heart. I think I can ofset my dream of Carrie with those "bewitching eyes" that call on you occasionly and so be about even. It was not my fault that I dreamt of her. I would much rather have dreampt of you. but I suppose I dream of you so much by day that my fancies refuse to be controlled at night and wander off on thier own account. Well my dear I must close this letter. Virgil McNeil is writing near me and says "give my love to her" I told him of course. so here it is. I shall be happy to hear from your father. but do not wish to trouble him while he is so buisy. It gives me pain to hear of the bereavement and trouble of others. I heartly sympathise with them and earnestly hope that their afflictions may work out for them the end for which they wer sent and be the means of leading them into the new and higher life. My heart grows sick when I think of the misery and suffering this war is the occasion of. I cannot bear to contemplate .it and turn from it with the earnest prayer of "How long. O Lord. how long" God grant that it may have a speedy and glorious end. I hope I may always be able to do my duty whatever the danger may be. and prove myself what your father cautions me to be a "good soldier" Good by Darling. accept much love from

Ned

CIVIL WAR LETTERS 1861-1865

No. 44 Rec'd May 5th 1862

Fort Pulaski Gg April 25th 1862

My Dear Sara

Now for a letter on a sheet of secesh letter paper obtained of one of the sick rebel prisoners. Not Bath Post exactly. but its of home manufacture and probably the best they could do. I must plead guilty to not having written a letter to you in over a week. it is getting rather difficult to find time for writing. We have roll call at six in the morning. breakfast ten minutes after. guard mounting at half past seven. drill commences at eight. and lasts till half past ten. dinner at twelve. drill agan from three until half past five. supper immediately after. dress parade at about sun down. tatto at "half past eight. taps at nine. About the oneley time for writing is between drill hours. as we do not have light very convenient in the evening. We are waiting impatiently for a mail. if we do not get one once a week we begin to scold. We are soldiering once more oneley a few men from a company are on fatigue each day. the rest are all drilling. Yesterday afternoon we wer allowed to go up on the parapet and look around. a privalige that has been denied us before. There are thirteen guns in all mounted there. all but two of rebel manufacture. These two are of English make. rifled pieces. that came in the <u>ship Fingal</u> which ran the blockade some time before we came to Tybee with a load of cannons. Enfield rifles and other munitions of war. She has not been so fortunate in her attempt to escape the blockade. with the load of cotten and return home. and is still lying at Savannah. The rebel cannon are made of a poor quality of Iron. and are very heavy. the maker trying to make up in quantity what they lack in quality. They are made after old patterns and are very inferior to our heavy guns. the carages are made of southern pine. There are two ten inch morters on the parapet one of which was dismounted. A vessel came in the other day loaded with fresh beef and ice. There is an ice house in the fort and a large quantity was landed. also about five ton of beef which is kept on the ice. So that we are to have fresh meat occasionly. which will be an acceptable exchange. The ice will be for the officers and hospital use. We have also got the oven at work. and today have received our rations of fresh bread once more. There are several men in our company who will probably be discharged soon. They have visited the surgeon this afternoon for an examination of thier cases. among the number is afternoon for an examination of thier cases. among the number is one of the Brinton boys. not the one whose father makes enquires of you as to his welfare but his cousin. sweet is still with us he is at present staying with the Captain. I presume he will go home with the others. but that will probably be some time yet. Well my dear I find I have but very little to write about today and fear you'l not get a very interesting letter. I suppose you are back in your school agan before this. I should love dearly to call and visit the school. or the mistress. I really think if I could be home tomorrow I should make a desperate effort to get as far as Lime Rock. on a pleasure trip. I send you a ring made of Tybee live oak with a small piece of gold let into it. I do not like it exactly and if I can find a prettier one will send it to you. I should love dearly to visit you all once more. by you all I mean Lizzie. Julia. Sara B. & Josie. I sometimes think if I could just make you one good visit I could return to my duties with renewed zeal and better courage. Mr Landon and, Wolcott are still sick in the hospital. George is well without a doubt. for he is as fat as a seal. There is a report here that Mr Robbins is engaged to Anne Bostwick. do you hear anything about it: I do not know what ails me. I cannot write today. I beg you not to think it is because I am thinking of and loveing you as ever. I will try agan in a day or two and hope I can then do better. Be assured that I love you my darling. that I think of you much. too much I am afraid for my contentment of mind in my present position. and am liveing on hopeing that some time in the distant future I may be premitted to realize the desire of my heart in calling you my wife and together treading the rugged path of life until God shall call us home. Praying that He will watch over. preserve. and guide us safely. and whatever. may be our fate on earth unite us at last in Heaven. I remain your own

Ned

No. 45 Rec'd May 11th 1862

Fort Pulaski Ga Apl 29th/62

My Dear Sara

    I sent you an apologie for a letter a few days since. and am not sure that I shall make out much better this time. We are still waiting impatiently for a mail which I some expect will be here today or tomorrow for we can see a steamer comeing over from Hilton Head to Tybee. It is very hot this morning makeing one feel both faint and unwilling to stir. luckily for me I have nothing to do and it is shady and cool in the casemates. If I could onley write good this morning I might turn off a number of letters during the day. but something is the matter with my head lately for I cannot write if I try ever so hard. It is real dull here now although we are all buisy either on fatigue or drilling. but we rather expected that after we had taken Pulaski there would be an immediate advance made on Savannah. but it was not done and there is no sign that one is contemplated. So we have come to the conclusion that we are to stay here awhile at least. It is said that there is to be a balloon ascension here shortly. a part of the apperatus and materials for makeing the gas have come already. A vessel loaded with brick is lying at the wharf and is being unloaded. Workmen are employed in removeing the rubbish and tearing away the loosend and broken brick and morter from the injured parts of the wall and getting ready to repair them. Some of the guns that we used in takeing the fort are being brought over here but have not been mounted yet. Last night heavy fireing was heard up the river. but the reason of it we have not yet learned. Yesterday. and this morning. the rebel gun boats wer in plain sight up the river with thier long trains of thick black smoke streaming far behind them. Significant I think of the withering and blighting effects of secession wherever it spreads. they are supposed to be placing obstructions in the river to prevent our going up. I hear that the mortars we used on Tybee are being removed and loaded on board of vessels but hear nothing as to thier destination. but we guess it is Sumpter of course. I suppose it begins to be quite Spring like up home by this time. I remember how excited we all wer last year at this time. and I believe you wer a little afraid once or twice that I would go out with the three months men. I did think seriously of it. but did not suppose I could stand it. I have often wished that I had gone with them as I could probably done much better when I came to go agan. I suppose there is not as much excitement this spring. though none the less interest. for by this time you have come to look at things as of every day occurance and matters of course that wer then new and startling. God grant that ere another Spring shall come the war shall be ended. peace restored. and the soldiers scattered to thier homes. Mr Landon and Wolcott are still in the hospital quite sick. Barnes is rather unwell on the sick list but not in the hospital. the rest of us are usualy well. I suppose by the time the "measel mania" has subsided. I hope no more deaths have occured from it. It has been a great affliction. I have still a number of your letters on hand that I had done up once to send home by Sweet but as I am not to have that opportunity I shall slip one in whenever I can. So you need not be surprised to meete one of them occasionally. In my last letter I mentioned that I enclosed you a ring. which you probably did not find on looking for. I took it down to put in the letter when it slipped from my hand onto the floor and the gold setting dropped out. I shall either have it fixed or else get another one made that will suit me better. I am sitting up in my bunk trying to write where it is so dark I can't see the lines half the time. Just because I am to lazy to get down and go to the light. My bunk mate is one of our sergants. a quiet pleasant fellow rather singular sometimes. but a good chum, on the whole I am as well pleased as I was in the tent with our own boys. Reid and Wells have thier bunk by the side of ours and they are all of our boys that we are in

our casemate. but its just like all being in one room after all and the noisest place when we are all in you ever saw. whistling. singing. talking. playing cards. swearing. danceing and almost everything else that man can do to make a noise is being done with a will. The profanity is awfull. I thought I had got used to it. but I find that then I had not herd it all. So much and such reckless swearing I never heard before and this from men who are liable at any hour to be called to face death at the cannons mouth. and prehaps be ushered into thier Makers' presence without time to cry "God be mercifull to me a sinner." I have just been to dinner. would you like to know the bill of fare: fresh meat: potatoes. new bread and butter. the last article furnished by ourselves at the low price of forty cents a pound (I have paid fifty) not quite half salt but near enough to it to ensure its not being spread to thickly. we have new bread every day now. We do not always fare as well as this noon by any means. We are agan disappointed about the mail as the steamer brings word that there is none at the "Head" for us. prehaps we shall get good news enough by and by to make up for all our waiting. Prehaps I mentioned to you that I had received letters from Mrs Randall and Carrie by the last mail I think it was we received on Tybee. I have answered them. Carrie is going to Bridgeport to school the first of May. I expect to hear from her occasionly. And now darling good by once more. be assured of my continued love and devotion and my earnest desire to be with you and do all in my power to make you happy. Praying for your continued health and happiness I am as ever your own
Ned

No. 46 Rec'd 19th May 1862

Fort Pulaski Ga May 7th 1862

My Dear Sara

The long looked for mail has at length arrived. We received a small one on Sabbeth morning. by it I received yours of the 26th. last night another and much larger one came. and yours of the 14. 19th & 21st came to hand. the last written comeing first. The reason for· it was that the "Massachusetts" with a large mail stopped at Fort Monroe while another steamer leaving N.Y. two days later came right along. I also received a short letter from Charlie. short enough, onley half a sheet. he did not write anything hardley. stated that he had left College for the present. but ment to graduate if he staid in P. three years. at present is in a drug store. said that things looked dark for him in Salisbury but offered not a word of explanation. said I would probably be posted up from there. did not acknowledge the receipt of either of my letters. but said he had written to me a while ago but did not know whether he directed it right or not as I have received but one from him since comeing away. I presume it is lost. he was very sorry he had not enlisted long ago and should not care if he had never returned. on the whole he appears to be quite down hearted and did not seem to care much for me or any one else. I have been thus particular in referance to his letter as I thought it might interest Lizzie. I do not hardley know what to think of his agreeing not to write to her. it is so contrary to what he has always asserted. that is that his mother should never come between him and Lizzie but I will make no further comments as I may do him injustice. like yourself I would not like to advise either of them what to do. time will show what is for the best. I am on guard today and mean to finish a letter if it is oneley a short one. and send it along. George is working at carpentering but where or what at I do not know. We have been at work for some time. all the men who could work to an advantage in clearing away the rubbish from the breach made in the walls of the fort during the bombardment. and in tearing away the cracked and loosend brick and morter that had not come down during the fire. Yesterday the masons commenced rebuilding the wall. Other parties of men have been removeing the earth and timbers that the rebels had placed at a protection for the walls inside. We have not drilled for several days and probably shall not for some time to come as ther is enough work to keep the regiment employed a good while yet. If I ever get home I shall have to hire out to shovel or roll wheelbarrow or attending mason or some such Irishmens work. I shall know how to do it all. A day or two ago. I sat on a pile of bricks that had been wheeled from the breach and with a trowel cleaned off the morter from the old bricks so they could be used agan. Yesterday I did not work much as I did not feel much like it. and besides was put in charge of a small party to see that they kept at work. The balloon that I mentioned in one of my letters is being inflated and I do not know but they are going up today. I am glad Josie is to have a new piano and hope to have the pleasure of hearing it. I conclude she is going to be at home a few days now. Am not agoing to answer any of your letters this time. but mean to write tomorrow and do it. I enclose one that of the 14th. I prefer to send them all back to you. the postage is of no account. I will make that all right very soon but am sorry that you have been thus annoyed. I will look out in future. I think the postage has been charged many times when it ought not to have been. Good by. much love from

Ned

MY DEAR SARA

No. 47 Rec'd May 19th 1862

Fort Pulaski Ga May 10th 1862

My Dear Sara

    Am going to improve a few leisure moments this noon in commencing a letter. There is another mail for us at Hilton Hd which we expect to get tomorrow. A paper of the 6th brought by the steamer on which the mail came has arrived here. and the chaplain read it yesterday to the men. in the chapel tent. I did not go in to hear it read. The news is very cheering indeede and I begin to think we may expect a speedy termination of the war. All is quiet about here as yet. A flag of truce was sent up the river yesterday but for what purpos is not known by us. no answer was obtained to the communication and it has gone up agan today. The rebel prisoners are still here. and it may be that the buisness relates to them. but as the dispatch came <u>from Gen Hunter</u> it is supposed to be of more importance. We are still at work clearing out the inside of the fort drawing out and pileing up the timber and filling up the holes and leveling off the ground. The breach is being rapidly repaired and we are in hopes that we shall soon be through with the work and get to drilling agan. Day before yesterday. the sixth Regiment Band came over and paid us a visit (they are <u>stationed on Dawfuskie</u>) they came just before noon. spent the remainder of the day. played for us on dress parade and returned home in the evening. It was quite a treat for us and seemed to enliven us all up. The prisoners wer out to see and hear them and seemed to enjoy it very much. they wanted to hear "Dixie" and the band played it. The same day. in the morning. the balloon was taken on board of a small steam boat and taken up the river the intention being to go up as far as our batteries and make an ascension. but the wind blew to hard for it. and the boat returned a little after noon. several ascensions wer made as pleasure trips by the officers and also by a couple of ladies belonging on board of the cosmopoliton. which lay along side of our little wharf. they went up but a little ways. being held down by ropes. Provisions for two months have just been added to our commissary stores makeing in all a supply for four months that we have on hand.

    11th Sunday. A very warm day. we have had our usual inspection and I have seated myself to finish this letter. The flag of truce returned agan last night. we know nothing of its buisness. the bearers of it brought word that news of the commencement of the battle at Corinth had arrived at savannah. I hope and pray we may be success full. We have beautifull moon light nights now. it seems most to bad to go to bed and leave thier beauties unenjoyed. Last night the boys wer quite lively. in one of the kitchens a party wer having a one sided dance as I should call it. when men have to take the place of ladies as partners. the music was furnished by <u>a secessh</u> violin operated by a live yankie soldier. After the men wer tired of the sport themselves. they went out and brought in three little contrabands about a dozen years old prehaps. who gave us some specimens of real negro shake downs. it was real laughable and the little darkies wer cheered on untill they nearly melted with the exercise. In the chapel tent Col. Hawley was leading our choir in practiceing singing. in various parts of the yard little groupes of men wer gathered singing such songs as pleased them. altogether it was quite a musical time. I should have enjoyed being one of your company the evening the "Waners" visited you very much. You know I use to ask you to visit there with me but you refused. for good reasons I suppose. but some how I failed to appreciate them. You mistooke my meaning when I spoke of my fear of being to sentimental. It cannot be wrong for two persons talk often and a great deal of what most nearly concerns them. I had no fear that you would weary of the story of my love. but you know that you discovered traces of the blues in my letter. and I feared that I might have written to much even of a good thing. So my dear do not get in a fit. but please continue to entertain me in your accustomed way. You know my dear that I told you long ago that I wished and expected you to receive whatever attentions from gentlemen you thought proper. I have all confidence in you. that your own good sense will direct you in all matters that may concern any relation to each other. A thousand thanks for your photograph dearest. I think it is an excellent one. I like it better than <u>the ambratype</u> I have. and shall keep it by all means though it may get somewhat ambratype soiled by the time I get by around home. I though could not think of sending it back to you. I know it must be pleasant for you to get back to Mr Barnum's in your own room once more. You have a splendid situation. I think. And enjoying as you do the esteem and favor of your patrons and the comforts of an excellent home. It is a query with me how you can be willing to sacrifice them all and consent to become the wife of a poor fellow like me. Mr Hawthorns cap came with our last mail. He is much pleased with it and wishes me to return you his thanks for it. Sends his compliments &c. (Editor's Note &c was the accepted abbreviation for etc.) My cap is not soiled near as bad as George's. and as weather is so warm now that we always wear straw hats except when on parade I do not use it much except to sleep in. so I shall not need another one yet awhile. Thank you. I will settle the postage bill on the cap with you. If Lizzy haves her photograph taken tell her I shall claim one. I think Frank is as good as one of the Green family already. it is probably onley a question of time. I wish them joy and hope Frank will be as happy as he <u>deserves</u>. I think the Robbins and Bostwick affair will meet the approbation of the old folks. and unless the young couple can manage to get into a love quarrel. I do not see John anything Holley. to I hinder thier course from running smooth. As for do not pity him a bit. I know I would be willing to have the rheumatism <u>awfully</u> if I could onley have my <u>lady love</u> to take care of me. I am sorry to know that I cannot have the privalige of deciding as to the beauty and merits of the new dress. I fear I shall not have a chance to find fault with it. I am very glad to hear of the decline in prices of cotten goods. it is of vastly more importance that they should be cheap. than any other kind of goods. for it makes a differance with a vast number who are the least able to sustain a high price. And now darling I must bid you good by agan. I am today in high hopes that another six months will find us at darling and much everything love from looking fair for our future happiness. good by darling   much love from

Ned

No. 48 Rec'd May 23rd 1862

Fort Pulaski Ga May 12th/62

My Dear Sara

The expected mail arrived last evening. and I received your two letters of the 27th & 2nd. I finished. and mailed a letter to you yesterday. but as I have spare time this morning I concluded to improve it in commenceing at least an answer to your letters. The steamer also brought some express matter. and George. Mr Landon and Pitt received each a barrel of "comforts" all I believe in in good order. although I was not present at the opening of them. Mr Landon is still sick in the hospital. does not gain very fast. although I do not think he is at all dangerous. his barrel was taken up to him by the boys this morning. Wolcott is better. and has left the hospital. but is not able to return to duty yet. All the rest of our boys are usualy well. I should have liked to much to have shared your ride to Church on the plesant Sabbeth morning you mention. I often think when one of those beautifull Sabbeth mornings come. how if I wer home I should be on my way over the mountain. catching a ride prehaps on the other side with some of the Lime Rock folks. or what I loved still better. climbing the rocky path over Sugar Hill. often stopping so long to view the beautifull scenery. that Mr R. would be well along with his sermon before my tardy steps crossed the churches threshold. Some how I never felt guilty for thus spending my mornings. for one who has all the week been closely confined plodding over his work. a pleasant sermon from smiling nature herself is often of more benefit than the dry theorizeing of man. Not that I ever failed of being interested or instructed by Mr R's sermons. but you know ones mind somethimes feels the need of a change or of fresh inspiration even in its faith. My sister S.B. is great on comparisons I must say. I wish you would just tell her that I am looking for a letter from her. Am sorry that I could not be present to take notes of the conduct of the happy bridegroom so that if possible I might be able to do likewise should kind Providence ever premit me to be come so blest as to be a husband. Many thanks my dear for your kind hint. I will surely try to profit by it if ever an opportunity is afforded me. I fear you will have yet many onely evenings to spend if my absence makes them so. before I can return to assist in enlivening them. True. to all appearances the war is being rapidly brought to a close. but many months must necessarily elaps yet before we can hope to get home. If the battle at Corinth terminates in our favor. I have not much doubt but that there will be no more fighting for us to do down here. But I do ache to have a chance at Charleston but it is not likely I shall be gratified. I should like it much if I could send some little trophy to your friend Mr P. but such as could be sent in letters are very scarce. Of course there was not the chance to obtain things here that there is on the battle field. for of course the prisoners haveing the liberty kept all thier things. I wish I was competant to give a description of the fort. but know that I am not. but at the risk of a failure I will try to give you some idea of it. The pictures in the papers that I have seen fail as they usualy do (in such cases) to give any very correct idea of the shape or construction of the fort. I have seen in one paper a diagram showing the shape and ground plan of the fort that is very correct. The fort is five sided. and so placed that two sides bear upon the river channell. and two others partly on the river and partly on the channell betwen Cockspurr Island and Tybee. while the fifth and longest side is faceing the river and in it are the officers quaters. kitchens and magazine. no guns are mounted in it but slits. or open places about three feet long and the width of one brick. are left in the wall {which is about five feet thick) and serve the double purpose of admitting light and as loop holes for mustetry in the event of a land attack on that side. The entrance to the fort is in the middle of this wall and directly over it on the parpapet is the flag staff. The other four sides of the fort are al casemated for the mounting of cannon. The casemates are about twenty three feet long by sixteen wide and as many in highth in the centre. the roof being arched. In the end from which the cannon is discharged--a space is arched out in the wall for two feet and is about two thirds the width and highth of the casemate. this is to give more room for loading and working the gun. the thickness of the wall from thence out is five feet. the embrasure is about twenty by thirty inches in size. the wall there is the thinest being as before stated five feet. The end of the casemate at which we enter is closed by large. wooden double doors. The casemates are all connected together by an arched passage way about twelve feet in width and rather more in highth. which runs from one to the other throughout the whole length of the fort. The distance betwen each casemate is about five feet. A flue for the escape of smoke and for ventilation is arranged in the outer wall of each casemate. The thickness of the wall overhead is I should think about five feet. and then it is covered with from two to three feet of earth. This formes the parapet which runs around the entire fort and here the barbette guns are mounted. of course I need not tell you that any gun mounted on the parapet of a casemated fort is called a barbette gun. There wer thirteen guns and two morters mounted in various parts of the parapet here most of them being placed to command the cannel as the rebels expected the attack to be made principly by our fleet. The outside wall of the fort rises about five feet above the parapet and is of about as thick as it is high. this is the parapet wall and the protection of things thereon. Large piles of turf and earth have been built up on the parapet around the guns. they are filed up nice and square. and are for the protection of the gunners aganst pieces of shell bursting on the parapet. they are called traverses. The whole width of the brick work of the fort is I judge about thirty feet. cut up by the casemates and the arched passage which commences at one end of the wall and runs around the whole length of it connecting all together all the casemates. in the way I have attempted to describe. The thickness of the wall to (be) breached is from five to ten feet. The papers say that the wall was breached in seven places. I have looked very closely but have failed to discover but two of them. Have come to the conclusion some time ago that the Southern papers do not do all the lying. The open space or parade ground inside the fort called in the military parlance the "terra plane" is of about an acre in extent. Around the outer side close to the brick wall is a moat I should think thirty feet wide and ten or twelve feet deep. filled with water and soft mud. this moat is crossed at the entrance of the fort by a draw bridge. at each end of the fifth wall are bastians. formed by extending the first and fourth wall about eight feet beyond the fifth. in these bastians guns are mounted so as to sweep the most of any assailing party on that side. From the fifth side which as I before said was the side faceing up the river and of course towards Savannah. runs out another fort in the shape of a letter A. the broad end to. and seperated from the brick fort by the moat. this is of course part of fort Pulaski. but as I do not know its proper name I'll call it the open fort. it is of earth with a brick wall four feet high to keep it in place. it is calculated to mount twenty five guns but there wer none mounted. except four brass howitzers. they are pretty pieces. wer made in Boston. and belonged to the Georgia Military Institute. This outer fort is also surrounded by a moat which connects with the one around the main fort and is also crossed by a drawbridge. the water in the moat flows in and out with the tide. and can be shut in an out at pleasure. I should think that the whole surface covered by the fort and moat is about three acres. There are I think fifty of the casemates. I do not think you or Mr P. will be able to form much of an idea of the fort from what I have written. it must be seen to be perfectly understood. Still others might give a much better description. The parapet is reached by five stairways. three of them circular. built in the wall. one in each angle. As you suppose my greatest fear of not being able to return to you arises from fear of disease. not but what I am well now. but my liability to disease makes me somewhat nervous. I trust I am not afraid to face the battlefield and I hope do my duty there but have not meant to show the "white feather" in anything I may have written to you. but may unwittingly have done so. I have answered your allusion to the photograph buisness in yesterdays letter. I believe that you love me dearly and always look for the acknowledgement of it in your letters. please do not say any more about makeing your letters interesting. they cannot fail of being thus to me. and although I do not feel willing to tax your time and good nature. yet as I have before told you I am much disappointed if I fail to get three or four letters each mail. Then you want to see me in a passion and scolding do you: we'll wait untill we have been married let me see, well I can't say exactly how long, and prehaps you will be gratified. I think you are making a great many friends for me. my darling. I am greatly pleased with Mrs Phelps kind message. and how can I well be otherwise when she is kind enough to notice me thus. I have almost come to expect a message from old Mrs B in your letters. when there. please remember me kindly to them all. and thank them for thier kind messages. We soldiers have no occasion to find fault with those we have left behind. for want of sympathy and kindly feelings. God grant me none of us may abuse the confidance reposed in us by them. and that we may return to be usefull members of society. uncontaminated by the vices of the camps. May 14th. I left my letter to go in batheing and after my return had no opportunity to finish it. Yesterday I was on guard and had no good chance to write. But I have lost nothing by delay as there has no mail left since I commenced this. We buried two more members of our regiment yesterday. They have both been sick a long while I believe. one of them had nearly recovered once. but by overeating brought on his disease agan. and could not stand the second attack. A sad accident occured yesterday on Tybee. by which the <u>Lieut Col of the forty sixth N.Y. regiment (Germans)</u> lost his life. As near as I can learn it occured as follows. A party of intoxicated soldiers belonging to the twenty eighth Mass regt attempted to run by the guard. the sentinal resisted the attempt and the Lieut Col seeing the affair ran out to assist the sentinal and came up behind him just as he was throwing his gun backward to strike at the offending party. the bayonet struck him in the neck and passed clear through cutting the juglar vein. he lived onley about an hour. His wife was within a few rods of him at the time. She has been with him through the winter. and we used the to see them riding or walking nearly every pleasant day. seemingly enjoying themselves very much. Now my darling. I must close this letter. for I have another to write to my brother today. be assured dearest that I am loveing you faithfully and dearly and that I am impatiently waiting for the time to come when I can make you my own dear wife. good by

Ned

No. 49 Rec'd June 3rd/62

Fort Pulaski Ga May 17th/62

My Dear Sara

 I have nothing particular to write today. but I have a little time. and I wish to send you another letter this week. and so I will try to write a few lines. We have no news of any importance here. We learn by rebel papers of the destruction of the Merrimac and other vessels together with the navy yard at Norfolk. also that the rebels have gained a victory over <u>Gen Pope</u>. this last we are slow to believe for they also claimed the victory at Pittsburgh landing. and besides we do not want to believe it untill we are obliged to. We lost eight men the other day through the oversight of some of the officers. They wer two Sergants and six privates and wer sent up the river in a small boat with a flag of truce. to take up a young rebel who was wounded at the time we took possession of H. Head. and has been in the hospital ever since. They went off without a commissioned officer. or our national flag. -both of which are necessary when communicateing with an enemy under a flag of truce. The rebels seized them and hold them as prisoners of war. I suppose they had a right to. or would have if they had any right to be our enemies. but it seems to me that it is hardley decent. when we wer returning them thier own man. whome we had nursed and taken care of for six months. but decency is not to be expected of the villains. Heavy cannonadeing was going on up the river yesterday and this morning but we do not know the occasion of it. Col Terry has been appointed a Brigadier General and has been assigned to the command of fort Pulaski. Tybee island and for the present <u>Dawfuskie and has fixed his headquarters here</u>. Liet Col Hawley is now our Colonel. an appointment that gives great satisfaction to the men. Last night at dress parade

Genl Hunters order declareing the three states of South Carolina. Georgia. and Florida to be under martial law. and all slaves in those states to be henceforth and forever free, was read. It caused a great deal of swearing about the "d __ d nigers" and any amount of controversy and loud talk among the men. Some swear that. now we shall have to stay our three years out anyway. while others insist that it will have the effect to shorten the war. My own opinion is that it will not make much differance any way. At any rate I am satisfied that it is a just and righteous act. and a much deserved punishment to the rebel slaveholders. One of the Rhode Island boys fell off from the parapet yesterday. breaking his arm and injuring him other ways. I believe I have forgotten to answer your kind offer to knit me another cap. My old one is not very much soiled yet and as I do not wear it much in warm weather it will answer until next Fall when perhaps I shall want another. George has been unwell for a day or two but is better today. Norton and Reid are now very well but able to be about. I suppose before this you have made your expected visit to the bride and groom. Wonder if Frank is in any immediate danger. Should not be surprised if he should cease to exist as a bachelor before long.

18th Sunday afternoon. The men who I mentioned as being made prisoners by the rebels returned this morning. They wer treated as prisoners while in the rebels hands. and expected to be sent off to Macon. but it seems they thought better of it and concluded to return them. We think that we shall have to leave our present quarters shortly. our captain said this morning that we should not probably be here another Sabbeth. No doubt some some movement is contemplated the present week. but whether on Charleston or Savannah is the question. The men who moved the batteries during the bombardment are now being drilled on light artillery and what is more the first drill we have had on the Sabbeth since we enlisted has been today. so that we are very much inclined to think something is to be done before long. George and the other sick boys are better. The weather is very hot. and we hear that it is very sickley on Tybee. I learned that Lieut Gale resigned his commission as soon as he was recalled to his regiment. Prehaps I do him injustice but I think getting married was the death blow to his patriotism. I know that I have not made out a very interesting letter but beg you will excuse me once more. A little steamer has just come in from Hilton Hd. bringing Genl Benham and good news from all quarters. The mail for us is at the Hdqtrs and will probably be down here in a day or two. I wish I could spend this evening with you my darling. but it cannot be. accept much love from me dearest. and believe me now and ever. your own

Ned

No. 50 Rec'd June 3rd 1862

Fort Pulaski Ga May 19th 1862

My Own Dear Sara.

We received a mail agan this morning. and I am in receipt of your letters of May 4th & 11th for which as well as for all your other kindness I am under great obligations to you. I think I have indulged in much more foolish fancies than could have been occupying your mind on the evening of the 4th. You know I am a great hand to build air matter. And I do not think your imagination often runs as wild as mine. besides if you think you do all the wishings for and fancyings of our next meeting. you are entirely mistaken. I have no doubt but what you would know me. I can't see as I have changed much. for the better. any at least I am more fleshy than when you saw me last. have not shaved since then. but have not a very heavy beard or moustache. am turned brown enough. wear my hair short. and look bad generally. In the event of my returning home._ should not fail to pay a visit to the barbers. before I made myself visable to my friends. As to the "style" your father speaks of. there will be but little of that about us. unless we choose to put it on for the occasion. for we have been drilled so much more in the use of shovels. wheelbarrows. and drag ropes. than we have in Military that it requires thought and care. to enable us to be up to the mark when on drill. I do not think we have acquired any new "graces" yet. no one would suspect us of haveing been soldiers after we had been at home a week. I am sorry darling that you should feel so bad because you wanted to see me. I am causeing you a sight of trouble. I should have been very glad to have attended the communion the afternoon you wer denied the privalige. I have attended oney one since leaving home. Am happy to hear of any reformation in such "girls" as you and Sara B. but fear that it will not be perminent. Think Sara B. must have been sick. or in trouble. to have been so quiet for a whole hour. ask her if she could not possibly be troubled enough to write to me once more before she casts me off utterly. I hope everything is being done that can be to ensure the destruction of the rebel army in Virginia. it seems as though we wer getting the advantage very fast. I have found a good deal of fault with McClellan. but think he is doing well now. and am disposed to give him all the credit his due. Am glad that you could discover no disunion sentiments in my letter. for I did not think to mention myself when speaking of the health of the boys. for I did not think but what you would understand that unless I wrote you to the contrary I was usually well. My health is very good now. but I never could stand hot weather very well. and the oppressing heat of this Southern sun sometimes makes my head feel bad. sing heat of this Southern sun sometimes makes my head feel bad. prehaps I shall get used to it by and by. You must not speake so disparageingly about the letters you write me. please remember that they are <u>my property</u> and that I cannot bear to have any thing said aganst them. It is I who ought to apologise for the appearance of my letters. I am glad that you do not find fault with them. for I should find it difficult to write much if I took time to write properly. Prehaps though it would be better to write less and do it better. you might be more pleased. Of course my dear I could not fail to be interested in the report of the way in which you had spent your Saturdays' holiday. am glad my dear. that you spent it so pleasantly and hope you may continue to enjoy them so well. I had not heard what had become of Joe Bostwick for the last six months and have often wanted to enquire but could never think of it when writing. is he staying at home now all the while or has he a situation some where and onley home on a visit. I have written but few letters since we came here. and none of them with even my usual ease. I have no idea that my imagination would serve me any better should I attempt to write to Carrie. but could not be certain as I have not tried. My memory does not serve me well as to all the events of my last Sabbeth in Salisbury and most unfortunately I have forgotten our meeting in the porch of which you speake. but what passed in the evening and also the Friday evening before are well ... and always will be. remembered. Mr Wheatley has not yet arrived here. but probably he went to Washington first and will yet be around. I have no faith in his being able to get Samuel discharged. though of course I do not know what influential friends he has. I know that they are very slow to discharge men. even when it is plain to all that they can never be of any use in the army. There are men in our company who will never be worth a pickayune in the service. and some of them will not probably live long unless discharged. but they cannot get even a furlough to go home and try to get well. Mr Landon is recovering slowley. I have not seen him in a good while. I know I am wrong. but the hours in which we are premitted to visit the hospital are from eleven to twelve. just after we have returned from a two hours and a half drill. and though I know it is inexcusable. yet I feel more like lying down in the cool of our casemates than walking up to the hospital in the hot sun. it is not doing as I would be done by. but I try to ease my conscience by thinking that if he was dangerous. I should visit him in spite of the sun and my own weariness. I dare not say my dear that my own conduct is always right or that I am by any means free from the sin of evil speaking. but I am not profane nor do I intend to be. I thank you most sincerely for your timely warning and will endeavour to keep it always in mind and be more guarded in my own language. while judging others so severely. True indeede that none of us are free from the danger of being contaminated by the evil by which we are surrounded. If any one has the least doubt of its being easier to evil than good I wish they might spend six months in the army. I think they would be looking to far ahead. of course you will see that it would be useless for me to ever speak to any one about a situation now. I have not much confidance in home this year any way. not but what everything looks encourageing at present. but I do not think we can be discharged in a long while after it is ended. The West is undoubtedly the place especialy for a young man to enter into buisness. if however I was going out to seek my fortune unaded. I should go as far as Minnesota certain. but of course would stop this side if I could get a good situation. I have not answered your letters very faithfully. you are very kind to write me such long ones when you are so poorly repaid. We had a very heavy shower last evening so that I could not possibly I "walk over and see you and my carrage is out of repair". but attended prayer meeting so you see I did the next best thing. George is lying in my bunk just now having come down to make me a call. he sayes he ment to have written to Julia that she must go to you for news as he has not felt well enough to write much. but presume he has written more than I have now. I am going to end off this letter pretty soon and make him some toast for his supper. he has not much appetite but thinks prehaps some toast would taste good. his head aches badly today. The balloon is being filled agan. I think an ascension will probably be made tomorrow. There seems to be considerable activity among the officers and from what we hear we think a movement on Savannah is to be made very soon. prehaps within a week. I will try to write to you agan very soon. and shall hope to hear from you again soon. Good by <u>darling</u> much love from your own

Ned

No. 51 Rec'd June 16th 1862

Fort Pulaski Ga May 25th 1862

My Dear Sara.

    This is a rainy Sabbeth morning. it does not rain fast but has during the night and we have the usual amount of hard. sticky mud that makes it so unpleasant here after a rain. It is no(w) raining a fine mist just enough to wet one before they know it and make them uncomfortable. We shall probably be spared our usual Sunday inspection. but a large number of men have been detailed for fatigue work and I do not hardley expect to be allowed to finish my letter in peace. We have good reason to expect that we shall leave the fort this week. prehaps as soon as tomorrow. The baggage of <u>the 48th N.Y. Regt</u> has arrived. and a few of the men. They tell us they are to take our place and that we are to go to Edisto Island which you know is near Charleston. We know that our own officers do not expect us to stay here long but they give us no intimation as to our probable destination. The 48th is the regt that Gale was in. I hear that the commanding officers are nearly all gone. and that is the reason why they are sent here. and we posted off to another place. Two batteries that we had up the river cut off the communication of the fort with savannah have been abandoned. as we have no further use for them. the guns wer all brought down here. and are now being put aboard schooners for transportation to some new scene of action. our men have all been employed on fatigue the past week with the exception of a little drill on the heavy guns. I went on guard Friday morning had a harder twenty four hours duty than I have ever had in any two days before. the weather was pleasant untill the last three hours yesterday morning when it commenced raining. The reason why I (or rather I should say we for the corporals all took it alike) had so hard a time was because we had a fussy old bachalor for officer of the guard. not so very old either. but he is old enough so that he ought to know better (than to be a bachalor). He was so afraid that everything would not be done just right that he kept himself. the sergant. And us three corporals in a constant worry all the time. and on the run the biggest part of it. I slept just an hour during the night. After all. he was so good natured and pleasant and patient. and on the whole as much in a stew himself. as he kept us. that we could not get angry. and had a good many hearty laughs at his expense. Her Majesty's steam sloop of war Racer. came in and anchored near the lighthouse on Tybee that morning. and during the day. her captain and several junior officers visited the fort. The British Consul at Charleston was with them. The captain is a cousin of Lord Lyons. quite distinguished guests you see. on the departure of the vessel late in the afternoon she fired I a salute which we returned. fireing twenty one guns. Yesterday I expected to have to myself. to sleep and write letters in as usual. but was obliged to go on fatigue in the afternoon. and so my letters must be written today or not at all. After all it is about as well as I can employ any spare time I may have on Sabbeth. for I seldom write anything that I think is very wicked. and if I am not thus engaged. I cannot help listening or haveing conversation that is going on all around which is anything but pleasant or agreeable. A rainy Sabbeth at home is generaly a weary day you know. and here it is unpleasant from being confined inside with those who regard not the day but spend it in wrangling over a game of cards or in any other say they choose to amuse themselves. our boys are all on the gain. but they all find as I did that it is very slow work getting up after ever so slight an attack of fever. We are unusualy short of rumors and news just now and I shall hope that when our next mail arrives we may hear some good news. The British officers who wer here the other day said that we could not whip the South in five years. but then you know they are liable to be mistaken. and I certainly hope they will be in that opinion. Does Lizzie consider her engagement with Charlie broken up or is it merely postponed untill a more convenient time. give my love to her and Julia. I don't see but you three girls are all in the same boat now. I hope it may not be so long. I shall be very happy my darling when the joyfull time comes that I can call you my owne dear wife. I had a very pleasant dream the other night of you. I cannot tell you now what I dreampt. but may someday. I was right glad to dream of you. for it is seldom that I do. but it is unpleasant wakeing up to the realization of the truth. still I must say that the rememberance of that dream followed me through the day and I spent it much pleasanter than I otherwise could have done. How much better one is for the rememberance. even of the good and beautiful 1. it will make bearable and happy many an hour that would be tedious and unendurable without them. I must close my letter now dearest. but will be sure and write often so that you may not be long ignorant of our whereabouts. With much love dearest I remain ever your own

Ned

No. 52 Rec'd 28th June/62

Legaria. S.C. June 6th 1862

My Dear Sara

  Yesterday after our arrival here I received your two letters of the 22 & 27th. the mail having arrived here just about the time we did. I also received two letters from you just before we left Ft Pulaski but had no time to answer them before leaveing. I tore. and burned them up on our way here. as we expected to have a fight and I did not care to have anyone read them in the event of my getting killed. I cannot answer your letters now for we are momentarily expecting orders to march. I am onley agoing to write what I can and let you know that I am well and have received your letters. I think George will write all about our leaving the fort and our present destination and I presume you will have an opportunity to read his letters. I do not expect to have much time to write for a few days at least as we expect to have stirring times. But if George remains unwell he will be able to write. I have not seen him since we left Edisco on Sunday evening but I hear this morning that our baggage which was left there with the sick who wer able to come thus far with us are to be brought up here. if it (is) so they will probably be here in a day or two. I hope so far we left our knapsacks with all our things there. bringing nothing with us but our blankets. I have not even an envelope for this letter and I do not know as I shall be able to mail it after it is written. I brought one along with me. but as we all got thoroughly wet through yesterday it was spoiled. We have had a tough seige of it for the last five days ending off yesterday with a twelve miles march in the rain. This morning we are expecting to go over the creek and <u>join Gen Stevens</u> fore now encamped on the other side under the cover of our gun boats. then we shall expect sharp work as we are now within five miles of Charleston and our forces have skirmishes with the enemy every day. Of course no delay will be made about an attack on the city as soon as we are ready. This place is a place of summer resort for <u>the nabobs.</u> contains prehaps a quater as many houses as the street at the centre. with a couple of very small churches. I may have more time than I anticipate within a day or two. and if I do will try to give you some account of our journey thus far. but if I do not it will have to lie over untill I can give you a verbal account of it. at present spirits. I can onley tell you that I am in first rate health and spirits. a little sore from the effects of our march. I shall not be able to answer your last letters. and hope you will excuse me. for not doing it when you know the reason. I shall destroy all of them that I have with me when I think we are agoing into action. Be assured my darling of my continued and undiminished love and affection. I always think of you when in danger more than at other times. I still hope that we shall meet soon but you know it is uncertain and is always well to remember that we are liable to be disappointed in what we think we are almost sure of. The rebels will doubtless make a desperate fight here. Our regiment is in much favor with our generals and we shall doubtless have warm work to do. all I ask for is courage to do my whole duty without wavering. which may God grant me. I cannot write more now. give my love to all my dear friends. May God bless and protect you and in his own good time premit us to meet agan. Good by darling accept much love from your own

Ned

CHAPTER 3

JAMES ISLAND

From the book:

After some rumors and false reports the regiment embarked on the steamer "Cosmopolitan" May 31st and steamed away to the north. Reaching North Edisto, sixty miles away, on the first of June, the stores were discharged and the men crossed the river to Johns Island.

Taking up the line of march on the 2nd we marched about five miles, oppressed with heat and thirst. We bivouacked at Sea View plantation until June 5th. On that day we started at 10 A.M. in a pelting rain and marched about a dozen miles to a deserted village called Legareville.

We bivouacked that night and the next day. The boys complained of sore feet caused by the long march in the rain. Rations were scarce and the Chaplain made a forced march to the rear for coffee and sugar and brought good cheer on his return.

On the 7th we crossed the Stono River to James Island where we lay down in a muddy cotton field with the rain pouring on our devoted heads.

No. 53 Rec'd June 17th/62

<u>James Islands</u>. S.C. June 10th/62

My Dearest Sara

I improve the first opportunity to agan inform you of my whereabouts and welfare. We are near the rebels and within about five miles of Charleston. forts Sumter and Moultrie can be seen from our outposts and we have a slight brush with the enemy about every day. I am well. and with the exception of a slight cold and a little soreness never felt better or tougher in my life. and that to. in spite of hard marches through the rain and mud. being wet through to the skin several times. lying out down on the damp wet ground for more than a week with poor and scant fare. and last but not least being exposed to the shot and shell from the enemy batteries all day while on picket. and winding up with a skirmish with thier picket during which I discharged my rifle at the enemy for the first time. I will try to give you some account of our doings since we left Pulaski. We turned into our bunks as usual after roll call on Saturday night. the 31st. but had hardley got nicely settled for a nap when the order came for us to get ready immediately and fall in line to leave the fort. As we had been expecting the order for two or three days. and had everything ready we wer soon on our way to the wharf and in a short time wer on board the Cosmopolitan, where we wer obliged to stay all night at great discomfort, for there was hardly room for us to sit down, and lying was out of the question with a great many. We left the fort about eight o'clock in the morning,(Sunday, June 1st). which was as beautifull a June morning as we often see and the day continued fine throughout. Nothing of interest occurred during the trip. We reached Edistoe about one o"clock. but did not land until just at night. We went ashore unloaded our camp equipage. pitched our chapel tent for a hospital. as the sick ones could be taken no further. also pitched a couple of small tents for each company into which our knapsacks wer all put and a guard left over them. then in light marching order we agan went on board the boat and droped down the stream about a mile and landed on the other side of it on Johns' Island. We went on board about twelve o"clock but there was a good deal of delay about landing us which had to be done in small boats. and it must have been about two when we wer all landed and wer ordered to lie down and rest. We lay in the sand on the beach where we landed. for the rest of the night and I improved the time in sleeping as hard as I could in spite of the fleas. In the morning (Monday, 2nd) I arose had a splendid salt water bath. breakfasted on hard bread and coffee. and found myself pretty comfortable. The morning was spent in landing some horses. a wagon. some ammunition. and a small quantity of provisions. As the day advanced the heat began to be very oppressive. At about noon we wer ordered to fall in and in a short time we wer on our march. In anticipation of a hot. hard march and possibly some fighting. I had lightened myself as much as possible during the morning. I had worn my dress coat thus far and carried a <u>blouse</u> with me. but I threw away the coat and some underclothing. tore. and burned up my letters and disposed of everything in my pockets that I could possibly spare. As we marched along I had plenty of reason to be thankfull that I had done so. The day was very hot and though our road lay through the woods· yet there was not a breath of air stirring. As we marched along.

the side of the road became almost lined with blankets. overcoats. and whatever else of clothing men could possibly get along without. that had been thrown away as thier weight became oppressive to the owners. We had to march slowley. stopping often to rest so that it was about sundown when we came up with the rest of the army and stopped for the night having marched about five miles. Light marching order means besides arms and equipments. canteen. and haversack and our blankets I had with me my rubber coat in addition. I regretted during the day having brought it. but found subsiquently that I had abundant reason to be thankfull. We build up fires. made us a cup of coffee and with hard bread made out our supper. and then lay down with our arms by our sides. accouterments on. and rested. I had a good nights rest but found my shoulder rather sore where my blankets had been strapped on the day before. We expected to have to march very early and so arose at four o"clock (Tuesday 3rd). got our breakfasts and wer soon ready but no orders came to march. About ten o"clock it commenced raining and continued with short intervals. to pour down all day and night and the next day (Wednesday) untill about one o"clock when it cleared up and we had a chance to dry ourselves a little. We had no orders to move. and so we stood and took it as patiently as possible. My rubber coat now paid me for lugging it the preceeding day. the boys resorted to all sorts of expediants to keep themselves dry. by sticking a few sticks in the ground and tying thier rubber blankets to them. quite a dry little shed could be made under which the owner could crawl on his hands and knees and lay down. No orders came to march the next day untill about <u>four</u> in the afternoon when we buckeled on our truss and flattered ourselves that we wer agoing to leave "Camp Misery" as we styled our stopping place at last. But we did not get off that time. it proved to be oney a brigade review. and after bobbing about over the ridges in the cotten fields untill we wer well tired out. we brought up at our old camp agan and wer dismissed for the night. after getting orders to march early in the morning. At half past twelve in the night we wer arroused up and without any time to eat or drink fell in line and at last got a fair start. I rolled up my overcoat and strapped it on my back inside my blankets so that it was not easily to be got at while marching. and as I had no chance to put it on until I was pretty well soaked I let it alone and took the wetting. It commenced (Thursday) raining in about an hour after we started and continued with but few moments intermission untill noon. We marched along quite rapidly oney stopping to rest twice during the march of about ten miles. As we passed along the road began to grow more and more and more muddy and the rain fell in torrents. the road was so nearly level that the water did not run off. but soon became ancle deep and we marched through it for miles without finding a place where it was any less. We expected to have a brush with the rebels at a place where the road crossed a small creek. the bridge over. which they had destroyed. but they wer not to be found so we waded the creek in which the water was about two feet deep. and continued our march. We reached Legaria about nine o"clock and took up our quarters in the houses which the owners had kindly vacated for our accommodation. There we remained over night. got dried and partialy rested. cleaned our guns. exchanged our spoiled cartridges for dry ones. and managed to get a full meal of hard tack once more. I managed to write you a short letter from that place which I hope you have received before this time. We left the place about noon (Friday). wer ferried over stanhoe creek and landed on this island. the boat had to make two trips and that with the usual delay brought it down to about sun down when we reached this. our present camp. about a mile from the landing where we bivuacked for the night. we had occasional showers through the day but did not get much wet. There was a slight shower during the night. for I recollect partially wakeing up and feeling the rain dropping on my face but it was to small a matter to wake up for. so I just dropped off into a sound sleep agan. from which I did not awake untill the morning (Saturday) was well advanced. The day was cloudy with occasional showers. the morning was spent in careing for our guns. cooking. and eating. About two o"clock orders came for us to fall in and make an advance into the woods. the object being to make a reconnoisance and find out the situation and probable strength of the enemy in the direction we wer to take. Part of a company of cavalry went with us. when we came to the woods beyond where our pickets wer stationed. we wer divided up. one company going in one direction. a couple in another. and so on. our company was sent off to the extreme left about three quaters of a mile from the others. with a cavalry man for a guide and to bring orders. Our orders on starting. wer to advance as far a road that was known to cross the path we wer takeing if possible if we met a superior force to fall back or if we heard heavy fireing to the right to fall back. We advanced to the road specified without meeting anything. and stood there in the rain. waiting impatiently for something to turn up. At length fireing commenced in the woods at our right. and then we knew that our boys had found them. We found that we wer far in advance of the main body of our men and that the fireing was comeing nearer our horseman who had been sent back for orders now came up with orders for us to retire immediately which we did in good order. and wer soon joined by the remainder of the regiment and returned to camp which we reached about nine o"clock. having been about five miles and back and such marching as I never saw before for a long distance our road lay along side of a ridge thrown up to keep the water out of the fields. here we had to go in single file through mud over shoe deep and slippery and sticky so that it was all one could do to keep on his feet. then to we had to wade through water and jump ditches and last but not least got wet to the skin with the rain which fell most of the time we wer gone. The result of our journey was one man slightly wounded. one man captured from us by the rebels. and the desired information. so that our object was attained and with but little loss. the lost man belonged to CQ A. he was sent out one side a little and his retreat was cut off and he made a prisoner. We wer pretty much used up on our return but we built up a good fire made us some good coffee. and then I spread my rubber blanket down on the wet ground. laid myself down. covered myself up with my woolen blanket and overcoat and though wet through. and my shoes it seemed half full of water. slept soundly untill morning and got up not near as blue as I used to many a morning at home. This was Sunday morning and we flattered ourselves that surely we should not have anything to do that day more than to clean our guns and dry ourselves. imagine our disappointment then. when before we had got our poor apology for a breakfast eaten. orders came for us to clean our guns. roll our blankets. take a days provisions and be ready at nine o"clock to go out on another skirmish. There was some <u>tall growling</u> done you may rest assured. a large number went on the sick list determined to take medicine rather than go out again so soon. but we wer told that our Colonel was agoing to try and get the order countermanded. and we knew that he would do the best he could for us. So we put the best face we could on the matter. and got ready for the worst. The colonel succeeded in getting released from going out but was ordered to hold us in readiness to march at a moments notice in the event of the enemys comeing on us in force. This was some relief but we could not rest much and at last about about three o"clock we wer ordered to fall in and six companies wer taken out and sent on picket. our company was not wanted and we congratulated ourselves that we at least would have one good nights rest under cover for our tents had just come and we went right to work and pitched them. During the evening it commenced raining very hard and we wer thanking our lucky stars that we wer sheltered from one storm at least. But we wer not to be let off so easily. about one o"clock we wer roused up and ordered out immediately. we wer wanted for picket. we went out about a mile and stopped at a deserted house. placed a few men out a short distance from the house. and the rest of us laid down and finished our nap. I slept untill morning (Monday). Soon after daylight the rebels commenced fireing at the house as I suppose they saw us about there. and continued with but few intervals to shell us all day. Nearly all of thier shell exploded short of us. but they have one rifled gun that threw shot clear past us. There was an alarm during the day and we marched out near the line of our pickets, but there was no occasion for us, and we soon marched back to the house. At length about two o"clock an officer came up who wished to find out the exact locality of a rebel battery. so our company. with a few cavalry. and Co B wer ordered to support him in a reconnaisance. by some mistake Co B did not join us as we marched past them. but onley stood ready with arms in hand ready to march. the officer did not notice the mistake untill we had passed our picket and wer well along towards the enemy. he then sent back for them and they came on at double quick. but did not come up until the brush was over and we had been ordered to retire. After passing our pickets a short distance we halted and the officer was giveing some orders about our further advance, when we were fired upon by the rebel picket who wer behind a fence sheltered by bushes about thirty rods in front of us. We were immediately ordered to fire which we did with a will. and as soon as we could reload. immediately advanced upon the rebels who fell back upon thier battery about a half mile in the rear. we went forward to where the rebels had been and stood a short time. while the officer rode forward to reconnoiter. while we wer waiting the rebels commenced shelling us from the battery we wer looking for. and it quickly became a warm place to stay. and an equaly warm place to get out of. for the shells fell thick and fast along the way we had to go. but we accomplished it in good order and received the praise of our colonel who had come out from camp with the remainder of our regiment to support us if necessary. The shells fell to the right and left. before and behind us and burst over our heads. but no one was hurt by them. one man was slightly wounded in the arm while we were fireing at the rebels. he was hit by three buck shot but it did not prevent him from going forward with us. in the advance we had to wade through water waist deep and had to look out well and keep our catrages dry. We returned to our old station at the house. which the rebels commenced shelling with renewed vigor. but with out doing us any harm. We returned to camp about eight o"clock pretty well tired out. We had a good nights rest. and after cleaning my gun I set myself about writing this letter. (Tuesday 10th) our knapsacks haveing arrived yesterday so that I am agan supplied with paper. Today is a splendid day. and I think it has at last cleared off after raining every day and night for a week. Our field pieces have gone forward today and there has been fireing all the morning and is now getting quite brisk. We have just had orders to have our arms in readiness. canteens filled with water and ready to march at a moments warning. so my darling I must hurry and finish this letter without writing near all I wish to. Excuse the miserable manner in which I have scratched this off. I could not stay to half write it. for I have feared every moment the orders that we have just received. I should love to write much more. but cannot now. There is no concealing that we are now in danger hourly. for we may be called at any hour to engage in deadly conflict with the enemy. who are both as strong and determined as we are. But we will remember that the same God still has us in his charge who has thus far cared for and protected us and if it be his will no power can harm us. As I said at the beginning of my letter I am in excellent health and have strong hopes that if I escape harm on the field. I be able to return much healthier than when I came away. Let me shall agan assure you my dear one that I think of you much and love you more than ever. Please write me all the news. and do not be angry if I fail to answer your letters properly and promptly. I will do the best I can but cannot have the time to write that I have had. May God bless and protect you is my earnest prayer. With much love I remain as ever your own

Ned

P.S.

I have time to add a time to add a few lines. so I enclose a few geranium leaves. a couple of flowers and a couple of leaves of cape Jassamine. which I picked out of a flower garden in Legaria. I have a nice cape jessamine but it is to bulky to send this time. I thought I had a geranium flower but cannot find it. George is not with us. he is at Edistoe unable to come. I am sorry for I would like to have him here. but I suppose Julia will not be very sorry. Please remember me to Mr Barnum's people. once more darling good by

Ned

No. 54   Rec'd June 28th/62

James Island S.C. June 12th/62

My Dear Sara

On the evening of the day on which I wrote you my last letter we received a mail and I was gladdened by the receipt of yours of the 30th. I wrote you that we had just had orders to be in rediness to march instantly. luckily it was a false alarm. and we wer not called out. though there was a sharp skirmish. and a good deal of cannonading that day. Yesterday and today have been days of rest for us though we had to fall in line yesterday afternoon. stack our arms. and remain with all our accouterments on ready for action untill sundown. then we thought we wer sure of another nights rest. but just as most of us had dropped to sleep we wer roused up and had to go out and work until nearly two in the morning drawing heavy seige guns up to near where they are to be placed in battery. which will also have to (be) thrown up in the night and the guns placed in position under the same friendly cover. If we did lose our nights sleep we wer partialy at least compensated by a beautifull sight. which we do not often see. a total eclips of the moon. The evenings are splendid now. The clouds and rain which made last week so uncomfortable for us soldiers have given place to clear sunshine. and although the weather is most to hot to do much in. in the middle of the day. the nights and mornings are very comfortable. Today has been a quiet one with us. We had a kind of regimental inspection this morning but it was made as easy as possible for us. and caused but little fatigue beyond the two hours labor in our tents necessary to put ourselves and equipments in order. The inspection was made by Col Fenton of the 8th Michigan regiment who is an acting brigadier general and in whose brigade we are now placed. This afternoon we have been agan ordered under arms and I am now writing with my equipments all on and my gun in the stack ready for action at a moments warning. We expect to have to go out and work tonight but whether at drawing cannon or throwing up batteries we know not which. The ground where our batteries are to be placed. must be pretty well ironed over by this time for the rebels have been shelling the place for several days thinking I suppose that we wer at work there. The work will have to be done in silence. for the rebels have pickets within a short distance and they are keeping a bright lookout expecting us. A little unusal noise at the place will be sure to bring down a small cargo of iron about the heads of the workmen. I suppose you will think that so much excitement of the kind is anything but agreeable. I do not like it because you know that I am not naturaly a fighting man. but. we do not mind it much. we just stack our guns and then turn into our tents and commence our reading or writing where we left off not worrying ourselves as to what may come next. Today we received another mail and I am agan favored by the receipt of your two letters of June 1st and 5th. Many thanks once more my darling for your continued kindness in being so punctual in writing. I know that before this you will have looked anxiously and often for letters and as often been disappointed. but I think the reasons I have already given will be received as sufficient excuse for the long delay.

June 13th. I was broken off from my writing yesterday afternoon by the order to "fall in." In a few moments the regimental line was formed. the colors displayed and the regiment on the march. It proved to be another false alarm. however. and like the famous king of old who "marched up the hill. and then. marched down agan". we marched out about a mile to the headquaters of our picket and after waiting untill about sundown. the cause of the alarm was found out. and we returned to camp. As I expected we had to go out to work last night. We had oneley time to drink a cup of coffee and nibble a hard bread after our return to camp when we wer called out to work. Part of the regt of which I was one. went out to work at the battery which had been commenced the night previous. It was much nearer the rebels than I had any idea that it would be. In fact as we marched along with our shovels. axes. and pickaxes. on our shoulders in single file. keeping in the shade of a hedge that skirted the bank of a creek. I began to think we wer going clear to Charleston. Arrived at the scene of operations we immediately set to work. and worked with a will. The digging was bad for the soil was composed of oyester shells and roots of trees with a little soil mixed in. but we succeeded in finishing it about twelve. Then we returned to the house which I have before mentioned as a sort of headquaters of our picket and where there is kept a small force under arms all the while as a reserve in case of an attack on the pickets. Here a quantity of timber. plank. stakes. etc to be used in the batteries. had been left by the wagons. which are not allowed to go further on account of the noise they make. Each man loaded himself with whatever he could carry. and returned to the battery. As the distance must be a mile certain. and the path directly across the ridges in the cotten fields. and the plank so heavy that it required four men to carry them. you can imagine that it was no easy job. Mean time while we had been shoveling. another part of our regiment had been buisy drawing up a part of a gun carrage. and two thirty two pound Parrot guns. which with a James rifled gun are to constitute the battery. Others had been employed all the time in carrying the planks. When we came away we met still another party of men belonging to some other regiment. loaded with boards. timbers etc and I presume the work was continued until daylight. In fact it did not lack more than an hour of that time when we got back to camp and turned in. for it was then three o"clock. I slept soundly from that time until after sunrise when I was awakened by the cry of "breakfast". after eating which attempted to finish my nights rest. but although tired and sleepy enough in all conscience. yet it is always difficult for me to sleep during the day unless very much fatigued. So after trying a spell. I gave it up for the present. went down and took a good swim. and then concluded to finish this letter. but think I shall manage to get a nap yet before night. An accident occurred night before last when we wer drawing the cannon by which an officer in the 46th N.Y. regt had his leg broken. We wer takeing a rest sitting along on the ground near the drag ropes which wer dropped on the ground just as we stopped. the officer was riding a wicked runaway. secesh horse. which had started into a run and the officer could not manage him. On he came with the speed of the wind tearing along through the line of men who scattered right and left in a hurry. As he was passing, the cannon he caught in the drag ropes and in an instant the rider was thrown far over his head. the horse was thrown all in a heap upon one of our men who luckily escaped with nothing worse than a severe brusing. Reid and Barnes rejoined us yesterday afternoon. They had been left back at Edistoe to guard our baggage left there. and assist in takeing care of the hospital. Reid tells me that George is recovering slowley but is most discouraged because he cannot get well immediately. After he has been sick as much as I have. he will know better than to fret because he cannot recover as fast as he wishes to. He wishes me to write to him. and I am agoing to try and write him a few lines today if I have time. I hope he will rejoin us soon. I burned a large number. some twenty five or thirty. of your letters to me the other night when I thought we wer agoing into a fight. I do not wish anyone to read them and presume my comrads would not do it even if I did not return to take care of them. but I did not feel quite easy until they wer all consumed. I was sorry to loose them for I wished to save them very much. I shall hereafter destroy them as fast they are received except such as I can readily return to you. I shall not be able to answer your letters in detail for want of time. be assured dearest that they are most welcome and interesting and that the contents of each is carefully noticed although not properly acknowledged. Please continue to give me any little items of news that you think of for it is the little occourances of every day life that the soldier loves most to hear about. they seem to bring one nearer home and to bring us into closer communion with friends left behind. whome may God grant that we may soon be premitted to rejoin. I must close my letter now for I am getting decidely tired and sleepy so hoping to hear from you agan soon. I remain with much love your own

Ned

CHAPTER 4

SECESSIONVILLE

From Wikipedia: At about 4:30 a.m. on June 16, the Northern troops attacked the Confederate fort at Secessionville where Colonel Thomas G. Lamar commanded about 500 men who had a number of very heavy artillery guns and a good field of fire. Marshy terrain to the north and south would constrict any Union advance. In the lead was the 8th Michigan and behind them was the 7th Connecticut and the 28th Massachusetts. The 8th Michigan were "mowed down in swaths" from "a shower of musket balls and discharges of grape and canister" from the Confederate cannon, according to one Union officer. Yet, some of the Union infantrymen made it into the fort fighting the Confederate artillerymen hand to hand before Confederate infantry reinforcements arrived to help Lamar's decimated men. These were Lt. Col. Alexander D. Smith's 9th South Carolina Battalion, up from Secessionville. Lt. Col. Peter Gaillard's Charleston Battalion soon followed and the battle became a rifle match along the battery wall and swamp lines. Lt. Col. Joseph Hawley's 7th Connecticut's advance halted when their left flank became mired in the marsh mud and their right received canister and grape. The 28th Massachusetts followed the 7th into the same mire and both regiments became intermingled as the Confederates continued to shoot and shell the confused mass of men. In the meantime, Lt. Col. John McEnery's 4th Louisiana Battalion advanced to reinforce Lamar's garrison, while Simonton's Eutaw Battalion advanced along Battery Island Road to face the Union left flank.

A Union battery, the 1st Connecticut under Capt. Alfred P. Rockwell, finally started firing on the Confederate garrison as the Highlanders of the 79th New York under Lt. Col. David Morrison advanced. Confederate artillery fire forced the 79th to the right flank of the fort where they joined the remnants of the 8th Michigan. The 79th mounted the top of Tower Battery and went over the wall. In the end however, they were repulsed, as had the 8th Michigan before them, when reinforcements failed to appear. The 100th Pennsylvania Roundheads, under the command of Maj. David Leckey, tried to support the Highlanders, but their attack stalled as did the previous ones with Confederate canister and grape. Col. Rudolph Rosa's 46th New York tried to line up on the 100th's left, but some retreated with the fleeing Irish 28th Massachusetts and the 7th Connecticut, while the remainder received Confederate canister. Finally, Col. Daniel Leisure ordered a general retreat. Isaac Stevens ordered the 28th Massachusetts, 100th Pennsylvania, 46th New York, 8th Michigan, 79th New York, and the 7th Connecticut to retreat back towards the hedges. The attack had lasted less than 45 minutes.

From the book, History of the Seventh Connecticut Volunteer Infantry: "The casualties to the Seventh were nine killed, sixty-nine wounded, and four captured or missing, making an aggregate of eighty-two — the aggregate of casualties to the whole command was 683, of whom 107 were killed.

The whole number engaged on the Union side was about 6,600 men. The number of the enemy is not reported. There were five regiments and five battalions. As they fought largely behind entrenchments, their aggregate loss was only 204."

Editor's Note: This next letter is written the day after the Battle of Secessionville, a Union defeat attempting to capture Charleston, S.C.

MY DEAR SARA

No. 55 Rec'd June 28th/62

James Island S. C. June 17th/62

My Dear Sara

It is with feelings of sorrow and regret that I set myself at the task of writing to you this morning. For the first time since leaving you has it become a task to write to you and certainly would not attempt it this morning did I not know that news of our yesterdays' disaster would quickly reach you and doubtless in a greatly exaggerated shape causing you a deal of anxiety until you should be able to learn of the welfare of those you love. our regiment has seen its first battle. and been severely repulsed. with considerable loss. Today we are mourning for the brave men who fell and fell without accomplishing the object for which they sacraficed thier lives. Thank God our Salisbury boys are unharmed except Dexter whose right arm was slightly grazed by a musket ball. Many of them wer in the thickest of fight. and had very narrow escapes. Norton was knocked over by the wind of a cannon ball which blew away his hat and left him half crazed. for a moment. but he was quickly on his feet. eger as ever for the fight. he's a brave fellow. every inch of him. Pitt was struck on the catridge box by a piece of shell that but for that obstruction would doubtless have finished him. Others of the boys had very narrow escapes for they wer in the advance some of the time of the colors where the grape and canister wer ploughing through our ranks. I will try to give you some account of the fight. though I know I shall fail to convey a very correct idea of it. The attack was ment to be a surprise. and in that it was partially successful. We wer aroused about one o"clock and immediately fell in line without stopping for any breakfast. The distance from our camp to the rebel battery is I should judge about three miles. A long time was consumed in awaiting orders. and in getting the other troops in order and advanced slowly stopping at intervals until we got beyond our pickets when the remaining distance nearly a mile was made mostly at a double quick. Until we came in sight of the battery none of us privates new anything what was required of us. soon after we left our camp we wer ordered to load. then shortly after came the order "that not a shot was to be fired but that we wer to go to our work with cold steel". All this sounds very well or would if we had been successfull but the storming of what is about the same as a casemated fort without a scaleing ladder or anything to assist men in crossing the ditch and climbing the stockade is worse than "requireing bricks without straw". After passing our outposts we advanced rapidly and soon our skirmishers came upon the enemys outposts which they put to rout. captureing a few prisoners. Here a few of our men wer killed or wounded. I could not tell which. we passed them being carrried back on litters or lying on the ground where they fell. This occurred at a house and the outbuildings belonging thereto. which was situated at the edge of a large plantation we had crossed. This place was afterwards used as a hospital by our folks and after we wer repulsed we burned it. first removeing our dead and wounded so that it could not agan become a cover for the enemys advance. After leaving the house we went through a narrow piece of woods and came out on the battle field. As we entered the open field we formed in line of battle as well as possible. It came very hard on us who wer on the left of the line for as the right kept on at double quick. we had to run in order to bring our line out straight and by the time it was accomplished we wer very much out of breath. for the ground was a deserted cotten field and they are always the most fatigueing in the world to march over even slowley. When our line was formed we continued our advance at a double quick. The skirmishers in advance driveing before them the flying rebels from the outpost. As soon as they could shelter themselves in the fort. the rebels opened thier fire upon us. Musketry. grape. canister and scrapnell. came plunging through our ranks makeing a clean sweep wherever they struck. As we neared the fort our left came upon a marsh which at high tide must be covered three or four feet deep with water. there was no water in it at the time. but the soft mud was very deep. the bank. as lined with fallen trees and brush purposley placed there to obstruct an advance. a howitzer was placed so that its fire completely raked this bank so that many who sought protection there. met the fate they wer trying to avoid. My position was in the centre of our company. and. and one other CQ. B. was to the left of us. As the advance continued my course was straight ahead through the marsh but I could not see what good we wer to do that way as by this time onley the right and centre of our column was faceing the fort. but as my onley buisness was to obey orders. we kept on. we sank in the mud nearly knee deep and it was worse and worse the further we went. An order came for us to come out of the marsh which we made haste to obey. Norton was with me in the marsh but nearer the bank and got out sooner. many others wer in there struggling in the mud. just as I gained the brush which lined the bank a terriffic charge of grape and canister came crashing amongst us strikeing all around us and wounding others near me. I never shall forget the ring with which that charge came about my ears. I gained the bank soon as possible. and looked around for the rest of our company. I could find but few of them. Our captain with a part of the company had gone ahead into the thickest of the fight where he was soon struck down by a ball. The regiment was now sadley disorganised. Three companies wer all huddled together on the bank of the marsh into which grape and canister of the enemy was makeing sad havock at every discharge. The ground was beginning to be quite thickly strewn with the fallen in spite of the constant labors of those who wer all the time at work carrying them from the field. A desperate attempt to rally the men in the colors around which stood a group of as brave men as ever drew breath. thier number was being rapidly thinned out but they never flinched until the order to retire was given when they moved off the field in good order. In the mean time other regiments came up cursing us for our cowardice. crying out to give them room. but I notice they got no nearer the fort than we did and then broke much worse. many of them throwing away thier guns in thier retreat. I do not think a man in our regiment who was able to come off the field alone failed to bring his rifle with him. About this time our light field batteries came up and began to play upon the fort. and Wrights brigade made an attack on the left of the fort. on the other side of the marsh that I have mentioned. Our regiment had now fallen back to the edge of the woods that I mentioned our comeing through on the advance. here protected by a ditch we rallied what we could and rested. Soon <u>Gen Stevens rode up</u> and wished us to make another advance and support a regiment that was then trying to storm the fort. Editor's Note: (On September 1, 1862, BGEN Isaac Stevens was killed instantly while leading his men at the Battle of Chantilly. He was posthumously named a Major General on July 18, 1862.) Our colonel reorganized us as quickly as possible dividing us up irrespective of companies into four divisions or companies under as many captains. that being all that wer left unharmed on that field. we agan advanced in good order. I should think half way to the fort until we came to an intrenchment thrown up for the protection of infantry I should think. here we wer ordered to lie down and keep under cover. Soon a section of the Conn battery consisting of two pieces came up and unlimbering opened fire on the fort. fireing right over our heads. this soon drew the fire of the fort upon it and our situation was anything but pleasant. crowded down into a ditch. our battery fireing over our heads deafning us with its thunder while we wer obliged to hug the ground to avoid the shot and shell which was falling around us. Our battery kept up its fire untill its ammunition was all gone and two of the horses killed. then they wer ordered off the field and when they were safely off. the order came for us to retire also. which we did in good order. General Wright had been repulsed from the attack he made on the rebels' right and it was evidant that nothing more could be done towards takeing the fort by the plan of operation then being pursued and our forces wer called off. The rebels did not attempt any pursuit. and weary and sad we took our way back to camp. which we reached about ten o"clock. The attack was made soon after daylight and lasted about two hours. and ended with a complete repulse of our forces engaged and with great loss and no advantage gained. Our captain was shot down after the action commenced. He was brave even to rashness and with a few men of his company. among whome wer Dexter. Norton & Hawthorn have got in the advance of the regiment and partialy sheltered by the bank of the marsh and some bushes they wer fireing away as fast as they could get sight of a rebel. The captain had fired several times and had raised himself to fire agan when he was hit by a musket ball it is supposed in the mouth. The boys attempted to bring him away and got a short distance when Dexter was hit and a sergant who was assisting was shot in the arm. they could not help any more and after going a short distance further. another man who was helping got frightened and left onley Norton and Hawthorn with him. they wer completely fatigued out. could carry him no further. but succeeded in getting him out of range of the fire and then placeing him in the best situation they could wer obliged to leave him. He was not dead at the time but was senseless and doubtless mortaly wounded. our folks have not been able to recover his body which has doubtless fallen into the hands of the rebels. It is bad. he was a good captain and we all liked him well. though of course he was not without his failings. He was engaged to be married to a young lady in N. Y. had no parents liveing. but has brothers and sisters. No one else was killed in our company. we have six wounded. The loss of the regiment is killed. wounded. and missing was ninty. onley four are missing and they are doubtless among the slain. Probably many of the wounded will die. at present the number killed is onley put down at eleven. The loss of our brigade consisting of three regiments. <u>the 8th Michagan.</u> <u>28th Mass.</u> and our own in killed. wounded and missing is three hundred and fifty nine. this is the unofficial report. how much other brigades suffered is not known yet. I am in hopes the whole loss in killed and wounded will not exceeded five hundred. but it may. we rested the remainder of the day after reaching camp. onley cleaning our guns. today it is raining hard. we have not yet been called out for any duty and are so sore and stiff that it would be difficult for us to accomplish much if obliged to turn out. I am makeing poor work writing and shall not try to write much more now. if the mail does not go soon I may add. George has not yet come from Edistoe. Mr Landon I suppose is in the general hospital at Hilton Head. the rest of our boys are all in good health. I have not given you a very minute account of fight (if such it can be called) yesterday and am not able or competant to do so and besides at present I do not like to dwell to much on the particulars of the occasion. One thing is certain whatever may be the report of the conduct of the seventh. our Colonel does not find fault with but praises us for our good conduct during the action. He is certainly as brave a man as ever lived. I do not believe there are rebels enough south of Mason and Dixons' line to scare him. We are sadley in need of field drill which I earnestly hope we shall have before we (are) agan called into an action of that kind. The rebel position was a strong one and it seems hardly possible to take it at the point of the bayonet. It is placed across a narrow neck of land on both sides of which was a marsh like the one I attempted to describe and both of which are impassable so that it is impossible to approach it except from the field I have attempted to describe and which narrows down as it approaches the fort untill the enemys works extend across it. I do not know but that I use the word fort wrongfully in relation to it. prehaps it is oney styled a battery. but that makes no differance with the strength of it. Near by is a lookout built up very high and must from the top command a view of the entire island. Of course we are not through with them yet. We shall have the works there. but shall go to work in a differant way to take them. There is just the nicest place to throw up batteries and shell them out. that could me asked for. I am sorry we could not have whipt them yesterday. they will make a deal out of it. It will be magnafied a thousand fold and spread broadcast over seceshendom to revive the courage and flagging zeal of thier soldiers. and they may even try to make capitol of abroad but it could never be done if the truth could be known. I believe that we can drive them out of that and every other strong hold they possess. but it cannot be done by sending regiment after regiment into such a den as that with orders not to fire a gun and expect them to carry by assault works that can oney be taken by siege. I will not try to write more now and hope if we are engaged in another fight I can give you better news. I am grateful to the kind Providence that preserved me in the hour of danger and hope I shall henceforth love and trust him more than ever. Leaveing you as ever and hopeing to see you yet my darling I remain your own

Ned

No. 56 Rec'd July 1st 1862

James Island S. C. June 20th/62

My Dear Sara

    I seat myself in the shade of the tent this morning to write you a few lines once more. The morning is most oppressively warm. with not a breath of air stirring. Last night there was a very severe thunder shower. one that is seldom surpassed. but it did not trouble me much for I was very much fatigued having been out all the night before. and slept but little yesterday. Al 1 has been quiet since the unfortunate events of last Monday. Not a gun has been fired on either side. I believe our line of pickets have been advanced a little. For two days nothing was required of us beyond the usual duties of the camp. But the third night. a detachment of fourteen men and one noncommissioned officer from each company with arms and accouterments was called for and we went out ready for action but not knowing what was to be required of us. we soon found out. for on reaching the picket headquaters we wer each provided with a shovel. ax or pickax. in addition to our other arms. and then we knew very well what was to be done. I mentioned in one of my previous letters our throwing up a battery one night in which three guns wer placed to bear upon the enemys works. This night we kept on nearly a quater of a mile in advance of that work and then stacking our guns close by. we went to work and by morning had a breastwork nearly completed of sufficient size to mount half a dozen guns. We wer obliged to be very carefull about makeing a noise and wer pretty successfull. I do not think our operations would have been detected at a distance of twenty rods off. The night was a good one for our purpos. though it rained most to much for comfort at first and most of us got nearly wet through. afterwards it cleared off. and was starlight and finely about midnight the moon arose. the wind blew. from the enemy to us which was also favorable. But it was hard for the men to work. for we wer almost all of us stiff and sore from our Monday work. and felt but little inclined for a nights job of the kind. It was day break when we arrived at camp. Yesterday guns. and thier carrages. wer being drawn past our camp. and I presume some of them wer mounted last night. George has rejoined us. he came into camp Wednesday night. he is getting better. looks quite well. but has not regained his strength yet. I was right glad to see him agan. I can assure you. I felt as though I wanted him here whether he was able to go with us into battle or not. I do not think there will be another attack made just at present and probably not in the same way of the other when it is done. But of course we cannot tell. It seems as though we could take the battery without anything like the amount of loss which we have already sustained. Our wounded men have all been sent to Hilton Head. the dead have been buried and we are ready for action agan. I have not been able to learn anything more definite with regard to our whole loss. than what I wrote you before. our officers have been out with a flag of truce to try and obtain some information about our missing. they told us that they had buried our dead and sent our wounded to Charleston. that when they could find out the names of our men in thier possession they would make out a list of them and give us. June 24th. Since writing the foregoing I have received your letter of the fifteenth and now have two unanswered ones to hand. and hardly know how to answer them. our company went on picket last Saturday afternoon and spent a most uncomfortable

Sunday watching for the secesh. We wer placed on the outposts. and where I was stationed with seven others was the most important post in the 1ine. We passed a sleepless night being continualy on the lookout from dusk till daylight. tormented almost beyond endurance by the mosquetoes which were in perfect swarms surpassing anything I ever saw before. The rebel line of pickets was just across a corn field about forty rod from us and wer frequently seen by our boys during the day. at the post where I was stationed the boys on the watch saw them prowling around dodging from one bush to another not over ten rods off. I did not see them myself. but Reid and Olin wer with me and they saw them. I think there will not be any fireing upon each others pickets. at any rate we had strict-orders not to fire upon the rebel picket and they certainly had abundant opportunity to shoot us had they been so disposed. Our people went out with a flag of truce that day. but I do not know its import. the bearers wer met by the rebel officers midway in the corn field that seperates the pickets. Everything is still quiet no fighting of any kind has taken place since the unfortunate 16th. That morning when you wer quietly seated at your table writing to me of the beauties of the morning I was on the battle field. in a position neither very pleasant or agreeable. For your letter of the 9th you refer to your correspondance and say that if I wish it. you will discontinue it. I do not wish it. I do not see any reason why I should. I have all confidence that your own good sense will always dictate what is right in these matters and I believe my dear you love me to well do me wrong in any way. As for a flirtation under the circumstances I believe you would think it to be beneath you and would never do me the harm or yourself the discredit to indulge in it. So you see my dear that I am pretty well fortified in my position and trusting implicitly in you have no fear that my happiness is not secure in your keeping. I received a letter from Sara B. at the same time I received yours of the 15th. they wer both marked the same day. She speaks of seeing you at church. wants me to talk to you about sitting with her in church. also about writing letters sundays. threatens to get you up there some Friday night and plague to her hearts content. thinks she can do it with impunity because I am not there to interfere. offers me some French gingerbread and cider if I'll call and get it. crows over her baby namesake niece. and makes out a good sisterly letter generly. please tell her it is duly received and shall be answered the very first opportunity. Tell little Lucy her message is gratefully received. <u>thankfully</u> accepted and cordiuly returned with the hope that I may yet be able to call and claim my <u>kisses</u> in person. to your wish. I heartily respond <u>amen</u>. I am so glad my dear that you are so pleasantly situated and can enjoy so many of Gods blessings. May his watchful! care ever be over you and his mercy ever strew your path with his choicest blessings. I am finishing up this letter on the morning of the 25th. have been a long while writing it and it will be time to commence another by the time this is mailed. All is quiet. but our folks are in thier turn expecting an attack from the enemy and last night we had to sleep with our equipments on and rifles by our side ready for fight at a moments warning. Night before we wer called out and after getting our things all on wer premitted to lie down. but with them all on ready for action. Prehaps you think it would be difficult for you to sleep under such a state of things but I assure you it does not interfere with our repose at all. we sleep just as well and just as sound as though we wer at home in bed. Our people have suspended operations so far as we can see aganst the enemy and we are wondering what is to come next. As usual under such circumstances the camp is full of rumors none of which I think worthy of credance. We have a new rumor that Richmond is ours. but have been so often deceived by the like that we place but little confidance in it. The weather is getting very warm and sultry especialey in the morning when it is all one wants to do to breath. in the afternoon there is generaly a little sea breeze. that makes it a little more endurabel. I had a good letter from Mr Randall with quite a long one from Mrs R. enclosed by our last mail but one. Mr R sent me quite a lot of postage stamps enough to last me quite a while so you neede send me no more at present. I do not care to have to many on hand. I lost a number that wer in my pocket during our march in that rain. we got so thoroughly soaked through everything in our pockets that could be affected by wetting was injured. My pocket bible was nearly spoiled with the rest. And now darling good by. give my love to Lizzie and Julia. remember me to Mr Barnums people. and believe as ever your own

Ned

No. 57 Rec'd July 15th 1862

James Island S. C.
<u>James Island S. C.</u> July 5th/62

My Dear Sara

Over a week has passed since I have written to you though I have had abundant opportunities to have done so but we have been kept in constant expectation of a move all the time and I have waited thinking that. when next I wrote. I should either be away from here or else have had a brush with the enemy. The weather is awfull hot here now and it requires a good deal of effort to even look at pencil and paper. One week ago today our regiment was on picket duty. when orders came to evacuate the island. Fatigue parties wer at the time at work throwing up batteries and drawing up cannon and morters from the landing but they immediately commenced drawing back and reloading those already placed in position and we have ever since been steadily at work returning the guns. ammunition and stores to the ships. and a greater portion of the troops have already left the island. We have been on picket once since the time mentioned and if we do not get off the island today shall have to go on agan tonight. Our pickets have been drawn in until now the outposts are scarcely more than a mile from our camp. Every precaution has been taken to conceal from the rebels that we are evacuating for if they knew it they would be likely to attack us which at such a time could hardley fail to be disasterous to us. We are now pretty well protected by our gun boats of which they are in great fear. and are not troubled much about an attack except at night then our gun boats could not render us much assistance. We have been obliged to remain under arms a good deal of the time and now while I am writing our arms are all stacked on the color line. and in case of an alarm we should be ready for action in three minutes. Yesterday was fourth of July. I suppose. but it was the most uncomfortable one I ever spent. We had nothing to do all day and as it rained hard most of the time. we could oneley crowd into our tents and contrast that day with those of former ones of the same date. The truth is I felt real miserable. just in the mood to make disparageing comparisons and more espicialy when I remembered the pleasant fourth that I spent last year. True you were a little vexed with me. but then it was infinately better to be with <u>you</u> any way than to be in our present situation. I hope you spent yesterday pleasantly. The gun boats around us fired national salutes and at noon Hamilton light battery camped for the night just below us. came out in order of battle and fired a salute after which the most of them left. probably to embark. Of course we do not know anything about our destination. but it will probably be either Hilton Head or Beaufort. to the latter place our sick have been sent several days ago. We may go to Edistoe. but I hope not. It is said that some of the troops here have been ordered north. but its not likely it will be our good fortune to go that way. But whereever we go it is more than likely that no more active operations will be carried on down here for two or three months at least. I dare not stop to write more now my darling but will try to write agan just as soon as we <u>light</u> somewhere. Many thanks dearest for your kind letter of the 16th which I cannot stay to answer now but will enclose it in this. I do not need any paper at present and most of the time our chaplin has it and supplies us cheaper than you can get it at home. when I do want some I will let you know.

    I enclose an Oleander I think it is picked from a bush growing at the house which has been our picket headquaters and from which fort Sumpter can be plainly seen and which is as near the fort as I am likely to get very soon. I wish you could send me another <u>ambrotype</u> of your own dear self. the one I have carried ever since enlisting is nearly used up the wetting we got during our march completely spoiled the case afterwards I carried it wrapped up in a silk handkerchief. the other day I threw my blouse down in the tent with it in the pocket and somebody setting down on it smashed the glass to pieces. The photograph I carry in my knapsack. but I want <u>something</u> to carry with <u>me</u>. Once more darling good by. I shall hope to hear from you soon. Ever your own

Ned

No. 58 Rec'd July 23rd/62

<u>North Edistoe</u> July 9th/62

My Dear Sara

I take the first opportunity to write. that has occurred since I sent you the hasty note from James island. We have got partialy settled down agan. but onley half our tents have come yet. We broke up our camp on James island late in the evening of the 5th as most or our camp equipage had already gone it was a short job. and we wer soon on our way to the landing. a distance of about a mile. Arrived there we stacked our arms and spreading ourselves around on the sand made ourselves as comfortable as we could for the rest of the night. A transport was lying at the dock. and fatigue parties wer at work all night loading her. In the forenoon of the next day she took on board the <u>28th Mass regt</u> which had bivouacked alongside of us during the night. and left. The Cosmopolitan came in during the afternoon and later in the day the <u>Deleware</u>. Several brigs and schooners wer anchored close by and wer being loaded as opportunity afforded. We spent the day (Sabbeth) on the ground. the weather was very hot but we kept as much in the shade as possible. hanging our blankets on our guns. and tying them to stakes driven into the ground so as to make a shade. We had no means of makeing either tea or coffee and did not as we usuly do on a march have coffee and sugar dealt out to us. but the <u>sixth Conn regt</u> which was in camp a little ways from us very kindly made us coffee and sent down to us. our company was supplied by them with a whole boiler full of excellent coffee for which we wer very gratefull. At night the Cosmopolitan came up to the dock and a party of our men wer employed nearly all night in loading her. This night we spent like the preceeding one. but it was much warmer. The dews at night are very heavy and the water will fairly run off our rubber blankets in little streams in the morning. This night the 6t<u>h</u> Conn and 97<u>th</u> Pa regts broke up thier camps and prepared to follow us off in the morning. About nine o"clock the next morning our regt went on board of the Cosmopolitan and dropped down the stream a ways where we waited two or three hours. Another transport the Ben D. Ford came in just as we wer leaveing the dock and she and the Deleware soon took on board the 6th and 97th and then we all bid good by to the James island and Stono river. Several gun boats wer in the river to cover our retreat so that the secesh did not molest us. A party of sailors from the Pawne went on shore in the morning with axes. sledges. and saws. to destroy and burn a couple of bridges. the last one was fired just as we wer leaveing the place. The Paul Jones a new iron clad gun boat came down a head of us. she passed us. and we wer favored with a good view of her. She is a saucy looking boat. sharp at both ends. a side wheeler and rigged to steer at both ends so that she can run one way just as well as the other without turning around. She had one hundred pound Parrot gun. and a large Columbiad or two besides smaller guns. Our boat had two schooners in tow besides her own loading so we did not make very fast time. The other two boats following us also had two vessels apiece in tow. Outside the bar at Stono a number of brigs and schooners wer anchored also a large ocean steamer. what they wer all doing there we do not know. We wer a long time in comeing into the Edistoe river. and did not reach the wharf here until about nine o"clock. I was on guard on the boat during the day. and was olbiged to stay untill all was off the boat. and when I got on shore the rest of the boys had all lain down and wer asleep. It must have been now about one o"clock and I was not slow to follow the others into the arms of Morpheous. The next morning we had a cup of coffee on the ground where we wer. and about the middle of the forenoon moved up to our present camp which is but a short distance from the dock. The remainder of the day was spent in clearing up the camp ground. bringing up what tents we brought with us and pitching them. George and I are at present staying in the same tent. with two others not of our boys. When the remainder of the tents come there will probably be differant arrangements but I mean to stay in the same tent with George if I can. This forenoon we have had a dress parade and inspection of arms. and afterwards worked at clearing up the parade ground. This afternoon I am trying to write you a letter. but it is very hot and I do not think I shall write much more. Five companies of our regiment are just going out on picket. as the distance is four or five miles the duty is rather heavy. The troops here now are the 55th & 97th Pa. the 6th & 7th Conn. two pieces of the 3rd R.I. battery and a few cavalry. The 55th Pa has been on the island all the time. the other regts came down with us. There was an alarm this morning. it was reported that the rebels wer landing a large force of artillery and infantry on the upper end of the island. a gun boat went up the river and we heard her fireing. she has returned but I do not know the result but we are told that it was onley a feint of the enemy. probably to find out whether we wer here in much force. The <u>gun boat Planter</u> taken from the rebels in Charleston harbour by her colored crew is here. She was here when we landed here before but I forgot to mention it. <u>General Wright</u> is in command here.

Friday. May 11<u>th</u> (He wrote "May" but meant "July"). I did not finish my letter the day it was commenced. and yesterday I was on fatigue nearly all day. and could not bring myself to write. A steamer came in yesterday morning from Hilton Head bringing a large quantity of baggage that we left at Pulaski together with any quantity of old rubbish that has accumilated in the quatermasters department since we came out. Also a number of our sick men that have been at Beaufort since we left Pulaski. Woolcott and Brinton who we supposed wer at home by this time made thier appearance. Woolcott is much improved in health but is not able to do duty. he thinks he cannot go home unless he is discharge(d) which he is unwilling to do. I think if he can get discharged he had better do it and think he will come to the same conclusion by and by. Mr. Landon is in Beaufort very sick. I am afraid he will not get well. Dexter went up there from James island as a nurse. I am very glad he (was) there as otherwise he would be alone. that is none of our boys would be with him. Dexter is a first rate nurse. kind and patient to all and will take good care of Landon. Today we had an hours drill from half past six to half past seven. some of the boys have since been on fatigue. we shall have another hours drill from half past five to half past six. and dress parade at sunset. I do not know but I suppose this is to be the order of things for the future. The mail came in yesterday and I received your two letters of the 21<u>st</u> and 29<u>th</u> for which as ever I am very thankfull. I knew that you would be in anxious suspense until you should hear of my safety and am glad that my letter reached you so soon after the news of the battle. I tried to give a correct account of the fight as far as my own observation extended. I see that the N.Y. papers with one exception fail to give us any credit. and more than that they do us injustice by mentioning and giveing praise to regiments from thier own State that took but little part in the action and did not begin to stand the fire as well as our regiment. I might have thought that prehaps you would be called on to read some part of that letter to others but I did not and so wrote as I always do to you just what came into my mind without trying to cover up or smooth over the rough edges. I was in a bad mood for writing any way and though I do not remember all or how I wrote. know that the letter did not merit the <u>enconiums</u> bestowed on it by your friends. but I am glad that you wer "well pleased" and am therefore content. I

think my other letter was circulated pretty fully. but will find no fault as it did not go out of the <u>family</u>. Am much obliged for the account of your visit home <u>and about town</u>. Hope you spent the <u>fourth</u> pleasantly despite my abscense. and that your vacation may likewise be pleasantly spent. Will Reid had a letter from his brother John by our last mail and allowed me to read it. John is at present Acting Assistant Adjutant General in <u>General Corroll's brigade.</u> Shields" division. he wrote that they had been severely defeated in an engagement with the enemy loseing eight pieces of artillery. and between six and seven hundred men and had been obliged to retreat about thirteen miles. He thinks the war has <u>just begun</u>. I do not think it so near the end as could be desired myself. I think very likely we shall stay on this island some time and think it doubtfull whether any active operations are carried on or at least commenced. during the hot weather. I think this will be a tolerably healthy place. we have a fine breeze nearly every day. the water is as good as can be found on any of these islands. and if we could onley be treated like men and <u>have</u> good medical treatment do not think we should suffer much from disease. I am glad to know my darling that you still love me as ever and that you think of and pray for me. I never doubted your affection for me but I must say that it gladdens my heart and gives me comfort to know that I am loved and my absence mourned by one so good and pure as yourself. Sara. I am getting awful wicked. the petty trials to which we are all constantly subjected to have proved to much for me and I have proved myself to great a coward to rise above and surmount them. I am loseing temper and begin to fear. am lacking in common sense. If I keep on at this rate. if I do not succeede in overcomeing my evil nature. I never shall be fit to be the companion of any woman much less of and so good as yourself. and loveing you as I do shall feel that I have no right to jeprodize your happiness for life by linking our destinies together. I know that <u>you</u> could not fail of exerciseing an influence for good over me. and thus make my own life happier. but would be right for me to be so selfish as to secure my own happiness at the expense of that of another and that <u>other</u> the one whome I most love. on earth. May God forgive me the past. and give me grace to overcome in future. so that if in his own good time we may be premitted to meet I may not be altogether an heathen in spirit but able to controll myself and live as becomes a christian. I must close this letter now dearest and hope that we shall now have a little quiet and be able to write more regularly. I have a few letters that I must write to my relatives and they ought not to be delayed longer. It is so very hot that it is almost impossible to sit down and write in the middle of the day. and morning and evening we are buisy other ways. With kind regards to all enquireing friends I remain as ever your own
Ned

No. 59 Rec'd Aug 1st/62

North Edisto, S.C.
North Edistoe, S.C. July 15<u>th</u> 1862

My Dear Sara

 Have just got through our morning drill and with writing materials am seated in the shade of a large tree at the end of our company street. The sun shines very hot this morning and it is almost melting out under its rayes. There is a little breeze and it is quite comfortable here in the shade. Sunday I wrote a <u>sort</u> of a letter to Sara B. and in the afternoon we went on picket. from which we returned just at night yesterday. It proved to be very pleasant duty. as the posts wer all at or near houses and from the negroes we could buy pancakes. fresh pork. eggs. milk and watermelons. and could help ourselves to tomatoes and cucumbers and green corn. Of course we lived well for one day. My place was at a planters house by the name of Hopkins. The government agent for this island makes his headquaters there. I do not know as I have told you. but presume you know that when the owners left the island last fall when we landed at Hilton Head they wer in to great a hurry to take away thier negroes. or any other effects. Our government took possession of everything. negroes and all. and at the proper time this Spring set them at work cultivating the ground as usual. At the present time large crops of cotten. corn. sweet potatoes and other vegetables are in the ground. and looking finely. All these have got to be left now. as orders have come to evacuate the island. The negroes are to be taken to H. H<u>d</u> and have been buisy now for three or four days. in lugging by thier effects to the dock. There is a large number of them on the island. I do not know how many. I suppose we shall now have to go into Virginia and reenforce McClellan. I learn that we are ordered North and presume that that is to be our destination. So all our hopes of a few weeks rest are disappointed. At the house we wer stationed yesterday. the grounds wer very prettily laid out. and shrubbery was trimmed up in all shapes. but the establishment was not as large or as fine as at <u>Seabrooks</u> just across a little creek and but a short distance from us. I visited the place in the morning. the house is not so very large. but it is high and airy. the rooms are large with large windows and folding doors. the stair cases are also wide and commodious. of easy ascent. most of them take a turn or two in rising from floor to floor. and you feel relieved on arriving at the top to find that you still have a little breath left which is not often the case on ascending our stairways at home. From the upper windows quite an extensive view can be had in every direction. but there is little to be seen. you know there is no hills in this part of the country. and the scenery is the same all around. open fields. belts.

and patches of timber. little creeks. bordered by wide marshes. the large white house of the planter. with its many outbuildings. and the long range of negro houses. called "the quaters" generly neatly whitewashed. with occasionaly a little church peeping out from among the trees. are the general outlines of these island landscapes. Most of the furnature has been removed from the house I speak of. a few pieces remain. but they are evidently old and have been pretty much demolished. Among them are the remains of an old piano made in 1803 in London. it is now pretty much demolished. but the boys say when they wer here a few weeks ago it was whole. but pretty much drummed out of tune. The house is not much injured. but the paper is much torn from the walls. the marble mantles broken. and occasionly a pannel broken or a lock torn off from a door. The grounds are uninjured. but haveing been untouched for several months. the walks are overgrown. and filled with leaves. the shrubbery. though it still retains the form into which it has been trimmed. is fast growing <u>wild</u> agan. the flower garden is large and beautifully laid out with large hot beds and glass house. Of course there are but few flowers. onley the most hardy. such as can take care of themselves and successfully battle for thier existance with the weeds. I saw a few fine roses and a few flowers of which I do not know the name. The Oleander grows here in profusion and to a size and beauty that would astonish the dwarfed specimens we have at the North. The place is beautifully shaded by live oaks. pines and magnolias. green oranges. lemons. peaches and other fruits are plenty. there is a fish pond with a little island in the centre covered with roses and flower beds. duck ponds. bird houses. arbors. rustic seats under the shade with many other things. that fancy could invent to beautify the place and make it delightfull. The owner or one of the same name ownes a large plantation on John's island. which we passed on our march a few weeks since and another on H. Hd. I picked a few flowers which I will try to send you. though I have no means of pressing them. After my return from visiting the place I lay down in the shade of some pines that stood in the yard on my rubber blanket. and whiled away the time by reading a book I had bought with me entitled "ten years of a preachers life" by the Rev. <u>Mr. Milburn the blind preacher</u> who you wrote me was to be the orater of the day at Milleston on the 4th of this month. I hope you heard him. I should like to know how you liked him and what he had to say in relation to the present unfortunate condition of our country. He has spent six years of his life in Alabama and speaks very highly of the Southerners and treats the "peculiar institution" very tenderly in his book. Prehaps however you have read it. if not I think you will find it quite interesting. We are agan hearing disasterous rumors with regard to McClellan. I do hope they are not true. It is time we received another mail. I dread the voyage north. my experience was so unpleasant on the trip down. we are not positively certain yet that our regt is to go. but it is more than likely. we are to leave this island in a day or two at any rate. I hope my dear I may (hear) from you agan in a day or two before we go North at any rate. for if we do not get the mail now due before we leave. we shall not get it in a long time. I am my darling with much love. your own

Ned

No. 60 Rec'd Aug 5th/62

Hilton Head. S. C.
Hilton Head S. C. July 21st 1862

Dear Sara

Seven months and two weeks ago we landed on this island for the first time. Yesterday. after a not very extensive but quite an eventfull wandering. we agan landed here. Today we are receiving the full benefit of a schorching Southern sun unprotected by the shade of either tent or tree. Partialy shaded by a blanket hung upon a couple of stacks of guns and seated flat on the ground. I am trying to write this letter. Even if my courage holds out till I finish it. I am much afraid you will find it will not pay for the trouble it will take to decipher it. Last Friday. the 15th. our company went out on picket duty on Edistoe island. we had then lain out two nights. haveing put our tents on board of a vessel expecting to follow them ourselves immediately. two days before. but as the transports did not come for us we wer obliged to remain on the camp ground without shelter. My post on picket was an unimportant one. and as I had five men with me part of them wer out scouting about. most of the time. George was with me. but not very well and kept pretty quiet. Saturday afternoon we wer not releived by a new picket but as transports had come for all the troops remaining on the island. we remained on duty until the picket was withdrawn. which was just at night. On returning to what had been our camp. we found that those of our regiment left there had already embarked on board the Deleware. whither we immediately followed them. We wer all on board by sun down and the vessel left the dock and anchored in the stream. The Cosmopolitan had already taken on board the 55th Pa regt. our regt was the last to leave the island. The Ben. D. Ford took our place at the dock and took on board a quantity of wagon horses. and a miscellanious assortment of stuff that had been taken from the differant plantations on the island. We remained at anchor during the night. On going aboard we wer allowed to go below and leave our knapsacks and arms. and then go on deck. as the air in the hold was fairly stifeling. Of course the deck was crowded. but we had pure air. I streached myself on the deck with my haversack for a pillow and somewhat fatigued with walking. and loss of sleep on the previous night. was soon sound asleep. I was partialy awakened about the middle of the night by finding myself becomeing quite chilly. and was soon wide awake by the occurance of two incidents. calculated to arouse a sleepy man on ship board. A number of horses wer on the lower deck and by some chance one of them got loose and commenced running about among the other horses and instantly there was a kicking and stomping that would have arroused the seven sleepers. The noise awakened all hands instantly. and as if by magic every man was on his feet and like all frightened men trying to run somewhere. I thought at once what the matter was. and did not feel inclined to stir. but quickly found I had got to get out of the way or be run over by the crowde. All this took place much quicker than you can read this account and all would have been quiet agan. but for the startling cry of a man overboard. he had been lying close to the edge of the boat outside of the rail and wakeing up suddenly rolled off before he could lay hold of anything to stay his course. The boat hands quickly sprang to the side of the vessel on hearing the cry and knowing I could do no good my first impulse was not look at the sight. for I feared he would certainly drown. but I finely did go to the side of the boat and look over. and there just under me as I leaned over the side was the man. strikeing out manfully and calling for them to fling him a rope. the tide was going out and in spite of his efforts he was drifting astern of the boat. but luckily he fell off near the <u>bow</u> and as he drifted by he caught hold of the wheele. and a boat being lowered. he was soon on board agan unharmed except by a good fright and a thorough ducking. All was soon quiet and going below and getting my rubber coat to protect me from the cold dew. streached myself in my old place and was soon sound asleep agan. Soon after daylight yesterday (Sunday) morning we took a loaded schooner in tow and was soon on our way here. The other two steamers also took each a vessel in tow and followed us. I kept very quiet on the passage down. for although not quite sea sick the foul air on these transports always makes me faint and it takes but little of the motion of the ship to upset me entirely. We arrived here at eleven o"clock. but did not disembark until about six. In the meantime a large mail was distributed to us and you can rest assured that the time was thenceforth well improved. I received your three favours of July 1st. 6th. & 10th. a letter from Frank French. one from Sara B. and one from Brookfield. from Mrs Randalls' sister. Wasen"t I fortunate: George received a number of papers. and after reading my letters. devoted the remainder of my time to thier perusal. About six in the afternoon we landed. not as on the eventfull 7th of Nov by wadeing ashore from the crowded small boats. but by stepping out on a long substantial dock which extended out far enough so that the largest ships can come up to it and receive or discharge thier cargoes. The place is altered some by the erection (of) large buildings. but it all looks familliar to me. We marched out about a mile from the landing. beyond the buildings and also beyond the intrenchments which we assisted in building when we wer here before. They are now completed. and are very formidable works. A heavy shower was rapidly comeing down upon us. and we had oney time to hastely stack our arms. and throwing off our knapsacks envelop ourselves in rubber coats and blankets. before it was on us. it came in big drops. and fast, for a few moments. accompanied by blinding flashes of lightening and heavy thunder. but it fortunately lasted but a short time and though the night came on pitchy dark with abundant promise of a rainy uncomfortable night. it passed over us and we escaped without the anticipated soaking. After the first shower was over. little fires wer quickly built and each man was buisy making his pint of coffee. Sugar and coffee sufficient for two rations had been dealt out to us on leaving Edistoe in anticipation of just such a time as this whereever we landed. You will probably think that our coffee would hardly do for a fashionable hotel. prehaps it would not. but I have drank much poorer in many places where it was called good. I can make a much better cup if I can have the material given me than is made by the company cook. A hard cracker or two. and prehaps a piece of cold pork brought with us constituted our supper. after eating which we made the best disposition we could of coats and blankets to protect us from the wet and lay down on the ground and wer soon asleep. I slept soundly and woke this morning with a good appetite which I have managed to satisfy with bacon and hard bread. washed down with another cup of "home made coffee." Today is the <u>anniversary of Bull Run</u>. how much of sorrow. suffering and death have we been called to bear since then. God grant that we may be spared from another such a year of suffering. Yet if it be necessary to purge our land of bad men to purify and exalt the mind of the nation. and place one government on a higher and firmer base. let it come. But I do hope that the army which is now being raised will be made strong enough to place the certainty of its success beyond a doubt and never agan while the war lasts. let the unpardonable blunder of refuseing volunteers be committed. Better enlist men up to the very last moment. and grant them all the benefits that those who beare the burden and heat of the conflict receive than that we should be agan so deceived and find ourselves in a critical position calling loudly for reinforcements. which are not to be had. By the papers received we are acquainted with McClellan's retreat and his new position.

All the efforts of the press to smoothe over and make comfortable the present position of that brave army. are unavailing to disguise the fact that it has been badley beaten. I have no fault to find. I feel satisfied that they and thier brave commander have done all that could be done to avert the calamity that has befallen them. A report is trying to gain credance that by some miraculous junction of other divisions with McClellan that commander has made some sort of a flying leape and gained possession of Richmond. But we have been gulled to often and the effort to make the story even plausible has hard work to succeede. Please express to Mrs Barnum the pleasure it gives me to be so kindly remembered by her. I should not dare have her read one of my letters for I am confidant that her good opinion of them must arise from some finishing touches put on them by yourself. I remember Mrs B. as one of my first school teachers when I was a little four year old. ragged. tow headed. urchin. and I fear if she should see one of my letters that she would think that I had neither profited by her teaching or any other I may have received. I certainly shall not attempt to write you a letter with the view of having her read it. for I know that it would be a complete failure every way. and would rather you would draw at a venture from your bundle for her perusal. I am obliged for the account of your fourth of July pleasures. am glad that you heard the oration for reasons that I have before written you. I do not know. I have no desire to deprive Emma of your society during your vacation. but it does seem to me that you could enjoy yourself well during a short visit to Lenox. and that it is a duty you owe to yourself. to <u>me</u>. and your friends to secure <u>all</u> the enjoyment you can. Always remember my love that I cannot be happy when you are unhappy and suffering. I know of course. that while you love me so dearly there will always be something wanting to complete your cup of happiness. while I am thus seperated from you. but always remember dearest that I wish you to look on the bright side as much as possible and be as happy as you can. Mr Barnum can rest assured that I shall surely <u>shoot</u> whenever I can but if he oneley knew how we "ketch it" sometimes for disobeying orders he would readily see why we withheld our fire so long that day. however I have the satisfaction of knowing that I did fire a few times orders or no orders and I shall never be so particular under similar circumstances agan. I am sorry to hear of young Hoffers death. I was not intimately acquainted with him. but my Sunday School recollections of him are very pleasant. I remember that I thought when he enlisted that if he lived he could hardley fail of being successfull and being promoted. He has died in a good cause and his friends may be as proud of his death as though he had fallen in the advance on the field of battle. The man who suffers for weeks in a hospital the results of fatigue. exposure and deprivation as was the case of young Hoffer is much more to be pittied than one who falls on the battle field often entirely unconscious of any hurt. It often seems to me that if I could go home for a few days onley and see my friends once more and have the satisfaction of a parting with you to which we are justly entitled. but of which circumstances have deprived us. I could return to my duty with a renewed spirit and more resigned to the fortune of war. But this may not be and I never allow myself to hope that it may be. But I cannot help picturing to myself the meeting and the pleasure we should enjoy. The boys are all usuly well. Dexter is at Beaufort assisting in the general hospital there and Norton is acting in the same capacity in our regimental hospital here. Pitt had a letter from Dexter yesterday. in which he forgot to mention Mr Landon so we conclude that he is better. It is not known how long we are to remain here. but are told we are awaiting transportation. there not being any large transports here now we may be here several days. I should like to write you much more but the heat is intense. and I am weary with dodging about trying to keep in the shade of my blanket. George has been trying to write and has given it up so I think I might be excused even if I had not already written you quite a letter. but then George is not very well and. no wonder. how any one but a salamander can stand such weather unaffected is a mystery. I expect to hear from you agan before we go North though this letter may not go untill we do. Still I shall write whenever I can and try to catch up with my correspondance in which I have fallen dear behind. With a kind remembrance as to all friends. I remain as ever your own loveing

Ned

No. 61 Rec'd Aug 5th/62

Hilton Head. S.C.
Hilton Head S. C. July 27th/62

Dear Sara

This is Sabbeth morning but O" how differant from our quiet Sabbeth mornings at home. No preparations for church going on here. but talking. laughing. cleaning guns. washing &c. I received your letter of the 13th on the evening of the 25th. the day you closed school. this is the first opportunity I have had to answer it though I was about writing you before I received it. We still remain camped where we stopped on our arrival here. It is still uncertain about our going North but we probably shall not go in a week or fortnight at any rate. Our sick have come down from Beaufort so we are together agan with the exception of the wounded who went home. It is quite cool and comfortable today for the first time since we came here.

I had much rather have remained at Edistoe during the hot weather as it is a great deal pleasanter. and I think healthier. You can hardly mention a place where we have been together. without the whole scenes recurring in my mind as plainly as when it was passing. I remember well the afternoon when we wer at Sara B.'s and the pleasant hour we spent under the trees by the spring. Mr Bosworths people wer always very kind to me. so much so that I could not help but feel at home there. I do not know why they should have been unsure. I am very far from deserving the good fortune that attended me during my sojurn in Salisbury and at the Falls. I do wish my darling that I could have occupied that near seat to you that Mr B. filled and also that I could have walked home with you from Sara B's but if our lives are spared we may yet enjoy many pleasant walks together. I shall not be able to write either a long or interesting letter this morning. but as the mail closes this afternoon I thought I would not let it go north without another line to you. I have forgoten how long a vacation you have at this time. but think it is a month. I hardley know where to direct this but as I suppose you have made arrangements with Frank to forward your letters I will direct to Lime Rock for the present. A small party of Gen Hunters' negroe soldiers have just passed our camp towards headquaters. they wer dressed in uniform and seemed to feel all the importance that usualy attaches to a negroe when he has a decent suit of clothes on. I believe I have no prejudice against negroe soldiers. but I do think if <u>Gen Hunter</u> would use the negroes to do the work which they can stand it to do much better than we white folks can and which they can do just as well. he would be advancing our cause in this section faster than he now is. and prehaps save many valuable lives. I believe I wrote you from Edistoe our daily routine there. it is the same here. The regiment has been supplied with light blue pants. our whole dress heretofore has been dark blue. the light blue pants look very well when new and clean. but as it (is) impossible to keep them clean long I like the dark blue best. I am not going to try to write more now my darling you are probably enjoying the Sabbeth under Mr Reids ministrations. we have been denied all kind of service even a prayer meeting since we left Pulaski. Hopeing for better days and and a happy hereafter. I remain as ever your own loveing

                                                                                                     Ned

No. 62 Rec'd at Guilford

Aug 14th 62
Hilton Head
Camp Hitchcock
Hilton Head S. C. Aug 1st/62

My Dear Sara

Your dear letter of the 20. & 23rd reached me two days ago. haveing been onley six days in comeing. the quickest I have known letters to come. I went on guard the day I received it and as I did not feel like writing yesterday I did not try to answer it. Sometimes when on guard I write a letter or two. and agan if I have ever so much time and do not feel like it. I do not attempt to write. You are probably at home now prepareing for your visit to the seaside. I am glad you are to spend the vacation so much pleasanter than you did the last. I do not think you could have enjoyed your vacation very well by spending it at home this time. and as you say waiting on company. Especialy as I happen to know that a part of that company at least. cannot be very agreeable to you. It will be so pleasant to spend a short time during the excessive heat of this present month by the sea side enjoying the cooling breezes. I feel real thankfull myself to Mr B's people for thier invitation to you. because I know you will enjoy it so much. to say nothing of its beneficial effects. I have been thinking since I commenced writing that I am not exactly (sure) where I was last year. at this time. I was in N.Y. a year ago today. and among other things that attracted my attention wer several regts of the three months men returning home. I little thought then that I should so soon be loaded with a knapsack and encumbered with all the traps that a soldier haves to carry and lugging them about under a burning sun. I remember that I wondered how in the world a man could walk. thus loaded and fettered. and must confess that I have never yet ceased to wonder how it was done after doing it myself. many times. Well time is hurrying along. we have spent nearly one year of our three in the army and thus far we have been mercifully spared from the hand of the enemy and from disease. It will be more than we expect if another year should pass and leave our number unbroken. still it is possible that it should be so. We all hope for the best. but few are bold enough to hope that the war will be closed before our term of service expires. Mr Moore has rather bettered his position by being appointed one of the division clerks at headquaters. he has for some time been our adjutants' clerk and when we wer on James island he was employed in writing for Col Fenton who was acting as Brigadier General. When we evacuated the island the brigade was broken up and Mr M. returned to his old place as adjutants clerk. now he is still further advanced. In his present position his pay is about doubled. and being detached from the regiment will not be obliged to follow us around further in our wanderings. The man whose place he takes was a seargent in one of the companies of this regt and has held the place since we first landed here. He now leaves it because he has been promoted to be adjutant in our regt. Our present adjutant haveing been promoted to the captaincy of our company. He came out as our first Lieut and was promoted to his present position when we wer on Tybee. Mr Landon still remains in the hospital with scarcely any preceptable change in his condition. I think if it wer onley cooler weather he would recover faster. Fred Brinton whome I have before mentioned as being so affected with the rheumatism that he is entirely unfit for duty and who we once thought had been discharged is now down with the same disease as Mr Landon and very sick. Brinton has never done an hours' duty since about a fortnight after we landed here last Fall. and there is not a shadow of probability that he will be able to do military duty agan if he is kept with us the whole three years. He has already taken medecine enough to ruin half a dozen ordinary constitutions. and to all appearance he is pretty nearly used up now. John Sweet who got a discharge last Winter. which <u>Col Terry</u> lost for him. has I believe got it arranged once more so that he can go home. and I believe is going in the next steamer north. he is at present in the hospital but not very sick. and recovering. There is at present no appearance of our immediate removal from here. <u>Col Hawley</u> and some non commissioned officers have gone home to recruit for the regiment. I doubt thier having much success in N. Haven and vicinity for the condition of affairs in the regiment is pretty well known among those likely to enlist about there. and I can hardly believe that many men will willingly and knowingly burn thier hands. I never have to you or any one else written that our officers wer anything but the very best. and our treatment by them all that could be desired. Nevertheless such is very far from being the fact. and George has done perfectly right in informing you to the contrary. Our friends at home can never be made to understand because they will never realize it. all that we have been and are continualy being obliged to bear. The hardships. deprivations. and dangers of a soldiers live are very far from being the most objectionable things in it for me. But however guiet I may have kept about the matter to you. I assure you I growled enough here to make it all up. it is no slight thing I can assure you to be in utter subjection to an unprincipled little upstart not yet out of his teens and who delights in showing his authority by punishing in some ignominious way any one he happens to dislike who crosses his path. I tell you my dear our young men at home would be still more loth to enlist than they now are if they onley knew the <u>whole</u> and how they are likely to be treated by those who they expect to see to and care for them and thier interests. Our town is certainly offering great inducements. and it would seem that if it wer possible to buy men. drafting would not have to be resorted to. in Salisbury at any rate. I see no reason why Mr Bostwick should object to Josephs' enlisting. with all due regard to the differance of our birth and fortune. I do not see why a soldiers' life should be any worse for him than for myself and others. I think that the larger part of the present levy of troops will have to be drafted. There are many reasons besides lack of patriotism that may conspire to make it necessary. I very much doubt if the war lasts. if the present evil will be the last. Many thanks darling for the renewed assurances of love and trust contained in our last letter. I am glad you are not afraid to trust your happiness with me and as you wish will give myself no more uneasiness about it. trusting that whatever may be my conduct to others. the love I bear you will ever prompt me to seeke your happiness and enable me to control temper and passion when they are likely to jeprodize it. You will probably be at the sea side before this reaches Lime Rock whither I shall direct it in the absence of directions to the contrary. I shall not be disappointed if you do not find even the little time you anticipate to write in during your absence. still I shall hope for short letters. at least for the time. and will then waite untill you have more time to give me full accounts of your trip. Good by darling with much love I am as ever your own dear

Ned

No. 63 Rec'd at Guilford Aug 16th

Hilton Head
Camp Hitchcock
Hilton Head Aug 5th 1862

Dear Sara

I am onley going to write you a few lines this morning. It is excessively close and warm without scarcely a breath of air stirring. We have not got to drill this morning but shall have more than an offset in going on fatigue this afternoon. It seems as though we wer bound to be laborers instead of soldiers. for work seems to follow us wherever we go. We thought surely if we staid here there would be nothing for us to do but drill. but it has proved otherwise. The extensive fortifications here wer never quite finished and we are now at work on them finishing up and strengthing them. This is being done in anticipation of an attack from the rebels. who are said to have completed a large and powerfull iron clad ram at Savannah and to be engaged in sounding and buoying out the channell evidently with the intention of comeing down and paying us a visit. It is also ascertained that they are accumilating a large force. and gathering means for transportation of it. So that it is possible that they may call down some fine morning and make us a visit. I do not worry much about it however. When we first came here it appeared as though we wer to have a compareatively easy time of it. But unfortunately our colonel has gone home and we have fallen into the hands of our Lieut Colonel. a merciless unprincipeled fellow who looks upon us soldiers as nothing but brutes or mere machines on which he can experiment at his pleasure and he has increased the hours of drill from two hours to four. and that of the hardest kind of drill. namely. battallion. And now comes this fatigue work on top of the rest. true we have not as yet had to work only half a day to a time but that is more than one wants to do in the hottest season of this hot climate. I received your letter of the 27th yesterday morning for which as ever I am thankfull. I hope that long ere this you have recovered from you fit of the "blues" and are "yourselfe" agan. I recollect aright. I have weighed you myself when your weight was the same that it is now exactly. Shall hope to hear that you are as heavy as <u>Maria Holley.</u> when you return from the sea side. If I wer asked I should be utterly at loss what to advise in regard to Lizzie and Charlie's matters. I do not see as anything better can be done than to let matters take thier course for the present. If he truely loves her it will be all made right in time and if they are mistaken in regard to the true nature of thier attachment it will be better to find it out now. than years hence. I went down yesterday afternoon with four other of our boys and took Sweet on board of the Star of the South which was to sail for N.Y. last evening and Sweet will probably be home by Saturday. We took him from the hospital and had to carry him most of the way down on a litter as he was not able to walk and there was not an ambulance or a wagon to be had. He is not hard sick. is able to sit up and walk about and in a week after he gets home will doubtless be as well as ever. I was very glad to see him once fairly on board the boat bound for home for it was wrong to keep him here amid so many discomforts when he could do no good. We shall miss him. for he is a pleasant companion and has always done his part cheerfully when able. There are others in our company who ought to be discharged and sent home and it is both a sin and a shame that they are kept in the army to suffer and die for nothing when they might recover thier health and be usefull at home. I do not expect to see our cause prosper until our rulers learn to close thier own selfish bickerings and unite all thier energies upon thier proper buisness and those in authority in the army learn to exercise judgement. justice. and humanity towards those under thier command. No cause ought to prosper when such great and crying abuses are allowed to exist which might be reminded. I tell you my dear Sara. I think there is a vast amount of inequity and corruption in our own government. and among us as a people at the North. for which the great Ruler is punishing us by allowing so many reverses to our arms. And not till we are sufficiently humbled and brought to feel our dependance on Him. and our duties to each other will he allow the justice of our cause to triumph and restore peace and prosperity to our at present distracted land. George. I am sorry to say. is quite unwell. he has not felt well for several days but as we are obliged to do duty as long as we can crawl. he has kept about until yesterday. he suffers very much from pain in his back from which he can obtain no releif except for a few moments by change of position. We can do but little for him. but you can rest assured we shall do all we can and I hope I can write better news from him when next I write. This weather and our present mode of life. being obliged to be exposed to so much. is very trying and there are but few but that feel its effects more or less. I enclose you a photograph of our lamented captain which I wish you to keep for me. It is a very fair likeness of the kind and will give you a very good idea of his general appearance when dressed as in the picture. I value it highly and therefore send it to you for safe keeping. knowing that you would be gratified by possessing the likeness of one with whome we wer so intimately connected and whose memory we delight to honor. Pitt tells me that James Deane talks of raising a company if he can get some one to supply his place in Canaan. Our people must hurry up and raise the new army quickly else the rebels who do not wait for volunteers but quickly increase thier army by conscription. will raise a large force and be down on us at all points. I can assure you my love that I am as sick of this war as you can well be and though as ready to fight as ever am heartily tired of being a puppet to be pulled and danced about for the amusement of a lot of fat. well fed and better paid. lazy shoulder strapped gentry one half of whome I should scorn to associate with at home. The fact is two thirds of our officers and rulers <u>do not want this war to close</u>. they wer never doing better in thier lives. or making more money and for the sake of that money. and the positions. they care not how long the country suffers. I have already written more than I expected to and all that I can think of at present likely to be interesting. Hopeing that you are enjoying your season of leisure and praying for a continuance of your health and happiness I remain as ever

Your own
Ned

MY DEAR SARA

No. 64 Rec'd Aug 19th 1862

Hilton Head
Camp Hitchcock
Hilton Head S.C. Aug 10th/62

Dear Sara

    I am so hot and faint this morning that I fear you'll not get much of a letter this time. but as I have not written since the first of last week I must write a few lines at any rate. I wonder where you are and what you are doing this morning. We have had our usual Sunday morning inspection and are now trying to make ourselves as comfortable as possible. Some are playing cards. some reading. some fixing thier tents. and a few like myself writing. It is one of the hottest of hot August mornings with scarcely a breath of air stirring. Poor George is still very unwell and is sick and uncomfortable. this morning from the effects of a large dose of castor oil. He is able to walk about and wait on himself some. but is very weak and is glad to lie down most of the time. if he wer at home he would be sick abed. He thinks he has scarcely slept at all for a week. I hope he will soon get some relief. but he is so weak it will be some time before he will be ale to do duty. but he is better than when I wrote before. Mr Landon remains about the same. The mail. I have just learned. closes at one o"clock. so I shall not be able to write but a few lines more as I wish to go up to the hospital and see the sick. and the onley hours we are allowed to visit them. are between eleven and twelve. I wonder if you have received all my letters lately. I wrote while on James island I think. requesting you to send me another ambrotype as I had unfortunately spoiled the case and broken the glass of the one I brought with me. You have made no mention of it in any of your late letters. and so I think you may not have received it. But prehaps you will yet. We have to go on fatigue now. every other afternoon. but do not drill in the morning and do but precious little work in the afternoon. every other day we drill about two hours and a half in the afternoon. with dress parade at six o"clock. There is no news here at present. The "ram" excitement is about played ot. the general opinion among the boys being <u>Gen Hunter</u> used it as a scare crow to prevent the removel of anymore troops from his commend. It is currently reported that <u>Hunter's negro brigade</u> is disbanded. I think it is so. I never can write when in a hurry so will close this apology for a letter. it will at least serve to show you that I am not forgetting you. Hoping that you are well and enjoying yourself. that I shall hear from you soon. and in good time be premitted to see you. I remain as ever you own dear

Ned

No. 65 Rec'd Aug 28th 62

Camp Hitchcock
Hilton Head S.C. Aug 14th\62

My Dear Sara

 Three days ago I received your letter of the 3rd and today the one of the 5th. I was intending to have answered the one first received today and am not at all displeased at having two to answer instead of one. I am heartily glad that our friends are at last satisfied as to our wheareabouts. While the uncertainty and anxiety was doubtless unpleasant and painfull to them. it has also been very annoying to us. as in the uncertainty many letters have been directed to us at Fortress Monroe. and many others have doubtless been witheld altogether. Of course no one is to blame except the heedless newspaper reporter. who from seeing the arrival of our late commanders at Fort Monroe. must needs jump at the conclusion that we wer there also. I think that you have not received two or three of my letters written on James island and Edisto. Unfortunately I stopped keeping account of the date of letters sent away. about the time we left Pulaski. so I cannot tell at what time those I think lost wer written. However it is of no great account onley like yourself. I do not wish any one out of "our" "family" to read my letters to you. I sent you a hurried apology for a letter a few days ago and hope you will not be angry when you read it. I fear I shall be obliged to write some short letters if any during this hot weather. George does not get along much. still I think he is on the gain a little. He is very weak yet. has no appetite at all. and eats but very little. he is not able to sit up all day and yet the miserable villain of a doctor returns him to duty each day. not excuseing him from any duty. Our company officers know his situation and do not require anything of him except to be present at roll calls. You will see some of the pleasant little features of a volunteers life. This creature calling himself a doctor has the lives and health of a large share of a regiment of men at his entire mercy. A position for which he is entirely unfit. he is a man of no character. and less soul. At home in our own town he could never get practice enough to keep him on bread and water. It would seem that the unavoidable hardships and dangers incident to a soldier's life wer all that a man ought to be asked to volunteer to encounter without recklessly jeprodizing his life by placeing it in the hands of an unprincipled scoundrel to experiment upon. You may think that I am useing harsh and ungentlemanly language. prehaps I am. but I feel indignant. and besides honestly believe what I have written to be true. My own health is still very good though I occasionally have a sick spell of a day or two. Mr Landon I think is slowly gaining and I think he will yet be able to be of much service.

but it is still uncertain. and will be a long time first. at the best. I wish darling that I could write you some good letters as you desire. but it seems impossible. Many thanks for the sample. and description of your new dress. The sample is very neat and pretty and I like it. I think it must look very nicely made up as you describe and sincerely wish I could get my arm around it with yourself in it. You need not have excused yourself for getting it. I do not think there is anything criminal in it. It is very true that people do not seem to realize the storm of destruction and misery that is rageing in our land. But dearest if that storm is not sooner staged. there will be call for few but mourning goods throughout our land by the end of another year. I am sorry one of my letters was so nearly illegable. I remember that it was written under rather greater disadvantages than usual and with a poor pencil. Young Knight may pretend to enlist as a private but I am loth to believe that he has any intention of going as such. I have seen to many of that kind of games played. to have much faith in them. Not but what there is many a man in the service now as privates. who are as good. and come from as good families as young Knight. But lately those kind of men do not like to go as privates. If they can get commissions. they are patriotic enough and are ready to wonder why others do not enlist. and in order to induce others to enlist they will do so themselves. but when it comes to being sworn in. they are not on hand. A man may enlist twenty times a day if he likes. and in as many different companies. but it will not bind him to go to the war. it is the swearing in process. that accomplishes that. I have no doubt but that if Knight goes he will be well provided for. It must be very laughable to see the chicken hearted ones who are going to be in danger of having to go now that drafting is about to commence. trumping up some disease or disability. to excuse them. I am glad of one thing. and if I live to get home it will always be a pleasure to me. to remember that I did not wait until I was in danger of being drafted before I volunteered. I learn that James Deane is buisy recruiting trying to get men enough so that he can get a captaincy. No doubt those who enlist now will gain some advantages over those (who) went earlier. but they are welcome to all that they get if we onley get home safe. Money or office never induced me to enlist and never would. so I do not envy those who waited to be bought by them. I do not get any _ale_ now days though I often wish I could get some. I do not think our soldiers are much in danger of falling into intemperate habits while in the army for there are no facilities for getting it at pleasure. as there is at home. It is true that generaly when we have been on fatigue, a ration of whiskey is served out to each man and it is drunk almost without an exception. but as it is not an every day occurance. I doubt whether there is one man in twenty in our regiment who did not drink as much or more spiritous liquers at home than since joining the army. I made a mistake in turning over this sheet while writing. and you will find that it does not hold together well unless you look well to the pageing. I am glad dearest that you wer pleased with my letter (I) fear that some I have written since will not be so interesting. I shall hope to hear from you from Guilford where I hope you will soon be enjoying yourself and gaining health and strength for new labors. You doubtlessly had an interesting tea party that Sabbeth eve. I have faint recollections of similar scenes myself. Never mind dear. if we live we will yet have opportunities to tell each other what we think without being obliged to put it on paper. I will write agan in a day or two. "till then. good by. As ever your own loveing

Ned

No. 66 Rec'd Aug 28th 1862

Camp Hitchcock
Hilton Head S.C. Aug 18th/62

My Dear Sara

I received your letter of the 10th on Saturday. It was oneley five days comeing. quite a quick trip. There has been a steamer in once in three or four days for nearly the last fortnight so that we have been more than usuly favored with frequent mails and thus far I have not failed of getting a letter from you on each arrival. There are now several transports in the harbor. and some of the boys think that prehaps we shall be called on to go North. but I do not worry much for I know that <u>the 1st Mass calvary</u> is ordered north and horses and all it will require two or three vessels for thier transportation. It is also said that the <u>Rhode Island light battery</u> of six pieces now here is also to go. and if such is the case. there is not probably any more ships in. than will be required for thier transportation. We have news that our forces under <u>Banks</u> have gained an advantage over <u>Jackson</u>. and I hope the report is true. Our prospects have been looking very dark for the last month. and do not brighten very much yet. We are just in the condition to be thankfull for small favors. I consider one of the brightest and most favorable symptoms of success. to be the prompt measures our government is takeing to increase our army. by drafting. Now if they will oneley take the men thus raised and put them into the skeletons of regiments now in the field. filling them all up to the utmost limit. before they allow a new regiment to be formed. it will save an immense amount of money. not oneley on the start. but henceforth while the war lasts. and the men will be fit for active duty and efficient soldiers in less than half the time that they would. formed into new regiments. My own opinion is that the whole three hundred thousand could thus be incorporated with regiments now in the field. without the necessity of forming a single new one. But I doubt whether this plan will be very extensively adopted as it will conflict with the interests of hundreds who are anxiously awaiting the chance to get a good office. I am very glad my dear that you are so pleasantly situated with so many acquaintances and friends and I believe you will enjoy yourself very much. I wish I could be with you but I should make a sorry figure amid so much fine company as you will doubtless find there. But I am glad you miss me. for it is a renewed assurance (wer any such necessary) of your love for me. Our present camp is such a distance from the shore (about a mile) that we have no opportunity for batheing. on <u>Tybee</u>. James Island. and Edistoe I bathed often and received much benefit therefrom. At Tybee there was a beautiful place for batheing. the surf came rolling in splendidly and one could walk out for more than half a mile before the water exceeded five feet in depth. We have not been out of sight of salt water hardly since we left <u>Annapolis</u>. and you can be assured that I am not anxious to make an incursion inland until cold weather at least. I thought likely the drafting would stir up the folks in Salisbury. they may just as well take it patiently for it is useless to try to avoid it. I onley hope that the draft will fall upon the rich and thier sons. also upon some of the lazy gentry. <u>Squire</u> Righter. for instance. or my estemable cousin C. W. Adams Esq. As for those men you mention as having run away. I am heartily glad of it. they are poor apologies for men any way and could not fail to disgrace any army. We have had three days now of quite cool weather. and it has been very refreshing indeede. Yesterday the weather was quite homelike. The air clear. cool. and braceing. quite a contrast to the hot sultry. debilitateing weather of the few past weeks. We had service in the open air on the parade ground in the afternoon. It is the first time the regiment has been out to church since we left Pulaski. We had a very interesting prayer meeting in the evening in the chapel tent which was well attended. We have a prayer meeting on Wednesday evening and also one on Friday evening which is especialy for the purpos of praying for the termination of the war. We have already had two or three for the same purpos. and they are quite well attended. as is the case I am glad to way with the others. The text yesterday was Romans 14 & 12th. I wish I could say today that George is better. but I cannot. I do not see as he improves at all. and think of the two he is getting weaker though he still gets out a little. he is evidently suffering for the want of proper medical treatment. A trouble that we are utterly powerless to remedy. He has no appetite. scarcely eats enough to keep a canary bird alive. I do not think he eats a slice of bread a day. It is very difficult just now to get anything to tempt the appetite of an invalid. Our chaplain who is also our regimental postmaster is makeing up his bundle to take down town so I will close and send this along as a mail may go North before this time tomorrow. Good by dearest. Much love from you own

Ned

No. 67 Rec'd 4th Sept./62

Camp Hitchcock
Hilton Head, S.C. Aug. 21st 1862

My Dear Sara

I do not know what in the world I am agoing to fill this sheet with this morning for I never felt less like writing. But as I have time I thought it best to commence a letter and prehaps by and by I may be able to finish it. It is a rainy morning. very close and warm. the cool weather which we have had for a few days past. seems to be over with. I have nothing new to write you. we are still jogging along the regular routine of a quiet camp life. Generaly a new story or rumor of our probable removal to some other place is started each day just to keep the nervous ones in a worry. By the present arrangements we have nothing to do until about three in the afternoon. But then there is a score of things to be done each day which help to occupy what at first seems to be spare time. Now that we are not in active service and having it comparatively easy. it is required of us to keep our arms and equipments in much nicer order than before. and consequently we are obliged to spend much more time on them. Then there is always some light fatigue work to be done about the camp requireing three or four men from each company every day. then there is guard duty to be done. and our own personal cleanliness and comfort to attend to. All which things help to employ the time when not on drill or at work. The first regiment of Mass. cavalry. and the Rhode Island light battery. have gone North the past week. I do not think it at all probable that we shall leave here. unless our folks conclude to evacuate Beaufort. which I do not think likely. George is a little better yesterday and today. and I am in hopes he will continue to improve. If he could onley get an appetite and then something to satisfy it. and be allowed time to gain strength before being returned to duty. I think he would soon be about agan. Mr. Landon continues to gain slowly. Frank Brinton and a young man named Root from North Canaan are very sick and it is not thought that Root can recover. I hear this morning that the surgeon has odered thier removal to the general hospital. I hope it is not so for I should not expect to see them ever come out alive. If I wer ever so sick. I should prefer to remain in my tent and lay on the ground satisfied with what attention my tent mates could give me rather than to be taken to the general hospital. In our regimental hospital the men have tolerable good care. as good prehaps as could be expected. and much better than can be obtained in ones own tent. and if I wer much sick I should wish to go there. When we first came here (this time) three of the sickest patients wer taken to the general hospital and from thence to thier graves in a short time. Of course it is very likely that they wer past recovery even by the best of care that could be given them down here. Since comeing here we have had our chapel tent pitched. and agan enjoyed our regular Sunday and Wednesday evening prayer meetings. To them have been added a meeting every Friday evening for the purpose of praying for the speedy termination of the war. The meetings are tolerable well attended though the tent is not filled. To me they are very interesting. We have a few christians who are able to take hold and make the meetings both pleasant and profitable. A debateing society has also been organized and the first discussion comes off on Saturday evening. A singing school has commenced operations. and on the whole. things look flourishing. It is generaly the case that just as we get nicely fixed and such matters as I have mentioned fairly started. we have to "up stakes" and move. A courier has just come in announcing that <u>Co. D of the 3rd N. Hampshire</u> regt stationed on outpost duty on Pinckney island was last night surrounded and captured by the rebels. The villains are all the while prowling around and will doubtless continue to annoy us in this manner. but I very much doubt whether they will dare to make anything like a general attack. We expect another mail tomorrow or next day. and I hope we may get some more encourageing news from the North. I have written more than I expected when I sat down and so think that I will enclose this without waiting to write more and send it along. Whether it can be justly dignified with the name of letter or not I must leave to you. I am loveing you as ever dearest. and still waiting. I dare not say patiently. in hope of a pleasant future. Excuse me darling if my letters are lacking in terms of endearment and affection. You know that I cannot use fine words fluently and well. and therefore I seldom attempt it. But I trust you will believe me that if my letters are not all that you could wish. or all that they should be. that it is the fault of the head and not of the heart. You have probably ere this enjoyed many of the anticipated baths. I expect you will derive much benefit from them. I expect to hear from you agan by our next mail. With love and best wishes. I remain as ever your own

Ned

CIVIL WAR LETTERS 1861-1865

No. 68 Rec'd Sept 4th 1862.

Hilton Head.
Spanish Wells
Hilton Head S.C. Aug 24th/62

Dear Sara

You see I date from a different place today though still on the same island. We have had quite a tramp since I wrote you last and the quiet routine of our camp life has been considerably disturbed. I mentioned in my last letter to you that while writing it a rumor had reached us of the surprise and capture of a Co of the 3rd N.H. while on picket duty on Pinckney island. That rumor proved to be true. The company was stationed on the island as a perminant picket. Within two or three weeks three men had deserted from the Co and it seems went over to the enemy. and it is to thier good offices in betraying the situation of thier comrads that we are indebted for thier surprise and capture. After dinner on the day I wrote you we wer ordered to get ready at a few moments notice to go to Pinckney island and cooperate with the gun boats in some operations against the rebels. We wer soon ready and on the march. Contrary to our expectations we took the road to Seabrook ferry. the same that we went on the day after landing here last fall in our pursuit of the rebels. We marched very rapidly. and it was very warm. but fortunately the sun shone but little. Arrived at Seabrooks we found general Hunter there. also the dead and wounded of the New Hampshire Co in the affray of the previous night. One of the dead. a Lieut. had six balls in his body. he would not surrender. but fought desperately. and it is supposed that his death was sought by the deserters who it is said had a grudge aganst him. We supposed that we wer to be ferried across the creek and landed on the island which was near by. but it seems that Hunter had made different arrangments. for we left three companies there and the rest of us took our way back to camp which we reached tired. hungry and wet just after dark. We marched very rapidly both ways and I was as wet with prespiration as though I had been out in a heavy rain. Our company with two others had orders to put ourselves and equipments in order and be ready to start the next morning on a cruse in a gun boat. It rained in the morning and for that or some other reason we did not go as expected. But at noon we had orders to prepare for a weeks picket duty on the opposite side of the island. The duty has been done for some time past by the 3rd N.H. regt and we wer now to relieve them. We left our camp standing and the sick to take care of it. About three o"clock we commenced our march for Grahams' plantation which was to be the head quaters of the picket. the distance satwenty fiveid to be about four miles.

but I think it much nearer five. This time we had our knapsacks on and though we did not march as fast as the day before. the sun shone more and it was much more fatigueing. Arrived at Grahams about sundown. we rested for about half an hour while the officers wer receiveing thier instructions and the various companies wer being assigned thier positions. Our Co and Co K wer assigned to this place. said to be distant from Grahams two and a half miles but in reality three good long miles. We started agan just at dusk and marched rapidly stopping onley two or three times to catch breath and arrived here about eight o"clock wet. sore. and fatigued. We rested as well as we could sleeping in the out houses. under the trees. on the piazza. or crawling into the tents of the N.H. boys. In the morning we releived the N.H. men from guard and they proceeded to strike thier tents and pack up to leave. which they did about the middle of the forenoon. There was but one Co of them. but in pursuance of an order to strengthen all the pickets along this shore two Cos wer sent here to take thier place. I was put on guard in the morning and was not relieved until this morning. My picket station was close by the house where we wer camped so that the duty was but little more than a name. During the day it was decided to move our quaters about three quaters of a mile back in the edge of some woods and accordingly a fatigue party has been out at work clearing up a camp ground. as we brought no tents with us we cannot move until they come. Our orders are that we wer to remain on this duty a week and as it was not known exactly where each Co would be stationed no tents wer brought. barely tents enough are to be brought to keep us dry. We are on Brainards plantation. and it derives its name of Spanish Wells from the existance of several springs of water which are supposed to have been known and used by the Spanish when they held possession of this island in old times. A spring is a great curiosity on these islands. and a natural rock still more so. I have not visited either of the springs yet as they are some distance from the house where we are. but the boys have been to one of them which they describe as being surrounded by a kind of soft red sandstone and the water very pure and cool. We are just about across the island from fort Wells. and distant about eight miles. we can see in a clear day the flag on fort Pulaski. Dawfuskie. Bulls island. and the main land are in sight and but a few miles distant. The negroes are still here. (I mean those belonging to the plantation). There is plenty of corn. and sweet potatoes growing. fish are abundant. also crabs. clams and oysters. And our boys have lived well for the last day at least. This is a pleasant Sunday morning. I was relieved from guard about eight o"clock. have been in batheing and have got this letter nicely commenced. I am writing out doors under the shade of two large pomegranate bushes. the weather has been fine this morning and it now begins to sprinkle with the prospect of a hard shower. at any rate I shall have to adjourn.

Augt 25th We had a heavy shower yesterday from which I took refuge in the house. it was some time after it stopped raining before the sun dried off my bunk so that I could set down and finish my letter and then feeling tired and unwell I lay down and went to sleep from which I did not wake up until afternoon. Then hearing that a negro meeting was in progress in a house a few rods distant I determined to go for I have always had a great desire to attend one. The meeting was nearly through when I got there. but I would not have missed it for a great deal. There wer prehaps twenty five men and women present in a little room in one of the cabins. Several other boys had got there before me. a few wer seated inside and the rest like myself stood at the door. They wer singing "A charge to keep I have" after the hymn one of the brethren made a prayer. then another old man arose and made some remarks and led in singing another hymn. after the hymn a prayer and the meeting was ended by shakeing hands all around. Much that was said I could not understand but I could not but be seriously impressed with the air of apparant faith. and sincerity of these ignorant people. and have no doubt that they could learn me many a lesson of faith. hope and patience. After returning from the meeting. a team came with our tents and we had to pick up our things and take our way back to camp. pitch our tents. and get things to "rights" for the night. Today we have been cleaning up a new camp ground and moveing our tents. and I am now trying to finish this letter. Am sorry to say that I have just got into a dispute with Wells in which angry words have been freely interchanged and the lie given by both of us. I was foolish and much to blame in the start for it was some of my noncense that first angered him and then I got angry in turn and it has ended unpleasantly for me at least. I do not know as he cares so much about it for I doubt if his self will and obstanacy will allow him to see that he was at all wrong in the matter. I was wrong in bringing up the subject and in calling him a liar. but as to the matter in point I was perfectly right. he certainly is mistaken as I think I can prove. but he will never own it. I am very sorry that the difficulty occurred and doubtless you will think me foolish for writing to you about it and I should not. onley that it took place while I was writing and rather upset the currant of my thoughts. You see I am not perfect. George was getting better when I left camp. and I have not heard from him since. It is said that the regiments are to relieve each other on this picket duty. each one doing it a week at a time. if so we shall go back to camp on Friday. It is getting dark and I have but little more to say. my writing was to unpleasantly interupted. Of course you will not mention what I have

written to you on any account. Pitt was present and can tell better than either of us which was most to blame. I enclose a few sprigs of "secesh" fennel that I picked in the garden at the house. where there is aplenty of it. I am hoping to find a letter from you when I get back to camp. and I will write agan in a day or two. Good by darling
                                                                  much love from
                                                                               Ned

No. 69 Rec'd Sept 8th/62
<u>Bainard's Plantation Hilton Head</u>
S.C. Augt 28th/62

Dear Sara

I have just been relieved from a twenty four hours guard duty during the greater part of which it has rained. and I feel both wet. sleepy and uncomfortable. But I received your letter of the 20th yesterday and thought I would try to write a short answer this morning. Our week here does not expire until Saturday and I hear that we are likely to remain another week. but hope most earnestly that it is not so for this is the most unpleasant situation we have been in yet. Our camp is in the woods. it has rained nearly all the time since Monday. the tents leake badly. our clothes and blankets are wet and we have no chance to dry them and worse than all the rest. the mosquitoes are so thick. and tormenting that it is almost impossible to get an hours undisturbed rest. One can hardly form a correct idea of the annoyance these little pests cause us. To escape the sting of the little torments we are obliged to make thick smokes around our tents until often the remedy is nearly as bad as the disease. I think by your account that you have enjoyed your seaside visit quite well. I shall expect to hear in your next letter how much you have gained in weight. You do not speak of the batheing. whether you enjoyed it or not. I have not heard from George since leaveing camp and am quite desirous to know howe he is myself. Also how two or three other men are who we left very sick. George was on the gain when we left. I presume Julia will hear from him direct before this reaches you. We do not hear much about the ram just at present. but the rebels are very active and have made attempts to land at two different places on the island since we have been out on picket. One attempt was made last night at Seabrook ferry. but our men wer not caught napping and the enemy left in disgust. They wer very buisy signaling last night which was a very dark and rainy one. and very good to scout about and capture pickets in. I learn that the rebels have thrown up a battery on <u>Dawfuskie island</u> and that they fired upon one of our gunboats day before yesterday. but with what success I do not know. I think they are rather presumptious for our boats can run clear round the island. and throw shot clear across it. While I am writing. there is heavy fireing in the directions of Pulaski and Dawfuskie. News by the last mail is very dark and discourageing. It seems we are nearly as bad off as we wer a year ago and have got to go to work and do it all over agan. Thousands of men have toiled and suffered. Thousands have been slain. and many thousands more are crippled or ruined in constitution and thier usefullness nearly destroyed for life. millions of treasure has been expended and today I think the rebellion more prosperous. the Southern Confederacy more likely to be established and its existance among the nations of the earth to be recognized. than when I enlisted nearly one year ago. When or what will be the end is harder to tell today than it was then. I think that there has yet to be battles fought. in compareasion with which those already fought are but mere skirmishes. But in this our time of peril and darkness. doubt and uncertainty we can onley look to Him who is the God of battles and the ruler of the nations. May He "who holdeth the nations of the earth as the small dust of the balance" speedily interpose and bring this our great national trouble to a close by granting us a peace by righteousness which shall endure to all ages. Give my love to Lizzie. tell her she need not be afraid to write to me. I hope school duties will not prove quite so unpleasant as you anticipate. Leaveing you as ever and earnestly desireing that you may ever be happy. I am your own

Ned

No. 70 Rec'd Sept 9th 1862

Hilton Head
Camp Hitchcock
Hilton Head. S.C. Aug. 31st 1862

My Dear Sara

We wer relieved from our picket duty yesterday noon. and reached camp about five o"clock. On our return myself and six others met with an accident on account of which we wer premitted to come directly into camp. while the remainder of the company wer obliged to wait at headquaters for something. Luckily the accident was a harmless one and its consequences nothing but a little wetting. Just before reaching Grahams' plantation. the road crossed a little creek. up which the tide rises four or five feet setting back over several rods of marshy ground each side of the creek. This creek was crossed by an old tumble down rickety bridge. in our hurry to get along. we crowded to many on it at once. and as a matter course a part of it "receeded".

takeing us down a distance of about ten or twelve feet to the water which was about two feet deep at the time. the tide being out. Had the tide been in. it would have been six or seven feet deep and we should have had to have swam for it. As it was. we picked ourselves out of the mud and rotten timbers. fished out our guns and made tracks for dry land. We wer fortunate in not any of us getting hurt. for it would have been very easy for one of us to have broken an arm or at least got a sore head in the tumble. This morning I had a hard job with my gun which was well rusted from the effects of its salt water bath. Our regiment has been mustered in today for our four months pay now due us. It has taken nearly all day. on account of its raining. which has hindered us. I find George some better than when we left but still far from well. and quite weak. Poor Root is dead. died a week ago today I believe. Fred Brinton in the general hospital is very low. Mr. Landon continues on the gain and is getting quite smart. Wolcott is also rather better. A large ocean steamer attempted to escape fromn Savannah last night by the river passing fort Pulaski. but she ran aground at high water just above the fort. and in the morning when discovered. fire was opened upon her from the guns of the fort when her crew set fire to her and left in the small boats. The commandant of the fort sent a small. armed. tug boat to ascertain what ship it was. and also dispatched another boat with the news to Genl Hunter at this place. Just one year ago yesterday dearest. we wer trying all day to go to <u>Mount Prospect</u>. Do you remember it! I spent a pleasant evening at your house. having first consulted Doct Knight about enlisting. He afterwards called at the house and assisted you in singing. I could not help contrasting the past with the present. as the events of about a fortnights time following. nearly as fresh in my mind today as when they occurred. came crowding upon me. A difference truely. and will another year find me in the service. or even alive! We hear that a steamer haveing a mail aboard went ashore some where near fort Monroe so that we shall have to wait awhile longer for a mail.

Monday. Sept 1st Was interupted in my writing yesterday so that I did not finish my letter. Oweing to our being mustered yesterday we had no time for service during the day. but in the evening we had service in the chapel tent. Our chaplain preached from Prov 14 Ch 12. A year ago today. I spent my last Sabbeth in old Salisbury. and the various incidents of the day are still fresh in my memory. I have been at work all the fore part of the day on my gun and equipments trying (to) renovate them from the effects of the salt water bath that they got Saturday. This afternoon I am to go on guard and as it is doubtfull whether I write any while on that duty. I will close. and mail what I've written. In rummageing my knapsack the other day I found an old letter from Sara B. that I've not sent home so I enclose it now as this letter will have a light load.

Good by dearest. as ever your own.
Ned

No. 71 Rec'd Sept 10th/62

Camp Hitchcock
Hilton Head S. C. Sept 4th 1862

Dear Sara.

Yesterday we received an old mail. and your letters of July 19th and Augt 14th & 16th came to hand. The July letter has been quite a round about as it was remailed at Washington Augt 13th. The receipt of it has restored the lost link in our correspondance which I had supposed was broken by the loss of one or two of my own letters. The most of the letter has already been answered in various ways. Your cousin Selina's kind message is gratefully recieved. please return her my thanks when you write. and say to her that it is one of the greatest comforts of the soldier. to feel assured that he is remembered and a kindly interest in his welfare felt by those he has left behind from whose society he is exiled. With regard to the recruiting service. if there wer to be a dozen men detailed out of our company. I should not stand a single chance of being one of them. The reason is this. you know that we enlisted in a N. Haven Co. all the officers are N. Haven men. and as all favors and most all promotions are given or made on the recommendations of Co officers. our boys stand a poor chance for anything of the kind. as the most barefaced favoritism is apparrent in the every day management of the most trivial affairs. So my dear you neede never expect to see me promoted on anything more than simply a corporal. the lowest office in the army. and from which I am liable any day to be reduced. either at the caprice of our company commander. or for an accidental violation of some petty rule or regulation. I would give twenty five dollars today if I wer not a corporal. If it wer not that it would lay me open to the charge of making a virtue of necessity I would say that I am very indifferent about going on the recruiting service anyway. I should like to come home and see you. if I could not stay overnight. and would give most anything for the privalige. But after I had made my visit I should want to come back. for despite a years service in the army. I have still to much regard for truth to make much headway at recruiting. for if I should answer truthfully every question most likely to be asked me by those whome I should try to induce to enlist. I should not get one recruit a month. I shall begin to look for your <u>face</u> soon. am very sorry that I cannot take the <u>original</u>. now but then there is comfort. and some consolation in knowing. that it is awaiting its claimant when the war is over. So I am exceedingly anxious to finish the war. Many thanks dearest for your particular account of your amusements and occupations while at Guilford. I could hardly think that the letter you wrote me after your return to L. R. contained all that you had to tell me about your visit to the seaside. but as you did not mention having written the two letters I have just received. I was trying to content myself with it and was therefore agreeably surprised to find myself favored with so good an account of yourself. The weather is getting quite comfortable here now. the nights are quite cool. I think the excessive hot weather is done with for this season. The Fall winds begin to blow. and of course the sand flies in clouds. Fortunately we have a good camp ground. pretty well turfed over. so that it is not quite as unpleasant as when we wer here last Fall. I went down to the general hospital yesterday to see Fred Brinton. I found that there is but one day in the week on which they admit visitors. and that is today. but by the kindness of the Doct having charge of the ward Brinton is in. we wer admited. Fred is very low. has failed much since leaveing our regimental hospital. I do not expect to see him alive agan. Still I should not think it impossible for him to recover. if he onley had a little more determination to get well. he has all the while presisted in thinking that he should die and all our efforts to make him give up the idea have been unavailing. and as not one in fifty who are as low as he is now recover. I have no idea he will get well. George is going on guard this afternoon. He is not fit to do the duty and I am confidant that he cannot stand it and that it will make him worse agan. I received a good long letter from Mr Edgar along with your three. He is one of the most valued of my correspondants. but I have been very remiss in writing to him. We have not drilled any since our return from picket. and have not had much fatigue work to do. We some expect to have to go out on the same duty agan on Saturday. but I rather hope not.

Sept 5th Did not quite finish my letter yesterday as I was occupied most of the day in putting up a bower of poles covered with palmetto leaves. in front of our tent. It was just as I expected with George. he could not stand it and was obliged to be relieved. he does not feel as bad this morning as he did last night. but I am afraid he will be worse for it. Another mail came in last night. and I received yours of 23rd and 24th. but think I'll not attempt to answer them now. but close up and mail what I have already written. and write agan in a day or two. With much love my darling I remain as ever your own

Ned

CIVIL WAR LETTERS 1861-1865

No. 72 Rec'd Sept 13th/62

Camp Hitchcock
Hilton Head S. C. Sept 6th 1862

Dear Sara.

I am agoing to commence a reply to your two favors of the 23rd & 24th received day before yesterday. though I do not much expect to finish it this afternoon. It is real pleasant weather now. except that prehaps the wind blows a little to hard some of the time. We did not have to go out today on another weeks picket duty as I thought we should. Our turn will not come until the last of next week. Col Hawley returned to us yesterday bringing with him about fifty new recruits. They were enthusiasticly received. But they did not know that thier troubles in this life had just commenced. General Hunter has been relieved of his command here. and has gone North. He sailed yesterday afternoon receiving a salute from the fort as the vessel sailed out of the harbor. The Adjutant and our second Lieut also went on the same steamer. they go on the recruiting service. We have recieved papers of the 1st giveing accounts of the late battles in Virginia and the precarious condition of our forces there. It is very humiliating. to think that after a year of hard fighting and at one time apparently almost successfull. we should find ourselves driven back over the hotley contested and hard won fields of strife. and put to such straits to protect our national capitol. We await with impatience another arrival earnestly hopeing that it will bring us better news. It is rumored that <u>General Butler</u> is to take command here. but I do not know upon what authority. I hope it is not so. for I think that he is now the right man in the right place and that he cannot be removed from his present command without doing injury to our cause. There is also another rumor and I think it a more probable one. that our next commander is to be <u>Brigadier General Brannan</u>. now I believe in command at Beaufort. S.C. I agree with you that it would be amusing to see the scramble for certificates of disabilaty wer it not so humiliateing to see the cowardice so plainly displayed. I never knew that our friend Frank was ever troubled with any disease of the lungs. and as you say never knew of his being sick at all. If he realy is troubled with the Bronchitis I can recommend the soldiers life to him as the medicine that cured me of quite an uncomfortable attack of it. after a years faithfull doseing of Doctors stuff. I think he had better apply to Dr. Knight for advice. Of course no man ought to come to the war but such as are able to stand exposure and fatigue. and each man whether drafted or volunteer undergoes a strict medical examination before he is sworn into the service and if found in the judgement of the surgeon to be unfit for military duty. he is promptly rejected. I suppose young Knight did a very romantic thing by marrying just before going to the war. but I see by the papers that that it is getting to be quite fashionable. I am sure I do not see why Dr. Knights people should dislike the match. I supposed that she was old enough to take care of him. I was not aware that your Uncle Benjamin lived at Millerton. but see from yours of the 24<u>th</u> that he does. I suppose when you get back to school agan and get settled once more. you will be more regular about going to church. Your pet Frank must be quite a boy by this time. I know the returned soldier you mention. have been to school with him in days long gone by when we wer both children. He is from Canaan. I saw him the day he left. His brother (Now in our regt) married a cousin of George's and frequently comes to our tent and spends the evening. I begin to think that there will not have to be any draft on the present requisition. At any rate Bert is safe enough for he would not probably be accepted wer he to volunteer if the true state of his health wer known. Let me assure you my dearest that my health is quite good. that I have not been better in a long time. George continues on the gain. all that he wants is a respit from duty until he can gain more strength. Before this time you are doubtless in school agan and quite "contented." Tomorrow one year of our time will have passed away. peace may be restored to our country and the soldiers to thier homes. Amid your unhappiness my dear you have much to be thankfull for. let us hope that the year to come may not be fraught with more of evil to us than the one that has passed. No my dear I am not an Abolitionist. Neither do I think with your Father and Uncle. that this should be made a war for the extermination of slavery. My idea is that every effort should be made and the last man and the last dollar if neede be. freely used to crush out the rebellion and then deal with slavery afterwards. I was always bitterly opposed to slavery and to its extension and certainly the present rebellion has not tended to increase my love for the institution but really I cannot see the way clear for the immediate emancipation of four millions of slaves degraded and ground down for ages. not one quater of whome have any definate idea of freedom. Whereever our victorious armies may advance by all means let the slaves go free if they are able to do so and take care of themselves. but do not saddle our alredy pressed Treasurey for the support of thousands of them in idleness. May I trouble you my dear to get me a few steel pens. The kind I like best your father generaly has in the store. they are <u>Gillotts' Number 303</u>. they are a small fine pen. You can send a dozen or so in a letter. Once more darling I am as ever your own

Ned.

No. 73 Rec'd Sept 25th 1862

Picket Camp near Spanish Wells
Hilton Head S.C. Sept 13th/62

Dear Sara.

Several days have passed since I have written to you. but it has not been through neglect or carelessness. But really I did not think that so much time had passed since mailing my last letter to you. I find by looking at my memorandum that it is a week today since I wrote. When it came time to write you agan I was sent down to the headquaters on a twenty four hours guard duty. when that was through with and I got back to camp. I found that there wer orders for us to raise our tent and make us bunks of some kind and raise them from the ground. So we had to go to the woods and cut poles and stakes and back them up to camp a distance of over a mile and then it took time to do the work. And besides we had commenced drilling agan. three hours a day. so that our leisure time was quite trespassed upon. During one of my trips to the woods. or rather while in the woods. I lost my watch and we spent nearly a half day in a vain search for that. On Tuesday I received a box of things from Mr. Randalls' people. Some of the things I had sent for and others wer nicknacks "from home." The box had come very quick. and every thing was in fine order. George received his box the same day. so that we had an abundance of goodies. We have been obliged to leave them nearly all in camp as we could not bring them out here with us: however they will be good when we get back. We put the things back in the boxes and left them in charge of Dexter who was not able to come with us. Dexter is not very sick. has a kind of slow fever which keeps him just comfortably sick. George is much better. but is still very poor and not very strong. He was on guard with me the other day and stood it very well and has come down with us today. By takeing his time. not trying to keep up with the rest of us. he made out to get here which was more than I thought he would do when he started. He is now on the seat with me and writing. Henry Brinton went down to the hospital yesterday to see his cousin Fred. he found him very low and failing fast. probably cannot live more than a day or two. We have all probably seen him for the last time. Henry feels very bad of course. They wer both brought up under one roof as it wer as thier parents live in a large double house. and thus thier relationship is nearer brotherly than otherwise. It seemed tough that he could not be allowed to remain with Fred to the last. but he was obliged to bid him good by. probably for the last time. The general health of the regiment still remains good although I believe this is considered about as unhealthy a month as any. We have lost two men lately very suddenly and when men are attacked with a disease now it terminates one way or another very quickly. The weather has been much hotter for several days back. than for two or three weeks previous. I had nearly forgotten one thing that occupied one half day of our time this last week. and that was an inspection on Wednesday afternoon by Capt Saxton of the regular army. He belongs to the artillery and what in the world he was inspecting us for puzzles us. The inspection was the most thorough of any we have ever been through and embraced not onley arms and equipments. but clothing. tents. cook tents. and cooking utensils: commissary stores and in fact the whole camp equipage. There is no news of any importance here at present. It is very quiet. We received orders to releive the regiment on outpost picket duty (The 76th Pa) this morning yesterday afternoon and wer excused from battallion drill for the afternoon on that account. We started this morning about seven o"clock although orders wer to start at reveille which was an hour earlier. This time we had our knapsacks brought for us. the morning was very warm with but little air stirring. and our colonel seemed to have forgotten but what we wer all mounted like himself for he came on at a tearing pace. scarcely giveing us a rest long enough to catch breath. The road is a sandy one. and of course very dusty and hard to travel on. The distance from here to camp is called eight miles and I should think it must be nearly that. About half way out the two companies that wer to come here (ours and Co K) left the rest of the regiment and came directly here. We came so fast that when we arrived here so many of the men had fallen out that there was not half a company of men left. The stragglers all came up in an hour or two. and most of the men thought it the toughest march we have had. I managed to stand it through without falling out but could not have gone much further. However after a good bath and swim in the salt water I feel much better and am doing my best to finish this precious epistle to you this afternoon as I expect to go on picket tomorrow. We are at the same camp we occupied when we wer here three weeks ago. The men we releived have built all sorts of queer little shanties and brush houses. to stay and sleep under. and although we brought along tents many of us prefer to occupy the outdoor quaters. I have taken possession of one of them. it is built of poles set one end on the ground and the other resting upon a ridge pole. and covered with pine boughs. it is open at both ends and has a good bed of boards raised from the ground. I share it with a comrad and we both unite our forces in disputeing the possession of it with the mosquetoes who bid fair to get the best of the contest. Our old tormenters of Tybee. the fleas. have commenced depredations agan and are very troublesome. Fifteen contrabands absconded from the main land ten miles above the village of Bluffton and came to this picket station yesterday morning. They report that the rebels are meditating a foray on this island to capture the pickets. So likely enough my next letter may be dated in some secesh prison. I am now writing in the evening. We have had a shower this afternoon and now it has commenced raining agan and bids fair to be a dark rainy night. just the right kind for a rebel raid. I am afraid we shall have another rainy week to be out in. it seems to be our fortune to have a large share of unpleasant weather when we are off on any such duty. It rained most of the time when we wer here before. The negros are holding a meeting in a house nearby and I can hear them singing. it is almost impossible to distinguish a word that they use when singing for they string it all together so. I can hardly understand them when talking with them. and when they are jabbering among themselves you might as well try to understand a lot of chattering monkeys. A mail arrived yesterday morning and I received yours of the 29th Aug. also a letter from Mr. Randall. I have now have two letters from Mr. Randall unanswered. and one from Mrs. R. and Carrie. The young fellow who took my place in the Bank has got through his time being out on the 7th of this month. Mr. R. has now a young man by the name of Northrop from New Milford. He comes there recommended by Mr. Conklin of the B"k of L'tfld Co. I guess that Mr. R. did not try to induce my successor to stay longer. Carrie was to return to Bridgeport to school after the fair. They have had quite lively times at home during her stay. I am going to try and write more of my letters with ink and have commenced already. but there will be many times like the present when it will be very inconvenient to carry ink along with me and yet I cannot lose the opportunity of writing so I shall still write a good many with a pencil. but I will comply with your wishes whenever practicable. I guess brothers Bostwick and Landon had better enlist as privates and try to be of some service to thier country instead of fishing for office and good salaries. Carrie wrote me that she thought her father must be about the onley one who had not obtained a certificate of disability. Canaans' share of the men are raised I believe. and five to spare. I learn that <u>James Deane has obtained a first lieutenants</u> commission. I am glad of it for I think he is thoroughly patriotic and willing if necessary to sacrifice something for his country which is more than I am willing to believe of Bostwick or Charlie Landon. I suppose Libbie is to buisy flirting to think of brother Ned now. I used to have a sister Sara B. up there some where. but begin to fear that she has seceeded.

<u>Sunday. Sept 14th</u> I did not suceede in finishing my letter last night. am now writing at our fartherest picket station about two miles from camp. we have a little shantie but whether we shall be able to keep dry or not I do not know. We are having

a real cold driving storm. I have six men with me and shall not get back to camp much before noon tomorrow. We came through a grove of woods on our way here which I wish you could see. If the brakes and underbrush wer cleared away it would be perfectly beautifull. The trees are mostly live oak. not very large. and from the branches hang long tufts of moss often reaching nearly to the ground and as the breeze sways them about I think it gives the grove a beautiful appearance. George is on guard at the camp. There was an alarm early this morning. and we hustled out of our shanties. fell into line. loaded our rifles and waited for something to turn up. but nothing appeared to interfere with us and so we stacked arms and turned in agan. I got cheated out of my nights sleep though for I had been unable to get to sleep the fore part of the night. and could not succeed any better in the morning. On our way out here we passed one of the springs from which this plantation takes its name. It is just under the bank which is ten or fifteen (feet) high. close to the edge of the beach. and I should think that when the tides are high the salt water would run into it. It is a large nice spring. the water is clear. quite cool. and altogether the best of any I have tasted down South. We filled our canteens and brought with us. and it is well that we did for there is no water between here and there. I am not attending church today by any means. two or three of the men are getting thier dinner. one of them is on duty. one is a fishing and another has gone off to get some sweet potatoes while I am fighting mosquetoes and writing. When we came from camp a steamer with a late mail was hourly expected in and I shall soon hope to hear from you agan. We have received no reliable information from the North since the third of this month and as there are painfull rumors afloat. we are anxious to have them set at rest and know the truth. which no doubt is bad enough. I fear I shall begin to think by and by that it is wrong to raise more men and place them under the control of our present leaders to be cut to pieces as our last ground army has been. I do not know but it would be better and more just to allow them to leave the country and save thier lives than to enlist them in an army to be butchered for nothing. I shall hope for better news darling and prehaps it may look brighter when I write next. Good by. love from

Ned

Editors note: Letters from here on are in ink.

No. 74 Rec'd Sept 27th/62

Camp Hitchcock.
Hilton Head S. C. Sept 21st 1862

Dear Sara

    I am going to write you a short letter this afternoon for I have not much time and I wish to finish and mail whatever I do write. It's a rainy gloomy day. and I feel very stupid. Your three letters of the 3rd. 6. & 9th. wer received three days ago. together with one from Lizzie and Sara B. but this is the first opportunity I have had to reply to either of them. We returned to camp yesterday reaching it a little afternoon. It rained nearly all day. and with the rain and prespriation together we wer pretty well soaked through when we got here. It has rained very steadily most of the time since yesterday morning. George has gone on guard today down at the headquaters. He will have a moist time of it. Our new general seems inclined to make us usefull. Our regiment received orders this morning to be ready to march tomorrow morning with three days cooked and seven days uncooked provisions. I understand that three regiments and the Rhode Island light battery are to constitute the force for an expedition. the nature of which of course we are in entire ignorance of. It is said that our camp is to remain standing and that we are to go in light marching order leaveing our knapsacks in camp. I am not particularly pleased with the idea of going so soon after our return from picket. and with the prospect of a wet time of it. It is our equinoxial storm we are now having. and it is likely to last a week. I had enough of wet weather campaigning on our James island trip. However pleased or not we have to march at the word. I guess that <u>General Mitchell</u> is a more active man than either of his predecessors in this department. But most unfortunately the time for action down here is passed. unless indeede it be at the head of a large force. Ten thousand men cannot do now what three thousand might have accomplished almost bloodlessly ten months ago. Fredrick Brinton is dead. He died on the seventeenth. Only two of our boys could attend his funeral. and they wer left in camp unwell when we went out on picket or otherwise he would not have had a friend to have followed him to his lonly soldiers grave. None but the heartless <u>Irish waiters employed in the hospital</u> wer with him when he died. It is sad. I pity his poor mother. for he was an onley son and it seemed as though she almost worshiped him. Fred always insisted upon it that he should never see home agan and we could never convince him but that his chances wer as good as any of our own. If he had fallen in battle we would not have murmered. for that is what we all try to prepare for. and expect. but to be obliged to see him dying by inches by a deadly disease which could have been easily and readily cured had he been sent home is almost to much for human forbareance. I think of nothing new to write you. If you do not hear from me agan in some time you need not worry for if we go off in light marching order I shall hardly be able to take along my writing materials so that I cannot write if I had opportunity. It is almost impossible to carry any paper about our persons on account of its getting wet and spoiled. But I shall write whenever I have a chance. And you must not be surprised if you get a letter directed in George's hand writing for he directed a number of envelopes for me one time when it was not convenient for me to write and I have never used but one of them. and when pen and ink are not handy sometime I shall probably use them. Please excuse me Sara dear for not answering your letters in detail this time. I have not time today and I wish to return them to you before we start off. for if I do not I may have to destroy them. I broke off a little sprig from the pomegranate bush as you requested and will enclose it. I have several times picked small wild flowers and intended to send them to you. but they would be forgotten when I wrote and when next I came across them they would be entirely spoiled. From a paper of the fifteenth brought along by some stray gun boat we recieve a little news of a more cheering kind than has greeted our ears for some time past. But we take it makeing great allowances and shall be very carefull about rejoiceing over any good news we may recieve for a long time to come. I am obliged to Martha for her kind oversight and labors of <u>love</u> in the <u>sparking</u> line. am grateful for her kind message which with <u>your premission</u> I will reciprocate. Am glad to be so kindly remembered by Mrs. Barnum. Please express to her my pleasure and tender to both her and grandma B. my kind regards. And now darling what new expression can I use to agan assure you of my continued devotion and unabated love: Words. I often find. are but poor expressions of our deepest feelings. but I feel that our attachment has alredy been so thoroughly tried. that it is beyond the power of words to increase or strengthen it. I was in hopes to have been able to have written often to you in the fortnight to come. but am likely to be mistaken. But you will know that I am loveing you as ever. and whether you hear from me or not. I am still your own

Ned

No. 75 Rec'd Oct 1st/62

Camp Hitchcock
Hilton Head S. C. Sept 23rd 1862

My Dear Sara

I have nothing in particular to write about this morning. but will write a line or two which my letter of Sunday last has rendered necessary. When at supper Sunday night word came that the order for us to start yesterday (Monday) morning was countermanded and that the expedition was indeffinately postponed. you can rest assured that the news was most welcome. for none of us liked the prospect of starting off in the rain. all wet as we already wer. not having had a chance to dry ourselves after comeing in from picket. It rained nearly all the forenoon yesterday. and was damp and cloudy. with occasional showers all day. In the afternoon. we had an inspection of the regiment. in light marching order. and some instruction with regard to our conduct in case we wer called into action. We also learned a few of the bugle calls. which are used in action. when the voice of the commanding officer cannot be heard. We still expect to go on the expedition whenever it clears off. We still keep three days rations cooked ahead. This morning it does not rain. but it is cloudy most of the time and the weather still looks threatening. It is now pretty generaly understood throughout the regiment that the expedition is for the purpos of captureing a battery which the rebels are said to have erected somewhere in the neighborhood of (as we suppose) Jacksonville, Fla and with which it is said they annoy our gun boats. Scarcely a man of us has any faith in the expedition. Not that we doubt our ability to take the battery. but we think it is not of importance enough to pay for the loss of live that it would probably cost. We cannot see how our possession or non possession of a single battery at a place that has once been held by us. and evacuated because it was not considered of importance enough to warrant the employmet of the number of troops necessary to hold it. can have any effect towards closeing the war. We do not think the whole state of Florida worth one good Union soldiers life. It is also thought that <u>General Terry</u> who has command in that section wants to put a feather in his own cap. and get the name of having done some great thing. whether it is of any real benefit to the country or not. It is said that he is to have comand of the expedition and direct its movements. And that does not tend to inspire us much for I have no confidance at all in his ability to plan or carry out any very extensive movements. I have said a good deal more about this proposed expedition than I need to or have any right to. considering that I do not know for certain that its plan and object is really what I suppose it to be. But I have not much to write about this morning and while on the subject I wrote just what I think about it and all that can be learned in relation to it. It is supposed that we shall not be gone over three days. but provisions are taken for ten. that we may not be short in case we do not get along as fast as we hope to. I did not fail to remember when the 9th of Nov came round. its corresponding day last year. The anniversary of each of those (to us) eventfull days.

last year is remembered and the great difference between them is much commented upon amongst us. But few. if any. of us thought we should be in the service this year at the time we left home. The 9th of this month. I and George wer on guard at the negro quaters. down at the headquaters. The day was very warm. and the night a beautiful clear moonlight one. On the anniversary of my last Sabbeth at home I was on guard here in camp and of the day on which we left N.H. for Washington (the 18th) was on picket guard at Spanish Wells. So you see my darling we do not forget these things by any means. In fact I think it more natural we should remember and think of them. than friends at home. For they are posetively the last impressions of home. while our friends are constantly surrounded by home scenes and pleasures which might well lead them to forget these little occurrances. Now that so many more of our friends and neighbors have been called into the field. we cannot expect to be remembered so exclusively by our friends as we have been heretofore. But we feel confidant that we each have a few very dear friends who will not forget us. though all the world should be at war. I feel very stupid this morning. dearest and not at all like writing. so please excuse me for not being more interesting. The men who wer to be sent home recruiting have gone. Our second lieutenant has gone from our company. though he had to take a low. underhanded. course to get the opportunity. However it made no difference with my chances. as a seargent of ours was to have gone. A corporal from Co F by the name of Dabull has gone on the same buiseness and I presume will get all the recruits that are to be had in our section which I think will be very few. The Adjutant of our regiment who was formily our first lieutenant has been promoted to be captain of our Co. He is now home on a furlough. Our first lieutenant is sick. so that we are without a commissioned officer at present. It makes but little differance with us though now. For since <u>Capt Hitchcock died</u>. we might as well have been without any as those we have had. Good by dearest. much love from
Ned

No. 76 Rec'd Oct 4th 1862
Camp Hitchcock
Hilton Head S. C. Sept 26th/62

Dear Sara.
  Our expedition has at last, to use a very expressive, if not polite phrase. "blowed out." Yesterday morning we wer arroused about day light and had an early breakfast. packed up everything snug. and with blankets slung in light marching order wer in line about eight o"clock. We marched directly to the dock. and with but little delay

wer embarked on board of the Steamer Ben. DeFord. The commissary stores wer not quite all loaded but wer fast being put on board. At length just as all was aboard and the steamer about leaveing the wharf. and we wer beginning to be a little reconciled to the trip. seeing that it could not be avoided. and we wer fairly started. orders came for us to disembark and return to camp as the expedition was to be abandoned. We arrived in camp agan a little before noon. and passes the remainder of the day very quietly. The steamer Erricson arrived in the morning. and as we wer going aboard we met the mail which she brought. two quite large wagon loads. A gun boat is reported to have come in while we wer on board the steamer. bringing dispatches to General Mitchell. in consequence of which it is said that the expedition was given up. Two other steamers. the Cosmopolitan and Boston came down from Beaufort loaded with troops. but I could not learn for certain who they wer. but as I saw the Colonels' of the 8th Maine and 48th Pa regiments on board of the Ben DeFord I conclude that those regiments wer on board the other vessels. Our mail was distributed just before night and I recieved your two letters of the 11th and 14th. also two others from a brother and a sister. This morning it is clear and cool. and I think our equinoxial storm is over with. By our mail yesterday we recieved a little better news than has been our fortune to recieve for a long time heretofore. Still I do not see as there is anything very decisive gained. About the old story after all. a bloody battle. and pretty near a crawn one. our side claiming to hold possession of the field. I say claiming. for I find that our people lie about as bad as the rebels in the regard to thier success in fighting. Our news is up to the twenty first. I recieved the paper you sent me. also the one from Martha. I am ever so much obliged to you for the paper you sent. I wanted it very much. but did not expect (to) get it. You have confered a favor upon us all. for it is the onley one of the kind that has come. and all of our boys wer anxious to see it. Please return my thanks to Martha for agan remembering me. I am glad to hear from her if she does get her letters printed in New Haven. And so you had a dull time of it at the Fair: Beaux wer scarce. and had there been a few gentlemen there of the right kind you think thier presence. and attentions. would have been tolerated do you? Well. it would be quite flattering to our vanity. poor fellows that we are. to know that our abscence is mourned. wer it not for the bitterness of being separated from those we love. I never could find much to be interested in myself at our county Fairs excepting of course the "fair" lady attendants and thier own deportment. There is generaly a fine show of flowers. and those you know are always beautifull. I think if the Fair was so well attended it has probably been a pecuniary success at least. I well remember last year when on our way to N.H. for the last time. on the first day of the Fair. how much we talked about our probable whereabouts this year at the same time. But few of us thought we should be in the army. The principle question was. who of us would be alive. Thank God we wer all alive this year though not premitted to be with our friends as we hoped. One of our number has since left us. we hope for the better land. And the rest of us still hope to once more meet our friends. but at the same time remember that it may not be in this world. It is quite possible that you and Sara B. thought of me while sitting by the Spring. Hope you did not say anything bad. or tell any wrong stories about me. Hope that Mrs B. will not think I am missimproveing my time because I spend so much of my leisure in writing to you. because you see I should dislike to do anything to displease her. at the same time I could not think of writing less often when the opportunity occurs. True. we do have a good deal of leasure time which is denied to troops in more active service. but I do not feel guilty at all. It requires some troops to hold possession of this place. and it may as well be us. for aught I can see. as any others. we expected to have to fight when we came here. and if we have not had to. it has been our good fortune. not our fault. We are liable at any time to have our comparative repose interrupted. and to be ordered into active service. in event of which of course we could not have the time to write that we now have. Therefore you see if I improve present opportunities well. you will the more readily excuse me if the time should come when you could not hear from me but seldom. It was very hard to be separated so suddenly as we wer. It was no part of my plan to leave you so abruptly when our separation was to be of so long and dangerous duration. and believe me dearest. had I believed that you loved me so dearly. I would have stayed with you long enough to have taken a more comforting farewell even though I had lost the place in this company or been court marshalled for over staying my leave of absence. But dearest hard as it has been doubtless it was all for the best. Unpleasant as many things are. and have been in this regiment. I am glad that I came when I did. rather than have waited until the present time and then have it said that I enlisted for money. or to escape being drafted. I do not wish to boast. but Sara. if we ever do get home. and whether we are ever called into action agan or not. it cannot be said of us that we enlisted. for want of occupation. for money. or fear of being drafted. No one can ever say we wer driven into the army. I think you did not mention that Fred White called at Emmas' though you did mention his being in the neighborhood. Emma was very considerate. for which I presume you did not forget to thank her. There has been a man calling himself an artist at the Head who took what he called ambrotypes I believe. I saw several specemins of his work. which wer miserable in the extreme and for which he charged enormous prices. I would not allow him to take a picture for me. at any rate. He was a miserable imposture and I learn that his establishment was shut up and himself arrested for swindling the soldiers. It is expected that there will be a decent artist here before long. I know that a sutible room is being prepared. And if it is possible I will comply with your wishes and send you a "shadow" of myself. I wrote the letter you allude to before recieveing your request to write with pen and ink. I had become tired of useing the pencil and concluded that I had rather try the pen and ink for a while. It would have been a great pleasure to me dearest to have been allowed to have tried my "charming" powers in driveing away your pain. You seem to have much faith in the magic of my touch. I would that it might always be potent enough to drive away your pains. and shield you from all ill. I do not wish you. dearest. to make your letters to me finished compositions. it could hardly fail to make them stiff and formal. and thus detract greatly from what to me is the real merit of them. thier genuine unaffectedness. Please to write just as you always have done until I find fault. will you? Prehaps if I wer not a soldier I should not be loved so well. and my absence so much deplored. What do you think of that? I learn by way of George that the Salisbury Co started. as expected. quite a long procession accompaning them to the Falls. By the way you ought to have heard George scold yesterday when he found that Julia had rented his house. afterwards on opening another letter he found that the lady was not going to take it thinking the rent too high. he felt better but is still in a twitter least she should alter her mind agan or Julia should rent it to some one else. Now Sara I should like to fill out this sheete but am afraid I shall get out of material. I have six letters to write beside this but they will not get written today. I must write to Lizzy and Sara B. and the others not being to lady friends are not of so pressing a nature but what they can be put off a while. Another small bag of mail has just gone past. but I guess it is onley some papers that wer not brought up yesterday. How does Franks' bronchitis come on now days: expect to hear from him before long: Suppose he tends out to Miss Greene well of course: Does he really mean to marry her think: or will it turn out as did his connexion with Josie. I see that Tom Orton goes with the nine months men. It does not seem as though he was old enough. Mr Moore very seldome visits us now. he is employed at the headquaters you know: has a very pleasant situation: Tom Norton is still employed in the hospital. Dexter has not done any duty since he joined the regiment on his return from Beaufort. He is not very sick. just comfortably so. I should think. Mr Landon has so far recovered that he was returned to the company two days ago. but he will not be fit for duty in weeks yet. Our first lieutenant has been sick for three or four weeks. and is going home. as both our other commissioned officers have gone home. we have a lieutenant from another company detailed to take command until thier return. Have managed to use up this sheet you see. and with much love will bid you good by. Ned

No. 77 Rec'd Oct 13th 1862

Camp Hitchcock
Hilton Head S.C. Sept 30th 1862

Dear Sara.

I am on guard this morning so you must excuse me for useing the pencil. That expedition that we hoped was given up. it seems was oney postponed after all. and has agan turned up. Our regiment is now under orders to leave this afternoon at two o"clock. I cannot stay therefore to write you much of a letter. but will just say a few words as it may be some days before I shall have another opportunity of writing. I have a hard cold and do not feel very good natured but I guess that is the worst feature in my case. But I do not feel a bit like takeing a sea trip on a crowded transport where the men are stowed away like herrings in a box. It is a fine clear day for the first time in over a week. The cool damp weather of several days past has given a good many of the men a touch of the fever and ague so that there has been considerable shakeing in camp. There is no news to write you. but I hope to be able to write you good news on our return which we hope will be in four or five days. I suppose we shall go in light marching order. leaveing our camp standing. I should love dearly to kiss you good by before we start off but that is impossible. But you will know that in whatever circumstances I may be placed I shall think of you and pray that all may be well with you. Excuse this short note. dearest. Good by darling. much love from

Ned

No. 78 Rec'd Oct 20th 1862

Camp Hitchcock
Hilton Head Oct 9th 1862

Dear Sara.

Once more I seat myself to write to you. I believe that so long a time has not passed before in a long time if at all. since I left you without my writing. to you. but you know the reason. if a Post Royal mail has arrived of late and you have recieved no letters. I do not feel like writing a long letter today. for we landed from the boat wet and hungry this morning and reached our camp at half past ten. I have just got partialy straightened out and improve the first spare moments in writing to you. Our expedition on the whole has been successfull though we did not capture the rebels and did not come in contact with them. They left thier works and cleared as soon as they found they wer to encounter a fire in the rear. as well as in front. They left eight guns. seven of them mounted and loaded. Thier tents wer left standing with all thier cooking utensils and small quantities of rice. hominy and flour. Also a plenty of ammunition. All thier small arms they had taken with them. Of the guns. two of them wer <u>eight inch columbiads</u> of the new pattern. and good guns. two eight inch columbiads of the old style and I guess of not much account. two wer rifled pieces but I do not know thier calibre. and two wer heavy eight inch sea coast howitzers new and good. The guns and shot and shell had been taken off when we left the battery. and the powder all placed in one magazine it being the intention I believe to blow it up. It was reported that another gun. a <u>ten inch columbiad</u>. was found about a mile from the battery. the day before we left. I have not been able to learn to my entire satisfaction whether it is true or not. though a captain of the <u>47th Pa</u>. which regiment accompanied us. told me it was so. We have lost two men from sickness during our absence. One of them. Mr Woodford is one (of) the new recruits. He was to have been our Quater Master. and has been in that department since he joined the regiment. He was a middle aged man. said to be wealthy. leaves a family. He was a christian man. not afraid to die. but I understand begged hard to have his body sent home. but it was not. he was burried at Jacksonville. The other man was a private in Co B whome I did not know. His name was Hubbard. He died on our way down the river homeward bound. He was buried in the sea. Neither of the men was sick long. Mr Woodford onley two days. Will Reid had the fever and ague before we started. was sick on board the boat and did not land at all except to change boats at Jacksonville. He is far from well now. but thinks he will get along without being much sick. George stood the trip well. I got along first rate with the exception of my old friend sea sickness. I believe I will not try to give you a very minute account of our trip as it was very much like others of which I have before written you. I have for the last month kept a kind of a diary and have noted down most of our movements in it. I intend to send it to you as it may be interesting to you though hastily written with no regard to style or grammar. It is written just when I can catch a spare moment. often just at going to bed while momentarily expecting the order "lights out". or in the morning while waiting the breakfast call. I am useing little books for the purpose. so that I can fill them often and send to you. The day after landing at the mouth of St. Johns river. we marched about seven miles in the direction of the rebel battery but on account of going around marshes and the like we did not advance more than three or four miles from where we landed. At night we bivouacked on the bank of a little creek that branched out from the river. At this place our artillery (Two pieces of the Conn battery) joined us and the cavalry was expected to but the boat that was bringing it up ran aground and we proceeded without them. In the morning we withdrew a short distance into the woods where we formed in line of battle. stacked our guns and rested by them until about four o"clock in the afternoon. The other regt did the same. forming in another direction a short distance from us. About three o"clock the gun boats moved up and commenced shelling the battery. which did not respond. After fireing a few times without recieveing any reply the boats ran up to the battery. landed and found that the birds had flown. They immediately raised the stars and stripes on the flag staff which was seen from where we wer lying. We immediately commenced our march for the battery. three miles distant which we reached between seven and eight o"clock. We wer obliged to march slow. for we expected to meete with the rebels in some good place and examined the ground carefully as we advanced. In the advance wer two large boat howitzers drawn by men. while the two field pieces wer in our rear and we often had to wait for them to come up. as part of the way they had to cut thier road through the trees. It was a beautifull moonlight night and the march was very pleasant. Not a rebel was to be seen. Our Co and Co B wer ordered into the battery where we remained doing the guard duty until Monday morning when we embarked on the steamboat Boston and went up to Jacksonville. Two companies from our regiment and two from the 47th had gone up the day before. The ride up the river was a dull monotonus affair to what a ride the same distance up any of our rivers would be. the banks are heavily timbered nearly all the way. not a sign of a village. and only semi occasionly a small clearing and a little house and outbuildings. We saw plenty of allegators and large cranes which was the only signs of life visable on the trip. Jacksonville is twenty three miles from the mouth of the St Johns river on the right bank of which it is situated. I should think it about as large as the village of Pittsfield. When our forces held possession of the place last Spring. considerable union feeling was manifested. and when our people evacuated it the rebels returned and satisfied thier vengance by burning the property of the union men. A large and fine hotel said to have been the largest one in the state. together with a large dock belonging to it. was burned. Several large lumber mills wer burned. and in fact the most business part of the village or city I suppose they call it was laid in ashes. The ruins. consisting of broken and blackened walls and chimneys. wer visable from the deck of the boat. and wer scattered all over the place. We arrived about nine o"clock I should think. and wer very much disappointed to find that we wer not to land. This occasioned a burst of a little the tallest kind of swearing. but it was of no use. The boat hauled out into the stream and we lay there all day. Finely at sundown we wer landed and immediately put on picket on the outskirts of the village. Early in the morning we wer called in. having had a quiet night of it. and wer quatered in the empty stores on the main street in the village. We wer forbidden to cross the street. and a strong guard was put on to keep the men in thier quaters. while the officers roamed about the village at will. gathering all relics they could conveniently carry away. Of course the men grumbled at such treatment and they had a perfect right to. The place seemed to be very nearly deserted. A few families wer about as usual but most of the houses wer empty. I saw a few white women which was quite a treat even though they wer secesh. There wer quite a quantity of negros whome we brought away with us with thier worldly goods. also a few white people. At four o"clock we recieved orders to go on board the Boston agan and start for this place. We wer aboard in a very short time. and then the negros piled in. till there was scarcely standing room to be found on the boat. We left about sundown. and dropped down the river to its mouth. where we found the Steamer Cosmopolitan stuck in the mud. a hole in her bottom and the water riseing and falling in her with the tide. We lay by her until morning and at daybreak started agan. getting over the bar at the mouth of the river and fairly started about half past seven. We found the wind blowing quite fresh and a considerable of a sea running which caused our boat to dance about in a very unpleasant manner. I did not like it exactly. for although I did not fear for the time. yet I was afraid that the wind would increse to a gale. in which event I considered it a poor chance for us. But the wind did not rise any more. We had a shower just at night which moistened those of us who wer on deck. We arrived off the quarantine ship in this harbor about half past one at night and wer obliged to anchor until morning and be boarded by the health officers. It commenced raining agan at daylight and for a while fairly poured. I was on deck and so recieved the full benefit of it. It must have been eight o"clock before we got premission to come in. and as I said it was half past ten when we arrived in camp. I forgot to mention in the propper place. that. during our first days march we came onto two rebel camps from which the secesh fled at our approach. the first one we came to they had burned everything. in fact wer just throwing the last tents on the fire as our skirmishers approached. At the next one they left everything including knapsacks. rifles. cartridge boxes. canteens with a quantity of carbines for cavalry use. After we reached the battery another camp was found about two miles from it in another direction. This one they had also left standing. I believe I have given you the outlines of our trip. the <u>filling up</u> was about the same as on other trips I have before described. Just as we wer leaveing the dock here at starting a mail was distributed. I recieved your letter of the 22nd. and one or two others from other friends. On our return to camp we found a small mail awaiting us. I did not get a letter. but recieved the Home Journal containing the pens. Am much obliged for them. Have not tried them yet. but they look like good ones. I did not find any thing on our trip to send you except a flower or two. There is a little white rose and another small flower. with some geranium leaves that I picked in the front yard of a house in Jacksonville in front of which I was on duty during the night. An oleander from <u>St. Johns Bluff</u>. where the battery was. and another from <u>May Port Mills</u> at the mouth of the river. where we landed and remained over night before commencing our march. I am finishing my letter on Saturday the third day after our arrival in camp. but it will reach you as soon as though it had been mailed the first day. for there has no mail gone North since. and will not until the arrival of the Arago which is now hourly expected. George recieved his last box last night. it had been opened but nothing taken out. A whole schooner load of Express has come. and for two or three days the Provost Marshal and his sattellites have been buisy ransacking the soldiers boxes. and takeing out every little bottle of liquer they could find. They pretend that it is turned over to the hospital. but it is pretty generaly understood that but very little of it ever finds its way into the hospital. One of our boys stood by while several boxes wer opened. from one of them a bottle of liquer was taken and the officer opened it and takeing a glass treated his friends who wer standing by. Night before last a lot of officers at the headquaters got beastly drunk and made night hideous with thier revelings. We heard them clear up to our camp which is over a mile distant. You can judge whether such proceedings are likely to increase the good feeling between the men and officers. Mr Landon has been detailed from the Co as a teacher of the negros. He is not teaching yet. but is employed at present in writing at the office of the negro headquaters. Mr Moore calls up and makes us a call occasionaly. I find that I had picked up a letter and a torn shinplaster in Jacksonville which I also enclose. I think I will close this letter now. as I may not have time to finish it by and by. Let me agan assure you my darling of my love. and that I often thought of you in my absence. Be assured dearest that I am longing to see and embrace you agan. Will that time ever come! Good by my loved one. Ever your own

Ned

No. 79 Rec'd Oct 20th 1862

> Hilton Head
> Camp Hitchcock
> Hilton Head S.C. Oct 13th/62

Dear Sara.

    Our mail came today. and I expected my usual number of letters from you. but instead recieved Martha's letter of the 2nd informing me of your sickness. It grieves me darling to know that you are sick and suffering. and I cannot be near you. to alleviate your pain and cheer you through the long dreary sick hours. It seems as though I could not stay here. that I must come home to you. but the utter impossibility of even takeing one step towards you my dearest. makes me almost crazy. I was utterly unprepared to hear that you wer unwell. Though I have often thought of the possibility of your being sick. and prehaps taken away from me. yet I could never bear to indulge such thoughts. and I have cast them aside until I had almost forgotten that you could be taken from me. But the sad truth is forced upon me that I may be called to mourn. That you who are the dearest of all things earthly to me. may be called to your heavenly home and leave me friendless and a wanderer in the world. God grant that you may be spared me. for how can I spare you. you who are my earthly all. who are all the world to me and without whome the world would be a dreary blank to me. It is useless for you to tell me not to worry. how can I do otherwise when you my own loved one are sick and in danger far distant from me. and I must wait many days and very likely weeks before I can hear from you. O. It seems as though you <u>must</u> get well. How can I ever live in this cold. careless. unfeeling world. without you. with no one to love me. no one to care for me. no one to look anxiously and loveingly for my return if I live and willing to trust thier future earthly happiness in my keeping. Sara. I believe that I can this day realize something of that bitterness of soul that you have often described to me as being your experience. at my departure. and abscence from you. Certain it is that though I have suffered much on account of our seperation all that has past is no comparrison to my present feelings. Martha writes me that you are better and tries to assure me there is no danger. God grant it may be so but I shall not be at ease until I learn from your own dear self. that you are out of danger and able to resume your accustomed duties. But darling you must not try to write to me to soon. I cannot premit that and I should derive but little comfort from reading one of your dear letters if I wer not sure that you wer well able to write it and that the effort had caused you no ill or weariness. No you must let Martha or some other friend write for you until you are thoroughly well. And then I shall almost want some ones endorsement of your ability to write. How earnestly I hope that you are now better and rapidly recovering. But then I do not know but that it is directly the reverse. The "terrible uncertainty" which you have so often experienced. it has now come my turn to fell. and I must endure it for long and weary days and prehaps weeks. You must return my sincere thanks to Martha for her kindness in writing to me. and say to her that I will gladly answer her letter. but she must excuse me for a short time. I will answer it soon as possible. I have onley time to write you this short letter tonight as our mail goes out in the morning. My own health let me assure you is good. Yesterday. the Sabbeth. was a rainy day so much so that we did not have our accustomed Sunday inspection. or even dress parade at night. We had service in the chapel tent. but it rained so hard in the evening that concluding there would be no prayer meeting in the evening I did not go to see. but remained in our tent listening to the wind and the driveing of the rain aganst the canvas. talking with Reid (George was out on guard) about old times. and friends left behind until I was real blue with home sickness. Had I then known of your illness I should have felt a thousand fold worse. I recieved a letter today from Mrs Jones. a real good and welcome one. she tells me of your sickness but her letter was written three days before Marthas'. Darling. I do not know how to close this short letter. I want to cheer you with brave words of hope and a bright future for us here on earth. but I cannot when my own heart is so sad and burdened with grief. I can only pray to Him. "who heareth the young ravens when they cry" that he will be very mercifull to you. to _us_. and speedily restore you to health and thus remove from me the anxious load of grief and gloom that this days news has thrown upon me. He is _Our_ Father. In him is _our_ trust. He has said in his word that "the prayer of faith shall save the sick and the Lord shall raise him up. and if he hath committed sins they shall be forgiven him." Henceforth it shall be my constant endeavor to raise that prayer of faith to the throne of divine grace. And may the good Lord hear my prayers and answer them as in his infinite mercy he shall see best. and give us both grace to be prepared for and resigned to his holy will. And now dearest once more. good by. my hand lingers as I pen the words. for I dislike to close this letter. it seems so much like takeing a last farewell. God grant that it may not thus be. but that I may soon recieve the joyfull intelligence that your health is restored and be agan premitted to indulge hopes of being once more premitted to see you. to clasp you in my arms. and tell you in words. of my earnest undying. love. Once more darling good by. remember that I am thinking of. and praying for you. Ever your own

Ned

CIVIL WAR LETTERS 1861-1865

No. 80 Rec'd Oct 31st 1862

Camp Hitchcock
Hilton Head S.C. Oct 15th 1862

My Darling Sara.
I have been highly favored today. I have been all anxiety since our last mail arrived bringing me the sad intelligence of your illness. I could not reasonably expect to hear from you agan in less than a week or ten days. But a steamer came in last night bringing a small mail and I was gladened by the receipt of two letters from Lime Rock. and one of them from your own dear self. A thousand thanks for it dearest. You cannot know how dear those few lines are to me. Doubly dear to me. for I know that you did wrong in exerting yourself to write them. You <u>must</u> not try to write when you are so ill. <u>Dear</u> Sara. do not I beg of you jepordize your health by trying to comfort me. For remember. if you <u>should</u> be taken away from me. I should never cease to regret that you exerted yourself to write to me. No darling. dear to me as assurances of your welfare are from your own pen I would rather recieve them second handed from some of our mutual friends until I can feel assured that writing will not injure you in the least. I also recieved Sara B's kind letter of the 10th with the postscript on the morning of the eleventh. and have also heard from you by the way of George. I beg of you darling not to try to decieve me with regard to your sickness. I <u>know</u> from what I have heard that you not onley have been but still <u>are</u> in danger. And if the <u>worst should</u> come. the blow would be all the more severe for being unexpected. But I <u>cannot</u> believe dearest. that that great evil will come. I am praying for you. Sara. and if the prayers of one so unworthy as myself reach the throne of grace and have any power there. you will recover. How I do wish I could be with you. Sara B. asks what I would give to share her watch with her. She might better ask. what would I not give. But the stubborn fact that I cannot come. almost maddens me. I know that you have plenty of kind friends with you. and good medical advice and that all will be done for you that can be. and it is of course a great comfort to know that it is so. And prehaps I could be of no earthly use wer I with you. but there is no satisfaction in that thought. I want to be with you if I can do no more than to tell you how necessary your existance is to me and ask you to live for my sake. It is possible dearest. that I am more than necessarily alarmed for your safty. and earnestly hope it is so. But some how the news of Miss Hollisters death reaching me at the same time with that of your sickness. has had the effect of makeing me look upon the dark side prehaps more than I otherwise should. I think you decieved me a little with regard to the real state of your health before you wer obliged to give up

entirely. this will not do darling. You insist upon knowing the exact state of my health all the time. and you ought to deal fairly with me. And if you are unwell. I wish to know it. You have a little the advantage of me in that George is sure to let you know if I am a little unwell sometimes when I think it will not pay to write about it. We have had very cloudy. wet. cold weather for several days past. and. it has fitly compared with the sadness and gloom of my own mind. Today it is clear and bright with a fine braceing breeze. and in spite of my fears. my spirits will rise as all looks so much more cheerfull around. and hope is more firm. I have now to wait another dreary period before I can agan hear from you. Return my thanks to Sara for her kindness in writing to me. and tell her I will answer it as soon as possible. Though she may have to write several times yet if you do not recover. soon. I have been interupted many times in writing this. and it is now near drill time agan and I am going to close and mail this for it may have a chance to start north before I am ready to write agan. Keep up good courage dearest. I shall hope to hear good news from you by our next mail. May the good Lord keep you. and in his own good time restore you to health. and if it be his will restore us to each other and to happiness. is the prayer of your own dear

Ned

No. 81 Rec'd Oct 31st/62

Camp Hitchcock
Hilton Head S.C. Oct 19th/62

My Own Dear Sara.

This is a very pleasant Sabbeth day. it was rather cool this morning. but the sun has come out fair and bright. and the day resembles a fine Oct day at home. We have had our usual Sunday inspections and the men are disposeing of thier spare time in thier usual manner. Our regiment is being paid off today. We have been without pay nearly six months. and shall get four months pay. It will come our Cos turn to be paid this afternoon and I presume I shall not find much time today to write. Still I am going to write a few lines. And how is my darling today: if I onley knew that you were well or at least on the gain. I should feel quite happy. Somehow I cannot feel so much alarmed about you today. It seems as though you would certainly get well as I earnestly pray that you will.

Tuesday morning. Oct 21st.
Did not finish my letter Sunday. the men got to talking in the front of the tent and I was just fidgetty enough to not be able to keep on with my writing and so lay down and took a nap. We had no service in the afternoon. but the usual prayer meeting in the evening. which I attended. As there has no mail gone North in several days.

you have not lost a letter in consequence of my nap. I recieved Sara B's letter of the 13th this morning containing your dear message. and the welcome intelligence that you are on the gain. I am glad that they would not let you write any more until you are better able to. We are expecting to start on another expedition this afternoon. Our regiment is very badly off for officers at present. many of them being home or sick. So the companies have been consolidated for this trip and there are to be but six companies. Our Co is to be joined on to Co B. It is represented as going to be a hard trip. and only strong and well men are required to go. those who think themselves able to march ten miles and then be ready for whatever they may meet at the end of them. Onley twenty six men. three corporals and two seargents are to go from our Co. George is not expected to go. and I am glad of it for he is not able. I am going. It may be that we shall not get away as soon as we expect. and the present arrangements be altered a little. Our Adjutant shot himself last evening through the heart. killing himself almost instantly. Cause: <u>rum</u>. His name was E. Lewis Bull. He came out as a seargent in Co C. was detached from the regiment in Annapolis and made a clerk for Gen Wright. On our arrival here he was placed as clerk in the Division headquaters. a very pleasant situation for a soldier and remained there until after our return to this island in July. He was then promoted to be adjutant of the regiment over the heads of a dozen men who stood before him. and Mr Moore succeeded him in writing in the office. His prospects wer as fair as those of any officer in the regiment but the habit of drinking was to strong for him and he either could not or would not abandon it. and he has been beastly drunk several times already since his promotion. Yesterday he went to the headquaters as usual and returned very drunk. So much so that it was not possible for him to disguise it. At dress parade almost every man in line noticed his condition. and it was freely commented upon. He did not read some orders that wer to be read to the regiment at that time and which it was his duty to read. It is probably that he was repreminded by the Colonel for his situation and he was just crazy drunk enough to kill himself. It is sad. I cannot help pitying him. much as I despised him yesterday. he was a young man not older than myself I should judge. I may not be able to write to you agan for several days. but you will know the reason if a southern mail arrives and there is no letter for you. George will probably write if he hears anything from us. I must bid you good by darling. for our orders have come to get ready. May the blessing of the good God rest upon us both and may I return safely and learn that you have quite recovered from your illness and are well agan. Once more darling. good by

Ned

CHAPTER 5

POCOTALIGO SOUTH CAROLINA

From Wikipedia: On October 21, 1862, a 4200-man Union force, under the command of Brigadier General John M. Brannan, embarked on troop transport ships and left from Hilton Head, South Carolina. Brannan's orders were "to destroy the railroad and railroad bridges on the Charleston and Savannah line." Under protection of a Naval Squadron, they steamed up the Broad River, and disembarked the next morning at Mackey Point (between the Pocotaligo and Coosawhatchie Rivers), less than ten miles from the railroad.

Colonel William S. Walker, the Confederate commander responsible for defending the railroad, called for reinforcement from Savannah and Charleston. He deployed his available forces to counter the two Union advances, sending 200 of his men to guard the bridges, and dispatching the Beaufort Volunteer Artillery (CS), along with two companies of cavalry and some sharpshooters in support, to meet the main Union advance on the Mackey Point road. The Confederates encountered Brannan's Division near the abandoned Caston's Plantation and the artillery opened fire with their two howitzers. The Confederates retreated when the Union artillery responded.

With Brannan in pursuit, Walker's men slowly withdrew, falling back to their defensive fieldworks at Pocotaligo. The Union troops encountered the Confederates on the opposite side of a muddy marsh, and their advance stalled. Brigadier General Alfred Terry, in command of the Second Brigade, ordered the nearly 100 Sharps rifleman of the 7th Connecticut Infantry forward to the edge of the woods where the Union forces had taken cover. The rapid fire of the repeating rifles quickly suppressed the fire from the Confederate battery and associated infantry across the marsh, and they were soon ordered to cease firing to preserve ammunition.[9] The opposing forces blazed away with cannon and musket fire at intervals for more than two hours, until Confederate reinforcements arrived.[3][8] By then it was late in the day, and the Union troops were running low on ammunition.

As dusk descended, Brannan realized that the railroad bridge could not be reached and ordered a retreat up the Mackay's Point road to the safety of the flotilla. The Confederate Rutledge Mounted Rifles and Kirk's Partisan Rangers pursued, but the 47th Pennsylvania Infantry Union rearguard held them off. Brannan's troops reembarked at Mackay's Point the next morning and returned to Hilton Head.

Editor's Note: The following letter follows the Battle of Pocotaligo, S.C.

No. 82 Rec'd Nov. 6th 1862

Camp Hitchcock
Hilton Head S. C. Oct 23rd/62

My Dear Sara.

Am agan safe in camp after another expedition and a fight. I will try to give you some idea of it so far as I went. which I am sorry to tell you to start with was not clear through. We left camp on Tuesday at one o"clock and marched down to the dock where we remained until about sundown when we embarked on the Steamer Boston together with the 3rd N.H. regt. as it was dark when we left the dock I took but little note of our course. for we seemed to take no particular direction at first. but wer going back and forth about the harbor. I managed to squeeze myself in among the hundreds of others who occupied the decks and soon got to sleep. When I awoke about daylight we wer going up what appeared to be quite a broad creek or river. Several boats loaded with troops accompanied us. some of which wer ahead others behind us. Several gunboats wer also along. Shortly after sunrise we landed on a point of land formed by another narrow creek or river entering the one we had come up. Part of the troops had already landed. We went a short distance. then halted until the remainder of the force wer landed. We had with us two twelve pound Parrot guns. said to belong to the Hamilton battery. and three light howitzers from off the gunboats maned by sailors. These last pieces wer drawn by hand. and a part of one company wer detailed to assist in drawing them. We wer in motion about the middle of the forenoon. I do not know exactly how many regiments there wer in all. several preceeded ours. and two wer in rear of us. The enemy wer encountered about four miles from the landing by our advance. and after makeing a short resistance fell back and took up a new position. prehaps half a mile in the rear of the first. When the enemy wer first encountered. we wer ordered on a double quick. the day was quite warm. and it took the <u>starch</u> out fast. On comeing up to where the enemy wer first met we began to recieve the fire of the rebel battery which had obtained a new position and had a good range on our line of advance. Agan it was double quick. at first over sweet potatoe fields. then over cotten fields. for a while we lay down in the ridge. though this was back in the potatoe fields. and the shot and shell played over and around us beautifully. occasionly a shell would burst in the ground near us covering with mother earth those who happened to be near. As we advanced through the cotten field unable to keep on I sat down and the regiment went on without me. The rebels wer driven from this position and after a short resistance. from the third one also. both of the two last positions wer in a piece of woods and the places very favorable ones for disputing our advance. After being driven from thier third position. they retreated for nearly two miles through the woods. then across a marsh some five or six hundred yards wide. through the marsh ran a small creek the bridge over which the rebels destroyed as they crossed. on the opposite side of this marsh the rebels took up thier position and being heavily reenforced maintained it. although our boys kept up such a well directed fire with thier rifles that the rebels could use thier artilery but little. but sheltered in a house and by some woods. they kept up a continual fire of musketry. The ammunition for our artillery was all expended and it was taken to the rear shortly after the enemy wer engaged in this last position. Had there been a supply of ammunition I have no doubt that our men would have driven them from this place. bridged the creek and followed them still further. but whether we could have accomplished the object for which we went. with the force we had I think doubtfull. The object of expedition was to reach the railroad and burn a bridge. destroying as much as possible of the road and property. I am told that the bridge was still some three or four miles off through the road must have run much nearer the battlefield for we could hear the whistle of the engines and very shortly after the cheers and yells of the fresh rebels that they had brought on to help thier bretheren. After the rebels wer driven from thier third position. our regiment was ordered to the front and followed the enemy until they had made thier last stand when they engaged them across the marsh with musketry until night put an end to the conflict and our men wer withdrawn. At this place our regiment did thier fighting and did it well to. recieveing much praise for thier conduct from Gen. Terry. as well as from all who saw them. They repetedly silenced the enemy guns. driving them away from them. Had the marsh been passable I have no doubt they would have given the rebels another run and quite possibly have captured thier guns. After resting myself awhile I went on but did not come up with the regiment until just as they wer withdrawing from the edge of the marsh. The boys wer all in fine spirits. none of our boys hurt. I am very much disappointed at not being able to keep up with the boys and shared in thier work. If we ever get in action agan I'll keep up some way if obliged to crawl on my hands and knees. Three men from our company wer wounded. one of them very severely. I have not been able to learn yet how many our regiment lost but think it will not exceede fifty in killed and wounded. I do not know either how great our whole loss was and have no means of judgeing. Some of the regiments lost more. some not so many as we did. I do not know how great our force was. but have heard it rated as high as five thousand. but there was no such number engaged. After the fighting was over and we had commenced to fall back they commenced carrying off our wounded. This was a hard job. There wer not over four or five stretchers to a regiment and onley two ambulances in all. and not a wagon along. or to be had. and the wounded had to be carried by thier comrads. in thier blankets. and on pieces of boards laid across small poles on thier guns. six or seven miles to the landing. After we had come back a mile or two. we halted for some purpos a few moments and a surgeon came through the line asking for men to go back and help bring up the wounded. he seemed to meet with poor success as but few volunteered. Feeling as though I might still help a little. though very much fatigued. I went back with him and assisted five others in bringing one poor fellow to the landing a distance of about five miles. At first we had nothing but a rubber blanket but finely made a rude litter by knocking of a few boards from a building and laying them loosely together. In this way we brought him the long weary distance. we wer all much fatigued ourselves and could only carry him a little way at a time. then lay him down and rest. The poor fellow was in great pain and almost every jolt forced a groan from him. For a while the road was full of men carrying off the wounded. but finely most of them past us. and the last of the way we had but few companions. it was nearly two o"clock when we got our man to the hospital and delivered him to the surgeons care. One regiment that was not in the fight was sent out to bring in the wounded and I think they wer all brought off though some of the dead wer burried on the field. The night was a very cold one. so much so that I could not sleep not withstanding my fatigue. In the morning the troops wer embarked as spedily as possible and by noon we wer agan in camp. I did (not) write at all yesterday but should have written a few lines at least had not George written to Julia. and I knew that you would thus hear of my welfare. I do not feel well today and cannot write. I earnestly hope that you are fast recovering. but am still very anxious about you notwithstanding Sara B's assurance that you wer so much better when she wrote. God bless you dearest and quickly restore you to health and grant you a long and happy life. Good by dearest. I am as ever your own.

Ned

No. 83 Rec'd Nov 15th 1862

Beaufort. S. C.
<u>Beaufort S. C.</u> Nov 2nd 1862

My Dear Sara.

 A long time has elapsed since I have written to you. and I fear you will worry and be distressed on account of not hearing from me. but it has not been because I have been to sick to write though I have been unwell since our return from the last expedition. but have been on duty untill yesterday. There is nothing serious ails me and I shall probably be all right in a day or two. I had commenced a letter to you three days ago and you would have doubtless recieved it at least one mail earlier than you will this had not our hasty removal interupted my writing. We left our camp on Hilton Head Thursday evening the 30th of Oct and went on board the Steamer Ben DeFord. Nearly all night was occupied in brakeing up our camp and getting the baggage on board the boat. We left the dock at sun rise on the morning of the 31st and steamed up the river to this place a distance of fifteen miles. We disembarked without much delay and arrived at our present camp ground between ten and eleven o"clock. it is about three quaters of a mile from the landing. The remainder of the day and all of yesterday (Saturday) was occupied in getting our tents and baggage together and in pitching our camp. We have an unpleasant place for a camp. it being as is generaly the case down here. a deserted cotten field overgrown with weeds and bushes. which when removed and the ridges leveled down will leave no nothing but the bare sand without any shade and when the wind blows we shall have the repetition of our delightfull experience on Hilton Head last Fall. Our removal to this place was quickly done. and without any idea of the intended movement getting abroad fifteen minutes before we wer ordered to strike tents and break camp. The reason assigned for our removal was that the rebels threatened an attack on the place and we wer comeing here to reinforce the troops already here. The real reason of our comeing was to escape the yellow fever which is prevailing to some extent at the head. Two Captains on Gen. Mitchell's Staff. Col. Brown of the 3rd R.I. and lastly General Mitchell himself have all died within the past week. I believe they claim that the officers died of <u>billious intermitent fever</u>. but it was doubtless so near the yellow fever that it would answer to call it that. We are not loseing many men out of our regiment. but sickness is somewhat on the increase I think. The 97th Pa which was camped on the same ground with. and but a few rods from us. at the Head has been suffering very severely for several weeks. They are to remove across to Bay Point I am told. Col. Hawley is unwell and has not been in command of the regiment since our return from the last expedition. He is not dangerous and is able to be about. Our Lieutenant Col is also unwell and has gone home on a furlough. Our Major is also home. so that the regiment at present is under the command of one of the senior Captains. It is a very pleasant Sabbeth day and while I am writing the church bells are ringing after the fashion of our own loved New England. and I need not tell you that they are forceably reminding us of familiar scenes in like occasions at home. By virtue of being a little unwell I have escaped the Sunday morning inspection. The regiment is to attend church down town this afternoon and I think I shall go with them. Gen. Mitchell was burried here the day we came. He was at the Head when first taken sick and came up here thinking it more healthy. but death claimed him here as well and none may escape his grasp. It is reported that several more of his staff are sick. but it may be nothing more than a camp rumor. I hope it is not. I have recieved your letters of the 16. 19. & 21st. I cannot stay to answer them properly for if I do am afraid I shall miss another mail.

Monday Nov 3rd/62

I am afraid I never shall get a letter written to you. I left off writing yesterday to go to church. after returning from church had dress parade and supper to attend to which took 'till dark. then I completed my bunk so as not to have to sleep on the ground. then a friend from another company came in and the remainder of the evening was spent in chatting. Today it is very warm and I should think good fever weather. We are employed in digging wells and clearing up our camp ground. Yesterday for the first time since leaveing New Haven I was favored with the privalige of attending divine service in a church. At half past two our regiment was formed in line and marched down to the church. every man not on duty. or on the sick list. was required to attend. which of course occasioned a deal of swearing. Our service was held in the Episcopel church. our own chaplain officiateing. Text Isaiah 3rd 1 & 2nd. It was the poorest attempt at preaching I ever heard our chaplain make. It was nothing but a eulogy on Gen. Mitchell and a poor one at that. Evidently the preacher was out of his element. and made sorry work of trying to be at home. There was a very good melodian in the church which with the singing by the choir made one feel quite at home in spite of the unvarying blue uniforms that occupied every seat unrelieved by a single scrap of bonnet ribbon or point lace. There was an evening service and all who wished could attend. A large number went down but as I have before mentioned I spent the evening in my tent. Now dear Sara I am going to close and mail this. so as to have one letter on its way to you any hour. I hope to have time soon to straighten out our correspondance. and also to answer a number of other letters lately recieved. I am very sorry my dear that I have kept you waiting so long for a letter. believe me dearest it could not well be avoided. I do not think the communication will be quite so direct and rapid now as it was at the Head. Good by darling much love from

Ned

No. 84 Rec'd Nov 17th 1862.

Camp Palmer
<u>Beaufort S.C.</u> Nov.10th/62

My Dear Sara,

More than another week has slipped away since I have written to you. But by no means has it been because I have not thought of you. or less of you. I can hardly give any excuse for not writing for on looking back I can see that although it has been more than usualy inconvenient for me to write still had I really been determined to write. I could have done so. I have had on one of my almost uncomfortable fits of averson to writing. and as we have from day today been vainly looking for and exepecting a mail. I have tried to satisfy my conscience by saying that I would write after the mail came. But no mail has come and we are almost in despair of recieveing another so I have concluded to improve a few leasure moments this morning in writing. Today is the seventeenth since we have heard anything from civalization. and we have ceased to be tantalized by the usual rumors with regard to the arrival of a mail Steamer. We have become pretty much settled down in our present camp which has been named camp Palmer in honor of one of our best and most respected captains who died of fever just as we wer comeing away from James island. For the past three nights we have had quite severe frosts which will probably effectualy dispel any yellow fever that may have been lurking around in search of victims. Thus far we have onley drilled in the afternoon. but have been employed more or less about camp during the fore part of the day. so that really we have not had much spare time. The evenings here are quite long now. but so cool that it is difficult to take any comfort in the attempt to either read or write. We have been compeled for the most part to spend them crowding over little fires built in front of the tents or else wrapped up in our blankets and impatiently waiting for roll call so that we can "turn in". Yesterday was the Sabbeth. I went down to church in the afternoon with the regiment. Our chaplain preached. text <u>Exodus 12th ch. & 14th vs</u>. Am sorry to say that it was not much of a sermon. to my mind at least. I would like the privalige of suggesting. very respectfully. to our worthy chaplain that if he would pay less attention to the negros and great men and attend more closely when in the pulpit to the real spiritual welfare of his hearers and his own legitimate buisness I think he would be doing more good than I think has been done by some of his late attempts at preaching. Although I have always maintained the right of ministers of the gospel to preach against any sin whether it be individual or national. intemperance or slavery. yet I never liked to hear what are called political sermons. and do not think it right. or fair treatment to the hearers to be bearing on one particular sin all the time. But I fear that our excellent chaplain has become tainted with a little of the false independence which so soon attaches itself to those in authority in the army. and we shall have to take just what kind of provinde he may see fit to deal out to us as the pure milk of the word. Now do not think that I dislike our chaplain or that he is not a good man. He is certainly a good man. so far as we can see. But he cannot possibly let the negro alone and some how has become possessed of the erronious idea that all the men have volunteered to fight for the abolition of slavery which most assuredly is not the case however willing they might be to see an end put the institution. I am sensible that this subject may not be very interesting to you. but for the want of some better ideas I have rattled away at that in order you see to make out a letter. I spent a few hours one day last week in looking around the aristocratic city of Beaufort. I was very much disappointed in the place. I had expected to find it very beautifull. but with the exception of its pleasant shaded streets. and they wer a led of mud in consequence of the rain the preceeding day. I saw nothing to admire. The houses are mostly old structures and there is the greatest lot of old tumble down sheds. out houses and rookeries imaginable. I did not see a fine building in the place. though there may be some. Some little excitement has been caused in the regiment. by some of the men's leaveing the regiment and joining the regular army. By a late order of the War Department the regular army officers are allowed to recruit the ranks from any who may be willing to join them from the "volunteer" regiments. A few men have joined the regulars from our regiment. Our Colonel is in high dudgeon about it. but he cannot help himself. Some of the men have been very ill used and though it is very doubtfull about thier bettering thier condition one cannot blame them for trying a change. I do not know as the health of the regiment has improved much since we came here. but think it holds as good as when we came. We have not heard from the Head since we left. and so do not know about the present health of the place. Everything is as quiet as can be here and it seems as though we had retired into the country. It is rumored that we are not to stay here long. but nothing certain is known. I think myself that it is quite likely that we shall return to the Head before long. George Olin is unwell at present. is having the fever and ague. he has been unwell some time. but instead of takeing care of himself. has presisted in keeping about on duty. My own health is good. I am getting very impatient at not hearing from you as I am anxious to know that you have quite recovered from your late illness. With many wishes for your welfare and much love I remain truely your own

Ned

No. 85 Rec'd Nov. 24th 1862

Camp Palmer
Beaufort S.C. Nov 13th 1862

Dear Sara.

I learn that a mail is to leave here early in the morning. and so I improve a few moments in writing a short letter to send along by it. I have not one word of news for all has been. and still is. quiet with us since we came to this place. We have but little to do. nothing at all in the forenoon. and onley an hour and a half's drill in the afternoon. no picket duty. and onley the usual camp guard to attend to. We are still in ignorance as to what is going on in the wide world around us. for we have now been nearly three weeks without a mail. and have no immediate promise of one. Of course we live in hope. I suppose the yellow fever is at the bottom of it all. though of course I do not know for certain. We learn that it is still quite sickly at the Head and that in consequence thereof it has been found necessary to close the hotel. and the Express office. I think there is no yellow fever in this place. at any rate I hear of none. and we have had three sharp. white frosts. which are said to put an effectual quietus to its existance whereever it may be prevailing. The general health of the regiment appears to be good. There is some change to be made in the surgeons now on duty in our regiment. Our head surgeon Dr. Bacon. a fine man. a gentleman beloved by the whole regiment. and withal an excellent surgeon. went home on a furlough shortly after the James island fight. and has never returned to the regiment. and will not. as he has obtained a much better berth somewhere about Washington. Besides him we had two young men ranking respectively as first and second assistant surgeons. and it has been to thier tender mercies that we have been subject since Dr. Bacon's departure. The first assistant surgeon Dr. Porter by name has been in charge. and it is of him that we have found so much fault. and not without good reason. He aspired to be the surgeon of the regiment but it seems that he has been disappointed in this and that he is going to leave us. This gives us great satisfaction and we onley fear that the news is to good to be true and that he will yet stay with us. at any rate he has ceased doctering in the regiment at present. It is rumored that our surgeon is to be a Dr. Jarvis and as it is said that he is from near Hartford. and some connection of Col. Sam. Colts'. I think it likely that he is a relative of the Rev. Mr. Jarvis at the centre. Dexter has got a situation in the general hospital at Beaufort. and has left the regiment and entered upon his duties there today. I am glad he has got the situation for he is not very rugged and hardly fit to rough it as a soldier. he has been unwell. and has done no duty of any account since we came back to the head in July. We have two rumors in camp now. one is that Horatio Seymour has been elected Governor of N.Y. and the other is that the Southern Confederacy has been acknowledged by our Government. The latter of course we place no dependance on. as it has doubtless been got up by the officers for a "sell" for the want of something better to amuse themselves about. The other report reached us by the way of the secesh pickets at Port Royal ferry. who hallored across and told our pickets the news. I fear it may be true for desperate efforts have been made by the secesh. and Southern sympathyzers in that State to elect Seymour. If the news is true. it may prove a great hinderance to the government in carrying on the war. If it be true. I cannot think that it is because republicanism is on the wane in that State. but think that the change is easily accounted for by the great mass of republican and union democratic voters who are in the army. Several of the men belonging to this regiment have joined the regular army since I wrote you last. The Colonel is awfull mad about it and makes a fool of himself over it. very much to the amusement of the men. who are glad to see that there is one chance to head him off. where he cannot help himself or control the men. I do not think many more will go. the most of those who have gone are men from companies where the officers are very tyranical and the men have ill useage to complain of. None have gone from our company yet and I think none will. We like our new captain very well thus far. but we still have two or three contemptably mean non commissioned officers who help to make it unpleasant. I must bid you good by agan my darling for I am finishing my letter in the morning with one eye on my work and the other watching the chaplain who I expect momentarily to see start with the mail. It is a beautifull morning. how I wish I knew where you are. and whether you are yet quite well. Do not think of returning to school until you are perfectly restored to health. I hope you will be able to enjoy many pleasant visits while you are awaiting the return of health. If you visit your cousin Selina. do not fail to remember me to her. Once more dearest good by. with continued love I am as ever your own

Ned

P.S.
    Please excuse this writing for it has been done in an awfull hurry. and I see on looking it over that it is hardly readible.

Ned

No. 86 Rec'd Dec 1st 1862

Camp Palmer
Beaufort S.C. Nov 19th/62

Dear Sara.
    The long expected mail has at last arrived. Onley a part of it has yet come up here. about a third of it they say. What we have was brought up last night and distributed early this morning. By it I am in receipt of yours of the 26th & 30th of Oct. The remainder expected tonight or tomorrow morning. No papers have yet reached us. I expect to get more letters when the remainder of the mail arrives but having an opportunity to write I thought I would commence a letter. Our regiment is practiceing at target shooting this week. and our company has been out shooting this forenoon. When Col Hawley returned from home he brought with him a fine Sharps rifle and bayonet with cartridge box and all necessary equipments. and they are to be presented to the best marksman in the regiment. and I suppose before long we shall all have a chance to try our skill in shooting for it. I am no marksman at all myself never having practised at all so I am in no danger of getting the rifle. Oweing to the inaccuracy of our rifles which are but little better than <u>smoothe bore muskets</u> for close shooting. whoever gets the rifle may thank his good luck for it. for it will probably be a chance shot which wins it. It is very quiet here indeede and we have been so long without news from the North. that it began to seem very much as though we wer a small world by ourselves. Yesterday I went down to the village agan and spent an hour or two in looking around. My previous impressions of the place wer verrified by the visit. It continues very healthy here. occasionly I hear the solemn strains of the dead march and anon the three volleys fired over the grave which tell of another soldier gone to his rest. They bury the soldier here in the woods which are composed of large pines with but little undergrowth. and altogether I rather like the

spot. In one place onley a few rods from our camp in a little hollow free from bushes and shaded by the towering pines. are the graves of about thirty men. most of them belonging to the noble. but now annihalated. <u>8th Michigan</u>. ten of them fell in action at one time. and are buried side by side in one grave. with onley a narrow strip of unplained and unpainted board at the head of each. with his name written on it with led pencil to tell his last resting place. The pencil marks are now nearly illegable and in a short time it will be impossible for any anxious friend to find the spot where the loved one sleeps. Two or three of the graves have been provided with good grave stones by thier comrads' takeing the marble tops from off stands. cutting the proper inscriptions on them. and then setting them over the graves in a firm stone base. At the Head every grave is provided with a suitable head board painted white. and the name. regiment. and company of him who sleeps beneath painted on it in plain black letters. So that it will not be at all difficult two years hence for friends to find the remains of thier kindred. When we wer on James island one of our company died in the hospital here and no one knows the place of his burial. We are told that the yellow feaver is disappearing from the Head. One of <u>the Adams Express</u> agents died with it. There have been thirty five cases in all at the hospital. The 97th Pa regt is suffering very badly from sickness. but I do not learn that they have the yellow fever. I cannot think of any thing more of interest to write about things in this section. I went down to church last Sunday afternoon with the regiment. Our chaplain did better than the two preceeding Sabbeths in that he left the negro out of his sermon. It might have been oweing to the fact that he had preached to the negros in the fore part of the day. but he is not very much of a sermonizer anyway. Text was Gal 6-7. & 8th. You can believe me dearest that it has given me great pleasure to hear from you agan. and to learn from yourself of your progress in recovering from disease. But I am sorry that your recovery is not more rapid. and I still hope to get letters from you of later dates than those received this morning. in which I shall learn of your further progress "up hill".

Nov 20th. We went out agan yesterday afternoon target shooting and on our return found that the remainder of our mail had arrived. and I recieved your letters of Nov 2nd & 7th. Many thanks dearest for your efforts to write to me when you are so unfit to write. and it is so difficult for you to do so. I love dearly to hear from you often. but dearest you must not exert yourself to much. Do not try to write long letters until you are able to do so without fatigue. I will get along with short letters. a few words even. until you are able to write longer ones with your accustomed ease. I wrote yesterday all that I could think of that would be interesting. with regard to matters here. So I will try and answer your letters today. I think those "young soldiers" will yet find plenty of buisness. not quite so agreeable as strutting through the streets of our quiet village. so I am content to let them enjoy themselves while they can. I learn that Milton Bradley has obtained the situation of Quater Master. and that Charlie Landon is Captain. <u>Whew!</u> I conclude that your fathers <u>beer barrel</u> is not entirely empty. neither is it quite as full as previous to Mr Reid's call to <u>rest.</u> Wish I could stop in to Jones'es to rest. not on the beer barrel. but on the sofa. which <u>you</u> "do now so much inhabit." That same sofa on which you wer sitting when my arm found its way first around you. on a pleasant Sabbeth evening quite long ago. Will Reid thinks that John is more scared than he need be and thinks it is doubtfull about his having the consumption. But knowing something of John's former mode of life. and also knowing something of the way young men like him holding offices in the army. live. I am inclined to think that it is quite probable. that the fears of his friends have good grounds. As near as I can make it out. it is nearly Sixteen months since you gave me premission to say "must not" in circumstances like those in which I have used them and as to a rebellion. she who is to be my <u>future wife</u>. (God willing) has to much good sense. and <u>loves me to well</u> to do any such thing. As to playing cards I think I am in no danger of learning a bad habit. for I am to indifferant a player to ever risk gameing. and I very seldome play with any one other than my tent mates. or "our boys" for a few days past I have been learning to play <u>crib</u>. with George. it is a good deal of a game and I do not make much progress. I suppose that it is a poor use of valuable time. to spend it in this way. but let me assure you. my dear, that there are ways in the army in which a man may spend his time much more wickedly than in a quiet game of cards with a friend. But I do not deny. neither would I hide the fact that there is a sight of gambling done in camp. spite of stringent orders aganst it. and severe punishments for those who are caught at it.

I mean to write soon to Martha. and Sara B. I have not yet done so in reply to thier favors while you wer sick. I ought to have done it before. for I have had time enough but have felt so averse to writing that I have written to no one but you. By the last mail I have recieved eight letters (four of them from you) and they all require answering so that I shall be obliged to write. Your sister Carrie must be a young lady by this time. I do not think of her as a member of your family except when you mention her. You must tell me all you learn with regard to the "brilliant affair" at Gov Hollys. When down town a day or two ago I had a couple of likenesses taken. they are both miserable as as specemins of the art. though no doubt they flatter the original. I will send you what I think to be the best one. the other I had taken. with my cap off. sitting down and is nothing but a light shadow with two little daubs of very red paint on the place where the face is supposed to be. <u>Red</u> indeede when my face is as swarthy as an Indians. I would have kept trying until I got a possible picture onley we are obliged to take and pay for every picture good or bad. and I thought it most to much of an imposition. If you think to tell me about the "new goods" it might be well to mention a few of the prices of different things. We hear almost fabulous accounts of the present cost of what used to be the cheapest articles. I will try and write a little more frequently hereafter and hope you will not have to wait so long agan for news from us. It must have caused a deal of anxiety at home to know that the yellow fever was prevailing so. and not be able to get any news from us. I learn that Charlie Gaylord died of the same fever in New Orleans. With much love dearest. I remain as ever. your own

Ned

No. 87 Rec'd Dec 4th 1862

Camp Palmer
Beaufort S.C. Nov 25th/62

My Dear Sara.

    I am behind hard agan in writing to you dearest but I will tell you something of how I have been employed for the last three days. and then you shall be the judge as to whether I am to blame. Saturday morning I went on guard. the day though bright and clear. was very unpleasant on account of a raw. chilly. wind that blew all day makeing it entirely out of the question writing at the guard tent as I have some times done. and it was uncomfortable to sit down and read. George was with me. and was on my releif. a thing which has not happened before in a long time. We wer relieved Sabbeth morning about nine o"clock. There was a review of all the forces at this place that morning. by <u>Brigadier General Brannan</u>. at present commanding all the forces in this Department. Contrary to general useage we wer obliged to turn out and go on the review. This occupied us "till about noon. And by the way I will say that I think the review would be a nice thing to see. but being in the ranks I could onley see but a very little of it at a time. There wer five regiments of infantry. two light batteries and a company of cavalry. being reviewed. but a small affair compaired with the reviews on the Potomac. but it was all we had to show. The review was ended about noon. After dinner I lay down to read a little. but soon fell asleep. from which I was aroused by the church call. Though I should not have been obliged to have turned out at this call. I did not feel at liberty to let the opportunity to attend service slip by on so trifling a pretext. We had the service in the open air on our parade ground. Mr Wayland preached from John 19th and part of the 30th verse. "It is finished." Col Hawley's wife and her sister are at Beaufort and they did us the honor to attend our service. and assist in the singing. and for the first time in fourteen long months we wer favored by being premitted to hear a ladies voice in singing. The Col's wife is very fair looking. neither handsome or homley. her sister is quite pretty. and they both appear very well. After service came dress parade. then supper. Our chapel tent has agan been put up. and for the first time since we came here. we had a prayer meeting in the evening. I did not feel like attending but did so and was afterwards glad of it. as there wer but few out. Monday morning we had the inspection which is usuly had on Sabbeth morning. Directly after it we went out target shooting. from which we did not get back until about one o"clock. At two we went out on battallion drill. which lasted an hour and a half or two hours. at any rate long enough to tire us pretty much out. The Conn light battery of six pieces came out to drill on the same ground with us and we struck up a sham fight with them. which while it was quite exciteing was also very fatigueing. The battery fired blank cartriges at us while we would charge bayonets upon them until we wer within a few yards of the guns. then came the retreats. flank movements &c. until we wer quite waked up. The affair brought out quite a lot of spectators. It is nothing unusual for the battery to fire blank cartriges when on drill. they do it to accustom the horses to the sound. but as they never fire near as many times as they did yesterday afternoon the idelers in camps around turned out to see what was up. In the evening our mail arrived so of course writing was out of the question. I recieved your favors of the 11 & 17th and the Enquirer. for all of which favors I trust I am truely gratefull. This morning is a bright and beautifull one quite warm. I am doing my best to improve it writing to you. I am very sorry that I could not have written to you last Saturday. or Sunday at least. for that was about my time to write and then you would have got a letter by the steamer which I expect will leave for the North today or tomorrow. Our chaplain left us this morning. to go home for a short time. his principle buisness is to get regularly appointed (as) our chaplain in conformity with an act of Congress passed since we came out. he does not expect to make a long stay. Previous to starting this morning he called the regiment together and made them a short address explaining the cause of his leaveing and bidding us all good by. The best wishes of the regiment follow him. and his early return will be looked forward to with pleasure. Yesterday two of our officers. the Major and a lieutenant. returned to the regiment. they have both been home some weeks. A considerable interest is felt in regard the fate of the missing transport George Peabody which left New York. for Port Royal previous to the Steamer that brought our first mail received a week ago. Our new surgeon and quater master are supposed to be on board of her. and until last night we supposed our two lieutenants wer on board of her. but last evening one of our seargents received a letter from his little son in N. Haven saying that neither of them started on her. but that both wer safe at home. Am greatly obliged to you for the description of Miss Lizzy. I think I understand your description of her ladyships toilet. I believe I have not forgotten what white muslin is though it's an age since I have seen the article. Your speaking of practiceing your father and Lib. reminds me of an evening spent at George's years ago when you wer teaching school at Lenox. Lizzy or Julia can tell you how Liz and I promanaded and flourished about practiceing in fun what I fondly hoped you and I would do in reality long ere this. I can appreciate something of your state of mind after seeing the whole family fairly fixed and off your hands. I had myself often rather forego the pleasure of an evening party than attempt to dress for it. and after being fairly ready to start. generaly feel more like pitching my dress goods into a corner and resumeing my every day dress. go about some buisness than to go among the gentry. and yet I usuly enjoy myself quite well. when once fairly in company. I hope that troublesome cold of yours is long ere this fairly disposed of. they are anything but pleasant companions. aside from all danger that may attend them. Now as to what you regard as your probable "worth" to me as a wife. prehaps there is a chance for an honest differance of opinion. If I onley wanted somebody to scrub. and scold." and dig. and help me earn and save money. I will acknowledge that I have but little faith in you and should have looked elsewher for a better half long ago. There are plenty of big strapping. red headed. freckeled faced Irish "biddies" that would be worth a ten acre lot full like you. for those purposes. But they would never make a wife. For a wife I want some one to love. some one possessed of those qualities of mind and heart that compell the love of men. One who would enjoy prosperity or share a diversity with me. and assist and encourage me in all that was good. and boldly and fearlessly. chide and rebuke me in anything that was wrong. For your enjoyment of perfect health I shall always pray. but should not be detered for an instant from marrying you though I knew you wer to be a "house plant" for life and that my unremitting labor was to be necessary for your. our comfort. Now do not get in a fever agan. because you are not able to do the washing and ironing of large family before breakfast. I have not written to Martha yet. but am going to soon. tell her not to think that I have forgotten her by any means. About joining the regular service. I should have mentioned that those who enlist in it out of the volunteers. have the privalige of enlisting for the remainder of the time they have to stay in the volunteer service. or for three years. I think that nearly if not quite all. who have joined it from among the volunteers. have done so for the remainder of thier three years. There are but few but what think we shall have to serve our full time of enlistment. so that we look upon it as merely a transfer. Those who have left us have gone into a light battery where without a doubt the duty is much easier and pleasanter than in the infantry. Yes my dear (I had almost written "Madam") I do expect you will keep your promises to me. and have no fear that you will not. for I wear upon my finger a pledge given me by yourself that you will do so and that pledge is very dear to me. It will never leave me while life remains <u>except</u> at the <u>wish</u> of the giver. I have not written to Sara B. since you wer taken sick. and have an unanswered letter of hers in my portfolio now. You may tell your father that I am very far from finding fault with his agent but that if he chooses to employ an agent to attend to his correspondance he must not be surprised to find all <u>his</u> letters directed to that agent. and he must run his own risk of getting possession of the letters. I do not know that I have seen the book you are reading. we have a few very good books in our library. which by the way is pretty much used up. I got "Baxter's call" from it the other day. and am reading it as opportunity offers. I shall have but little time to read. for some time to come if I do justice to my correspondance. I received a letter last night from Mr Randall in which he expresses much gratification at the removal of McClellan and strongly condems him for his past actions in which opinions I heartily concur. Our chaplain told us this morning that from all he could learn Gen Banks was comeing down here. I am not sure but if such is the case we may stand a chance to see some of our brethren from the "<u>nutmeg</u>" State for I am told that a part of his force consists of Conn men. I knew Seargent Reynolds whose death occured at <u>Pocotaligo</u>. I see you marked the notice in the paper. He was a brave man. earnest in the conflict in which we are engaged and ready to face any danger in the preformance of his duty. At one time during the battle the men wer lying down and fireing. orders came for them to rise up and advance with his usual promptness he sprang up and as he arose he was struck by a bullet in the brest. before he could be got off the field he was agan hit recieving a severe wound in the thigh. I believe he lingered until the next morning. There is to be a man shot at Hilton Head on the first day of December. He is a private in the <u>fourth New Hampshire</u>. He deserted when the regiment was at <u>Fernandina</u>. to the enemy giveing them important information. he was to big a villain for even the rebels and after a short stay with them during which he perpretated a series of villainnies ending by stealing some money from a poor women the rebels delivered him up to us as a punishment. he was tried by a Court Martial of which G<u>en</u> Terry was president and sentenced to be shot. The sentence has been approved by president Lincoln. Now my dear I think I have done pretty well for one "sitting" though there is something else I was going to write about but for the life of me I cannot think what it is. Guess it cannot be of much importance.

    Good by dearest. I am with much love your own

Ned

No. 88 Rec'd Dec 16th 1862

Camp Palmer
Beaufort S.C. Nov 29th/62

Dear Sara.

    I am on guard today. and cannot hope to finish a letter. but as it is past my time to write to you I am going to write a few lines this morning and prehaps I can finish it tomorrow. It is a cold windy morning and I have a hard cold and consequently feel as uncomfortable as one could wish. Thanksgiveing has come and gone since I wrote you last. Many more homes are desolate than at the last observance of our national feast. For myself I little thought that I should spend the second Thanksgiveing day in Uncle Sam's service. but so it has been. We wer granted an entire holiday. no drill or fatigue work. not even a dress parade. An amazeing condescension on the part of our officers. for which we are trying to be sufficiently gratefull. Our company went to work and got up quite a supper. two good sized hogs wer bought and quickly butchered. and we got them nicely baked at one of the bakeries down town. A quantity of pies and cakes wer also bought there. At camp we cooked a fine lot of onions. potatoes. and turnips. We did not get all ready until it was dark. but you can be assured that by that time our appetites wer in a fit state to do the eatibles ample justice and we did. All pronounced it in soldiers <u>parlance</u> a "bully supper." After supper a barrel of apples was distributed and if we had onley had the cider to have gone with it I have no doubt we should have felt quite gay. I think most of the companies endeavored to have something extra, Co E had a fine supper. and after it had quite a variety of games. such as sack raceing. rolling the wheelbarrow blindfolded. jumping and the like. which created a deal of merriment. In the evening they laid down a platform of boards and had quite a dance. They had some good music but no ladies. A couple of miserable imitations wer manufactured. by rigging up a couple of boys in skirts made of blankets. bonnets rigged out of our old hats and false hair made of the long moss that hangs so plentifully from all the trees in this country. A party also rigged themselves up as guerrillas (don't know as I have spelled the word right but can't stop to see) and made much sport by thier comical appearance. Altogether the day passed off very pleasantly. I went down town in the morning and mailed a letter to you. and took a walk about the town until noon. At Fort Pulaski they observed the day in great style. Most if not quite all of the higher officers went down. takeing with them all the ladies that could be mustered. Our Colonel went takeing his wife and sister in law of course. They had quite a ball. and a variety of other amusements.

Sunday. 30th My guard duty passed off finely. It was a beautiful moonlight night. cool but still. My cold though unpleasant like all colds in the head is no worse. This is a lovely day. There has been a regimental inspection this morning which I escaped by comeing off guard this morning. Our chaplain being gone. of course we have no service today. Mr Moore and Mr Landon came up from the Head last night to make us a short visit. They go back tomorrow. I have been visiting with them a little. I will write agan in a day or two but I have little of interest to write about. Believe me dearest. I am loveing you fondly as ever and that I am faithfully your own.

Ned

No. 89 Rec'd Dec 16th 1862

Camp Palmer
Beaufort S.C. Dec 8th 1862

Dear Sara.

It has been several days since I have written to you. and I ought to write you a long letter. but fear that I shall not be able to for want of material. It is real chilly today. though the sun is shining brightly. It froze quite hard last night. harder I think than I knew it to last winter. The ground was frozen quite hard. Last night was clear and perfectly still. and the full moon shone clear. and bright. I went on guard yesterday morning and was relieved this (morning). It is the first time I have been on guard on the Sabbeth for a long time. and the day passed stupidly. and unprofitable. enough. Had I not been on duty I intended to have got liberty to have gone down town to church. Since we left New Haven I have not heard any other minister preach than our chaplain and a change is sometimes pleasant if not profitable. We met with quite a misfortune a few nights ago. It was cold and wet. and our guard tent blew down. so the guard took up quaters for the night in the chapel tent. There was a stove in the tent and the boys wer not carefull enough with thier fire and managed to burn the tent down before morning. It is quite a loss to us for

we have now no place for holding evening meetings which of course will have to be given up. The chaplain previous to leaveing had got it fixed up with tables. benches. and a stove so that we could be quite comfortable there. We may get another in time. but prehaps before it comes we may be in battle. and half of us never have neede of it. We wer having three very pleasant evening meetings a week besides useing it for singing schools and a reading room. I have a hard cold and considerable cough. but am happy to say that it is some better today. We are awaiting another mail now as it is quite time that we had one. We are drilling about four hours a day now and it is a considerable like work. There is absolutely nothing new to write about. A part of Gen Banks expedition put into the harbor at the Head a couple of days ago. We thought at first that our reinforcements had come. The last war story that is afloat is that the rebel pickets at <u>Port Royal Ferry</u> told our pickets that there had been a great fight in Virginia in which the rebels had been badly defeated. I place no confidance in the report for I think it improbable that they would be in haste to tell of thier own defeat. The pickets do often talk to each other at the Ferry. I have just learned that the mail will close at eleven o"clock and so though I have not written half a letter. I shall close and mail this. I suppose you are having sleighing at home by this time. I shall have no nice rides with you this winter dearest. but shall often think of the pleasant ones I have enjoyed. I hope you are quite well by this time. It must be about time for the gay season to commence but I should think that the society in town wer pretty well thinned out. If this war lasts much longer the ladies will have to wait on each other or else take up with pretty <u>old</u> beaux. Good by darling for this time. be assured that I am loveing you as ever and that I am still your own
Ned

No. 90 Rec'd Dec 19th 1862

Camp Palmer
Beaufort S.C. Dec 12th 1862

Dear Sara.

By the mail which arrived this morning I recieved your letters of Nov 23rd and Dec 1st. If I have good luck. I mean to get a letter written in time to go in the mail which is expected to leave the Head tomorrow. I have nothing new to write of. or about. so think my epistle will not be very lengthy. By the way I have not troubled you with very long. or frequent letters. lately. I must be more carefull and shall get into trouble. for even your patience and long suffering may give out at last. and you demand as you have a right too. more frequent and interesting letters. We have been out target shooting agan this afternoon. the whole company did very poorly. and no good shots wer made. Tomorrow we try agan. I will tell you the name of the lucky winner of the rifle. if the final trial for it is ever made. I do not stand anything like as good a chance for it. as I do to get home. am no marksman. and never can make one. Our drills are rather severe lately. The Colonel seems to think because the weather is a little cooler. that he can put us through with impunity. He has taken it into his head lately to drill us on the double quick which is very hard work. the men scold and swear and wish him any where else but here. Our orders are to drill three hours a day. but the officers manage to make it nearly or quite four. Yesterday. General Terry with three ladies. and a number of officers. came on the ground while we wer drilling. and we wer kept at it nearly an hour over time. just for thier amusement. Most of the men wer pretty mad and did not care whether they drilled well or not. and so of course the Colonel got vexed. made some mistakes himself and on the whole I think our performance any thing but creditable to us. We very seldome drill worse. However the Colonel let us off without any dress parade which was a slight mitigation of our grieveance. Today we are signing the pay rolls. and so expect to get paid off agan in a day or two. Last night. our first Lieutenant returned to us. he has been home on sick leave. and returned on the same steamer that brought our mail. Our second lieut is still home recruiting. but we expect him by the next steamer. Things look as though we wer likely to remain here for the present. but rumors are afloat. that we have been ordered to Ft. Pulaski. Key West. and the Dickens knows where. Our new Surgeon has arrived. but I have not yet had a near view of him. His name is Jarvis. and as he is from Hartford and some connection of the "<u>Colts</u>". I presume he is a relation of our minister Jarvis. The Colonels wife is still at Beaufort and if her presence is the cause of the change in his actions lately I heartily wish she was at home. I wrote you a day or two ago that some vessels belonging to Banks expedition had put into the harbor at the Head. They are the Erricson which stopped to coal. and the <u>Bienvil</u> whose machinery needed repairing. and a brig which one of the vessels had in tow. They are all loaded with troops. but I do not know what State they are from. I have mentioned many things which you may not think worth the telling. but they are such little incidents as go to make up our every day life. and as I am writing to one who I think is interested in all that concerns me and to whome I am not to write stiff. formal. letters. I put all these little things in. <u>to fill up</u>. I knew you would not like those pictures. They may be much better looking than the man they attempt to portray. but they do not resemble him much. If I can ever get a chance I will have a good one taken if it is so true to live that it scares you. No my darling though I may have changed some in appearance. and character. and in neither for the better. to you I am still the same. loveing you fondly. trusting you implicitly. and earnestly looking and praying for the time to come when I can claim for my very own the to me dearest earthly being. I rather think you do me injustice when you infer that neither yourself or the sofa could tempt me to forego the pleasure of a visit to the beer barrel. Now I acknowledge that I do have good ale. and why should I not! Your own fair hand has often held out to me the tempting glass. and it is too much to expect of <u>poor frail humanity</u> that it will resist such inducements. but I submit that it is not fair to accuse me of loveing the gift more than the <u>giver</u>. I remember very well the time to which you allude. when we wer sitting together on the sofa. I know that you wer much worried for fear that <u>father</u> saw us. and also know that whatever anxiety on the subject I may have professed at the time. in reality I did not care a rush whether he saw us or not. at least as far as my part of the transaction was concerned. Do not worry dearest because you cannot send me all you wish to. The papers you occasionaly send me answer every purpos in that line. I can get <u>good</u> books whenever I want them out of our regimental library. and though they are not as interesting to me as some I could select. they are such as I ought to be interested in. I could not carry books with me you know if they wer ever so good ones. I have thus far carried with me a bible that I bought in New Haven and a prayer book. the latter more particularly on account the hymns in it. In our rainy march across <u>St. Johns island</u> the bible which was in my brest coat pocket got wet through and nearly spoiled. but I still hold on to it. but if I have a good chance I shall get another. Our chaplain haves plenty of testaments. but no bibles. By the way dearest do you still <u>take Petersons Magazine</u>. It is very ungallent in me to have forgotten to attend to that before. but I shall have to plead guilty to having entirely forgotten it. Please let me know how the matter stands. They must have queer ways of doing things in the 28th if they preform after the fashion of Capt Landon & Co. The gentlemen would be apt to get his walking papers if he did that kind of buisness in our regiment. or any other where they pretended to do things up in military style. I guess from all that I can learn that it is what we call a one horse regiment. No wonder the boys do not like Landon for a Captain. He is no more fit for one (than) I am to preach. Am happy to learn that you are able to get out in society once more and hope you will be able to enjoy the respit and pleasure that was denied you during your vacations. I must finish this letter rather abruptly as the drums are now beating the tattoo. and it will quickly be "lights out." and I must get it into the mail early in the morning. With much love. I am darling. as ever.

Your own
Ned

No. 91 Rec'd Dec 29th 1862

Camp Palmer
Beaufort S.C. Dec 18th 1862

Dear Sara.
Tomorrow morning we go out on picket duty. in the same manner that we did at Hilton Head. two and three companies in a place. The distance out is about ten miles and we stay ten days insted of a week as at the head. This morning there was another review of all the troops on the island. similar to the one I wrote you we had several Sabbeths ago. There was no notice given of it. our company was out target shooting. an orderly came out and told us that there was to be a review at ten o"clock. we had barely time to get back to camp and black our shoes and brush a little of the dust off when it was time to march out. This afternoon there is no drill and amid preparations for an early start tomorrow morning. and some time between now and then. I must manage to write a bit of a letter to you. We have had very cold weather this week. although it has been very clear and not very windy. writing except in the middle of the day when we wer buisy. or it was inconvenient to write. has been out of the question. Last night I think was the coldest we have had since we have been South. At any rate the ice was thicker this morning than I have seen it before. I dread going out on picket on account of the cold. I suppose as we shall be near the enemy. no fires will be allowed when on duty. Wer it not for the cold I should rather like the change from the dull routine of camp life. For it has been. and is. very dull here. we have not had mails as frequently as formaly and so not having news from abroad to talk about. and speculate on. we have been forced to supply the place of it by all sorts of flying reports and vague rumors. We have news of the <u>occupation of Fredricksburgh</u> by our forces. it was brought us by a gun boat. A number of men wer discharged from our regiment and sent home last Saturday. Among them was Gardner of Co F who came and enlisted at the same time that Wolcott did. Wolcot is on the sick list most of the time I guess. I seldom see him on duty. He does not look as bad as at one time. The <u>47th Pa regt</u> left here the other day for Key West. I hear that we have been placed in the brigade they wer in. in thier place. if it is so (and I think it probable) we shall be more likely to remain here for awhile than we otherwise should. Nothing has yet been seen of our expected reinforcements. and though large preparations are makeing for them at the Head. yet I think it very possible that they do not come after all. One Company of the <u>28th C.V.</u> are at the Head. the vessel they wer on being one of those that put in for repairs. It is not our Salisbury Co. but I learn that Judson Ham and a young French who I conclude is from the

Falls, is with them. I think it very doubtfull, Dear Sara. if I am able to write to you much while we are out on picket. What with the fatigue and cold weather I think we shall seldom feel able or willing to write. But I shall not forget you darling. Last night I dreamed of you. Dreamed you wer sitting with me and my arm was around you as of old. It was a pleasant dream. but was of short duration and I quickly woke to realize you wer hundreds of miles away. and that it would be long ere I could agan recieve your fond embrace. Trust me darling. I love you fondly as ever. I feel that time. and abscence. have but strengthened our attachment and that God willing I shall yet return to you. as your own

Ned

No. 92 Rec'd Jan 7th 1863

<u>Port Royal Ferry S.C.</u> Dec 23rd/62

Dear Sara.

As I wrote you in my last that we expected to do. we started betwen seven and eight last Saturday morning for this place. and our ten days picket duty. The morning was clear and bright and the weather beautiful. as has been the case since we have been here. though the nights are sometimes very cold. The distance out here is ten miles. but the road is excellent and we arrived here a little before noon. There are no buildings on this side of the Ferry. so we are liveing in our tents. The men for duty on guard that day wer immediately detailed. and at once set off to relieve those then on duty. It was George's bad luck to be one of them. but he came out the next morning all right. The rest of us set to work and pitched our tents. and proceeded to make ourselves as comfortable as possible. Saturday night was very cold. and the men on picket not being allowed any fires suffered considerably. Sunday I hoped to have had a day of rest. and a chance to write to you. but the officers took it into thier heads to move our camp a short distance to a more sheltered position. and so what with leveling off the cotten ridges. prepareing the ground and moving out tents. it kept us buisy all day. In the afternoon the mail arrived. and I recieved your letters of Dec 8 & 10th. and also one from Mrs. Randall. Monday morning it came my turn to go on picket. from which duty I was relieved this morning. so that I am now improveing the first opportunity to write you since leaveing camp. The city of Beaufort is situated on the lower end of Port Royal island. At the upper end this island is separated from the main land by a narrow creek or river. the channel of which at the Ferry is prehaps eighty or ninty yards wide. A marsh extends back for a quater of a mile on each side and at high tide is entirely overflowed giveing us then about half a mile of water betwen us and Secesh. At the Ferry however the road is built across the marsh on both sides clear to the channel. so that we can approach thier pickets within less than a hundred yards and have a good chat with them which is done about every night. In the day time our men do not go out on this causway. but remain close to the shore going out to the end of it after dark. George was on that post the first night and from the rebel pickets we that night learned of Burnsides' disastrous defeat. which was confirmed by our own papers the next day. My post yesterday was about half a mile to the left of the causeway. and of course there was the half mile betwen us and the rebel picket which was posted directly opposite. But the distance did not prevent us from talking across to each other which we did several times. No effort was made by either of us to keep out of sight. and when talking we would mount a stump or anything that would assist us to hear the better. Our conversation did not amount to much as the distance rendered it difficult to carry on a connected talk for very long at a time. I believe it commenced by thier asking us "what we had for dinner" upon replying. they invited us "to come over and they would give us some fresh beef and sausages and plenty of "old rye." meaning whisky. of course it was impossible for us to comply with thier hospitable invitation. notwithstanding the inducements held out to us. Then followed a number of questions from them such as. "Do you have any whiskey." "Do you have plenty of tobacco." What does it cost." How long since you have been home" "What regiment do you belong to" Don't you wish the war would end." "When are you comeing over agan to see us." To the last question we replied at random that "in about two months." They answered that is to long to wait. They saw us reading and inquired if we had a New York Herald. what the news was &c. This morning they wanted to trade overcoats. At the Ferry last night our folks had quite a long talk with them discussing the cause of and probable termination of the war &c. Our camp is about three quaters of a mile from the Ferry. there are three companies of us here. A. F & G. and about half a mile back on the Beaufort road is the headquaters of the picket. where the officer in command stays. Co. I. is posted there. and also one section (Two pieces) of a light battery which is also a part of the picket force. The other companies of the regiment are stationed at different plantations along the shore of the island. to the right and left of us. and they throw out thier pickets each way. meeting ours. and thus forming a line of sentinels around this part of the island. There used to be a little shooting between the pickets. but there has been none for some time until a week ago last Sabbeth evening when the <u>55th Pa</u> regiment being on duty. thier officers got <u>drunk</u>. and ordered up the battery and commenced fireing across at the rebels. It is said that they fired nearly a hundred round. but I guess that part of the story is stretched. but at any rate they blazed away in thier drunken stupidity. throwing indiscriminately shot shell. grape and canister. Of course they did the rebels no harm. and onley injured ourselves by disturbing the peacible relations betwen the pickets and giveing them an excuse for passing over any of our men whenever they got a chance. But I believe they have not fired in return. and probably will not as we have not scrupled to tell them that the fireing was done by a lot of drunken officers. without any authority to do it. The onley connection of Beaufort with the mainland by land is by the means of the road I have mentioned and this Ferry. There are no buildings near the Ferry on this side. but opposite of us there are quite a number of fine looking buildings. If it is not to cold. the duty will not be very tedious and the ten days will soon pass off. Our commandant was stupid enough to order that no fires would be allowed on the outposts. but since the first night I guess the order has been genarly disregarded. and I hardly think he will attempt to enforce it. It is a foolish and useless order. for the men have had fires all the while before. and the rebels keep up good fires along the shore line of thier pickets. But our officers think that they know more than all the rest of our military leaders put together. and the consequence is that we are certainly as badly harrassed. and fooled. around as any other regiment in the service. It is singular to me how any man. not an idiot. can grow up to man's estate so utterly devoid of common sense. as are the most of our officers. I am happy to learn from your letters. that you are improveing so much in health. But do not be in a hurry to get shut up in the school room agan. Spring will be plenty time enough of resumeing to your labors. Am much gratified with your account of the entertainment at Tuppers. and happy that you wer able to participate in it. I am surprised to hear of Annie's dissatisfaction. had Mary preformed so I should not have been. as I have heard of such preformences on her part before. I suppose Annie thought she had good reason for her conduct. though you have not given me a hint as to the cause of her "<u>miff</u>". I recieved a paper from you on Sunday. also one from Martha. I have sent you and Martha both copies of the New South and should like to know whether you recieve them or not. as I have been told that they wer very often stolen when done up and sent as papers. I must write to Martha. do excuse me to her if you can. I will try to write while we are here. If Mr Frank wants to hear from me. let him write. I am sure that I wrote to him last. The high prices of the commonly cheap articles of clothing I should think would make it difficult for poor people to clothe themselves comfortably. Have fine goods increased in cost in the same proportions? I am finishing my letter on the morning of the 24th. Tomorrow will be Christmas.

MY DEAR SARA

I may as well wish you a "merry one" now for I shall be on guard tomorrow and shall not have time. Last year I was on picket on Christmas down on old Tybee. and thought surely I should be home if alive before this time. But I ought to be thankfull that I am alive and that it is as well with me as it is. and I will try to be so. I recollect that I was writing to either you or Sara B. it was a nice warm day such as tomorrow promises to be. but the night before was cold as the mischief. The <u>46th N.Y.</u> regt was on Tybee with us. it is a Dutch regiment and you know they make a great deal of Christmas. Well <u>they</u> did. they had plenty of whiskey and I guess the whole crew got as drunk as was necessary to be. to pay proper respect to the occasion. I know they had large bonfires and made night hideous with thier revelings. By the way we have a half a gill of whiskey night and morning while out on this duty. I expect to be quite a whiskey drinker by the time I get home. But you can't preach to me. this time. for by your own account you have made the brandy and wine suffer severely lately. I think there will be a chance to send this letter to headquaters from whence it will be taken to the P.O. in Beaufort. very soon. And so with much love to you my darling. will bid you good by. As ever your own.

Ned

LETTERS FROM EDWIN JANES BARDEN TO HIS FIANCE
AND LATER HIS WIFE, SARAH MARIA JONES,
WRITTEN DURING THE CIVIL WAR

SECTION 3 – 1863

No. 93

Camp Palmer Beaufort S.C. Jany 1st/63

Dear Sara.

The first day of the new year. May it be a happy one to you and through it may our kind Father in Heaven preserve you from all evil and adversity from sickness. sorrow. and suffering. is the prayer of your own Ned. I was disappointed about going on picket on Christmas. as I wrote you I expected to. I began to feel unwell soon after finishing that letter. Attempted to write to another friend. but scarcely got through the salutation and gave it up. The next day I was not able to do duty. no better the second day. and the third I returned to camp. I caught cold nearly four weeks ago. I tried to be careful and get over it but it seemed that the more cure I took. the more cold I caught. I can"t begin to be troubled with a cough. which I have not yet been able to get rid of. While on picket I caught an additional cold. and for several days had a little fever. That I have pretty much disposed of. but the cough still troubles me. Am also quite week. though able to be out and around. The Docter sayes my cough is not one caused by any living difficulty. but by <u>Malaria</u>. So if that be true you neede not be alarmed. for I shall doubtless soon recover from it. It is probably as the Docter sayes. but it is differant from what the Malaria troubled me last winter when on Tybee. Doctor Jarvis has been doctoring me for a few days first. but I expect to return to the care of one of the other surgeons tomorrow. If it is to <u>Dr. Hine</u> I do not care for he has always been kind to me and done all he could for me when I have been under his care. but the other fellow. Porter by name. would scarcely be recognized as belong to the human family. except by his form. and that like his heart is of very small proportion. I like Dr. Jarvis very well. The little I have seen of him. I think there is in his accountments a family resemblance to the previous Mr. Jarvis whose cousin he is. but he is a much finer looking man. apparently fond of the good things of this world. and of convivial taste. Our new Quatermaster and our second lieutenant came on the last steamer which did bring a male. but said male did not contain a letter for me from you. Strange. but true. I onley hope that nothing has happened to you. for since your illness I am continualy fearfull you will be taken down agan. hence I am a little uneasy at not recieveing the usual letter. or letters.

You must not expect any news. for there is absalutely nothing of the kind here. The regiment returned from picket yesterday afternoon. Today the men are employed in fixing up thier tents and makeing them as comfortable as possible. Tomorrow there is to be a regimental inspection. which cause the men a good deal of labor to prepare for. When out on picket as for the last ten days. but little care of guns and equipment is taken except to keep the former in good shooting condition. So that on our return to camp there has to be a general cleaning up. and blacking of shoes and belts. and polishing of bayonets and brasses. The camp also has to be cleaned of the accumilated dirt and rubbish. The members of Company K. in our regiment presented thier Captain today with a fine and new sword. costing a hundred and forty five Dollars. it is a gift from the members of the company. They are exceedingly fortunate in having an excellent Captain who always treats them well and like men. The forces on this island are now under command of General Seymour who has lately arrived and assumed the command. no new General has yet come to take command of the 10th Army Corps and General Brannan still retains the command. The reinforcements we expected have not come either and I learn that the work on the large new store houses has been suspended. Mr Landon is here today having come up from the Head for the purposes of being mustered in with the rest of the Co for the last two months pay. He appears to be in good health and excellent spirits. He was quite unfortunate in losing his watch. which was sent him from home in a barrel of apples. He failed of getting either I believe. He valued it very highly on account of it having been given him by his Father on his death bed. President Lincoln's Emancipation Proclimation goes into effect today I suppose. where it can. The negros in this region are assembling at Beaufort for a general holiday. I am unable to determine in my own mind as to the wisdom. and propriety. and probable effect on the present struggle. of that proclimation. That it is right I do not for a moment doubt. with regard to its propriety at the present time I am not so sure. Wolcott has lost a barrel of stuff that he had sent for to speculate on. So I am told. I cannot feel much sympathy for him if that be the case. Quatermaster Tompkins has expresly declared that he could not trouble himself to transport things free. to the soldiers. for that purpose. But anything that was needed for thier own use and comfort he would willingly forward to them. So you see that if every one is going to speculating and smuggling his goods through in that way. by and by it will be found out. and we shall get no more things forwarded to us in that way.

Jany 2nd.

A male arrived last evening and I recieved your letter of the 20th also one from Rev Mr Edgar. your apoligy for not writing more and oftenmore is satisfactory and if it wer not I could not complain. for I am under the same charges to much of the time myself. Should love dearly to be present at the concert. but fear that my engagements are such that I should be obligated to deny myself the pleasure. I learned that the two Salisbury boys Ham and French wer on one of the boats that put into Port Royal for repairs. they landed and remained a few days. untill the vessel has made the necessary repairs. There was only one Co of the 28th on board. and that not Salisbury Co. the remainder wer Mess troops. Ham and French wer off over to Brooklyn when the regiment embarked and when they got back they wer onley in time to get on board this boat. not able to join their own Co. French had been sick on the way down. and fought it awfull tough. but I guess from the manner in which Mr Landon spoke telling me of it. that he did not think he had been very roughly handled for a soldier. Mr L. did not say that Ham was at all unwell. They must have queer works in that reg't, I should think from I hear. If they wer under our officers for a year and <u>ground down</u> as we have been. they might grumble. and I guess they would raise a hugh cry of distress. But they may catch it yet. It is pretty evident that Uncle Sam is going to get his moneys worth out of the nine month men if feassible. I believe there is but little sympathy for them by the older troops. as they are looked upon as being bought. and driven. into the service. This feeling is wrong. but soldiers are not always as considerate as they should be. The event may prove that the men last raised may do the best fighting. Several officers wives came down on the last steamer but one that arrived. It looks more and more. like our remaining here for a long time unless reinforcements should come. I presume if there is no forward movement made in this quater we should winter at Beaufort. Do not think that I am short of paper. but having spun this letter out much longer than I at first expected. I have added on. from time to time as "occasion required." This is a beautiful warm day. but it was very cold last night with a heavy white frost. Shall try to write you agan within a week at least. With much love.

Ever your own
Ned

No. 94 Rec'd 18th Jan 1863

Camp Palmer
Beaufort S.C. Jany 6th/63

Dear Sara.

If I could onley take my seat beside you this morning and have a good long quiet chat with you. I have no fear that we should be troubled for subjects of conversation. but to sit down and write you a letter with nothing at hand in the shape of materials to make it out of. is not so easy. I am still on the sick list on account of my cough. which I am happy to say is not so bad as when I wrote you last. and I am in hopes to be soon rid of it. Aside from the cough I am feeling very well. though not Strong. It is raining today. the first rainy day we have had in a number of weeks. Last Sabbeth I have obtained premission to go downtown to church and though the wind was blowing pretty fresh I put on my overcoat and went. I attend at the Episcopal Ch and heard a good sermon from <u>Eph 6 & 10th</u>. I think the clergyman was an Episcopalian but he did not read the service in Litany as usual. This was the first time since leaving New Haven that I have heard any but Mr Wayland preach. This man preached to. and about. the congregation instead of the negro. and the change was most acceptable. The onley other church in town is Baptist. and is used half of each Sabbeth for the benefit of the negros. I reached the church sometime before the service commenced. and employed the time in looking around the church yard and examining the inscriptions upon the grave stones. Many of them are very old. There are the graves of the Barnwells'. Rhetts'. Elliotts'. Stuarts'. Draytons'. and other names that are familiar in S. Carolina history. better had it been that the now living representatives of those once honored names had been laid in the old church yard in thier infancy. while yet they wer innocent of the great crime of treason. than that they should have lived to be the perpretratons of such crimes. and the cause of so much misery and suffering all to satisfy the cravings of thier damnable ambition. But He who seeith not as man seeith has decreed otherwise. and they are premitted to make thier names a loathing and contempt to all lovers of free institutions and thier country. In one side of the old churchyard close to the wall grows a venerable ceder. It is one of the largest I ever saw of the kind. Several graves are beneath it. I picked a small sprig which I send you. A soldier was gathering some seeds from what looked to be a kind of briar. but he said the flower was beautifull. He was going to send the seeds home he said. So I picked one of the large berries and will send you the seed. and if you care to you can plant them in the Spring and see what will come from them. I have answered several letters since I wrote to you last. and have been reading the life of Martyn and <u>General Havelock</u>. I dare not call myself a christian when I compare myself with those truly Godlike men. Georges' barrel of apples came last night they had rotted pretty bad. and the head had been broken in and quite a number taken out.

Dear Sara. I hope you are enjoying yourself and that your health is fully restored. I think of you very much lately. and the probability of our meeting again. It looks dark. not very dark now. and scarcly a ray of light is to be seen. I get almost discouraged some times. but then how much better off I am than thousands of others. I ought not to complain but it is hard very hard to be contented. amid so much that is unpleasant and disagreeable. With love to you my darling I remain every your own.

Ned

CIVIL WAR LETTERS 1861-1865

No. 95 Rec'd 22nd 1863

Hilton Head S. C. Jany 11th/63

Dear Sara

Here we are agan on the same old camp ground from which we fled a little over two months ago. to avoid the threatend pestelance. It is Sabbeth evening. I cannot sit by your side this evening and tell you all that I wish to. and so I will do the next best thing. write to you. It is not very quiet. in the next tent the men are having a sort of a miscellanious sing. That is singing anything they happen to think of. Across the street in another of our Salisbury tents the boys are singing sacred music. In our own tent. George is writing. Olin is reading. and Reid is out and in just as it happens. I do not feel exactly like writing tonight but as I am now much better. and able to do light duty during the day. I shall not have as much time to write as I have had for a short time back and therefore do. not feel at liberty to let pass the present opportunity to at least commence a letter. I really thought that we should probably remain at Beaufort during the winter. or at least during our period of inactivity. and wrote so to several of my friends. and I think to you. But alas for a soldiers calculations. just as we had settled it in our minds that we wer fairly in winter quaters. comes the order to pack up our worldly effects and prepare to leave for this place. We have been quite moderate in our move this time. oneley half of the regiment going at a time. and takeing with us everything that we had accumulated for our comfort and convenience. The right wing of the regiment (Five companies) left Beaufort on the morning of Thursday. the 8th. and we. the left wing. spent the day in packing up and prepareing for a start the next morning. Friday morning we commenced getting our stuff down to the dock. and on board of the Planter. this was not accomplished until two o"clock. at which time all being on board we bid good by to Beaufort and set our faces towards this place. arriveing at the dock here at four o"clock. We immediately landed and started for this. our old camp ground. The day had been fine though cool. but on arriveing here we found the wind blowing a hurricane. and the sand flying in clouds. as onley Hilton Head sand can fly. It looked pretty blue for us. and our nights lodging. for it was to late to attempt to unload our tents and get them on the ground. but by getting a few new tents from the Quater Masters and borrowing some old ones from the companies that had preceedid us and wer already snugly in camp we managed to all get under cover for the night. By the kindness of the same companies we wer also provided with coffee. and thus fortified we passed the night quite comfortably. All day Saturday our stuff was comeing from the boat and night found us with not more than half of it on the ground. However the tents came. and we wer sheltered for the night. Today has seemed like any day but the Sabbeth. The remainder of our things have all arrived from the boat. and we have been buisily engaged all day in pitching our tents. fixing our bunks so as not to be obliged to lie on the ground and makeing our tents as comfortable as possible. The star of the south arrived on the morning of the 9th. and when we reached here that evening we found our mail awaiting us. I recieved your two favors of Dec 27 & 30th. and two papers. for which as ever I am very gratefull. our chaplain also arrived on the same steamer with the mail. and I had the pleasure of greeting him before I got off the dock on the day of our arrival. He has brought his family with him. This afternoon we had a short service on the parade ground. The text was the <u>90 psalm 12th vs.</u> Mr Moore has been up and made us quite a visit. staying to service and until after dress parade. We are now nearly settled but it will be several days before all will be in "ship shape" agan. and until then we shall all be more or less buisy. We have moved many times more stuff this time than we ever thought of doing before. All our boards. bunks. and boxes wer brought for us. To see the stuff you would wonder how anybody having thier reason. could ever think of moveing it. but onley a soldier can realize the value of every bit of board. and old barrels. and the like. The bunks in our tent for instance are nothing but the staves of an old flour barrel nailed across a couple of poles. and would not be worth at home over six and a quater cents. but we took pains to tie them carefully together and have them brought with us. It is the want of the thing. not the worth. that makes such compareatively valueless things of so much importance. Had we not brought our rude bunks with us. weeks might have passed before we could have got a chance to buy barrels to replace them. and in the mean time we should probably have lain upon the ground. now we have them all in shape and flatter ourselves that we are quite comfortable for soldiers. It is not likely another such an opportunity for a "clean sweep" will fall to our luck agan while we are in service. It was onley the shortness of the distance and the having plenty of time to move in. that premitted it this time. Quite a number of new buildings have been erected here since we left. and more saloons and sutlers stores have been opened. Mr Moore tells me that there are a good many ladies here now. a great many of the officers having sent on home and had thier wives join them. Our camp ground which we left nice and hard with a good turf on it. we find all cut up and turned into a bed of sand. It has been used as a drill ground for the artillery which has done the mischief. I look for a second Tybee season with the fleas which already commenced a vigorus attack upon me. Thus much for our present situation. now for answering your letters. but just here I will say that the Star of the South. has sailed for the North today. I do not know when the next mail will go North. <u>the tatoo is now beating</u>. soon it will be "lights out" and so reserving the answer to the letters for another sitting. please consider yourself kissed. wished pleasant dreams. and a good night.

Monday Evening: Alas for a soldiers calculations. I think that we can realize the truth of the passage of holy writ. which says that "we have here no continueing city." This afternoon we received orders to leave this place for Fernandina Fla tomorrow morning. George and I have both been buisy all day fixing up and around our tent. we had just returned from the woods with each a back load of poles when we recieved the news. This time we shall move no such amount of truck as before. No boxes. are to be carried for the men. so that we can carry nothing but what we can put in our knapsacks. We are to relieve the <u>9th Maine Regt</u> which has been posted at Fernandina since its occupation by our forces about a year ago. There is also one company of engineers there. and that I believe is all the force that is posted there. Col Rich of the 9th Maine has been dismissed from the service. and as the regt is in command of onley an acting Lieutenant Colonel. I believe it was thought not proper to leave the command of the post in the hands of an inferior or subordinate officer. hence the sending of our regiment here. I know but little about the place. it is a village or I presume the Southerners call it a city. Fort Clinch is situated about the premises somewhere. and is garrisoned by part of the regiment. It is I believe a healthy place. there was no yellow fever there last summer and onley that it is an out of the way place. and that we shall always be several days behind the "Fair" in everything. I think it might prehaps not be the worst place that we could be sent. Am exceedingly obliged for the description of your Christmas. you must have had a very pleasant time indeede. by my last letters. you will see that I did not enjoy festive season quite as well. still I would not complain. I have great reason to be thankful to my Heavenly Father. that my unprofitable life has been spared through another year and that I am now under such favorable circumstances. I am sorry the concert was so slimly attended and am at great loss to account for it. Concerts are not want to be so poorly patronised by Salisbury people. and especialy when thier own singers give them. I know of one who would have been most happy to have been part of the audience. I have no doubt that Prof Taylors commendation was justly merited. for I have heard the same ladies (girls) sing. myself. and know them to be good singers. Am glad that you are not to return to school agan this winter. Enjoy yourself to the full extent of your opportunities. and the best of your ability until Spring and then will be time enough to begin to search for employment. if you are able to engage in any. Shall be glad to recieve the ambrotype. and am now considerably interested in the arrival of the box. Hope that you have told me the true reason for not having a new one taken. and that it is not for fear that I should see that you are not looking so well in health as formerly. Shall be quite satisfied with what you have sent me if I succeede in getting it. Next to your own dear self. what better could you send me than <u>your shadow</u>. Shall not try to write much more tonight. the mail leaves this place for the North tomorrow night. this is the last letter that I shall write you from Hilton Head very soon. <u>probably</u>. I feel as though I was going further away from you. though in reality the distance is of little account but I must think of you all the more and I know that I shall not love you any the less. I will write agan as soon as we get settled in our new quaters. and tell you all I can about them. Short as the trip is down there. I cannot scarcely step on board of a boat without being sea sick and uncomfortable while there. But it is much better to move as we do. than to be obliged to march and pack our knapsacks as our brethren in other places have to do. always providing that we have a pleasant voyage and arrive safely at our destination. Now darling with a good night and good by. I must close. May our Heavenly Father protect. preserve. and bless you is the prayer of your own.

Ned

No. 96 Rec'd Jan 29th/63

Fernandinia Fla
Saturday Evening Jany 17th/1863

Dear Sara.

Im pretty tired this evening but thought I would write a few lines and make a commencement of a letter. <u>The Steamer Delaware</u> on which we came here returned to the Head yesterday afternoon having on board the <u>9th Maine</u> which regiment we relieved from duty here. Many of the boys managed to write letters in time to send them back by the steamer. but I could not. and as it may be a long time ere we have a chance to send even to the Head. agan I do not propose to hurry this letter. but prehaps write in it as I have opportunity for several days. I must go back to last Tuesday morning. We commenced packing up and prepareing to leave directly after breakfast. At ten O'clock we struck the tents and several teams are buisy takeing the baggage down to the wharf. I got my knapsack which was pretty heavy carried down on one of the wagons. At noon I obtained premission to go down to the boat. not waiting to go with the regt. which I supposed would soon be along. We had a number of little necessaries which we wanted to take along with us and I put them in a water pail which I took in my hand and brought safely through with me. The baggage arrived on board the boat very slowly and it was not untill nine o'clock in the evening that it was all on board. On my way to the dock I called to bid Mr. Landon good space by. and much to surprise found him expecting to have to go with us. but not quite certain. He had recieved information that Col Hawly had required him with several others out of our regt who had been employed in various duties around the Head. to return to the regt. but the Superintendant of the contrabands did not want to have him leave. and had tried to retain him. While we wer talking over the probabilities of his being obliged to go. the order came for him to rejoin his company. That settled the matter. and I left him that he might pack up unhindered. I was sorry that he was obliged to come. for he was very pleasantly situated. and he is not fit to rough it around and do duty with the rest of us. Mr Moore did not have to come with us. he came down on the dock to see and bid us good by. Dexter has a pleasant situation in a hospital at Beaufort. and will not probably join us. It was dull work waiting around all the afternoon and evening. but at last all was on board and then the regement went aboard. It was about ten o'clock when the boat left the dock and anchored out in the stream. At three in the morning we hoisted anchor and commenced our journey. I spent the night upon deck as I usually do. rolled up in my blankets. slept a little but as I lay close by the engine room there was to much going on about me to permit me to sleep much. I prefer to stay upon deck during these trips unless it be very stormy on account of the air being so foul between decks. There was oney a slight breeze during the day. not enough to roughen the water. but oweing to the high winds of two or three days preceding there was a heavy swell on that gave the boat a disagreeable motion. I was as usual very uncomfortable and glad enough when the boat reached the dock here which was not until six in the evening. We landed immediately and wer quatered for the night in unoccupied buildings. Four companies of us wer put in the upper room of a large store house. While eight officers occupied nearly the whole of the room below not allowing a soldier to come in with them. We had scarcely room to sit down. Many of the men went out and slept else where. so that we were able to squeeze down after a fashion. We made our supper out of such things as we had brought in our haversacks. At daylight the next morning myself and twelve other men from our company joined a fatigue forty to unload the boat. which was accomplished by noon. and the fatigue of forty dissmissed. Coffee was prepared for us who wer on fatigue and with bread and butter we made out a passable breakfast. in that getting the store of the others. for coffee was not made for us all until late in the afternoon. The afternoon was necessarily a buisy one. pitching tents and making them comfortable as possible. The regement we relieved has been quatered in houses all the time they have been here. and have had things in excellent shape for soldiers. Our boys growled a good deal at not being allowed to go into the houses. but of course that was unavailing and useless. I believe that the Maine boys wer expecting to be obliged to go into tents soon. as it is said that the property they occupied was to be prized and sold. Fernandina is quite a village. nearly as large as Beaufort. It is situated on quite high ground for this country. Our camp is on the side of quite a hill just in the edge of the village. The place is quite new. Its principle builders I believe wer <u>ex.Senator Yulee</u>. and an <u>Irishman named Finnegan</u>. the same man who ownes a large interest in Jacksonville and of whose fine house there I made mention of in my letter to you some time ago. He is now a General I believe in the rebel army. He has also a fine house. beautifuly situated here. but at present I believe it is occupied by negros. Most of our officers are quatered in buildings. Our Colonel. Lieut Colonel. and Chaplain. and one Captain have thier wives with them. and other officers are expecting thiers by the next steamer. Ft. Clinch is situated at the entrance of the channel leading to the town and is about three miles distance. Prehaps it as our way up. I have not visited it yet but hope to get premission to do so after we get fairly settled. It is not finished yet. Mount at present twenty odd guns I am told. It is garrisoned by a company of the <u>N.Y. Vols</u>. Engineers and one company from our regiment. <u>Major Rodman</u> has been placed in command since our arrival. before it was commanded by the Captain of the engineer company. The rebels removed all of the guns that wer in the Fort on our approach leaving nothing but a couple of old guns utterly worthless and that would not stand fireing. Nearer the town they had commenced some earth works and had mounted two or three good rifled guns. These our men captured. and the works they commenced are being finished. They are close by our camp. and are to mount three guns. There is a gun boat stationed here all the time. The town is situated on an island and I believe that our boat can sail clear around it. There is a rail road termainateing. but where it goes to I do not know. it crosses the channel by a bridge and as there is a draw in the bridge I suppose vessels sail that way around the island. The rebels burned the bridge and escaped with I believe five trains of ours but we captured one train. the gun boat throwing shell at it as it was running off. one of which hit the locomotive and disabled it. This train is now used to run out to the bridge and carry provisions or whatever else is wanted to the company which is stationed at the bridge on picket duty. Co F. is there now. Co. I. is in Ft. Clinch. and Co. C is assigned as provost guard. and have pitched thier tents near the centre of the village. The town of St. Marys is in sight across the marshes distant apparently eight or ten miles it is situated on the main land. and at the mouth of St. Marys river. it is occupied by the rebels though I believe the 9th have made one or two excursions over there.

<u>Monday evening</u> Saturday was spent in working about camp tearing down buildings and carrying boards to camp to floor our tents with. Friday night was rainy with a high wind which has continued through the day the weather being quite cold. The 9th Maine left this afternoon on the same steamer that brought us here. first after they had left. a house that had been occupied by one of the companies was found to be on fire. It was a good sized two story house. The fire broke out in the roof. There wer no conveniences for putting it out. and no effort was made to do so. other buildings nearby wer in danger on account of the wind. but the tearing down a couple of outhouses standing near was prevented from spreading. Some think that the Maine boys set it on fire. and it is not unlikely. but still it would have been very easy for it to have taken fire accidently. Sunday morning we had the usual inspection which with the time spent in preparing for it occupied most of the forenoon. In the afternoon we had service which was held in one of the churches. The <u>text was Ex. 20. 3rd.</u> The wind and cold still continued unabated. The Maine boys said that they had seen no frost this winter. Had they been here Saturday night they would have seen some. I spent part of the day in reading Paradise lost. Today (Monday) has been quite windy but much warmer. and tonight the noise of the frogs in the marsh nearby remind me of a warm April night at home. There are gardens with peas in blossom and other vegetables growing finely. We have had four hours of drill today. and as it is the first day I have drilled in the long time I am pretty tired. Mr Landon has been detailed to issue the Government rations provided for the contrabands. Wells is detailed as nurse in the hospital. One of the Maine boys left sick in the hospital died last night. and was burried this afternoon. There are a few white families here. but I cannot learn anything near how many. A plenty of negros of course. There are several <u>Sutters Stores</u> and Saloons where most anything that a soldier wants can be had for about double what it is worth. There is a bakery. so that we are still supplied with fresh bread daily. I believe I have now told you all I know about the place and our present situation that will be likely to interest you. One thing I had nearly forgotten. Georges box came on shore at the Head the day we wer leaving. He was lucky enough to get it and put it aboard the steamer and it came down with us. The things wer in fine order. onley a few of the eggs wer broken. Many thanks for your ambratype. it is a very good picture and will last me a spell. but I fear the case will get broken in some of the rude knocks it will be likely to get in a soldiers jacket. but it must go with me. and if it cannot stand the rough travelling. I shall send for another when it gives out. Am much obliged to Mrs. James for the dried fruit it is

most acceptable. Please thank her for me. One or two things in your last letters require answering. I should be pleased to send Martha my photograph. but it will now be impossible for ought I can see. I was in hopes to have remained at the Head long enough to have had some taken for I learn that they take very good pictures there now. But you see I am disappointed in that. And there is no guessing when an opportunity will be afforded agan to have a picture taken agan. I will expect you to send me a Bible the one I brought with me has nearly given out. You could send one by mail at but little expense by leaving the wrapping loose at one end so that it could easily be seen that it was a book. But there is no neede of being in haste you can wait if you choose until someone is sending a box to one of the boys and put it in.

    Wednesday Morning   Have nothing new to write this morning but the mail closes at six this evening and I must finish my letter while I have the opportunity and have it ready. We have now been here a week and begin to get settled. I think we shall like it pretty well when we get accostomed to the place. The duty is going to be pretty hard on account of so many men being detailed out of the regiment on various duties connected with the post. The new duty neede not be hard. there is no necessaty for it but our officers will insist upon having more men as guard. on having more useless work done. and in drilling more than any other regiment that I have yet met with. One thing we are going to miss very much and that is our former frequent communication with home. The Maine boys said that they wer frequently a month without recieving a mail. and sometimes then two or three would come all together. There is probably one at the Head for us now. as a steamer was expected in a day or two after we left. We await its arrival with much interest. for the two great battles in the west were in progress and both undecided. when our last mail was recieved. I may not write you as many letters as formaly. but will try to keep the amount good. by lengthening those I do write. It is doubtless going to be pretty dull here. for aught I can see. nothing but the steady routine of camp life. It will be uninteresting work but will be better than bloody fighting. With love to you my darling I remain as ever your own

                                                                             Ned

No. 97 Rec'd Feb 10th 1863

                                                Fernandina Fl. Jany 26th 1863

Dear Sara.

    After an interval of Sixteen days the much desired mail has arrived. It came in late this afternoon and is not all distributed yet. I have recieved with great pleasure your favors of Jany 5. 7. & 12th. also one from a sister in Illinois from whome I have not heard directly in eight or ten years. and one from Carrie Randall. I may get another one or two yet. for in the hurry of distributing the mail on its arrival. mistakes are often made. letters. getting strayed away into the wrong company. None of the papers have yet been distributed. But I have read with delight the three letters and as there remains a few moments before roll call I cannot better improve them than by commencing a letter to you. And first of all about myself. I am quite comfortable. my cough still continues to trouble me some. mostly nights accompanied towards morning with a difficulty in breathing. not serious but enough to make it rather unpleasant. I apprehend no danger from it. but think that soon as it becomes more settled weather it will wear off. I have been very fortunate thus far in being under the care of Doc Hines who has always treated me kindly. My general health is good and for some time I have not asked to be excused from any duty except guard. which on account of the necessary exposure in the night. I did not wish to do. Doct Hines has readily granted me this. Yesterday (Sunday) morning. I attended the surgeons' call as usual. for some reason Doct Hines was not present. and Doc Porter attended our company. I told him my trouble and that I had been excused by the Doctor from guard duty. he prescribed for me. and did not excuse me from any duty. I had scarcely got my breakfast down. when the seargeant came to me and ordered me to go on guard. I thought first that I would not go anyway. that I would first see the captain. and if he would not excuse me I would hunt up Doc Hines who I knew would. But finely thinking that I thought as little of the captain. as I did of the seargeant. and dislikeing to ask any favors of him. and thinking that I would not trouble the Deeter. I put on my traps and started for the guard tent. a good deal more angry than I care to get every Sabbeth morning. The day was damp and foggy but I got along very pleasantly and do not think I am any the worse for the days duty. Today has been fine. I spent most of the forenoon in looking around the village. this afternoon we drilled a little while and have spent the rest of it in waiting for the mail. We are pretty much settled down. and begin to consider ourselves at home once more. It is going to be dreadfull! dull. We drill four hours a day except Fridays and Saturdays. on Friday we drill five hours. having a battallion drill of three hours in the afternoon at which all the cooks. ordelies. clerks. and extra duty men who are excused from all other drills. are required to be present. On Saturday we have no drill at all. This last is no favor for which we are indebted to our Colonel. for it is according to the army regulations. But we have always been obliged to drill on Saturdays except a short time last summer when we wer under the command of Lieut Col Gardner. The men find much fault with being obliged to drill to much and not without reason. For there is not a shadow of excuse for it. We are not raw recruits whome it is necessary to render effective as speedily as possible. but have been in the service over sixteen months. and there is nothing more for us to learn. except the bayonet exercise and that is of but little importance any way. Every man of sense knows the regiment will drill much better and learn more. drilled onley two hours a day than when drove through a four hours performence. Another of the unjust acts we complain of. is that men are obliged to drill in the afternoon after comeing off guard. This is not according to Army regulations. which provides that men shall have twelve hours after comeing off guard. to themselves being excused from all duty during that time. But it is not of the least use to complain. any attempt of a private to obtain redress for any insult or injury he may have suffered is utterly useless and onley makes his own situation worse. such a thing as justice between officers and men except so far as the officers please to allow it is entirely unknown in this regiment. On Sunday the pickets out at the bridge became alarmed. thinking that the rebels wer about comeing down on them. They sent in word to camp. and Co A was ordered out and went down there on the cars in great haste. but finding there was no cause for the alarm came back. A number of men in the regiment have been practiceing together for some time to preforme as a negro minstrel band. During the past week they have preformed two evenings in one of the small churches. I attended on Saturday evening. the house was crowded full. mostly of soldiers of course. but there wer quite a number of citizens. among them a number of ladies. and officers both of the regiment and gun boat stationed here wer present. The preformance did not equal those of the <u>Christies</u> of course. but all things considered was very creditable. Some burlesques on our soldier experiances wer quite funny. The music was very good. but we had to take up with old songs. George is not very well. and does not seem to recover from the attack that he had last summer. He is at present helping do the cooking for the company as it is easier than doing the regular duty.

Monday. 27th Two gun boats came into the harbor yesterday afternoon. they did not come up to the town. but this morning they have run up to St. Marys and have been shelling the town and the woods near. A part of the negro regiment is on board of the Ben Deford which steamer brought our mail. and is now lying at the dock. and we suppose that the gun boats have also some of the negros on board. What they are up to is a mystery. but it is probable there will be a great <u>splurge</u> in the papers over the brilliant achievements of the negro brigade. No more letters came for me. but I recieved three papers you sent me. also some from Mr Randall. They will recieve propper attention when the letters are disposed of. I have succeeded in commenceing a letter to Martha. and should doubtless have finished it this evening. had not the mail arrived. I should like to write her at least a <u>respectable</u> letter. but I the truth is I do not <u>know enough</u> to write to I those with whome am not intimate. except it be on buisness. I had anticipated the pleasure of sending those little papers frequently to my friends. but shall not be able to procure them now. I fear that I conveyed a wrong impression when I spoke about the apples. which I hasten to correct. We perfectly understood that they wer for both of us. and used them accordingly. I should have used the word <u>ours</u> instead of <u>Georges</u>' when speaking about them. Hope you have not mentioned the mistake for George would not like it. and I should regret it. It gives me great pleasure to hear of your good health. But please remember that "pain in the side" and be carefull not to over work. recollect that you are an invalid. and though you may feel able to accomplish considerable. you will find that the strength and ability are wanting. Those pains in the side are my especial horror. they are so often the precourses of fatal disease. that I dread them as I do the plague. I suppose Miss Lizzie anticipates a fine time in N.Y. in which I hope she will not be disappointed. I should I have thought Miss I Bidwell would have been married before this. but do not know as have any reason to think it strange. I am to apt to think that everyone is as anxious to get married as I am. but it will never do to judge thus. you know Dear Sara you must not think that I do not miss you. and long for your loved society. True I may not have time to reflect so much upon our separation as you do. but there are times. and they are not few neither. when the truth of our situation comes back upon me in all its dreary reality. and I long. o how earnestly. to take you once more to my heart ad never never more let you go. I have made up my mind that it must be a long time yet before I can see you. and having come to that conclusion I try to reconcile myself to a long. long weary waiting" until the period of my banishment is ended. But remember I love you none the less. nor is the separation less painfull. General Hunter has arrived at the Head. Two of the new Moniters are also there. and I hear that two or three iron clads are up at Stono. I do not know what they are here for I am sure. no land forces have come. and I should not think any could be spared from what are now here for any new operations.

Afternoon. The fireing from the gun boats still continues or rather has been renewed for it ceased for a while. we can see the boats. but cannot tell what they are fireing at. I think of nothing more at present to write. With much love dearest I remain

Ever your own

Ned

No. 98 Rec'd Feb 13th 1863.

Fernandina Fla Feby 2nd/63

Dear Sara.

Last evening we recieved a mail. two letters fell to my portion. but neither was from you my darling. But as I recieved three from you a few days before. l suppose I had no buisness to expect another so soon. This morning I was excused by the Doc from drill. and I flattered myself that I was going to have a whole quiet day to myself and that I could write several letters. But I had not got my breakfast half eaten before I was ordered to go out on fatigue duty. The Steamer Delaware came in yesterday with a load of commissary stores and we had to unload her. It has taken us all day. Am rather tired this evening. but it is several days since I have written to you. and I must delay no longer to commence a letter. I do not know exactly where I left off in my last letter. so if I happen to write the same thing twice over. you must excuse me. The gun boats which I told you wer up the river shelling the country. landed the negro soldiers the night previous and they marched about four mile inland. intending to surprise a company of rebel cavalry or whose whereabouts they had become acquainted. but the rebel pickets discovered them and gave the alarm. and the rebels came out to meet them. they had a smart skirmish in which one of the negros was killed and seven wounded. They do not know how much the rebels suffered. but they think that some of them fell. The rebels retreated into the woods and the negros did not pursue them. It is said that the negro soldiers behaved well. As the boats wer about leaveing St. Marys the next day. some rebels came out of a house where they had been secreted and fired upon our men. A women in the house had assured our officers that she had not seen a rebel in a month. Upon being thus fired on. our folks turned about. landed agan and set fire to the place. burning it all down but a few houses that some how escaped. That rebel nest is pretty much used up now. Two days afterwards. the two boats proceeded up the country. entering some very narrow creeks. Up one of these they came upon quite a body of rebels who liked to have proved troublesome to them for the creek was so narrow that a tree felled across it would have stopped the boat. when it would Fortunately have been the banks comparitively of these easy to creeks have are boarded. not wooded. and taken but her. Fortunately the banks of these creeks are not wooded. but the rebels followed them. peppering the old boats with musket balls. by one of which the captain of one boat. the John Adams was killed. The negros captured half a dozen prisoners. who wer for a while confined in the Provost guard house here. Yesterday (Sunday) we had a regimental inspection in the morning. I was not well all the forenoon and after the inspection I lay down and slept about three hours. a thing which I very rarely do. In the afternoon. I attended service. The Rev. Mr Hill chaplain of the 3rd N.H. regt was present and made a prayer. Our chaplain preached from <u>EK 20 & 12th</u>. There was a prayer meeting in the evening which I ought to have attended. for it is the first time that I have had an opportunity of so doing in nearly three months. but Peter Turner came in to see me as did also two or three more of the boys. and so I allowed the meeting to pass without attending it. Some express came on the Steamer which we have been unloading today and George recieved his box containing his boots. Did I tell you that our second Lieut George D. Sanger had resigned his commission and was going home! Well he has and we are all sorry. for he was a good officer and always treated the men well. It is true he loves liquer and is apt to get intoxicated. but that is the trouble with nearly all of our officers. onley the rest of them are more fortunate in concealing it than he is. I am told that is the cause of his resignation. It is said that he became intoxicated when we came to the Head from Beaufort and that an old goofe of a second lieut. belonging to another Co whome we all dislike. prefered charges aganst him and he was obliged to resign or be court martialed. which would be likely to end in his dismissal from the service. We do not know who will take his place. but we do not think that anyone who is likely to get it will be as acceptable to us. as he was. A seargent in our company wants the place awfully. and is creeping around.the officers and eating all kinds of dirt for them. in hopes of getting it. He is despised by nearly the whole company. but is in favor with the Captain. I hardly think he will get the place though. The first Lieutenant of CQ B has also resigned. on account of ill health. he was one of the finest officers in the regiment. We have heard lots of news from the Head but do not know how much of it to credit. A large fleet was met by the Delaware just entering the harbor as she was comeing out. A valuable prize is reported to have been captured while trying to run into Charleston harbor. The attack on either Charleston or Savannah prehaps both is not far distant. The iron clads. and large land force. being concentrated in this Department. can be ment for no other purpos. I had thought that we should have no part in the expedition. and it is still very doubtful! But Col Hawley is anxious to be with it with his regt and his clerk told me a day or two since. that the Col was promised before we came down here. that whenever the attack was made. he and his regiment should be along. So we shall not be surprised any day to see a boat comeing in with a regiment to relieve us and we ordered into active service. This will be more likely to take place if the attack is delayed some time. but if it is made at once. I hardly think Genl Hunter will take the trouble. A Debateing club has been started in the regiment. called the

"Joe Hawley Debateing Society." The first debate is being held this evening in the church which we use for service. The question is whether whiskey is a benefit. or an injury to the army! by the stomping and laughter which I can hear from where I sit in my tent. I think they are having considerable sport over it at all events. I have been repetedly asked to go over. but am desireous of finishing this letter quickly as possible so have refused. I think of nothing more to write tonight. it is now almost roll call. Shall leave this unclosed for the night as I know of now main about to leave and I may wish to add something more on the morrow.

Tuesday Morn. It was decided last night that." whiskey is no benefit to the army." The debate was quite spireted and passed off pleasantly. Well we do not get whiskey enough to benefit. or harm us either one. A few times since I have been in the army I have drank my ration when I was confident that it did me good. but probably nine tenths of what I have drank. has done me no good and might better have been left alone. I do not apprehend any danger of ever loveing it. for the taste of what is given us is anything but palatable. I think that if I live to get home. I shall be ready to stop drinking. according to the special order issued in one of your late letters. I am rather vexed this morning. for I dreamed last night of seeing Lizzy. but for the life of me cannot remember that you wer any where around. I do not see what buisness my imagination had to be running off after other girls. and takeing the advantage of me in that way. when I was asleep too. and so could not control it. I frequently dream of you both together. but this is the first time that I recollect of your giveing me the slip. and leaveing me alone with madam. Don"t you think it was rather unkind of you and when I had spent most of the evening in writing to you too. In passing the house occupied by the Colonel yesterday I heard some one playing the piano. and my thoughts wer at once carried back to the good old times when I could seat myself beside you. and ask you to sing for me my favorite songs. Alas. I can do it no longer. May kind Providence soon restore to me the ability to renew those precious visits. and to make one which shall be lasting as life itself. I think you will feel quite lonely with Lib. and Julia both away. I had like to have forgotten to tell you that at length I have written to Martha. I would tell you what I wrote if it was not to much trouble. I was sadly puzzled what to write and I guess Martha will think so to. I presume she will show you the letter if you wish it. If she does. and any part of it does not please you. or is improper. just give me a "Candle lecture." Now darling. good by. with love. ever your own

Ned

No. 99 Rec'd March 4th/63

Fernandinia Fla. Feby 10th 1863

Dear Sara.

Your letters of the 21st & the 26th have just been recieved and read. and having a few spare moments gladly employ them in writing. that I may have a letter to send you when the steamer returns to the Head. I shall be on guard tomorrow and have no chance to write and as it is probable the steamer will return tomorrow I must finish my letter tonight. It is a fine warm day most to warm to be comfortable on batallion drill this afternoon. However. the Lieut Col who was in command either had a good streak on or else had an "ax to grind" for himself. for he dismissed us after we had drilled about an hour saying that as the mail had come and the men wer anxious to read thier letters he would drill them no more. So that instead of drilling I am much more pleasantly engaged in writing to my darling far away. It is over a week since I have written to you dearest. So that you see I cannot give you a schooling for not writing more. There is so much irregularity about the mail that neither of us must be disappointed if a mail arrives without a letter for us. I mean to write so as to send you a letter by every steamer that sails for the Head. but then you know it might arrive there first after the mail for the North had left. and of course would have to lie over a week or more until the next steamer sails. I shall not attempt to teach you your duty for you are the best judge of it yourself. I love to hear from you whenever you have a word for me. it would be selfish and exacting in me to expect. or require you to write oftener. than opportunities or inclination prompted. I have no reason to fear that your love is lessening towards me. Shall never accuse you of that. until I have some better reason than a short interruption of our correspondence. for although that would be very unpleasant. yet I should content myself by believeing that all would soon be satisfactorily explained. I do not know of but little to write you. for it is dreadful dull here. A few days ago <u>the Boston</u> brought up from St. Augustine about a hundred seccesh ladies. who for about a year past. have been supported there by our own government. From here they wer taken in small boats over to the rebels. at one of thier picket posts about six miles from here. I am told that a couple of hundred more are to be carried to the Head and from there turned over to the rebels. I hope I am some better of my cough. but it seems bound to trouble me about so much. and as the Docters medicine does not seem to do any good. I left off going to him a day or two ago. and determined to do the best I could for myself by being as carefull as possible. and using such simple remedies as I can obtain. Last Sabbeth we had a communion service at two o'clock. after it the customary service. This was the third time since we came out that we have been thus savored. The first time was before we left Hilton Head for the Tybee. I presume I wrote to you of it at the time. The second I believe was when we wer on Tybee. but I am not positive for I did not have the opportunity of attending for I did not know of it. Our service last Sabbeth was of necessity very short. but I trust was prized and enjoyed by all present. the chaplin remarked that it was very encouraging that more wer present than at either of the other occasions. This was doubtless oweing in part to the fact that none of the regiment are now together than on either of the previous occassions. After the service Mr Wayland preached from Ps 17th. 13th. I also attended the prayer meeting in the evening. Yesterday Co H. of reg went down to the Fort to stay in the place of the engineer company. which has been ordered to the Head. I learned that Dexter has succeeded in getting a situation in the same office with Moore so that we shall not be likely to have his presence in the company any more. He wrote to Barnes that the 10th Conn. was there and that he has seen Dr. Knight. about twelve thousand troops are now there. I am now writing in the evening. since leaveing off this afternoon. the Cosmopolitan has come in from the Head. and it is currantly reported that the paymaster has come We all hope it may prove true. for we have nearly six months pay due. and most of us are getting decidedly short. Uncle Sam is beyond question a good. but not regular. paymaster. we have to wait his time. for there is no looking up a new job. am pleased to hear that Madam Lizzie is having such a fine time. I can rejoice in the pleasure my friends are enjoying. though it is denied me. True sometimes I am inclined to musses when I reflect that amid all the tumult of a great and terrible war and the suffering. misery. and death. that is constantly being caused thereby. people at home seem scarcely to notice it. but go on seeking thier own pleasure and apparantly caring for "none of these things." But it is doubtless a wrong thought. and even wer it true of the majority of the people. I find comfort in the thought that each soldier has a few faithful friends who will. and do. care for him. and look forward to the glad time when he can agan return to him. Then agan scarcely one of the many thousands so rapidly passing away. but what leaves a few true mourners in his loved home. who lament his hard fate. and will fondly cherish his memory. And after all. those few are all that we should care for. and. trusting in thier love and faithfulness. why neede we let what the thoughtless throng is doing trouble us. Have agan been interrupted in my writing. It seems that the Government Inspector has arrived to inspect our regiment. Col Hawley anxious to have the regiment in as fine order as possible. has ordered us to prepare for the inspection which is at nine o'clock tomorrow morning. this evening. so writing must be laid aside   If I have time I will add more tomorrow. but if not you must make this pass for a letter.

with love ever your own.
Ned

No. 100 Rec'd Mar. 4th/63

Fernandina Fla. Feby 18th 1863

Dear Sara.

    I see by a notice on our bulletine board. that it is likely that a mail will close some time this evening and I am desireous to send along a line. I have not written a word except what has been absolutely necessary for over a week. Nearly all of the time I have been very buisy. When closeing my last letter to you. I mentioned that the officers had come to inspect our regiment. I was obliged to be present during the inspection and immediately afterwards went on guard. After the inspection there was a long and tedious battallion drill. at which the inspectors wer present. They made it a point in thier examination to find out something of the fitness of the various officers for thier situations. and for that purpos took the drilling of the regiment out of the hands of the usual officers. and required each of the others to take thier turn at drilling it. In the company drills he tried the non-commissioned. and as our training has been sadly neglected of course he found us deficient. Col Hawley has set to work to remedy the defect and has commenced with us. where he should have commenced seventeen months ago. we now have to learn a lesson in Harde's Tactics every day. and as guard duty comes frequent. and we drill four hours a day besides dress parade. what with keeping arms and equippments in order. and ourselves. clean. you can calculate that but little time is left for playing. our lessons would be insignificant at home. where one could sit quietly down and give his attention to it and would hardly be worthy of the name but here where it must be learned amid all the adverse circumstances incident to a camp life. without a quiet moment. it requires a good deal of time and attention. I am in hopes the Colonel. will soon get over his notion and let it die out otherwise. if I attempt to keep my end up. I shall be obliged to give up my correspondance almost entirely. Do not think dearest that for a moment I think of writing to you any the less. I do not rank you as an ordinary correspondant. by any means. our communications are a part of my existance. not for an instant to be thought of parting with. or except as circumstances render it unavoidable. to be any way abated. The inspection was the most thorough of any the regiment has ever had. It is very quiet and dull. here. but if we are to be kept as buisy in the future as we have been for a few days past ther is no I danger of the time's hanging heavy on our hand. Last Sabbeth I attended church as usual. in the evening the prayer meeting was unusuly interesting. The chaplain took for the subject of a few remarks the parable of the prodigal son and urged upon all present the importance of at once returning "to thier father's house." Several fervent prayers wer offered. and two or three persons seconded the chaplain in his endeavour to impress upon the minds of all the necessity of liveing a new life. At length a man arose in a distant part of the house and declared his intention to at once seek his Father's face. he stated that by our last mail he had received the news of the death of his mother. whose last request was that he would prepare to meet her in heaven. Shortly after. another quite a young man arose and stated his determination to try to live a new life. a few moments later another arose and said that he had once professed religeon but had wandered far from the path of life. but that he was now determined by <u>Gods</u> grace assisting him to turn once more from the error of his ways and renew his covenant vows. All this I was, as it wer new language to us. it was the first time since have left home that had heard it. God grant that the good work begun may be bountifully blessed. his spirit poured out upon us. and a revival of pure releigon take place in this regiment which shall honor God. and save a multitude of souls. The little church that we use for worship is also used for several differant purposes. Monday evenings. The debateing society occupies. last Monday evening I attended the debate. The building was filled. The question was whether it would be preferable to have the south return. and everything go on as it did before the war. and we return home in two months. or the war continued until slavery was entirely abolished. There was the usual amount of Lyceum <u>bun</u>comb, beating all around the bush. with but little said that really bore upon the question. It was decided that we had better "<u>stay</u>." Last evening (Tuesday) our minstrel band gave another concert useing. of course. the church for the purpos. Our chaplain thought it best to appoint a meeting for enquirers last evening. and as our church was occupied. it was held in a little building used for the like purpose by the negros. I did not attend. for I had gotten behind in my Hardee and was obliged to spend the evening over that. This (Wednesday) evening our usual prayer meeting is held. and I can ill afford to lose it. so am writing with all my might to finish my letter in time. I have been employed two days this week in connection with two others. in takeing the census of the town. It was very pleasant buisness and me I onley regretted that I was not alone in it which would give longer We wer supplied with the proper blanks. and took down the names &c with pencil. when through the account was handed in to the Provost Marshall. by whome we wer employed. and by his clerks are to be copied and summed up. so that I do not know the number of inhabitants any better than as though I had not been out looking them up the same time. George is still in the cooks tent. and I guess is feeling comfortably well but to tell the truth I have not asked him in a week. Now dearest I want to go to church. The bell has rung and I know you will excuse me from writing. for that purpos. and would do it the more willingly if you onely knew how much I <u>needed</u> to. I am going to send this along (excuse) without all stopping imperfections. to look With it over continued and love correct it. so please excuse all the imperfections. With continued love darling. I am as ever your own dear

Ned

No. 101 Rec'd March 9th 1863

Fernandinia Feby 21st 1863

Dear Sara.

We recieved a mail day before yesterday. and your favor of my 29th came to hand. Three weeks on its way here. So you see my dear our communication is not very rapid. as I had hastily written you an apoligy for a letter on the preceeding evening. and was very buisy. I did not attempt to answer it so as to go by return of boat. yesterday I was out on picket duty. and today am very stupid for want of sleep. I have just come in from witnessing a matched game of baseball. played by twenty men from Cos G. and E. against and equal number from Cos A. and S. Our boys (Cos G and E) wer the victors. It is Saturday so you see the boys have had plenty of time for thier sport. The men have played a little along back during thier leasure hours. but this was the first matched game. One of our boys has first been promoted to be a Seargeant. in the Co. It is R. J. Hawthorn. the one for whome you knit the cap when we were in Pulaski. I am somewhat surprised to hear that your father has sold out. I fear that he will find that he has made a mistake. but do not know enough about the buisness he has been doing for the past year and a half. to be able to judge. It is no new idea with him. As long ago as when I was clerking it for him. he was thinking of. and talking about it. so as he has not done it hastely I presume he is confidant of bettering himself some way. One thing is very much against. in entering upon any new buisness. and that his age. And his health is not any to good. I do not know as I am at all surprised to hear of Frank's marriage. neither should I have been surprised had the whole thing fell through. for I know him to well to be surprised at his actions. Frank. you know. is looking sharp after the Dollars. and father Greene's purse is both longer and heavier than Mr Orton's (probably) and there are not so many to share its contents. This fact I think would be a powerful motive and have much to do with Frank's decision. other things about the two girls being equal. I think Josie had forgotten that consideration when she formed her opinion. Prehaps she did not think of it. or would not impute to Frank. so mean and sordid motives for his actions. I know the though must be repungnant to any lady. that while a gentleman is paying her his attentions and vowing unalterable and eternal love. he is at the same time secretly calculating how much better off in Dollars and cents. he would be by possessing her. Therefore I am not surprised that so many are decieved. and find when to late. that it was father's gold. and not themselves. that thier pretended adorers. worshipped. I do not say dearest. that this is true of Frank. and I would not like him to know that I have intimated any such thing. But you know that he will think of such things. and we are all so uncharitable. I remember having seen Frank Richardson, and have heard others than yourself speak of him as being a very smart intelligent young man. one who would surely make his mark. and doubtless be successful in life. We are not entirely free from Diptheria in the army. occasionally a man dies with it. and our Docters are unusuly attentive to a case of sore throat. Lizzie was fortunate in escapeing her acustomed dose of. Quinsy. As I have said before. I see no impropriety in a friendly correspondence with her Rosette if you wish it. but of course I do not urge it leaveing to your own good sense and judgement to act as you see fit. I should judge that he had not entirely given up Martha. and that. that. friendship is the main reason for writing to you. if you have not forbidden him to write to you. or given him some intimation that his letters are not acceptable. I do not see how you can avoid writing at least once more. if as you say his letter was a purely friendly one. but you are the better judge of what is right. and I leave it with you. The Bible came safely in the same mail with the letter. Many thanks dearest. I shall enjoy reading it all the more for it being your gift. You did not tell him how long Selina was going to remain with you. but presume she will have returned home before this reaches you. if so when you write to her do not fail to say to her that I am gratefull for her kind message and heartily return like for like so far as I can without Sara's being jealous. By the way has no one spoken for Selina yet. I think she is a good girl and wonder that some lucky fellow has not made a prize of her before this. I commenced this letter Saturday evening intending to have finished it then. but there was to much going on around. and my own nerves wer not very quiet. so I laid it aside. I have many other letters that I ought to write. but have got on one of my fits of aversion to writing and the letters lie unanswered. I should be out playing ball with the boys now. if it wer not for you. What a bother you are to be sure. here I am nearly a thousand miles away from you and yet I cannot go out and play ball with the boys. which I want to do real bad. on account of you. If this is the way you control me now under these circumstances what hope is there for me if I live to get home and you become my wife. Now see how I have blocked my letters in my frett. Oh dear. Yesterday (the Sabbeth) passed off as usual. if any thing I was more lazy than common. Today is being spent by the regiment as a holiday. yesterday was Washingtons' birth day and we are obeserving today as a holiday in honer of it. This evening Col Hawley is going to give an address and we are to have some music. and do the best we can to warm up and recieve what little patriotism fifteen months of petty tyranny and ill useage has left us. I heard something yesterday which I tell you in strict confidence because should it prove to be incorrect I should be very sorry to have reported it. I heard that the papers recommending Mr Moore for the position of Adjutant of this regiment have been sent here for Col Hawleys approval. and that it was more than probable that he would give it that the recommendation was already signed by Gen Hunter. it is therefore likely that Mr. Moore will be our next Adjutant. which is quite a step up. the rank being that of a first Lieutenant to which the position is preferrable. and the pay is that of Captain. I shall be glad to see Mr Moore so successfull. and I shall not incur the limitation of being envious. when I tell you that I think it is a piece of gross injustice. not oney to commissioned and non-commissioned officers whome he will out rank. but to hundreds of privates in the ranks just as capable for the position as he is. Ever since before the fall of Pulaski Mr Moore has had what the soldiers call a "soft thing of it." I will explain. he was made a company clerk some time when we wer on Tybee. in that position he had neither guard or fatigue duty to do. After we went into Pulaski he was made the Adjutant's clerk. then he had nothing to do in the company so he still messed with us. his oney buisness was writing. A few days after we arrived at James Island he was detailed as clerk to <u>Col Fenton of the 8th Mch</u> who was in command of our brigade. he did not participate in the fight there. on Edistoe where we remained a short time. he was again Adjutant's clerk. and continued in that capacity a short time after our return to the Head when he was advanced to his present position. where his pay is about doubled though any man would prefer the situation to that of doing duty in the ranks without any extra pay. Thus without being exposed to the dangers of suffering the hardships of hundreds of others equally capable. he is to be promoted over thier heads. As I said in the start I am glad he is to have the position (probably) but I have endeavored to give you some idea of the justice which is exercised in the promotions. Mr Moore ought to have had a commission when he came out. I think he will make an excellent officer. At any rate he is head and shoulders above three quaters of our officers in all that constitutes a man. While on this subject I may as well make a clean breast of it and tell you some other things. and I shall tell them partially because you may have wonder why I never recieved any promotion. and partly because you have a right to know my own feelings and whatever concerns me. The promotion of our friend Hawthorn was an injustice to me and also to three other corporals who stand before me on the list. Without egotism I may say that I am first as competant for the position as Hawthorn. The Captain made the promotion to gratify a mean, petty, pieque, of his own the cause of which did not concern me and which did concern another of our boys. but I may not mention it for it was told me in strict confidence. I am satisfied that I am not fit to be an officer. I have but little taste for military anyway and have not enough of the tyrant in my nature (thank God) to qualify me for a good officer. Another thing. I have never saught the favor of the officers. and have never conducted myself in a manor to gain thier good will. I am to fond of expressing my likes and dislikes. and whenever anything has gone contrary to what I thought right. I have not failed to express my opinion freely and condemn in language more expressive. than elegant. and complementary. the conduct of my superior officers. Of course this has not been to thier faces else I should long ago have been in <u>confinement at the Tortugas</u> with a ball and chain attached to my leg. But the things said have doubtlessly reached thier ears and I know to expect no favors at thier hands. I cannot stoop to court favor with a man whome I utterly despise. and will not though I wer to remain in the army all my life. The paltry position of Corporal. which I have held since we came out. I am bound to get rid of I have a note now in my portfolio written last evening to Capt. Mills and which I am waiting an opportunity to hand him. requesting him to relieve me from the position. if he refuses I shall appeal to the Colonel and if neither of them will grant my request I shall do something for which they will reduce me as a punishment. So you neede not be surprised to hear that I have been reduced and disgraced. My fellow Soldiers understand my situation and will not blame me. you will believe what I tell you and neede not care what others may say. I can do my duty as a private soldier with a clear conscience. and what neede I care for aught else. The position of a corporal is no sinecure by any means. being the lowest office in the army it is respected by neither officers nor men and at the same time those who hold it are expected to enforce respect for an obedience to themselves from the men. the pay is the same as that of the privates. A non commissioned officer is liable to be reduced and disgraced at any time for any little real or fancied offense. A case in point occurred during the last week in our own Co. A seargant and a corporal from our Co wer on picket duty. they wer on separate posts having each three men with them. Now a man on guard or picket must not take off his equipments during the whole twenty four hours except it be for a most urgent necessity. The officer of the day makes the rounds visiting the different picket posts during the twenty four days. once in the day time and once in the night after twelve o'clock. On the day of which I am speaking the officer of the day visited the different posts and came upon our seargant first as for good and sufficient reasons he happened to have his equipments off. Without asking or accepting an explanation he at once reported him to the Colonel "for neglect of duty". Without the form of a trial. without a hearing even before the Colonel. or any chance to excuse himself. and though the men on picket with him wer ready and anxious to testify in his favor they wer not allowed the chance. and he was reduced to the ranks. apparently in disgrace but every soldier who knows the circumstances will exonerate him from all blame and look upon him as he is. and injured man. We all very much regret it for he was the best seargant we had. one of the best men in the company. and one whose word alone would be perfectly good to exonerate him even though there wer no witnesses. The same day the officer visiting the post at which our corporal was station found one of the men who had been into camp after thier bread and just returned. with his equipments off. and the officer reported him also for neglect of duty. and he to was reduced without a hearing. though if he could have had one. no man with a spark of justice in him. would have reduced him. Visiting another post. the officer found the corporal absent from his post. this under no existing circumstances the corporal had a right to do. and he paid for it by the loss of his position. Last Friday I was on picket and the same man was officer of the day profiting by the lesson just learned by my brother soldiers I steered clear of all trouble with though ordinaraly on inside picket I am as easy and careless as any one. but the seargant who was on the next post from me came in for a share of his displeasure. by having on his post a little fire. he to was reported. but as the captain of his company haves things his own way. and the Seargant is a favorite with him. I think he will get off free. So you will see dearest that everything is not as agreeable as it might be. I am sorry to have written so much on this subject. or to have troubled you with it at all. but I thought that prehaps you might wonder at some things and like an explanation which you did not wish to ask for. I shall strive hard to get along without any trouble and go through the remainder of my soldiering as I have thus far done without subjecting myself to a reprimand. do not worry at all about me. for there is no cause for it. I hope the Captain will make no objection to my request. we have no news here. the <u>gun boat Mohawk</u> which is stationed here. has taken the alarm from our late disaster at Galvestone, and keep themselves in constant preparation for a like attack. I do not think there is any reason to expect it. but it is well to be prepared. I know that our officers think we are likely to be attacked here. but I very much doubt that we are. an old negro made his escape from thirty miles up the river. and came down in a little Dug out canoe. reaching here yesterday morning. he sayes that there is a small steamer up there loaded with cotton waiting for some favorable night to run the blockade and escape. A schooner loaded with cotton left the same place two or three weeks ago. as we have taken no such schooner she has doubtless escaped. He reports that but three companies of men are any where near here. but that they had sent to Savannah for more. Good bye love. Ever your own

Ned

No. 102 Rec'd Mar. 28. 1863

<u>Camp Starr. Fernandina Fla</u> Mch 5th 1863

Dear Sara.

 I am a long ways behind the time dearest. but you have not been forgotten or thought of any the less. My time has been so fully occupied that this is the first opportunity I have had for writing in several days. and I purchased this by staying up late last night assisting Reid to correct the company muster rolls. As I have before told you. we are engaged in learning the "tactics." and it occupies me. all my time which would otherwise be leasure. besides the labor of learning the lesson I am further hindered by being obliged by the scarcity of books. to watch my opportunity when the fortunate owner of a copy is not useing it. to borrow the book and copy off the lesson. When I wrote you last I was in hopes that ere this I should be relieved from the necessity of thus useing up my spare time. and be at liberty to use it as formaly. in writing or reading. But I have not yet succeeded in my design. and am particularly desireous to keep my "end up." while I remain a non-commissioned officer. After finishing my last letter to you. I handed the note. I mentioned. to Capt Mills' servant. to give to him. Immediately after dress parade I was summoned to the Cap's quaters. There I found somewhat to my surprise. all of the rest of the non-commissioned officers of the company who wer not on duty. I knew at once that something was ill but did guess the truth. I onley thought that the Capt was probably more angry than I had anticipated and that he was going to <u>come down</u> on me (to use a common phrase) harder than I had anticipated. But I was just angry enough not to care for all the officers in the regiment just then. I was however mistaken. for although I could see that the Capt was that he had called us together. for the purpos of finding out how many of us wished to be reduced. as he wished to make a clean thing of it. and not be troubled with frequent applications. he then went on with quite a long speech regarding matters in the company and the non-commissioned officers. putting on the air of injured innocence he pretended to be sorry if he had not treated them properly. and expressed himself willing to listen to any complaints and to do whatever he could to make matters pleasant for them. When at length we got an opportunity to speak. I found that two other corporals besides myself had applied to be reduced. and that if we wer. two or three other non-commissioned officers would apply to go down with us. This then. was the cause of what surprised me when I found we wer all assembled. Still I was thrown somewhat off my balance at the new aspect of affairs. and particularly as I found that the Capt was very unwilling to have us presist in our applications. Moreover I had some things to say that I wished to say to the Cap alone. and though we all spoke freely. yet the Capt. managed to occupy most of the time and as it was nearly time for the services which wer to take place in the church in honor of Washingtons' birthday to commence. and we wer anxious to attende. he dissmissed us with fair promises and desired us to say no more about it. The fact was that in some respects we had unconsciously got the better of the Capt. I say unconsciously. for neither of us was aware that the other had applied for reduction. this he was slow to believe at first but was at length obliged too. Besides one of our Seargeants had applied for. and obtained. his reduction a few days previous. on the ground that the Capt did not treat him with the respect that he considered his due as a non-commissioned officer. Added to this. two of our very best non-commissioned officers had just been reduced to the ranks by order of the Colonel for some triflieng (alleged but not proved) neglect of duty while on picquet (one of them was a seargeant and the other a corporal). If we presisted in and obtained our requests therefore. it would make a change of nearly all the warrant officers. and beside this. so much change. and so marked an evidence of discontent. among those who ought to be the best men in the company (but is not true that they are) would not be very flattering to his pride. and prehaps excite remarks from his brother officers. He reported the matter to the Colonel. as he told us he would. The next day while the company was out drilling. the Colonel came and calling one of the applicants aside scolded him soundly. and I guess thoroughly scared him into submission. he enquired for me also. but unfortunately I was out on picket and so lost the much desired chance for a set too with his lordship. I presume it was for the best however for I was in such humor then that I presume I should have got myself into a guard tent. if not a <u>Tortugas</u> scrape. The third applicant has I hear been quieted with the promise of the next Seargents' berth that becomes vacant. As for myself I have decided to let matters rest just as they are for the present until the matter has a little blown over. But I shall yet have a talk with Capt Mills. and after the other corporals are pacified shall insist on his complying with my request. In the mean time I shall have to keep a sharp look out. for not withstanding his fair promises. I have no doubt he will gladly sieze the first fair pretext that occures to reduce me dishonorably. But as I told you before it will matter little to me. onley I would not like him to outwit me. Mr Moore's promotion is no longer any secrete. Dexter has obtained a situation in the office at Hilton Head with him. Ried is company clerk. Norton and Wells are employed in the hospital. and in the whole matters are very favorable as regards the Salisbury boys. The exercises on evening of the <u>23rd</u> in honor of Washington wer very interesting. Col Hawley made a good speech. in which he took occasion to compliment. and flatter the men of the <u>7th</u> and I think he succeeded in produceing for a time at least. a better state of feeling than has existed in the regiment for a long while. Col Hawley is quite noted for his ability as a stump speaker. and certainly what I have heard of his powers. have not been overrated. The officers spent a very gay night of it afterwards. they had a kind of a ball or party. or a little of both at the house of one of the Assistant surgeons. and between dancing. feasting. and playing. passed the time merrily until morning. Our Major took it into his head to double his existance a few days ago. and accomplished his purpose very suddenly and unexpectedly to all. A young lady from Groton Conn. has been South here teaching school for some time. a short time since during one of the Majors visits to the rebels under a flag of truce he found this lady desireing to return home and she came here with the Major. where she has remained staying at the house of our Lieut Colonel. whose wife is not well. and has been for sometime waiting to get well enough to return North. when the young lady above referred to was to accompany her. But the gallant Major has succeeded in persuadeing her to postpone her return North rather indeffinately and after a short acquaintance in induceing her to change her name to that of Mrs. Rodman. We have been mustered for six months pay which we are in hopes to get some time during the present month. I think of no other incidents likely to interest you. The weather is getting to be generally pretty warm. and the mosquetoes and gnats. very troublesome. they will be a terrible annoyance here this summer. not quite so dangerous as bullets. and bayonets. but I had almost said. as much feared. Last Monday a schooner came in from the Head bringing us a mail. I was favored by the recept of your letters of Feby

6. & 11th. and also a paper. I get but few letters now from any one but you but so long as you do not fail me I shall be content. I know that it is principaly my fault for I write but few letters now. and correspondants are more particular than they once wer in expecting answers to each letter. I cannot blame them. "Out of sight. out of mind." is a time honored and truthfull saying. We have been absent a year and a half. the times are very stirring. everything to call off the mind from friends long absent. what wonder then that affection should fade and interest should slacken. I am not surprised. Enough for me the "faithful few" to whome I have before alluded. fail not in thier love for and interest in the absent <u>one</u>. My last letter to you was dated the 21st of Feby though not finished until two or three days later. it is now the 5th of Mch. and I have not written a line to any one during the itervening time. I hope that ere this Carrie has entirely recovered her health. She must be quite a young lady now. I remember her well. as she appeared during her visit at her fathers. when I was liveing with you but she has doubtless changed much since then and probably has forgotten me. It has made it quite unfortunate for Selina's visit hasn't it? It's a pity. her visit once before was nearly spoiled (I mean your enjoyment of it) by my comeing off to the war (Egotistical). But we will hope that maybe premitted to one day return her visits. and yet have a happy time together. Tell you father that I hope. that before he gets an appointment in the army. he will get into some pleasant and lucrutive buisness. from which he would not be willingly absent a single day. One of the greatest curses of the army. and I might say the nation. is the fact. that the officers. ninteen twentieths of them. are men who wer out of buisness or never had any. and therefore it is for thier interest to have the war last as long as possible, for when it ends thier occupation is gone. However if he really wants to get an appointment in the army -tell him to be sure and not take up with any thing short of a Major Generalship. because if he takes any office beneath that it will be absolutely necessary for him to know something about military matters and movements. and besides he might be required to do something whereas if he takes up with a Major Generalship. he can easily avoid them both. particularly the latter. Carrie does write to me occasionaly. of which I am very glad. She is widely differant from you dearest. but she is a very good sister to me nevertheless. Knowing her so intimately. I can easily account for what to you might seem to be grave faults. and though like others she has her faults. yet she has a kind heart. is yet young. and years will doubtless do for her what they ought to do for us all. make her better. She writes but seldome. and her letters are interesting for she tells all the little news. which is so welcome to the absent ones. Thank you darling. for the renewed assureances that you are wholly mine! those

word are very dear to me. I sometimes feel as though I did not appreciate all the glad truth which they express. It is more of happiness that I have a right to. God grant that the time may speedily come when distance. nor the want of the conventional ceremony shall any more separate us. Gratefull for the privalege you have granted. I shall not hesitate agan to claim full possession whenever speaking of you. Not knowing exactly how well informed Martha was of our affairs and not understanding her perfectly. I hesitated about being as free as I should have otherwise I do not know how Martha will construe my allusion to a further correspondance but if she should write occasionly I should not be sorry and of course should answer to the best of my ability. thank you for your caution and will observe it but presume she will show you whatever letters she may recieve from me. and if I am not sufficiently carefull you must tell me when I have done wrong. I ought to write you a much longer letter dearest. but it is most time to drill. and I do not know when I shall have an opportunity to write agan. so I shall close this at once. Pitt has just run in and sayes "give my love to her." I enclose a flower which I picked on my way to old town a few days ago. it looks the most like a dandelion of anything I have seen since leaveing home. I call it Fernandia dandelion. to venture. There are not many flowers around here but there may be more by and by. With kind regards to all the dear friends. and love to you my dear one. I am as ever your own.

Ned

P.S. Guess you will laugh when you come to see how this letter is folded. and you would have laughed all the more if you could see my attempts to get the awkward sized paper into some shape so that the envelope would cover it. Shall try to get some better before I write agan.

Ned

No. 103 Rec'd March 28th/63

Camp Starr
Fernandinia Fla Mch 14th/63

Dear Sara.

Yesterdays mail brought me your letter of the 23rd & 25th Feby. I neede not say how welcome they wer. and the more especially as the previous mail did not favor me with a letter. I am rather tired tonight having spent the day in visiting Ft Clinch returning in time for dress parade. But if I am not interrupted I hope to write you a letter in time to go by a boat which is to leave for the Head tomorrow. The day has been fine. would have been uncomfortably warm but for a fine breeze. We have burried a member of our regiment today. the first who has died since we have been here. The burying ground is on the road to the fort. and I overtook the funeral train just as they reached the grave and of course stopped while our brother soldier was lowered to his last resting place. As I believe I have before described a soldiers funeral I neede not stop to do it again. The chaplin read some passages from the Bible made some very appropriate remarks. and an earnest prayer. then throwing a spade full of earth upon the coffin. he stepped back, the three volleys wer fired over the grave. the drums and fifes struck up a lively tune. and the funeral escort marched off in quick time, leaving the grave to be filled up by the contrabands. By the way did I ever tell you that the tune played for the dead march is <u>the Portuguese hymn</u> that you sing in the choir at home. Fort Clinch is not nearly finished yet. It is larger than Pulaski and will be a much stronger work if ever finished. which will take at the present rate of progress three or four years. Comeing home I got a ride part of the way in a sailboat. A few evenings ago there was a meeting held in our church for the purpos of adopting resolutions expressing of the sentiments of the regiment in a view of the approaching Spring election in our State. The quiet of the meeting was shamefully disturbed and its proceedings interrupted by a few shameless scamps. who <u>true to loco-focos principles</u>. and practices. got into the galleries. and backseats. and vented thier spite by hissing. shouting niggar. hurraing for Tom Seymour. and interrupting and black guarding the speakers. Strong resolutions wer brought forward upholding the Government. and deprecateing the election of men to office who wer known to be traitors. and often empathizers with the rebels. They wer voted upon. and declared to be adopted by the secretaries who counted the votes. The buisness was now nearly concluded when an unlucky remark from one of our Union officers. gave our Lieutenant Col George F Gardner who was present. an opportunity for which I am uncharitable enough to think he has been waiting to break up the meeting. He arose and declared himself a Seymour man. Whereupon that clique set up a tremendous cheereing. and after putting. and stamping until they wer tired left the house with Gardner. yelling like a pack fiends. The more orderly part of the audiance remained and concluded the buisness. But the matter did not end there. The copperhead thus so suddenly exposed in our very midst. having shown thier hideaous features. could not be contented without making a further demonstrations. So they have been at work and written a set of resolutions to which they have included a large number to sign thier names. Many, in fact the large portion of the signers knew not what they wer signing. and many who heard them read, or read them themselves, are not capable of comprehending the meaning of what they so readily endorsed. I have not been able to see the paper during its circulation of to find out exactly, or with any definateness its exact purpose. But I am told it is highly treasonable. The matter came to Col Hawleys ears. and he is highly exasperated. He called out the regiment this afternoon and gave them to understand that there must be no more such work. that if he hears of any man uttering treasonable sentiments or trying to disseminate the same, he will at once send him. or them. to the Head in double irons. to be tried by a General Court Martial and recommending that they be dealt with without mercy. I thoroughly and heartily endorse his course in the matter. tomorrow afternoon he is to address the regiment upon the object for which we are fighting. I learn that he has since been examineing some who appeared to be the leaders of the treasonable paper. all but three professed not to be fully aware of what they had done. or rather of what might be the consequences of thier actions. those three I am told are to be sent to the Head tomorrow in irons but I am not sure that it is so. I am much afraid from what I learn from home that the traitor Seymour will be elected. I hope not. Much as I am disgusted with the manner in which the war is conducted. and earnestly as I desire to return home a free man once more. I am not yet prepared to give up everything for the sake of a dishonorable peace. I cannot bear to have the rule of the North fall into the hands of those contemptable traitors and rebels. whome the leniancy of our government has premitted to live. and even go free and liberty to disseminate thier infamous and disloyal sentiments while every day they have threatened. and endangered the stability of that government whose mistaken mercy has spared thier worthless lives. Every day I grow more and more satisfied that had the Government ruled with a more firm hand and hung without fear or favor a number of the vile traitors who infect the North. at the beginning thousands of valuable lives would have been saved and the war before this brought to a successful and honorable close. I am very sorry my dear that you burned up the letter you commenced. Sorry did you not finish and send it along. Am I not to know of your provocations. You ought. I think to let me know what they are. I do not suppose you are in Paradise exactly. I should like to know what troubles you. and what you would say when you are vexed. it might be interesting. You know I am somewhat acquainted at your home and I think would be able to understand. and appreciate your troubles. and then you know prehaps I might be able to comfort you. Now my dear it is more than probable that, that little fret of yours cheated me out of a letter by one mail. for I did not recieve a single letter by our next to the last mail. I fear I shall have to consider myself an injured man. I cannot aford to be deprived of my comfort in that way. Well darling you are very kind in being particular to write cheerfull letters. but you should remember that I am interested in all that concerns you. Your trials and temptations. and shall always be glad to know of them. Now the other day I wrote you a real blue letter. I did not hesitate to write because I was angry but went right at it and told you what the matter was. Sunday afternoon Have been to church this forenoon <u>Text Math 6 & 10th</u>. am going this afternoon to attend the negro Sabbeth school that is if I can pass out of camp without getting a pass I have written all of interest that I can think of. By the way I was going to say that we are having some very interesting prayer meetings and I enjoy attending them very highly. I have great need of all these external helps to enable me to hold fast my profession. May the good God continue to bless and protect you is the prayer of your own.

Ned

No. 104 Rec'd March 31st 1863

Camp Starr.
Fernandina Fla Mch 23rd 1863

Dear Sara.

I have been looking at your ambrotype this morning and as I looked I asked myself if it was true that I had ever held that dear form in my arms. ever pillowed that head upon my breast. or recieved the kiss of love from those lips! It has been such a long. long. weary time since the happy hours we spent together. fled away. that I almost doubt sometimes whether the rememberance of them is real. if it is not a pleasant dream that I am indulging in when my mind wanders back to them. I am lonely this morning dearest. and my thoughts are of home. and you. If I wer with you. you would accuse me of having the "blues." I should hardly dare contradict you. I am sick at heart. and disgusted with my situation and its attendant circumstances. You need not tell me I am wrong. that I ought to be contented for that. "Godliness with contentment is great gain". that will not mend the matter. Don"t I know all about it! Well dearest I must write you a letter any way but I am sadly in want of the proper materials for it. Over a week has passed since I wrote you last. In that time I have not written a letter though I have nearly a dozen unanswered ones in my pocket. I could hardly look at my portfolio with any patience. The week has been dull. and monotonus as usual. I have been on guard twice. acting as seargeant. Most of the week has been damp. cloudy. and very windy. with but little rain. We have been furnished with a new second Lieutenant in the place of the one I wrote you had resigned shortly after we arrived here. I know but little of the new Lieut. he appears very well. He was Orderly Seargeant of Co H. which by the way is considered one of the <u>scaliest</u>

companies in the regiment. But <u>Lieut Wood may be a fine man</u> for all that. he was much liked in his own Co. It remains to be seen how well he can bear prosperity. He was wounded in <u>the fight at Pocotaligo</u> and has but just returned to duty. On Friday the Boston came in having on board the 6th Conn. on thier way to Jacksonville. They lay here a few hours and then went on. On Saturday a small steamer came in having on board a part of the 8th Maine also on their way to Jacksonville. They had had a very unpleasant time coming down as the sea was very rough. they a landed and spent the night on shore. and reembarked early Sunday morning and left for their destination. Sunday afternoon the Delaware with the remainder of the 8th Maine aboard came in, made a visit of two or three hours and then went her way. They had their band with them and favored us with some good music. which we enjoyed if it was the Sabbeth.

 Yesterday I went to church as usual. Text. the last part of the 25th verse of the 1st Chap Acts. In the afternoon attended the negro Sabbeth School. and tried to teach a lot of little sooty damsels from four. to ten. years old. their catechism and letters. If it was not wicked I should say it was <u>fun</u>. It was awfull hard work for me to maintain my gravity. It was a droll sight. over a hundred wer present. of all ages from four to <u>thirty</u>. Of course it is utterly impossible to keep them still. but after all they are as quiet as can be expected. The classes are to large. There ought not to be over half a dozen in a class. but there is generally a dozen. The want of teachers makes it necessary to have the classes so large. There are three or <u>four</u> lady teachers. among them the wives of our chaplain. and Colonel. and the Colonels wife's sister who is teaching one of the negro schools here. I noticed two or three negro men. teaching. The most of the teachers are volunteers from our regiment. After the school. I went to the hospital in company with Mr Landon. and Wolcott. to see Tom Norton. who. in playing ball a fortnight since accidentaly sprained his knee very severely and is a cripple. He is getting along finely. but it will doubtless be some time before he will be well agan. He is in good spirits and otherwise in good health. The Colonel did not give us that-talking to that he promised us last Sabbath afternoon, of which I made mention in my last letter to you. The matter has quieted down. I believe the three men wer not sent to the Head. We learned by one of the last boats that came in. that a pretty story had got started at Beaufort and the Head about us. to the effect that seven companies of us under Col Gardner had mutinied and refused to do duty. Of course there is not even the semblance of truth in the story. I have already acquainted you with all there is of it. so much for having a "Copperhead" for Lieut Colonel. One morning a week ago a dwelling house occupied by a white family took fire and burned down. It was at daylight in the morning. and it is supposed that it was set on fire. but by whom and for what purpose is not apparant. the family escaped uninjured and most of their effects wer saved. The man who occupied it is a strong Unionist and has already lost largely by the rebels. The Boston has just come in on her way back from Jacksonville. I shall close this in order that it may go to the Head on her. Mr Moore has not yet joined the regiment but is expected on the next boat. With unchanging love. I am dearest. your own.

Ned

No. 105 Rec'd April 8th/63

Camp Starr
Fernandinia Fla Mch 27th/63

Dear Sara.

Three days ago a mail arrived by which I recieved your favor of Mch 8th. I also recieved letters from Mrs Randall and Carrie. they wer most welcome for I had not heard from them in a long time. Mrs R's letter I have answered. They wanted to send me a box of "goodies" if I thought it would reach me. and requested an early answer. that they might know if there was anything in particular that I wanted. and that they might hurry it on. I have taken advantage of thier kindness to send for a few things which the tardiness of the Pay Master has rendered it difficult to procure here. and of which I stand in neede. I hope you will consider that I have sufficient excuse for answering her letter first. Today I am agan made happy by the receipt of your letters of the 12 & 16th Mch. I hasten to commence a reply to one. prehaps all of them. and will begin by telling what little of interest I can think of that has occurred here of late. Mr Moore came on the same boat that brought the first mentioned mail. he had quite an unpleasant trip. The boat was an old. and small one. and I guess pretty well loaded. it is about a fifteen hours trip from the Head here. The boat started in the morning and having some things on board for the blockading vessels <u>Ossabaw Island</u>. expected to visit them. and then reach here the following morning. But it was a week lacking half a day. before they arrived. I cannot tell you all the reasosn of thier lengthened trip. but most of the time the weather was very rough. and they attempted to run down part of the way. by what is called the inner passage but had much difficulty for want of a good pilot. they got aground several times, and one time wer obliged to lie twelve hours before they could get off. They had some ladies on board. who were occupied nearly all the limited accomodations the boat affoarded for passengers. So the gentleman had to lie on the floor. Mr Moor is looking finely and shows off the broadcloth gilt buttons. and shoulder straps to good advantage. he was much like in the Co and I think he will be a very acceptable and popular officer. he commences his duties today. Last night the <u>gun boat John Adams</u> came in from Jacksonville. she brings word that the forces there have little skirmishes with the rebels nearly every day. The rebels are very much incensed to think that a negro regiment should be sent there. They have a sixty four pound rifled gun mounted on a platform car. and they would run in on the railroad and shell the town. Our folks tore up the track for about three miles out. and then placed a light battery in ambush. when the rebels came down to the break in the road they of course came to a "halt" then our light battery opened on them and they cleared out as fast as steam could draw them until they got out of range of our light pieces. when they opened on us in return. It was now our folk's turn to back out. and while retreating a shell exploded among them killing two and wounding two of the 8th Maine. The object of the John Adams visit here. is if possible to have Colonel Hawley send down one of the three guns mounted in the redout here in the village. The Col complied with the request and JA. left this forenoon with the gun. and men to work it. You really had quite a waiting spell before you recieved those two letters. I regret it but it can"t be helped. I know just how to pity you. I see it would have made no differance if I had written one a day between the 10 & 18th. as you recieved them both at once. But I shall write whenever I have opportunity just the same as when I knew the mail went every day. I ought to write much and often to you. if my letters are of so much importance. and comfort to you. but Ill not complain to you more of my lack of brains. for you must be weary of that complain ere this. I am sure I never was made for a ladies man. I can neither "talk like a book" or write a love letter. much less verses. I don"t know what you can find loveable about me I am sure. I am neither rich. witty. good. or handsome. Are you quite sure you love me? I often find myself wondering if this be true. I am sure I love you more than I can tell. but why should you love poor me! I do not deserve your love. hard as I have tried to win it. Sometimes when I do not have the blues. and think that I may yet return home. I almost shrink from the thought of makeing you my wife. fearing that I am unworthy. and that I could not make you as happy as I should wish my wife. As I should wish you to be. But I love you dearest and I think my selfishness would surely claim you for its own gratification. After all your love for me for so many years. after all the efforts that I have made to secure that love. and to lead you to trust in me I should be a villian unfit to live if I did not love you all my life. Yes darling God helping me I will love you all my life. and if it be his pleasure that our lives shall ever become one. I will not. in becoming a husband. cease to be a lover. Yes dearest it is enough. I had no buisness to complain as I did. you love me. it is enough. But Sara. it is so tedious. so dull. so trying. nothing pleasant and agreeable. everything coarse. rough. vulgar. Yesterday the same as today and tomorrow will be but the repitition of hundreds before it. what wonder then that the weary soul sometimes murmers. and cries out for a change. and wonders why others should be premitted to enjoy so much. and it so little. To complain and find fault is the rule. To be contented is the exception. I see nothing amiss in your letter to excuse. you had been reading the Bible you said. it is full of love. God is love. and he requires of his children their first and best affections and oh how unwillingly and grudgingly we yield him a little of our hearts. From loving Him freely and fully we come to love all his creatures. We can do no otherwise. then. if we love all how natural to love much. those dear friends that God has granted us. You see my dear I find it perfectly easy to account for your feelings on firmly Scriptural grounds. Thanks dearest for your economical suggestions. but first at present I do not neede them. Why? Because the Pay Master has not come yet. and it no more probable that he will in the four months to come. than that he has in the six that have passed. But what does my Sara think I do with my money? She must remember that I am working for almost nothing. and boarding myself almost. Not quite so bad as that its true. but then she must know that it is almost impossible to live on Government rations alone. and that everything a soldier wishes to buy to live on is only to be had at an exborant price. Let me see. Since I have been out I have not lost a cent gambling. have spent no money for liquer. and not about half a dollar for cigars. Then those habits have not been very expensive. have they? True I have spent money for things I did not neede sometimes. but I believe in the majority of the cases Sara herself would not find much fault. True I have not saved anything. Oh! I shall have to laugh at her a little when I see her. But seriously dearest. I neede lecturing on economy and I am not displeased at your kindly suggestions. Money was never of any good to me except for purposes of spending. I never loved it for any other reason than that by useing it I could be benefitted. If I wer onley rich I would not listen to any economical lectures. but as I am not. I shall have to be gratefull for a great many of them if I live by the way. how stands affairs between Charlie and Lizzie (there now what a blunder I put Charlies' name first) I have not heard since you told me that he had concluded on account of his mother to let matters rest for the time being. There are no new instances of conversion in the regiment. but the prayer. and class meetings are well attended. and are very interesting. There are more professing christians in the regiment than I had supposed. They seem to be "thawing out." I remember when we had been out but a little time. I used to attend prayer meetings when there would not be over twenty present. oftener not over a dozen. The comeing Sabbeth our chaplin is to baptize by immersion (Have I told you that Mr Wayland is a Baptist!) one man who has experienced a change of heart since we came here. We have a regular class meeting one evening in the week and three regular prayer meetings. I attend as many of them as I can. and enjoy them much. I think my dear. that prehaps your fine sleigh-ride and subsequent enjoyment was partially to compensate you for delaying your visit. once at the call of duty. not that you suffered. or that you were called upon to do anything unpleasant. but when we have set our hearts on attaining. or doing a thing it is not pleasant to be opposed. thus it is ever. The cross before the crown. Oh I can preach but I am a miserable practioner. I see that I was uselessly cautious with regard to Mr Moor but you would not wonder at it if you knew how many utterly truthless responses we are continually hearing. many thanks for your kind advice with regard to my own actions. it is good. and it would be in part of prudence. and discretion. and if my misearble temper does not get the upper hands of me I shall follow it. But it is hard to be treated with contempt. and as an inferior by men with whome one would not associate at home. I shall not try to conceal from you that I should very much like to secure some one of the many positions which I know I am more competent to fill. than are the present incumbents. I am ambitious. as well as lazy. and I should love promotion for your sake. but you loved me first the same. so what matters it. it is worse than useless for me to think of ever being anything but a private. so I will dismiss the subject. The reason of <u>Mr Daboll's</u> being home is this same time ago there was an order issued for a certain number of men. Non-commissioned officers and privates. from each regiment to go home recruiting. I think it was one from each Co. Corporal Daboll succeeded in getting the appointment from his Co. That recruiting party under the command of a Lieut from our regiment are still in home. ostenseably recruiting. in reality lying around home in the same way Daboll is still in the R.I. service as much as I am drawing his pay just the same as though doing duty in the ranks. and would get a ceratin allowance for every recruit he would get. He is liable to be ordered back the to regt and is eaqually likely to remain enjoying himself as he is for six months to come and for aught I know until the end of the war. In short Daboll has got what the soldiers call a "soft thing" of it. This is just the way the case stands and Daboll tells the truth when he sayes he expects to rejoin the regiment again. but when he does it will be when he is obliged to. Mr Daboll understands how to work his cards and get the right side of such men as our officers. hence his success in getting home recruiting.

Not but what he is a good soldier and has. and always would probably. do his duty. but Mr Barnum if acquainted with the man will understand my meaning. Pity you and Mr Barnum could not end the war. I still greater pity that those who have had the power to do it. have not done it long ere this. by the way if this reaches you while at Mr Barnum's I wish you would try to find out what Mr Barnum thinks of McClellan. I really wish that if able to teach this Spring you might get back the old situation though I have no doubt you will be able to engage elsewhere. yet it seems as though you wer at home while at Mr B's. and I love to think of you there. at the same time I would not counsel you to urge your claims if they seem at all disposed to savor or wish for some other teacher. but of this I have no neede to speak. Your own judgement will guide you aright. Many thanks to Miss Martha for her kind offer to "hold on" for me. but she must not require me to "double quick" or I might get out of breath and so lose the race. I am finishing my letter on the evening of the 28th. in the forenoon I was on fatigue helping to unload the Burnside. the vessel that brought our mail yesterday. She was loaded with commissary stores. nearly all of which wer intended for the horses that have lately been sent to Jacksonville. a small part of her loading was for St Augustine to which place she has gone tonight. After she had her loading on board it was decided to withdraw the troops from Jacksonville. and she was ordered to discharge the cargo intended for that place here. At noon her party was relieved by a fresh one. and this afternoon has been principally spent in cleaning up and preparing for the inspection tomorrow. and in writing a little on this letter. This evening there is a concert by our minstrel band but I prefer to finish my letter to attending it. We learned that the iron clads have left the head. and that the attack on Charleston is about to be made. It is expected that four or five companies of our regiment are to leave in a day or two. either to form a part of the expedition or to do guard duty at the Head and thus relieve a regiment there. that it may join it. We are all feeling rather unpleasant over it. We do not know for certain what Co's will go. if any. but it is very likely our Co will be one of them. it is also more than probable that some will have to go. Three steamers have passed today. doubtless going to Jacksonville after those troops. I am sorry to close my letter to you amidst such uncertainty. for I know you will be worried. and in suspense. until you hear agan. I think I will keep the letter by me until I know the mail is about to leave. and I may be able to let you know in a postscript whether we go or not. May our Father in Heaven. watch over. protect. and bless you. is the prayer of your own dear.

P.O. Tuesday Morn

    We are to go. A boat came in this morning after us. and we are to start first as soon as we can get ready. Five Co's of us. we do not know for certain our destination. prehaps it is the Head. prehaps Charleston. The weather is cold damp as it has rained hard all the time since Sunday morn. A very unpleasant time to start. Good bye dearest. I have but little aprehension that we are to be engaged. and will write at the first opportunity. May God give me strength and courage to discharge my whole duty whatever it may be. and may his blessing rest upon us both. and we be prepared for all his will concerning us. Once more my darling with love I am as ever your own
Ned

No. 106 next - dated Apl 10th was received after No. 108

No. 106 Rec'd Apr 15th 1863

camp Houghton
Hilton Head S.C. Apl 10th/63

Dear Sara.

    Your letter of the 29th was received by the Arago's mail two days ago. In consequence of large fatigue parties being called for to unload that vessel. we are having no drill today. so I'll commence answering your letter. I have just finished and mailed an answer to Martha's letter and mailed to you a copy of the "New South" of Apl 4th. It contains some verses entitled "Off for Charleston." which are quite good for the kind. The Captain alluded to as "Black David" is <u>General D.R. Hunter. You may not know that "Black Dave"</u> is a very common nick-name for him among the soldiers. "Robbers Row" is the street occupied by the sellers down town and is so called on account of the exorbetant prices they charge. I hope to be able to send you a copy of the next issue. also a copy of the Free South published at Beaufort. the latter is <u>the "Gideonite"</u> organized under the patronage of General Saxton. I received the N.Y. Tribune. but have not yet had time to read it. So far as I can learn we have not yet made much headway. in the attack on Charleston. Comdr. Dupont is confidant of success. One of our iron clads is sunk. the Keokuk. She was also called the "Whitney Battery." after her inventor. she was considered the weakest of the ironclads and I understand regarded with disfavor by Dupont. Eleven wounded men went down with her. A couple of the rebel water batteries have been captured and the Ironsides is said to be doing good service on Sumpters walls. The roughness of the water has prevented much being done for the past two days. one of the Moniters came into the harbor two days ago for the purpos of makeing some slight repairs. some strong piece of iron had got knocked into her machinery. and prevented the turret from revolving. She returned to Charleston this morning. Yesterday morning early. one of our small gunboats. the George Washington. ran aground at, or near, Port Royal Ferry. The rebels seeing her in the fix hurried up some field pieces and opened fire on her. the boat carried nothing but light field pieces I believe. and being aground the rebs had her at a disadvantage. After a sharp fight a rebel shot entered the magazine of the boat and she blew up. I have not been able to learn how many. if any. wer lost. All is quiet here. We have not heard from Fernandia since comeing away. A boat left for there two days ago. I presume many of our convalescents will come back on her. I think likely George will come but hope not. as he is not fit for the duty here. One (year) ago today. we commenced the bombardment of Ft. Pulaski and I was in for differant buisness than writing to my loved one at this time on that day. It is nearly two years since the rebels took fort Sumpter. how earnestly I hope that on the anniversary of that day. the glorious old stars and stripes may agan be planted in triumph upon its ramparts. never agan to be replaced by any other ensign so long as its walls stand. Now dear Sara you are entirely mistaken in putting such a construction upon my remarks about Carrie. I never for an instant thought that your remarks wer "uncharitable." I did not write what I did about her. in defense of anything you said about her. for the very good reason that I do not remember that you wrote anything particular. concerning her. I knew that you wer not acquainted with her. and I merely thought to give you some little idea of what I thought of her. I know that she has faults. and I know too that you are far superior to her in those good qualities which I admire and which make me love you so dearly. You have given me no occasion to think you uncharitable. why my dear. what put that into your head. I meant to have written you a longer letter. and been more particular in explaining myself. as I do not remember exactly what I wrote. and cannot stop to recall it to mind. but I have just learned that a mail leaves tonight. and I have but few moments to finish this in. Now my dear believe me that you are entirely mistaken in this matter for I am entirely innocent of the motives you impute to me. I cannot write more now for my tent is full of men talking and smokeing and I cannot hear myself think. With much love. ever your own

Ned

MY DEAR SARA

No. 107 Rec'd April 16th/63

Hilton Head S.C. Apl 2nd 1863

Dear Sara.

    Here we are back at the old starting point once more. and on the same ground we once before occupied. Not the same we left when we went to Fernandinia as that was out side of the intrenchments. our present camp is inside of them. but some distance from the Headquaters right in a sand heap of course. and the other blowing breeze is distributing it plentifully about. We left our own camp at Fernandinia at four o"clock on the afternoon of the 31st. and went immediately on board of the boat. which left the dock at first about five o"clock. The sailors on board of the Mohawk manned the rigging and gave us three cheers. which we returned with a will. as we pass the Fort the two companies there mounted the ramparts and cheered us lustily. The storm of the three preceding days had been all day in cleaning up. and at our starting the sun shown brightly. while the wind had lulled to a pleasant breeze. The moon shown brightly as the evening danced. and everything bade fair. for a quick and pleasant trip. Gradually as the night advanced the wind increased. at first it was in our favor and we made good progress. but finely it changed around and blew dead ahead. So that we made but little progress. The wind towards morning became quite violent. Our ship the Cassack had on board a cargo of ammunition and shell which in their hurry to get us here. they had not given her time to unload. This was quite fortunate for us as it helped very much to steady her and prevent her rolling. But it did not prevent me from being. as usual. dreadfull sea-sick. The night was very cold and as the boat had besides our five Cos. of the 7th N.H. on board. many of us had to lie upon the hurricane deck. which swept from end to end. by the wind. made it difficult to keep our blankets around us. Luckily for me. I got into a large boat that lay on this deck. and covered with a sail. there I lay as snug as a "bug in a rug" and warmer than I should have been in my tent. I did not sleep much. for the boat lay near the main mast to which it was lashed. and the sailors wer there frequently raising or lowering the sail. and the noise they made kept me awake. The crew on board were mostly green hands. and negros at that. So that it took an unusual amount of cursing and swearing on the part of the officers on the boat to get things done to suit them. We reached the dock here first at twelve o"clock and landed within an hour afterwards and marched directly out to our present camp which is about three fourths of a mile from the wharf. Our tents wer packed up and brought down to the dock at Fernandinia and left there under guard. to come on after us on the Burnside. New tents wer lent us to use until ours arrive which will doubtless be today or tomorrow. Today we are as usual immediately after moveing all stirred up. a small guard is on duty around our camp. a fatigue party has gone down to the wharf. probably to <u>unload the Burnside</u>. and bring up our camp equippage. there is to be an inspection of arms some time during the day. I have taken advantage of a few unoccupied moments to talk with you dearest as best I can with so much distance between us. We suppose. and it is more than probable that we are to remain here and do guard duty during the absence of the expedition. If so we shall have plenty of duty. but it will not be dangerous like that of the expedition. One of the regiments here which has done the duty which we think will fall to us. are cursing us to thier hearts content. but we are not to blame. They are a heavy artillery regiment and could not expect to be left behind in such an expedition as this. It is not certain but we may have some outpost picket duty to do. but I think not. if we do not know how to do it. I am told that Col Hawley used to be the commander of the post during the absence of the expedition. I know that he remains at Headquaters instead of coming to camp with us and that <u>Capt Skinner</u>. the senior Captain is in command of us. Our Lieut Col and Major are left at Fernandinia. George did not come with us. he went to the Docter and got excused I am told. I am glad of it for he is not fit to knock about with us. I expected that he was coming though and so did not see him when comeing away. Our Capt had to leave his wife behind. he mounted one of the paddle boxes of the steamer glass and handkerchief in hand and waved an adieu to her as we left the town. She standing on an upper piazza. and returning the salute. As we wer coming into the harbor here we met two of the iron clads. the Montauk and <u>the Keokuk</u> going out each in tow of a tug boat. the Keokuk is shaped something like a big turtle. in fact they call her the turtle back. she has two turrets. The Montauk one. I was prepared to see some odd looking craft. but they rather go ahead of my expectations. They wer prehaps half a mile from us. and the deck of the Montauk appeared to be level with the water. in fact she looked exactly like the pictures of the Monaters which you have often seen. The navy feel confident that they shall succeede in the attack on Charleston. I cannot learn how many. if any. of the troops have yet left. But few of the new troops have been landed on Hilton Head. Nearly all of them landed on St Helena island. Gen Hunter is to command the troops in person. Generals" Terry and Seymour and I do not know how many others. also accompanying the expedition. Dexter gave us a call last night. he is well. white faced and soft fingered like any clerk. will doubtless get his commission in a few months. He expects to go with the expedition but as he will go as a clerk at Headquaters it will be a desirable trip. far different from shouldering a gun and roughing it in the ranks to say nothing of the differance between perfect security and danger. Now dearest I have told you all I know and some that I have guessed at. of course rumors are plenty. but I will never write any that I do not think have the appearance of probability. After Dinner. Our tents have come and we pitched them in place of those borrowed for the night. The guard left with our tents at Fernandinia has come into camp. and by him I learn a few items of news. of which I will write. A large hotel called after its proprietor. the "Patton House" was burned yesterday about noon. supposed to have been set on fire by its owner who was reported to have come from St. Augustine on the Burnside. The house was quite a large one. built of wood. and occupied by several negro families. Mr Landon and two or three others. in charge of the contrabands. also had rooms in the building. It is supposed that the owner was mad because the building was occupied by the negros. a little negro child was burned to death. for two nights before we came away some one in the village had been signaling a cross to the rebels by means of lights. the same signals wer also repeated the night after we came away. Some rebel there is trying to make us trouble. Jacksonville has been agan evacuated and this time burned to the ground. The Col of the negro regiment would not allow them to burn it. but the 8th Maine burned it. The negros captured fifteen secesh prisoners and brought them up with them. I learn that the detail for guard out of our own five Cos. for tomorrow. is one Capt two Lieuts two Seargeants. six Corporals. and eighty privates. in all ninty one men. This will bring the men on. once in three nights. Pretty good. I also hear that Mr Moore is to act as Post Adjutant for the present. and our first lieut Townsend. as adjutant of the regt. Two more of the iron-clads went out this morning. and it is said took thier course towards Savannah. I wish you could see this place first now. the wind blows almost a hurricane. and the sand flies in perfect clouds at times so thick that the buildings at the Head are hid from view. While writing in my tent I have to keep blowing the sand off my paper. You can imagine what a time we have of it trying to keep clean. No floors to our tents. and at present not a board to be had. If this is to be our stopping place we may remain here three weeks. and prehaps two months. there is no telling. We expect to see Fernandinia agan but we know the uncertainties of soldiering to well. to be either surprised or disappointed if we are mistaken. A very small mail arrived here two days ago. brought I believe by <u>the Adams Express Co.</u> but the mail for the whole of our regt was not as much as generally comes to our one Co. So I am not disappointed at not recieveing anything. I shall hope to be able to write to you often while we are here. and will tell you what is transpiring as fast as I can learn the truth. but you will learn. facts with regard to the expedition from the papers before we shall know of them here. You must excuse my assortment of paper this time. I have got entirely out. have sent down after some. but having the opportunity to write thought I would use up the scraps and not wait. with love dearest. I am as ever your own

Ned.

No. 108 Rec'd Apr 16th/63

Camp Houghton
Hilton Head S.C. Apl 5th/63

Dear Sara.

I have nothing in particular to write you now. and there is but a few moments before dress parade. but my portfolio being handy. and myself well occupied and last but not least. having thought of you in my mind. I concluded to commence a letter thinking that prehaps fortune would favor me with material to fill it out during the evening. It has been a very pleasant Sabbeth day. the wind having condescended to give itself and the sand a resting spell for the day. Last night there was a sharpe frost. something we have not had in some time before. I find there is quite a difference in the temperature of this place and Fernandina. though there is but a little over a hundred miles of distance between them. The last of the Charleston expedition left the harbor early this morning. A fatigue party from our regiment wer down at the dock at work nearly all night helping load the vessels. About forty vessels of all kinds started this morning. besides all that have gone before. The present week will probably tell the result of the nearly a years preparation for the attach. This place looks quite deserted. all that could be spared have been sent away. and there are not men enough here to protect the place aganst anything like a determined attack. But there is no fear of our being disturbed. the rebels will have all they can tend too elsewhere. A mail arrived yesterday and I received Martha's letter of the 22nd and yours of the 23rd. I am very sorry you have to waite so long for letters. for as I have before told you. I know how to pity you. I keep account all the letters I write and when they are sent. also of all that I recieve and the date of them. I find I had sent you a letter on the 5th and another on the 15th. and hope you have recieved them ere this. I wish if it is not to much trouble. you would keep a little track of the letters you send me. and see if I acknowledge them all. I do not think it likely they will. or that any have been lost in the mail. but I am a little jealous of one or two seargeants who occasionly bring our Co mail from the Chaplain's and distribute it. We have had no service today. and there is no prayer meeting this evening to attend. our chaplain has gone with the expedition as aid to Gen Terry. He is very enthusiastic and brave. was very anxious to be along when Charleston is taken. and as our regiment was not to go he could not very well go in any other capacity. We are now engaged in learning all we can of the heavy artillery drill. but as the men are nearly all either on guard or fatigue most of the time. we make but little progress. Every night after dress parade we have to march out to the intrenchments and each squad necessary to work a gun is assigned to its particular piece. and we retire with orders. that in case of an alarm and the long roll is beat. the men appointed to each gun will immediately put on their equipments and takeing their rifles with them at once and without delay repair to the piece assigned to them. Col Hawley is fearfull of an attempted raid on the island during the absence of the troops. but I think the rebels have enough to do elsewhere without troubleing us. If our outpost pickets do their duty well. in the event of a rebel clash. we should have sufficient time to prepare for them a warm reception. and I have but little fear of the result.

Apl 6th. Did not finish my letter last night but as a mail is to leave for the North tomorrow. must get it ready for its journey this evening. I have been on fatigue nearly all day. with a squad of twenty men I have been employed near the general hospital in an old camp clearing the ground of rubbish and leveling it off. tearing down old tents. and loading and unloading wagons. The ground is being prepared for a convalescent camp hospital. <u>commodious</u> tents are to be pitched. and those patients in the large hospital who are recovering. are to be put in them thus makeing room for wounded men in the general hospital. I understand that large preparations have been made for the accommodation of wounded. at Beaufort. There are plenty of large houses there which are easily converted into excellent hospitals. No tidings of the fleet and expedition have yet reached us. Sara I have a favor to ask of you. You must have been often shocked at my frequent blunders. and mistakes in orthography. It was always my failing to be very deficient in that. and the careless manner into which. partly from necessity. I have fallen of writing without takeing particular pains to correct this error. since I have been in the army. has tended make it much worse. Now as I presume you read my letters carefully. you must notice these errors. and if you will note. and correct them when you answer the letters you will do me a great favor. I know it is rather a singular request to make. but you know we are never to old to learn and it may yet be of much use to me to be able to be correct in this particular. Having answered your queries as to our whereabouts. and welfare in a preceeding letter. I have onley to tell you that I am tolerabley free from a cough just at present. but do not consider myself rid of it yet. I catch cold much easier than I did a year ago and then my cough returns. hope it is nothing permenant. I am happy to hear that your health is so good that you are looking so well. No my dear it will make no difference to me whether you are "fresh" or "faded" if the joyful time ever comes when I can claim you as my own. I shall love you all the same. Though I by no means despise personal attractions. yet I know that they are not indispensable. You rate me very highly dearest. in comparing me with Mit. Robbins. you must not make that comparison aloud. for I am vain enough to think that I had been blest with Mit's advantages I might have known as much as he does. I think Mr Barnum meant to be a little ironical when he spoke of John Holley and Mit. as going into business. he must know that neither of them are hardly capable of takeing care of what money may be left them. much less of adding thereto by entering into any buisness. John is much the smartest. and is not by any means lacking in intelligence. but as a financier I think he would make a very poor show. Rich mens sons are good to spend money but are seldome good at makeing it. Marthas' kind letter was gladly welcomed. as she showed it to you I neede not notice its contents. I shall answer it at the earliest opportunity. With regard to the war question. she and I are widely at variance and as I am more confirmed than ever in my views it would be useless for us to argue the point. I regret having averted to it in the way I did in my former letter to her and would not have done so had I thought we differed so much. Now I remember that we had talked the matter over before. and had I stopped to think. it would have occurred to me. However she agrees to drop the subject and I ought to be thankfull that I am let "down" so easily. Please excuse the pieces that go to make up this letter. It has grown beyond what I originally planned its. length. and I have added on pieces as they wer required. happening to have a quantity of them just at present in my portfolio. Please ask Lizzy if I am owing her a letter. I cannot tell. If I am I shall pay her as quickly as possible and if not I want her to write. Remember me to Julia. I cannot write anything about George this time. It is after roll call. I have spent the evening with you to the best of my ability and must bid you goodnight. Loveing you fondly and praying for your welfare and happiness I am your own.

No. 109 Rec'd April 30th/63

Camp Starr
Fernandinia Fla Apl 15th 1863

Dear Sara.

After an absence of thirteen days we arrived at this. our old camp. last Monday afternoon. My letters written to you while at the Head have doubtlessly ere this time reached you. and now while you are thinking of me as being there. behold: I am safe back in Florida. Your long and interesting letter of the 5th reached me last Saturday. and I was anxious to answer it from the Head. that you might get my letter quicker. but I received it the same time. another letter. that it was necessary to answer immediately. and as we left the Head the next day I barely succeeded in writing one letter. On Saturday some of the regiments that had arrived the day before. landed. and we made up our own minds that our time there was short. During the day Genl Hunter arrived. and transport loaded with troops. kept comeing in. Meanwhile the camp was full of rumors as to the cause of the return of the expedition. As we are entirely in the dark with regard to the true reason it is useless for me to notice any of the reports circulated. You doubtless know all about it before this time period. On Sunday the troops continued to land. and when we left just at night there was quite a force on the island. We recieved orders at noon to be ready to leave by three o"clock. and orders to embark might be expected at any moment after that time. We had a short service on the parade ground in the afternoon. Mr. Wayland officiated. and of course was obliged to refer to the proceedings of the past week. and did so. but in such a back-handed manner that we wer just as wise after he had spoken about it. as before. and not a bit more. About Five O"clock we struck our tents and wer quickly en-route for the boat. on board of which bag. and baggage. we wer safely stowed before nine o"clock. I was on fatigue with a party. to load our stuff on board of the boat. and as we wer anxious to be on board quickly as possible. we worked lively and I managed to start the sweat pretty freely. We wer put aboard of the Delaware. and there being nothing but our five Cos on board. we had lots of room. but with the exception of the upper deck which had been scrubbed off just as we wer coming on board. she was awfull dirty. A regiment had been on board of her eleven days. and they had. had no opportunity to cleane her out. The smell on the lower deck and hold. was anything but savory. I preferred takeing my blankets up on the deck and making my bed in the open air. to sweating it out in the close. hot. hold. We left the dock and draft out in the stream where we anchored until about two o"clock in the morning when we commenced our trip here. where we arrived at two in the afternoon. The day was fine. the sea as smooth as a mill pond. and but for the misery of sea sickness. I should have enjoyed the ride exceedingly. Under any and all circumstances at sea. I am sea sick. I cannot get over it. I am provoked at myself. but that does not help it any. We got all our luggage off the boat and up to the camp on the afternoon of our arrival. and by night wer about in our old tracks left a fortnight before. Yesterday morning I went on guard. from which as usual. I was releived this morning. This forenoon was spent in cleaning up gun and equippments. in being mustered in and. in a short visit downtown. and to Tom Norton at the hospital. The object of said visit. being to help him dispose of sundry lots of goodies. received by him yesterday in a barrel from home. am pretty sleepy. and stupifid. this afternoon. but hope. if not obligated to drill. to finish this letter between now and tatoo. The object of the muster this forenoon. as I learn. is to find out the number of men in the regiment in obedience to a late law of Congress. it is said that the old regiments are to be filled up by draft. If any new men are raised. I hope that by all means they will be incorporated in the old regiment. as it will not onley save a vast expense to government. but at the same time the new men will thereby be rendered effective at once. Whereas. if new regiments wer formed of them. much time would be lost in drilling. Of course the number of men in each regiment could be told at any time without mustering. by applying to the Adjutants office. but I suppose that would not be "military". The <u>old "tub" Neptune</u>. (The same boat on which Mr Moore had such an unpleasant trip) started to come down here five days before we left the Head. several of the men came down on her. it being them supposed that we should likely remain at the Head some time. and they wer to bring up to see us articles of clothing and the like. that we had left behind. She also took along the mail for the men here. Well. the fun of it was. that she never arrived here until yesterday noon. a day after we did. The joke was decidely aganst the boys who came on her. I did not find George so well as I had hoped too. although I believe of his principle trouble he is no worse. but he had caught cold and rheumatism troubled him. I have got so that I write on the Sabbeth whenever I have opportunities. and so cannot school you. or anyone else for it. Doubtless when at home. I was so scrupulous with regard to it. I might much better have been quietly writing to friends. than to have passed the time as I did. it is generally hard work to realize the Sabbeth in the army any way. true there is no drill and except in urgent cases. no fatigue. but an inspection must be gone through with. which with the buisy preparation for it takes up half the forenoon. Then the dress parade at sunset is never omitted. unless the weather forbids. And then some of the men have always some little thing to do for themselves. which they improve the leasure of the Sabbeth in doing. So that. although the camp is somewhat more quiet than on other days. We can by no means have the still New England Sabbeths of our homes. by the mail which brought your letter. we received the glad news that the traitor Seymour had been defeated in the election. and that <u>Wm. A. B. is to be Governor</u> of the old state for another year. We could learn no particulars with regard to the election. and impatiently wait another mail that we may get them. Our news is. "that Buckingham is elected by three thousand majority. I am anxious to hear with regard to the Congressmen as they after all are really of the most importance. I cannot bear to hear that John H. Hubbard is defeated. and hope I shall not have to. Should like to hear that Geo Burrull is elected. but am not so particular about Mr Landon. The Republicans done very wrong in my humble judgement to nominate him. and wer the question before the people of a less serious import should be glad to see them well beaten for their folly. tell your Father that I cannot see how a man who has been in the political field as long as he has. could think of nominating Horace Landon. Politically he cannot go alone. if run for office. everything must be done. by those whome circumstances compel to support him. Of himself alone. he could not fall twenty five votes outside of the Canaan family. he is a curse to any party. and a drag. and a dead weight. to any project that may be unfortunate enough to have him along its would be supporters. Put him in. and shake him up. with the two nomanies of the copperheads. and I would not bet a pin on who would come out first. I may be egotistical but I beleive I know Horace Landon better than most of his townsmen who have been acquainted with him all his life. I would not say so much about him. but your Father cannot have forgotten Mr Landons despicable course. and the trouble he. (or his son rather) caused the Republicans. three or four years ago. As to the copperhead candidates. I must say they are "awfully" matched. I think that T.riz has got the inside track completely. He must be a big load for even Mr Richardson's broad shoulders. I earnestly hope they will both be elected to stay at home. Am much obliged to you for thinking to write me so much political news. I am aware of the "recruiting" story to which you allude. Some such law is said to have passed Congress. but some of it is said to apply only to the Virginia army. I do not know. But one thing I do know. and that is that Col Hawley is so averse to having any of the men go home. that he will let no more go. than he can help. and none at all unless he is absalutely obliged to. And in the event of the number mentioned. being allowed to go home. for reasons that I have before given I think I should stand the poorest of a chance to be one of the fortunate ones. I might possibly buy out the chance of some one more fortunate and go home in his place. and you can rest assured that if there even is a chance. I shall try every means consistent with honor to obtain it. But darling much as I wish to see you. I never will lick dust for our Captain to obtain favors of him. During our absence at the Head. we have heard nothing about the furlough buisness from the troops here. and but little is said about it now in our own regiment. From the first I have had so little faith in it that I forebode to allude to it. because I thought it might arouse hopes in your breast that there would be no feasibility of realizeing. Do not think dear Sara. that it is to me a matter of very little importance whether I see you or not. It is not so. I love you dearly. I want to see you more than tongue can express. I would do anything. would do anything consistent with honor for the privalige of spending one week. one day even with you. and though the parting would be bitter and sorrowfull. I feel as though I could then return with a better grace and discharge my unpleasant duties more cheerfully and faithfully than before. I doubt not. you suffer more than I do. You are not continually confused with different scenes. frequent changes of place. little adventures of oneself. on others. to talk about and wonder over a thousand and one little annoyances. petty spites. and mineature indignation meetings to attend. where each new. real or fancied grievance is talked over. commented upon. almost momentarily surrounded by some one talking. so that during the day a quiet moment is a rareity. and generally weary enough at night so that sleep quickly overtakes one. leaveing but very few moments for one to think of the pleasant and loved enjoyments of the past. You are at home at that home our love commenced scarcely a corner of it but reminds you of pleseant. prehaps stolen interviews. when our love was young. and it was not necessary for others to know of its existence. That existence which for so long you doubted yourself. you have many quiet moments and your mind goes back to all those delightfull meetings. and sighs in vain for the presence of the absent one. Oh. Sara it almost maddens me at times to think that I must be so long separated from you. believe me darling. nothing but absalute impossibility prevents me from comeing to you. I try to hope. let us hope. that all will yet be well. that there is happiness in store for us. and that we may yet be able to see. what we now try to beleive. that all is for the best. I think that Irene Stoddard has made a very poor choice. have no doubt but that Bostwick will quickly run through his property and Irene may one day know what it is to want a home. I hope not. Bostwick may. get turn from his present disasterous course. become sober and steady. but I fear he has got a fearfull start downhill. should dread to have a sister or lady friend of mine make choice of such a one for a husband. Dawn here we do not get a mail or an average. oftener than once in a fortnight. At the Head. we get about two mails a week. I am catching up a little with my correspondence and hope to be able to answer some long neglected letters soon. Will see to it that I do not defraud you. At present I am keeping all your letters by me. I am hoping that something will turn up by and by so that I can send them to you but I have no idea what that something will be. I keep them in a package in my knapsack with the request written on the wrapper that in the event of anything occuring so that I could not see to my things myself. the letters may be burned without their being read. I know fear but what such a request would be instantly complied with. I intend to send them to you if fessible. Should be very happy to indeede to see a lot of your girls come in and take a class of little ducks some fine Sabbeth afternoon. but should be a great sight happier to take a seat beside you some fine Sabbeth evening. Wouldn't that be an improvement think! I beleieve that I wrote you that Dexter had gone with the expedition. he intended to have gone had his things all packed. and was to have gone aboard the ship in the morning but the General. changed his mind during the evening and concluded to have the writing done at the Head instead of takeing a clerk along. So Dexter missed the chance. I have at last <u>written to Sara B.</u> I was almost afraid to. it had been so long since I had received a word from her that I began to fear she would turn me a cold shoulder. it was wrong for me to answer Martha's letter before Sara's but I presumed upon our intimacy and Saras' friendship to put her off. Prehaps there would have been no harm in it if I had not put her so far off. Sara I do beleieve I do not know enough to keep good friends when I have them. It is now evening. I should have finished my letter this afternoon. but that Olin, full of nonsense hindered me. Father wonders what you find so much to write about. does he? let him change places with me. with darling far away. as mine is. prehaps you would not wonder much longer. I can oneley assure him that his daughters' letters are more than interesting to me. There is a prayer meeting in the church tonight but I am to sleepy to go. I should surely get to sleep. I think of no more to write tonight dearest. I do not know where you are. or what you are doing. my thoughts are with you. I love you. I hope you are well. I hope you are happy. and enjoying yourself. Good night my love. I am ever your own.

Ned.

No. 110 Rec'd May 11th 1863

Camp Starr
Fernandina Fla Apl 20th/63

Dear Sara.

I want to talk to you tonight dearest. I would love to sit by your side this evening and tell you of all the little incidents of the last five days. of too little importance to pay for the trouble of writing them down. but you would doubtless be willing to listen and let me do the talking long enough to tell you about them. Now "firstly" the weather is getting to be decidedly hot. and it is all one wants to do to breathe during the middle of the day. Fortunately. for us. we have not had to drill but once since our return from the Head. but we have been buisy raising up and flooring our tents. and makeing them comfortable. Saturday a steamer arrived from the Head. with orders for our two flank Co's. A & B which are armed with Sharpes rifles to return on the boat to the Head. also bringing news that the Charleston expedition was about to start agan. I suppose the companies that have gone will be likely to go with it though they may be left at the Head as we wer before. Last Thurday a fatigue party went out to a plantation about eight miles distant for the purpos of takeing down buildings that we might have lumber to use about our tents. They wer not to have returned until this (Monday) morning. but a courier was sent out to them on Saturday evening to have the men belonging to A & B come in to camp that they might go on with their companies. The messenger sent did not half do his buisness. he must have been fearfull of being taken. for he did not approach near enough to the house to find out whether they wer there or not. but stopping some distance short he <u>hollared</u> to them. no answer being returned. he turned about. put spurs to his horse. returned to camp and reported the party captured or at least missing. Co D was at once sent out to skirmish the island and find what trace of the captors they could. The company reached the plantation at sun rise. and found the party quietly eating their breakfast. So much time had been lost however. that the departure of the men was delayed until four O"clock in the afternoon. instead of going as was first intended at the same time in the morning. On Sunday I was on guard. in camp. and passed a dull quiet day. We wer disappointed in not receiving a mail by the boat on Saturday. and it may now be a fortnight before we receive one. I think we shall not drill much more any way. Guard duty will be pretty hard during the abscence of the two companies. the men will have none too much time for rest when off duty. By the time the companies return it will be to hot to do anything. Luckily for us our surgeon is opposed to having the men drill and he has a good deal to do about it. George complains that he did not receive any late letters from Julia. at least not as late as I did from you. 21st. Afternoon I meant to have made out more of a letter than this. and can hardly dignify it with the title. but a gun boat has just come in. it is said she is going to the Head and will take a mail. In these times of unfrequent communications it will not do to let the opportunity to send even a line slip unimproved. so I send this along. It will tell that I love you. that I think of you much and often. and that I long to see you. to be with you. to live with and for you. I am very well now. dearest. With love. and a kiss. I am

Your own
Ned

No. 111 Recvd May 11th/63

Camp Starr
Fernandinia Fla. Apl 25th/63

Dear Sara.

Your letters of the 8th and 15th were received yesterday morning. Also two from Mr Randall. One of which contained the sad news of the death of one of my cousins in So Canaan. A boy of fourteen. Or fifteen. years of age. bright. active. intelligent. the only son of his parents. who are quite poor. I was brought up in the family. and he was like a brother to me. though for the past few years we have been together but little. His death was very sudden. A <u>severe attack of brain fever</u> carried him off in five short days. We corresponded quite regularly and I was answering his last letters to me. written the second of the present month. while at the same time the poor boy was struggling with death. His Father has been sick nearly all winter. is still quite low. and I fear the affect of this sudden blow upon him. You may recollect mentioning having heard of his sickness in one of your late letters to me. You may also remember that I lost a cousin while in the Store. She was the eldest child in the same family. Possessing good natural abilities. a kind heart. and energy and activity beyond his years. with more favorable advantages I have no doubt had he lived he would have made his mark. In the limited sphere he would have been compelled to have lived. had his life been spared. I have no doubt he would have been usefull. and respected the support and stay of his parents. and beloved by his friends. But <u>He</u> who seeith "not as man seeith" has thus called him home. and though I mourn his loss it is a comfort to believe that it is for the best. Although on guard yesterday I tried to write a few words of sympathy. and comfort. to the afflicted family. but succeeded but poorly. how very. very few there are that know how to comfort the afflicted. I also received the paper you sent. Thank you dearest for two such good long letters. I wish I could answer them as they deserve and shall try to have an answer ready to go by the mail which will probably leave tomorrow. We have the promise now of a mail once a week. hope we shall not be disappointed. but have seen too many military promises to place much confidence in the present one. One word about the burning of Jacksonville. it may not be the right spirit to manifest. but I cannot say that I regret very much that the place was burned. The most I care about. is that copperheads will make a great hue and cry about it. calling it an act of negro vandalism. while the truth is it was burned by the white troops. Do not waste your sympathies upon the "Union" sufferers. I have no faith in their pretensions to loyalty. and believe that all the genuine union feeling there was there. could be put in your eye without injuring it. There are so called union people here. who have taken the oath of allegiance. but in reality they are as great traitors as live. I tell you Sara there is no Union feeling down this way except among the negros. Now about the "Economy" matter we"ll <u>use</u> that up this time. My dear girl do you know that I am both surprised. and delighted? Who would have thought you could give such good lectures. I had no idea that you had any such serious notions of economy. Young ladies now days are not supposed to be fond of either studying or practicing of economy. I think my dear you must be an exception. or else the young ladies have been abused. My dear. Do you know that the very thing you mention. my want of means. to enable me to wed you if a kind Providence should premit me to return. has. and does worry me greatly? It's an ugly subject and gives me pain but I cannot avoid it. Neither can I avoid the truth which is so unpleasantly forced upon me. I never would mention it to you. I thought you did not think of it. and that the truth would only give you pain as it does me. But now the ice is broken I am glad of it. A thousand thanks. my love for meeting the question so squarely. and showing so firm a front to the ugly obstruction. Angry with you indeede. I adore you. I loved you before. You are priceless. But you are indebted to my absence. for the opportunity to show off your abilities as a lecturer. I should have stopped your mouth with a kiss before you could have fairly commenced your fine speech. Well seriously my dear I will try to do better. but I fear I shall make sorry work of it. for all that I get in the army is so little compared with what we <u>must</u> have. that it hardly seems as though it would help any towards making up that sum. So it seems that money could not buy a fine day for the wedding. Well if they only have pleasant weather through life. it will be of little consequence whether. or not. the sun shone upon their bridal. it is to be regretted that Mr Burrul could not have been elected in the place of the senseless blockhead. and the senseless blockhead Friz. It is a disgrace to Salisbury to allow such a fellow to <u>mis</u>represent the town in our State Legislature. He would deserve hanging as a traitor. but for the fact that "the sin of ignorance is to be winked at." I was much gratified with the success of the unionest Canaan. That has ever been a <u>hot bed of loco foco-ism</u>. but they wer this time nearer lipped than ever before. I was rejoiced to hear of the success of Mr Hubbard. that alone is a good victory. On the whole. we have been more successfull throughout the State than I had dared to hope. I am rather stupid today dearest. my thoughts are dull and wandering. and hand quite unsteady. I am feeling usuly well. but am always more or less dull after my guard duty. I am sorry I am not able to make my letter more interesting. I fear you will not feel encouraged to write me good long letters. By the boat that brought our mail yesterday we learn that our two companies are at the Head drilling skirmish drill. I begin to doubt again whether an advance on Charleston will be made after all. The long looked for Pay-Master. has at length arrived and we shall be paid either tonight or in the morning. The members of the 19th Regt are thus far quite fortunate. I hear of any quantity of them being home on furloughs while we who have been out twenty months. could not get leave of absence though the necessity for it be ever so great. Still we may be the better off in the end. Milton Bradley has lined his pockets and got home much sooner than I expected he would. Now I wish that he might be drafted and obliged to serve through the rest of the war as a private. It would I think be but justice. His disgraceful quibbleing to avoid the draft. and his precedings since I think entitle him to the supreme contempt of all honorable men. Thanks dearest for so readily complying with my request to rectify my mistakes. I will try to be more carefull. but in looking over what I have first written I fear you will find plenty to do to correct all the errors. Though I shall not make the request in each letter separately or prehaps ever allude to it again I shall expect you to continue your labors. for which I shall be gratefull to you. Now while I think of it. the story has reached home that I was sick in the hospital. I am puzzled to know how the story got around. I have never yet been obliged to be an inmate of the hospital and pray that I may not be. We have a very pleasant hospital here. a good lot of nurses. and a kind Surgeon. and the men seem to think that most of those now in it. are not at all anxious to leave their snug quaters and return to duty in their companies. You may recollect my telling you a short time since of one of our Seargents. and one corporal being reduced. for a trifling offence. and as I thought very unjustly. A few days since they wer both restored to their former positions. which act of justice has given much satisfaction. I send you a copy of <u>the Peninsula</u> a little paper just started here. together with two copies of the <u>Free South published at Beaufort</u>. One of them is for Lizzy. she must excuse me for not sending it to her direct as I have no penny stamps. Am going to send to the Head for a supply. Tomorrow is Sunday. I cannot go to church with you. but I shall think of you dearest. I must now try to get a letter ready to send Mr Randall along with this. as he particularly required an early answer. Have been so much engrossed with my own affairs for a few days past that I have hardly spoken with George. but I see him many times a day. and think he is about as usual. both my tent mates are on guard duty today. so I am alone in my tent. We have had but one drill since our return from Hilton Head. this gives us a considerable leasure time. and we are quite good natured now days. Good night darling. and love from

Ned.

No. 112 Rec'd May 25th 1863

Camp Starr
Fernandina Fla May 2nd 1863

Dear Sara.

I do not know what I am to write you. I am sure. But it has been a week since I have written and I do not feel at liberty to let a longer time pass without writing to you when I am at leasure to do so. Tomorrow will be the Sabbeth and I expect to be on guard. I am sorry for it will be our communion Sabbeth. Last Sunday we wer highly favored. The Rev. Dr. Eustice of New Haven is on a visiting tour to the various Conn Regts. Sent I believe by the state or Gov. Buckingham to ascertain their condition. and circumstances. He preached to us in the afternoon from Luke 15-20. At the close of the service our chaplain requested him to preach to us in the evening a sermon that he had already preached twice in his own pulpet. once on Fast day. and once on the evening preceeding the Election. The Dr. consented though he said it was rather lengthy and he would not preach it all for fear of wearying us. The little church was crowded in the evening. and the preacher might have lengthened. instead of shortened. his sermon had he sought to please his audience at that point. As it was. we listened with unabated interest till some time after roll call and then regretted that the sermon was so short. The text was the last verse of the third. and the first verse of the fourth chapter of Malachi. The Revern gentleman handled copperheads without gloves. while at the same time his language was as faultless in style as Dr. Reids' "best." He seemed to have "picked out" the best words. This sermon was largely made up of accumilated facts regarding rebel atrocities. and barbarities. going to prove the tendancy of slavery to brutalize and debase the society in which it exists. It also called strongly upon all to sustain the Government. and not to premit the blood and treasure already expended. to be wasted. by deserting our armies and so crippling our government as to force it into a dishonorable peace which could not be lasting. At the close of the sermon Mr Wayland called upon Col Hawley to express to Dr. Eustis our thanks for his kindness in visiting us and for his labors of the day. and also through the Dr. to Gov Buckingham for being so mindfull of us. and so solicitous for our welfare. "Old Joe" was taken a little unexpectedly but of course would not refuse. and in a few moments was bringing "the house down" as usual. He is a good speaker and talks well whatever may be his actions. I have before told you that our chaplain is but an indifferant preacher. and our visitor had therefore a double advantage. On Monday the Revd gentleman. with a party of ladies. and officers. made a pleasure trip over to Cumberland island which is <u>opposite Ft. Clinch</u> and six or eight miles from here. There are several plantations on it. and one or two fine houses with large pleasure grounds surrounding them. A small party of soldiers accompanied the party as a guard. One of my tent mates went with them and brought back a fine boquet of beautifull flowers of which he said the grounds wer full. I have tried to press two or three of the smaller ones which I will send to you. The place they visited is called "Dungeness" owned by a Mr Nightengale. I send you a ring made of tortise shell. it was made by one of the men in our CO. I think it rather neat. but do not know how strong it is. or (how) well the bit of gold will stay in. The frequency of guard duty. and heat of the weather. together. make us rather dull but little inclined for exertion. now days we can hardly muster up courage enough to write. There was a negro ball down town last evening. Of course I could not attend. but I hear that it passed off very quietly. The ladies were all dressed in the whitest of white. with the usual profusion of gay colored ribbons and trimmings. Shan"t try to write any more today dearest. for it's up hill work. I am sick of writing. I want to talk to you. With love. and a kiss on both cheeks. Ever your own

Ned

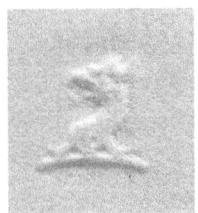

No. 113 Rec'd May 25th 1863

Camp Starr
Fernandinia Fla. May 5th/63

My Own Dear Sara.

  Here comes back to you your sheet of paper which enclosed your letter of the 19th Apr. received this morning. I cannot improve its looks. I fear I cannot improve its value. Will leave the latter for you to judge. It is just getting comfortable so that you can go away by yourself and sit without a fire is it? Well its so warm here. that without coat and vest off. seated in a draft of air in the shade. I can hardly keep the prespiration from blotting my letter as I write. Quite a differance. Well we are off again. Still going from home. O when shall we ever turn our course homeword in peace. Our next move is to St Augustine. We are to exchange places with the 7th New Hampshire which has been there since early Fall. I cannot imagine the reason for the change if indeede there be any. It is said that the duty is lighter there. but I cannot accept of that as a reason. Our two companies have not yet returned to us. and it may be that they are not to rejoin us in some time. and therefore they send us to St Augustine that we may not have so much to do. But I think it is more than probable that it is a part of the Generals policy. not to allow any one regiment to remain very long at any of these posts. I hate to move so often it is always so. Just as we get nicely fixed up any where so as to be tolerably comfortable we must pack up and be off. Beyond the trouble of moveing I care but little about it this time. as I suppose the place we are going to. is as pleasant and healthy as this. I have grown quite contented with this place as we have been so much better off here than elsewhere since we have been out. But believeing that all is for the best I shall not grumble much. We shall not get away in two or three days yet. Boston comes in this morning bringing us orders. and our mail. she staid but an hour or two. and has gone on down to St. Augustine. will return with five companies of the 7th N.H. and exchange them for five companies of us. return again to St. A. with us. and soon until we are all exchanged. I am glad we are to move so moderately. and only a part at a time. we can be so much more comfortable. It would be our Co's turn to go out to the Rail Road bridge on outpost picket next Sunday. I think I have told you before that each Co goes out there and stays a month. Well I guess we shan"t spend our month there. Last Sabbeth I was on picket. the morning was beautifull I lay in the shade of our little shantie gazing at the beautifull clouds. bright flowers. and green woods. and thought of you. and the friends at home. of how you would be spending the day. for I remember it was our communion Sabbeth at home. I finely fell asleep and had a good nap from which I was awakened by two of the men with me coming and sitting down in the same shade to play cards. They were gambling. one won from the other in a short time Eighteen Dollars. There is a vast deal of gambling going on now. and in our Co too. In the tent next but one to mine there is a party at it all day long. and a few nights since unable to sleep I got up in the middle of the night and on looking out of the tent found two men playing by the light of the moon. on a little bench near the tent. I love to play cards. and do very often. but I do not fear ever acquireing the habit of gambling. Well, but to return to the picket. A little after noon a shower came up and though we did not get wet. yet the sunshine was over with for the day. and sky-gazeing out of the question. It rained a good deal the latter part of the night and yesterday morning. I did not get much wet. and as it was warm. cared but little about it for a good wetting does no harm if one can keep warm. After comeing into camp and cleaning up I went down town went aboard of the gun boat. that lay up to the dock takeing in coal. spent a little while looking at the players in the billaird room. and ten pin alley and returned to camp in time for dinner. The afternoon was spent in reading and writing. this forenoon I have been arrangeing my things a little in view of the expected move. I am obliged for the "wedding" items and all the more because it was unpleasant for you to mention them. I should be a brute indeede if I fail to appreciate your desire to please me even though you are obliged to write upon distatefull subjects. Now about that white satin cravat. Don"t you see that it is absolutely necessary that Mine should be different in some article of dress from the poor vulgar crowde of bridegrooms who are every day being married? So too with regard to the invitations. Something a little odd must be done in order to differ from others so that people may be made to talk and wonder. and thus have the affair more noticed. Thats' my idea about the wedding now good bye to it. A little worried about a letter you sent me are you? Well I am not much. I have answered the letter. What do I think of what your Father said? Think he told the truth. Why should I be angry with either him or you for that? It would be foolish even though the truth be not very palateable. I cannot blame him for looking out for your interests. and certainly cannot blame you for seeing to your own. Tell Lib. to send along those two pages she has got written and may be I will finish out the letter for her and return it. pretty story. that she can"t find anything to write about. Does she think she must write an article as carefully as though she intended it for publication in the Atlantic Monthly? I"ll bet she never once troubled for something to write to Charlie. Let her just imagine she is writing to her lover. tell me who she is flirting with. what she wore to the last sociable. what is the Spring fashion for bonnets. whether they are worn on the head. hung around the neck or worn on ones shoulder. What lucky fellow helped her last. and when and where. and whether she helped him in return. The chance. if I was only a lady i'll bet I could manage to write nonsense enough to every fellow that wished to bother me with a correspondance so that he would soon get sick of his bargain. and leave me in peace. Kiss her for me. and tell her to write. How are matters between her and Charlie? Do you have to pay extra postage on any of my letters lately? Now my love I shall not probably write to you again until we are settled at St Augustine.

That is if we go as soon as we expect to. I wish to write two or three letters more to go by return mail. and we are so near through with Fernandinia that it will not pay to write much more to you from here. I love you darling. You know it. and that I am your owne

Ned

CHAPTER 6

ST. AUGUSTINE FLORIDA

From the book:

The importance of the operations of the Union Army against the defenses of Charleston during the summer of 1863 is little realized in popular history.

The reasons are evident. Three great armies were then operating at Virginia, Port Hudson and Vicksburg. Their very magnitude overshadowed the quiet work that General Gillmore, with but 10,000 men, was laboriously prosecuting at a point where newspaper correspondents were not encouraged and sometimes not permitted.

For this reason it seems proper that we should not confine the story of that campaign to the part in it taken by the Seventh, but should sketch briefly the plan and execution of the general movement.

It is not too much to say that in that campaign more ingenuity was displayed, more devotion and bravery exercised, and more knowledge gained in the science of ordnance and fortifications than at any other place or period during the Civil War.

Of this Major General Halleck, General-in-Chief, wrote: "General Gillmore's operations have been characterized by great skill and boldness. He has overcome difficulties almost unknown in modern sieges. Indeed, his operations on Morris Island constitute a new era in the science of engineering and gunnery."

Toward the close of May 1863, General Gillmore was ordered to Washington, and informed that the Navy Department wished to make another trial of the iron-clad gunboats against the defenses of Charleston Harbor. He was asked what part the small land force available for the purpose could efficiently take in such an operation.

His opinion in substance was, that Fort Sumter could be reached and its offensive power practically destroyed, without any material increase of the land and naval forces then serving in the Department of the South; but suggested that there should be a "cordial and energetic cooperation between the two branches of the service," and that the naval commander should be one who had "confidence in the efficiency of the monitors and their adaptation to such work, and was willing to risk his reputation in the development of their untried powers."

He also suggested that the most that the land forces could accomplish was the demolition of Fort Sumter. A land attack against Charleston was not contemplated. The naval authorities then at the seat of the government regarded Fort Sumter as the key to the position.

The final result of the conference was that General Gillmore was placed in command of the Department of the South and Rear Admiral Dahlgren in command of that portion of the navy which was to co-operate with him. No written instructions whatever were given by the War Department; everything connected with the operations of the land forces being left to General Gillmore's discretion and judgment.

He assumed command of the Department of the South June 12, 1863.

No. 114 Rec'd June 9th/63

St. Augustine Fl May 12th/63

Dear Sara.

Here we are in the "Ancient City." A delightfull place I think. I hardly know where to commence writing. or what to say first. I would not write at all. yet to any one but you I had much rather streach myself lazily on the green grass and give myself up to the enjoyment of the delicious sea breeze. the fragrent air. and to dreaming of the probable past of this old tumble down place. We reached here on Saturday noon (The 9th). Six companies came. two more we expect today. along with the remainder of our things. The two companies that went to the Head a few weeks since have not yet rejoined us no(r) do we expect them at present. On our arrival three companies went into the fort (Ft. Marion). the other three pitched their tents on a level green just outside the town. Sunday two of the companies left and went into the barracks at the other end of the town leaveing only our Co in the tents where we shall remain until the rest of the regiment arrives. when we shall go into barracks too. Sunday morning I went down to the "Plaza" or park. to see the guard mounting and got cheated out of the privalige of spending the day down town in conseqence. for when I got back and asked for a pass. so many had got the start of me that the Lieut in command refused to give any more passes. so I was obliged to spend the day in camp. I was real vexed for I wished to attend the catholic church. and I could have more liberty then. then I shall be likely to have another time. However I made the best I could of it and visited the Fort. an old Spanish structure commenced nobody knows when and finished in 1753. It is built of concrete as are also most of the houses in the town. All the guns are mounted on the parapet "en barbette." There are sixteen in all I believe including several howitzers. I should not think it would offer very much resistence to even our wooden gun boats. but it is a good defence aganst any land attack likely to be made by the rebels now in this vicinity and no gun boat is stationed here. It and well ventilated forts of more modern build. It is said that the N.H. boys opened an old cell. or dungeon. like in the corner of it. and found one or two skelitons. The companies that occupy it have quaters on parapet which is very pleasant. as there is always a fine breeze there which keeps at a distance the mosquetoes and gnats while at the same time it commands a fine view of the town. the bay. the distant ocean. and the surrounding country. At five o"clock Sunday afternoon we had an inspection. then went down to the Plaza to dress parade. then returned and the whole company went out on picket. our lines are close in to the town. the furtherest picket post out not being over a mile from the out skirts. In the day time the company on picket go out to a deserted plantation about four miles distant. but at night they return inside the lines and act as a reserve to the pickets regularly posted in the morning.

May 19th. I was interrupted in my writing a week ago by an order to start at the moment on a boat picket to be gone twenty four hours. It was a pretty hard duty. but I liked it pretty well. and should have liked it first rate if I had not had a couple of arrant cowards with me who were afraid of their shadows that they dared not go out of sight of the town. I had four men with me and my orders were to go down by the town and cruise around in the bay and creekes. overhaul all boats and if I found any without a proper pass to bring them in. I took a sail boat and used during the day. and had a fine sail. but found it was not going to answer my purpos at night when there is seldome any wind. so I came back just at night and took a row boat. During the day I landed on Fishers' island which lies in front of the town and went up to a plantation a little way from the shore. and got a lot of oranges. They were the wild or sour kind. quite bitter and not very palatable. but quite healthy I am told. and am inclined to believe it for I ate six or eight large ones during the day and so far from hurting me they seemed to do me good. These wild oranges are beautifull looking ones. larger than any we are accustomed to see at home and of a rich deepe yellow color. prepared in the same way as lemons. they make an excellent drink. nearly equal to lemonade and quite as healthy. I went out among the trees and gathered what I wanted myself. The orange trees all around here are hanging full of the greene fruit giveing promise of an abundant supply in due season. Orange trees are more abundant here than currant bushes are in our gardens at home. every little yard is full of them. I can give you no idea of the luxuriance and beauty of the vegetation here. It seemes to me I never saw such a such a beautiful green as envelopes every shrub and tree. Some of the houses are almost entirely hid among the oleanders. and orange and lemon trees. The oleanders are now in full bloom and their perfume is delightfull. The miserable little dwarfed specimens of that beautifull shrub which we see at home would shrink into insignificance if placed beside their monster brothers here. Both varieties. the white and red. are very plenty. and grow to the height twenty-twenty five. and even thirty feete. forming a shade which for beauty and perfume can scarcely be equalled at home. I neede not add that all kinds of flowers are in great profusion. The streets in town are very narrow varying from ten to twenty feete in width. Most of the houses. particularly the old ones (and they are most all old) are built of concrete. are rather small and low. and are built right in the streete. thier front walls forming the sides of the streete. Some houses of a later date are built in a more northern style at a distance back. with a yard filled with trees and flowers interveneing. St. Georges' is the principle streete. it running the whole length of the town and nearly in the centre. it is not over twenty feete wide in any place. except where it passes the Plaza. from wall to wall of the houses which line it. its general course is nearly straight. but it is full of short turns. At the Plaza it is crossed by King streete which divides the towne nearly in the centre in the other direction. King streete is about three times as wide as St. Georges and is a very pleasant streete as it is lined its whole length on both sides with fine shade trees. The Plaza is a park of about half an acre. enclosed partially by an open fence and shade trees. it fronts the water on the east side or rather its east side extends to the streete which separates it from the water on the north side. a short street which runs up from the dock (there is but one in the place) separates it from the catholic church or cathedral. at the head of this streete where it intersects St. George is the convent of st. Marys which is only separated from the cathedral by St. Georges streete. The Plaza is bounded on the west side by st. Georges streete on which stands the Court house facing the Plaza. on the south side the plaza is bounded by King streete on stands the Episcopel church facing the cathedral with the Plaza between them. Ft. Marion is at the North end of the town and the barracks at the South end. I use the terms North and South. though it is not exactly correct to do so. but as the town stands in nearly that direction they are near enough for my purpos. **I do not know** as you will be able to get any idea of the situations of the places I have mentioned. I am a very poor hand. at giveing descriptions. st. Augustine bay on which the town is built is a long and narrow sheete of water. more resembling a Sound. than a bay. Its course is about parallel! with that of the coast. nearly Northeast and Southwest. just opposite the town it runs more nearly North and south. It is entered by a shallow. narrow. channel. running between a long narrow peninsula which juts down from the North. and Fishers island another long narrow strip of land on the South. Thus far I am charmed with the place. The weather is rather to warm to exercise much. but there is a delicious sea breeze nearly all the time which tempers the extreme heat of the sun. I have been on guard three times within the last week so that you will not wonder that I have not written more to you. But for this boat picket I should have had a letter ready to have sent on the Boston when she came with the remainder of reg"t which she did on the afternoon that I first commenced this. There has been no boat in since. so that though you will be a long time without hearing from me. I shall have missed but one mail. and that unavoidably. The men from our reg who were to go home on furloughs went on the Boston. Hawthorn. who is from Canaan and one of "us." with three others went from our CQ. the other three were all New Haven men. You will of course hear of Hawthorne's arrival. prehaps see him. before you will get this. Prehaps you will wonder that I sent nothing by him. not even a letter or a word. I can't help it. Some time may be I will tell you why. I am sorry that I could not have sent you a letter by the mail which would have reached you the same time he did his home. Its of no great consequence. I only wanted to have asked you to keepe your ears wide open to anything you might hear that he had said about not only myself. but others. and about the regiment. without your appearing to notice that you did so. and without leading any and to suppose that you heard them. Its not my ways likely my darling that I shall be permitted to see you until my time out. I <u>shall not ask permission</u> to go home. and it is not any ways probable that the opportunity will be offered me unasked. Hawthorn is at present the Captains pet and he cannot do enough for him. so he not onley promotes him over the heads of half a dozen corporals who stood before him and a dozen of equally worthy privates. but allows him to go home at the very first opportunity though there are plenty of men equally deserving. who have families that require their attention. But of all this dearest not one word. to any one not even to your sister. If our friends wonder why some things are as they are you must let them wonder. If they could have those things explained they would only wonder still the more at the causes of them. I wish you would write to me anything that you may hear that you think may be interesting.

About the middle of last week we moved our camp from the upper end of the town near the Fort. to the lower end below the barracks. We are still in our tents. waiting for the barracks to be fitted up. the carpenters are at work makeing the bunks and the rooms are being cleaned out and white washed. It will be a fortnight yet before we get into them. I wish we might not go into them at all. would much prefer to remain in the tents if they would give us boards to fix them as we had them at Fernandina. We shall be crowded in the barracks. the ventilation is poor and such a thing as a quiet moment I fear will be a rareity. Co's & C. & H. are now in them. Co's F. & I. are in the Fort and D. E.G. K. are in camp. Co's A. & B. have not yet rejoined us. Guard duty comes plenty often enough particularly to the non-commissioned officers. There will also be some fatigue work until we get everything cleaned up and arranged about the barracks. Drills have not yet been heard of. and hope they will not during the summer.

May 24th. Sunday Afternoon. our Co has gone on picquet today. Did not go with them. Have been to catholic church all day to the neglect of our own service. What do you think of that. As I attended merely out of curiosity I do not think it a proper way to spend the Sabbeth. Have now attended three services there and as my curiosity is satisfied. shall not go there any more on the Sabbeth when we have service of our own. You know that all of the catholic service is in Latin and of course utterly unintelligible. the sermon which was preached in the morning was meant to be in English but the speaker being a Spaniard it was out of the question to understand him. Today is "Whitsunday" or the day of Pentacost. and therefore a high day with them. A week ago last Wednesday was the Ascension day. and I attended the morning services. they were nearly the same as those of this morning. Nearly all the people are catholic. but the congregation is not very large now. so many of the inhabitants being absent on account of their secessh principles. The cathedral will seat prehaps a on third as many people as our own church. It is nearly as large on the ground and as high. but there are no side galleries and a large space is taken up by the alter and shrines. Besides the principle alter. there are two others. one on each side. the one on the right being st. Josephs'. the one on the left st Marys'. Today the principle alter and St. Marys' were beautifully adorned with flowers and at vespers as the afternoon service is called. when all the candles were lighted. the scene was beautiful! indeede. I cannot bring myself to realize the services are devotional. it looks to me like a "preformance" and in spite of prostrations. kneeling. crucifixes. and images. there is nothing about the ceremonies that excites a devotional feeling in me. but the music. There is of course an organ and though the words of the music are all Latin. yet the music is fine. and I enjoy it. The service at vespers is almost exclusively music. The Episcopel church is about the size of the one at the centre. it has also a small organ. There are some very fine windows of stained glass in it. it is to be occupied as our chapel.

May 25th. Thought I should finish this letter last night but was interrupted. and put it by for a more convenient season. Today we have received our first mail in this place. Our last one came when at Fernandina the 5th of the month. Your letters of Apl. 26. May. 3rd. 6th. & 10th. are received. also the papers. I have read them (the letters) once but have not time to notice them particularly and attempt to answer them now. The boat returns tomorrow and I have but time to finish up what I wish to say in this. I hope to have more time. and more inclination write when we get fairly established in our quarters in the barracks. We are now the only CQ in camp. and expect to go into the barracks in a day or two. Co's A. & B. are not to return to us at present. They are posted on Bulls island. a small island separated from the Head by Scull creeke. and they are probably doing outpost picquet duty. I received five letters from other correspondants by this mail. so you see that I have some writing to do if I do my duty. I am still two or three letters short of what I expected. I have one letter from Martha. it is somewhat different in its tone from her former letter. or else her late conduct has made me suspicious and critical. I shall send it to you as soon as I have answered it which I am puzzled to know how to do and not give offence. Many things in her letter are anything but pleasing to me and I am taxing my brains to remember if I can. anything that I have written her which would lead her to suppose that such sentiments. or the utterance of them. would be pleasing to me. You must not mention this criticism of her letter. I am not at all anxious to continue the correspondence very much longer. but desire to drop it so as not to give offence. You can rest assured that Sara B. will never again be neglected for Miss R. I do not consider that I ever have exactly done so. I presumed upon Saras friendship. and ventured to answer Marthas letter first. Do you think my dear that for a moment. I place the two girls side by side. in my estimation? By no manner of means. One is <u>day</u>. The other is <u>night</u> to me. If I desire the society of rebel ladies, I can try to ingratiate myself into the good graces of some of St Augustine's dark eyed beauties. They at least are consistent in their sympathies with their own section. As Carrie once wrote me in answer to a remark that I made about seeing the ladies at Fernandina. "She should think I would rather not see any ladies. than that they should be rebels." For Southern ladies as well as southern traitors there may be some shadow of excuse. But for Northern ladies and Northern men who sympathize with the South. none. This is the first and only letter I have written since I have been here. most three weeks. Isn"t it shamefull! I see you have not received (or had not) my letter telling of our expected transfer to this place. News from home is going to be pretty old before it reaches us down here. I have just resumed my pen after attending dress parade. eating supper. and takeing a hasty look at the Tribune. our dress parades are held up at the Plaza. Its a nice place after we get there. but is quite a walk. sometimes we are favored with a few lady spectators. but not many. The sentiment of the people is to strongly secessh to allow them to manifest much interest in us. The residence of the rebel General Kirby Smith's mother is faceing the Plaza so that she is daily favored with a view of her son's enemies. She is also the mother-in-law of General Hardee. whose wife is burried in the army cemetery here. She is an old lady. eighty four years old. I have seen her several times. She is a Litchfield CQ lady. I am of course much surprised at Mr Fosters sudden death. But to him death was no surprise. he had so long been prepared to meet him. that he could feel his approach without alarm. I do sympathize with the afflicted ones. but they have the blessedness of receiving consolation from the fountain head. To them. the departed one is not lost. only gone before. Oh! what a blessed consolation. To know. to believe. to realize even the glorious truth. that our loved friend is safe in the arms of a merciful and loving saviour. If. as you seem to think. Auntie would be pleased to have me write to her. I will try to do so. but fear I shall not succeede very well. Mr Foster was always very kind to me. and I have reason to believe. my friend. but you know he was never in a hurry to express his preferences. choosing rather to wait and prove men before he espoused their cause. Poor Lizzie. doubtless feels as though she were left almost friendless. and truly she has lost an earthly father. Mr Foster has not been premitted to live to see the consumation of what he has so long and manfully battled for. I mean the abolition of slavery. but he must have been much cheered. and his hope and faith strengthened by the favorable prospects of the ultimate triumph of the cause for which he has so long contended. I fear my dear that you will think that I pay no regard to your corrections. I ask your pardon. In writing this I have done it so hurriedly that I could not stop to think whether I were right or wrong. I have much more that I wish to write you. but must stop now. Be assured darling. that I am loving you fondly. May our heavenly Fathers' choicest blessings rest upon you. and may you ever be happy. Accept many kisses love. and believe me ever your own

Ned

MY DEAR SARA

No. 115 Rec'd June 23rd/63

St. Augustine May 30th/63

My Dear Sara.

    I will try this afternoon to answer your letters received by our last mail. But first let me tell you I was on picket last night and so do not feel very bright. I received a box from Mr Randalls by the steamer that brought our mail. and I have just finished a letter to Mrs R. acknowledgeing the receipt of it. It was only a small box. most of the things I had sent for. Mrs R. writes me that I must not be backward in sending for any thing I want. that they are glad to do anything they can for my comfort. They are very kind indeede. Mr Randall has another new clerk. the third since I left. We moved into our quaters in the barracks last Tuesday. we are fortunate in getting as pleasant and comfortable quaters as any in the barracks. still for many reasons I would prefer to be in our tent. if we could have them fixed as we did at Fernandina. Where there are so many in one room. even the little quiet that we used to have in camp. is almost impossible to be had. Then too many little conveniences that we used to fix up. have now to be dispensed with. Still I do not complain. it is vastly better as it is. than to be marching and fighting. as our brethren on the Potomac are doing. scarcely stopping two nights in the same place. our officers think there is a probability of our going North. It is reported that Gov. Buckingham has requested that the Conn Regts in this Department may be sent into one of the Northern Departments. His request is made upon the ground that the second summer at the South is more fatal than the first. The Gov intended his request for our benefit. doubtless. but I think it would be difficult finding a more healthy place than this. to either summer or winter in. But as we can never know what is for the best when laying plans for the future. I shall try to be content to go wherever ordered. I should hate to leave this pleasant. and comparitively save place. where the duty is light and go into more active service. But it would be no worse for us. than for others. and no doubt we ought to exchange places with some of the toilworn and decimated regiments that have borne the brunt of many a fight. I went into an eating house kept by negroes this noon and had a good dinner of new potatoes. string beans. squashes, onions. cucumbers. tomatoes. fish balls and a kind of "chicken pie made of birds" and fresh bread. all for the small sum of twenty five cents. I forgot one item. nice fresh fish. The things were all well cooked. and though the furnature and crockery. wer rather "scarce." yet every thing was perfectly clean and neat. Green corn is to be had. but not very plenty. and cheap. yet. All other garden vegetables are in profusion. We are now having a good deal of rain. which the inhabitants tell me is unusal at this time of year. It is very favorable for the crops. I have no news to write today but as I have many letters to answer I concluded to write to you first. then you would be sure of yours any way. I do not know whether I am sorry you are not going to teach in Lime Rock this summer or not. I believe you are entitled to a resting spell during the summer if you choose to take it. Onley. you will be apt to work harder at home than if you were in school which you see would just be no resting spell at all. I cannot but wonder that Mr Barnum should do both so ungentlemanly. and unbusinesslike an act as to fail to answer your application. and in proper time. All's for the best. and something as good. or better. may be found by next Fall. You will now have time to look about and not be obliged to decide in a hurry. I know that you prefer being engaged about something rather than to remain quietly at home. and it would have given me much satisfaction to know that you were pleasantly ensconced in your old quaters at Lime Rock for the summer. But I am heartily glad you did not ask them but once. I can scarcely forgive Mr Barnum. such conduct to a lady is unpardonable. in such a case a <u>gentleman has no buisness to forget.</u> It seems he does not even deign to offer an apology himself. Very well. carry your head as high as you please Mr Lofty. You are rich and whose business is it. I am obliged to you for sending me the copies of the notes. I approve of yours. decidedly. presume that Mrs. B. is sincere in what she sayes. As for Martha. she is as much my confidant now as she ever will be. I am sorry that we have ever become so intimate. But you are my witness. that I never was egar for her friendship. I mentioned that I had received a letter from her in my last letter to you. I intend to answer it as plainly as I can and not give offence. but I am not going to allow her to suppose that sentiments that express sympathy with rebels are pleasing to me. I do not wish to make her angry just at present neither am I going to court her friendship further. I am forced to conclude that there is to much Richardson in her composition to permit her ever to become a myrtar to truth.In the letter she wrote me sometime since. she spoke of her intention of going west. as almost a sure thing. No doubt you would have spent a pleasant summer with your friends in L.R. I hope you will enjoy yourself equally well in Lakeville. wish I might be with you while the family are absent on their visit. I have no fear that we should be lonley or that the time would hang heavy on our hands. And then you know I should have a chance to find out what kind of a housekeeper you are and whether you <u>scold</u> any or not. I guess if I were there that Carrie would have most of the work to do for you would be <u>otherwise</u> engaged. I was just going to try my luck at getting some photographs taken when we left Fernandina. Some of the boys had very passable ones taken there. I had no opportunity to try. I hope you will not put the miserable thing you have in your album. I can never furnish you with a handsome picture. but may be able to give you a good one. You can put the photograph of Capt Hitchcock in if you like. and think proper.

Tuesday. June 2nd. One year ago today we commenced our march across Johns island. The two days are very much alike. both exceedingly warm. I have been down town this morning. and saw one Thermometer indicating the heat at 92 degrees. it stood in the shade and a light breeze was blowing on it. The year has flown very quickly. and as a regiment we have been very highly favored. Will the expiration of another year find us favorably circumstanced? Last Sunday I went to church in the forenoon. As there was to be a review and regiment inspection in the afternoon. the men were not required to go to church unless they chose to. There was quite a respectable audiance present. but Mr Wayland did not seem to think us entitled to a sermon. and so put us off with a prayer meeting. Guess we did not lose much. The afternoon was very hot. 91 degrees. At three o"clock we fell in with our arms. accouterments. and knapsacks on. marched about half a mile to the upper end of the town. near the Fort. There the line was formed. then we passed in review. and finely the Colonel and Adjutant inspected the arms of each man. As our CQ comes last. we did not get through until near six o"clock. there was no dress parade. In the evening I attended prayer meeting. a cold dead affair. Monday I was on guard at the barracks. a very easy post. The guard here consists of a seargant. a corporal. and six men. two men on duty at a time. There is no walking to post reliefs. and the seargant and corporal divide the night between them. each sleeping half the time. Today has been so very warm that I have hardly had life enough to move. The Colonel is having all the people here take the oath of allegiance. they must either take it. or go over the lines into rebeldome. The Provost Marshall's office is filled with women today waiting their turn. All over fourteen years of age are required to take the oath. They make some wry faces over the matter. but not in the Colonel's presence. I do not believe there is a spark of Union sentiment in the place. and if not. those who take the prescribed oath perjure themselves. I have been giveing Martha's letter a second reading and do not find it so exceptional as I thought it to be upon the first hasty reading. I suppose I was vexed at her treatment of you and ready to take offence at anything. Like every one else. I should prefer to have her friendly than otherwise. but cannot afford to pay much for her friendship. as I could never feel safe to rely upon it. I do not feel competant to advise in regard to the settlement. of differances of opinion between ladies. but if I were you I would avoid the subject of the war. when in conversation with Martha. unless she crowds it upon you. in which case do not back down one bit from what you think is right. Though many and grievous have been the errors in the present prosecution of the war. Though we have been cursed with traitorous and know-nothing. do-nothing Generals. Though speculators and human sharks have fattened and feasted upon the miseries and sufferings of thousands of noble patriots. Though the war is not yet ended and none but He who knoweth the end from the beginning can tell when it will be. or how much more blood and treasure must be poured out upon the alter of Liberty. Yet we know our cause is just. that we are in the right. and we firmly believe that by the blessing of God we shall ultimately triumph. And Sara. we also believe that the time will come when every northern <u>man</u> or <u>woman</u> will blush to own that they ever sympathized with this rebellion. and when rebel sympathizers will be looked upon and execrated as bitterly as were ever the Tories of the revolution.

Friday of June the 5th. Was on picquet yesterday. The weather has been very hot all the week and shows no signs of cooling off yet. Pitt is a little unwell. with a touch of the fever and ague. is better today. What do you think of shaking with the cold. when you are wrapped up in two good thick blankets. and thermometer stands at 92 dg. in the shade? on Tuesday a member of Co B died. the first man we have lost since we came here. he has been sick a long while. and all who knew about him. knew that he could not live long in the army. Probably if sent home when he should have been. he would have recovered. Another Military murder. He was buried on Wednesday in the soldiers burying ground just below the barracks. Two men came in from rebeldome on Wednesday. one of them has been a rebel soldier. was discharged for disability. came home. recovered his health and the rebels wanted him again. but he concluded not to stay. so made tracks for Yankee dominions. This is a queer letter you are getting this time. I catch it up and write a little just when I get a chance. You will have but this one letter from me in a long time. It's a poor apology for a fortnight. or three weeks production. Some way I do not know what to write or how to write when we do not get a mail only once in a great while. You must not think I love you less. I cannot bear to have you do that. I might borrow from the future. and build air castles and paint bright pictures of happiness which <u>we</u> are to enjoy <u>sometime</u>. But when I think of what a failure my whole life hitherto has been. and how all my plans have failed and how uncertain life is even for a single day. I have not faith enough to allow myself to dream bright dreams of pleasures <u>to be</u> enjoyed. Do not think. dearest. I have a fit of the blues. for I have not. I may just now have an eye to the <u>practical</u>. and inclined to <u>prove</u> everything before I accept it. fearing least it should be false. and find to late that I had grasped a shadow instead of the substance. nothing more I assure you. It is better to look things square in the face. figure them right out and if they be stubborn. unpleasant. facts that must be overcome. or they will overcome us. grapple with them manfully and clear them if possible from our path. Pshaw. How I am preaching. I am almost sure to fall into the first temptation that besets me. What I am driving at after all is simply. that it is best to deal with the present. the today. Enjoy what we can reasonably of it and not trust that all our earthly happiness is away somewhere in the dim future. The future is treacherous. It promises us fair. but too often when we attempt to taste its promised pleasures. we find them like the dried and bitter apples of Sodom. It will not do to promise ourselves happiness in the future. we can only "Hope on, hope ever."

Wednesday June 10th. Last Saturday I wrote a letter to my friends in So. Canaan. Sunday we had the usual inspection at nine o"clock. At eleven. went to church. Mr Wayland preached from 1 John. 3rd ch & 20th verse. I believe I did not go to sleepe but for the life of me cannot now recall anything he said. There is a small organ in the church and one of our men played it. but a little negro that worked the bellows made as much noise. and almost as much music. I intended to write in the afternoon. but somehow I did not get about. Attended the prayer meeting in the evening. Monday. I was on picquet. The morning was very close and sultry. and to lighten myself as much as possible concluded not to take my blanket. or overcoat with me. for the last few times I had been out they were nothing but an incumberance to me. so took nothing but a rubber blanket with me. Well. the wind blew up fresh. damp. and cool at night. and though it kept off the gnats and mosquetoes beautifully. I caught a beautifull cold in consequence. I spent the day in reading "Adam Bede." a novel. presume you have it. did not finish it "till yesterday. It is a very good story but has no other merit as I see. will do very well to while but away idle hours with doing the reader no positive damage prephaps other than encourageing a taste for fictitious reading. and certainly doing no good. Yesterday could not bring myself to write as my head felt large on account of my cold. For two days past there has been a great slaughtering of the dogs in the regiment. All but three or four have been poisened and I am heartily glad of it for they had got to be a regular nuisance. CO D has a large dog that they got when he was a puppy on Hilton Head when we first landed there. He is called "Jeff". Our Co has a middleing sized black and white dog which we found in fort Pulaski. he was wounded in the leg I suppose by a piece of shell. he has been with us ever since. and answers to the name of "Bill." These two are to be passed over in the general massacre. being considered as "dogs of the regiment." So much for the dogs. I have just been and had a good shave. and my head thoroughly rubbed and well dressed with bay rum. it feels a little better but is pretty large yet. The weather is a little cooler this week. and more comfortable. The oranges on the trees behind the barracks are growing fast. they are as large as peaches now. Watermelons are getting ripe. and all the other fruits and vegetables are coming to maturity very rapidly. We are getting impatient for the arrival of the boat. Officers and men go back and forth to the lookout on the top of the barracks from whence a good view is to be had of both sea and land for many miles. and strain their eyes in hopes to catch the first glimpse of the expected boat. Somehow I can"t bring myself to feel so anxious for the mail to arrive as usual. I don"t know why. prehaps it is because I fear I may hear bad news. prehaps because I fear we may be ordered away. If we are not ordered into more active service soon I fear I shall lose all taste for fighting. In our present safe quaters we are apt to forget that anything more dangerous than guard duty is likely to be required of us as soldiers. which would be quite a wrong impression to get. you see I am not spoiling for a fight nor anxious to leave by any means. only when we do have to go to work in earnest it will come all the more tough after a long season of quiet. Should like to be sure all the <u>hard work</u> that we were required to do was done. then I should feel at ease about it. But then it is better. as it is I have no buisness to worry about the future. On some accounts a more active life is pleasanter than this. The mighty barriers between officers and soldiers get very much lowered when we are all equally exposed to danger. when on the march. the officers must foot it along with the men. and at night spread their blankets on the same ground with them. Then too there is less grumbling and fault finding among the men and there is not the time to indulge in pettie jealousies and strifes which are often very unpleasant after we have been awhile in camp. There is nothing like common danger. and common hardships to unite. and make men of one mind. In a few days more it will be time for the return of the first lot of furlough men. but then it's not at all probable that they will be back on time. if they return by the first of July it will be as soon as I expect them. Another anniversary of our national independance will soon be around and I shall not be able to spend it with you. I shall have to comfort myself with thinking of the enjoyment of the past. I rather doubt whether there will be very great demonstratons made. as a general thing this year. people will of be more inclined to wait I think untll they can celebrate the day of our <u>second</u> independence.

June 14th Sunday. The steamer came in yesterday. I received your four letters of May 11. 24. 27 & 31st. Cannot answer them in time for the return of the boat. but shall set myself about it soon as possible. I have nothing more to <u>add to</u>

this letter without commencing to answer your last ones. I am very well now. my love. shall go to church in a few moments and while on my way there shall take this to the Post Office. I remember walking over to S. church one fine morning and stealing a chance to drop a letter in the box for you. that was when you were at Lenox. Wish I could walk to church with you this morning dearest. <u>We must not hope</u> for so much pleasure till fifteen months more have past. Libbies letter of May 24th is received. tell her I will answer directly.

With love. my darling. I am ever your own
Ned

No. 116 Rec'd June 29th 1863.

St. Augustine Fla June 15th/63

Dear Sara.

If you are a housekeeper now. you would hardly find time to sit down and go to writing early on Monday morning I imagine. But then there is no comparison between houskeeping and soldiering. We have leisure when we least expect it. and often when we have said to ourselves "now I will have a good quiet time to write a letter or to read" we are ordered off on some disagreeable duty. Its a cloudy morning but very warm. The steamer started for Hilton Head about six o"clock. We did not get any very important news by our mail. Papers of the date on June 4th were received. A schooner came in the day before the steamer did and set the boys all to talking over the probability of our being releived and sent North. The 7th N.H. which went from here to Fernandina. has been releived and itis said is to go North. The 11th Maine take their place. The boys that looked to see a regiment on the steamer to take our place were disappointed. on the. contrary it is reported that General Hunter said in speaking of Col Hawley. that "He was the right man in the right place." Well. its a monotinous treadmill kind of an existance here. but I suppose is better than being shot at. or shaking to death with chills and fever. in Virginia swamps. A number of stout able bodied negroes who have been waiters. and hangers on. to our regiment for some time. were taken on the boat this morning to the Head where they are to be "conscripted" and placed in the negro regiments. Quite a number of officers came down on the boat and returned on it. I suppose they came for the pleasure of the trip. Guess they are not troubled with sea sickness as I am else there would not be much pleasure about it. After finishing my letter to you yesterday morning. I went to church. The men were not required to go unless they choose to. the reason being that as the mail was to leave so soon. it was desireable that all should have as much time to write as possible. The opportunity was very generaly improved. pens were buisy and frequent packages of letters passed on to the office. For one Sabbath the cards had a resting spell. for all not writing were buisy with the papers. Mr Wayland preached from Isaiah 53rd ch. & 4th verse. The audience was small. and as our organist was not present we were not diverted by the performance of his limp with the bellows handle. In the afternoon I attended vespers at the Catholic church. I go there to hear the music which is very good though of course I cannot understand a word they are singing. If I could get hold of a Catholic prayer book. I could follow the service. At vespers. the service is all music. and burning of incense. The congregation are very quiet and devotional. I did not suppose my dear that you were so short of funds as to be unable to pay postage bills. but you know it is not pleasant for a lady to be troubled to pay additional postage. and a gentleman should always be careful not to subject her to the annoyance. I must confess it seems strange to me that our people can be so easily gulled. They do not seem to learn anything after all the miserable cheats and impositions that have been practiced upon them for two years past. Any story. no matter what. or how absurd it may be that the N.Y. papers publish. they will swallow at a gulp. without waiting to see whether it is confirmed or not. It does seem as though they ought by this time to have found out what a contemptable. miserable. time serveing. money grabbing. lying. pack the N. Y. dailies are and never allow themselves to believe any story they may tell. until it is confirmed by some reliable papers from other places. They have been a great curse to our cause. and still are. My sister wrote me from Illinois. May 12th. that they were then having a great glorification over the capture of Richmond. I do believe that people love to be gulled and fooled. else why does not an indignant people raise an outcry and visit the authors and publishers of these miserable frauds with such summary vengeance as will effectually deter them from all desire to continue the practice in the future. There is much reason to believe that there is truth in the saying that "men like a lie better than the truth." Several men in our CQ are acquainted with Bostwick. the Col of the 27th c. V. he served with them in the "Greys" during the "three months." From their account of him. I should expect some just such result. as occured to his regiment in the late battle. I find that those who are acquainted with him are not at all surprised at the rediculous figure he cut in the engagement. I have not learned the number of their killed and wounded. but guess when the "missing" are accounted for that the loss of the regiment in them is very small. One regiment properly officered and well disciplined. would arrest a host of "Flying Dutchmen" and not be run over by them. I have no confidance in the nine months men any way. There are undoubtedly some good fighting men among them. who will do their duty anywhere and be an honor to any regiment. but the majority are men who were bought or driven to enlist by fear of the draft. This, enlisting men for nine months. or nine years. is all wrong. Every man that is wanted. should be drafted. and drafted for the war. There should be no volunteering about it. I hope that the government will never accept of another volunteer regt or company. I expect that if I am alive and well I shall have a chance to enlist again in this war. after my three years time has expired. The Copperhead party are determined that this Administration shall not have the credit of crushing this rebellion and I think they will suceede in their infernal design. They mean to keep the war alive for political capital and upon it elect the next President. I like Sara B's grit first rate. and hope she will not take back a word that she said to Martha. If madame M. wants to make "much ado about nothing." let her choose some less important and tender subject than the present national troubles. And she is right about not wishing the war to close until slavery is entirely driven out of our country. I was in hopes to get a letter from her by our last mail. but perhaps she means to keep me waiting as long as I did her. I have told you. I believe in a previous letter. that you must not expect me home on a furlough on no account. So long as we are both well I shall never ask our Captain for such a favor. I cannot bear to ask. and court favor with a man I despise. I would remain a private all my days first. to say nothing of asking to go home. However. aside from that the War Department has settled the question effectually. by ordering all furloughs stopped for the present. Some are in hopes that they will be granted again shortly. but it matters little to me anyway. still I would be glad to have all go home for a short visit that possibly can while we are inactive as at present. I have thought much of you my love for several weeks past and how much we might have enjoyed together during the absence of the family. In imagination. you have set with me in the easy chair. my arm has been about you on the sofa. and I have stood by you at the piano while you sang for me the songs I love. But this pleasure we may not hope to enjoy yet. I cannot kiss you under the "white jockey" unless it lasts you a year and a half. I am glad the white ones are in fashion. for I like them best. I always disliked the colored ones. The ladies down here were them. they braid them themselves out of the palm leaf. They also braid card baskets and a variety of fancy articles. also hats for the soldiers. I have not purchased one because I do not care to pay a dollar and a half for a secesh hat though both the hat and its maker are pretty. The order from the War Department granting furloughs to enlisted men premitted five per cent of each regiment to have leave of absence for thirty days. Thus if there were a hundred men in a company. five could go at a time.

We had that number when we came out. and though but four have died. our number is so reduced by various causes that but four men were allowed to go home from our CQ. The men to go are selected by the Captain of the Co and by him recommended to the Colonel. who then grants the furlough. All who go home do not have to curry favor with the Captain and ask him for the privalige. but it is generally his favorites that are selected. Mr Hawthorn <u>did ask</u> for his furlough. but do not mention that I told you so or that you know that he did. One of the men who went did not ask. nor expect. to go. he was very diserving and I am heartily glad that he was premitted to go. I have not allowed myself to hope for so much joy. as to be premitted to pass a few days in your loved society. for I knew there was no probability of it. But I have thought much. how we could enjoy ourselves and happy we would be for the time. for I should not allow a thought of the sorrow to come when we should have to part again to mar the pleasure of the few days that were ours. No. I have the thought to myself. if I did go home I would settle it in my mind that I would enjoy the little time allowed me to the utmost. and no shadow of sorrow to come should cheat me out of a moments pleasure. I would not write so much about this dearest. only to show you that if I did not come home it was not my fault. that I would love as dearly to come to you as you could possably wish to have me. Often, often have I thought of our sad and hurried parting. The incidents of those last few days at home are as fresh in my mind today. as when they occurred. I frequently think if I could only return and see you once more. though the parting would be dreadful. yet I could feel better satisfied. and more reconciled. It always seems as though I had run away from you and had thus added to your sorrow. I am glad my love. that you are feeling more satisfied that I did right in enlisting. I had rather be as I am today. with your commendation and my own approving conscience. than in the places of any of the young gentlemen of whom you make mention that were at the wedding party. They are too useful. and ornamental. of too much importance to the welfare of the community and the good of the nation, to admit of their jepradizing their valuable lives in the army. So be it. I am content to be a "mud sill." and become food for powder if I can be of any service to my country. If I live through it. I shall always have my own private opinion of some of our <u>loyal young</u> men at home.

June 16th. One year ago this morning we were engaged in deadly strife on James island. The weather is almost precisely the same that it was then. warm and cloudy. Many a brave man gave his life for his country in that fatal hour. But it was nothing but a skirmish compared with other conflicts of the war. I have no opportunity of knowing to what extent the good things which our ladies provide for the sick and wounded are appropriated by the surgeons. nurses. and hangers on about the hospitals. but I have no doubt that in many instances. those for whome the things were sent never either get a taste or sight of them. My opinion is that not one tenth of the wines and liquers that are sent for the use of the sick are ever appropriated to their destined use. I may be mistaken. but I know that probably nine tenths of our officers and surgeons are drinking men. and I know further that a good many of our officers would not hesitate about the means by which their liquor was obtained. But I do not wish you to speak of my opinion in this matter. I may be doing injustice to the majority of surgeons and hospital attendants. Our hospital is at present under excellent management. and I do not think that anything sent it would be misapplied. I think that at present it is well supplied with all things necessary. our hospital steward. and most of the nurses. are fine fellows. and surgeon Jarvis is liked by all. But I imagine that but few hospitals are so well cared for. and in so good hands as ours. The larger part of the surgeons in the army are young ones. just from collage many of them. of course entirely without practice. Many of them would starve if dependant on their profession for support at home. Very many of them are reckless. unprincipled. wretches. who care for nothing but to gratify their own base passions. On the night of the Pocotaligo fight. I saw some of the surgeons myself. so beastly drunk that they acted. and appeared. more like a lot of truants from Dr Knights school. than like men. There was no mistake about it. they were drunk. I should have as much confidance in the sanitary commission. as in the surgeons. rather more I think. Wherever there are any ladies engaged. I should have no fears but what the things would be properly used. The truth is dear Sara though I have been in the service twenty one months. I have had but little real experiance in war. at least not as I used to imagine it to be. and in fact as it is in other parts of our country. I repeat what I have often said. we have been very highly favored. If we have no harder experience in the months to come than in those that are past. we cannot claim a very great debt of gratitude from our country. A schooner is to leave here in the morning. and I am just informed that she will take the mail. so I shall put this in the office today. I mailed a little paper to you on Sunday. I did not receive the papers that you spoke of doing up for me. George is pretty well now. he is still cooking. I think his work is harder than it would be in the company. but the exposure is less. and that is a great item to be considered. With love darling. I am ever your own.

Ned

                                                    We had when we came out.

Enlisted men	98
Commissioned officers	3
Have received new recruits	3
Total	104

Of this number we have lost by death Commissioned officers. 1.
Privates 4. Discharged 5.

Promoted	6
Present number of the Co	91
Of these there are	
Commissioned officers	3
Musicians	3
Sick and wounded	10
On Extra Duty	18
total	34

Leaving for duty in the company...including non-commissioned officers   57

Taking out non-commissioned officers. the men on furloughs. and some on extra daily duty. it leaves us at present but 40 privates to do guard duty. Our detail for guard each day varies from eight to twelve men. so that the men come on duty about once in four and five days. I have made this little statement thinking it might interest you.

I must ask you to excuse my many blunders in this letter. I do not know what has got into me. for the two days past I have scarcely had an idea of what I was about.

                                                    Ned

No. 117 Rec'd July 7th 1863.

                                        St. Augustine Fla June 19th/63
Dear Sara.

The schooner by which my last letter was sent did not leave until yesterday. A steamer attempted to come in the harbor this noon. but the tide was running out and the wind was blowing violently. and she could not succeede. She has gone back probably to Fernandina. perhaps to st Johns river as the tide will not serve her until about eleven o"clock tomorrow. As she only left here Monday morning we are a little curious to know the reason of her returning so quick. Last Wednesday the 17th being the anniversary of the battle of Bunker Hill. we had a meeting in the evening which was held in the Presbyterian church. Speeches were made by <u>Lieuts Dempsey</u> and Wildman. Mr. Wayland. and Col Hawley. some attempts at music were made by a Quartett club. but did not amount to much. One man sang the "Sword of Bunker Hill." very well. all united in singing "America" to conclude with. A few lusty cheers and the "day" was celebrated. The church was very neatly trimmed with evergreens. and flowers. and decorated with flags. swords. muskets etc. A few citizens were present. including some ladies. one promising young Lieut succeded in finding three of the latter to accompany him. though much I doubt if they would have done so. did they but know him as well as we do. The speeches were decidedly tame. even "Old Joe" was very sparing of his ammunition. but he made himself popular with some of the men by his remarks. He patted the boys on their backs a considerable. and called them good fellows. A deserter came in the night before. and the Col got him to make some remarks. His name is Wright. he is a telegraph operator and has been employed by the rebels at a salary of Fifty Dollars per month and expenses paid. his board cost Seven Dollars a day. Two Dollars per meal. He confirmed the reports we hear of the high prices of all the necessarys of life in the confedaracy. A pair of shoes that he had on cost Sixteen Dollars. I did not take particular pains to remember the prices he named. you have seen them quoted in the papers frequently. He concluded by saying that he had come to offer his services to the Government in any capacity they might see fit to employ him.

<u>Hilton Head S.C. June 29th\63</u>
Dear Sara.

    I concluded to piece out this letter rather than to throw it aside as what I wrote several days ago will do no harm. I should have sent you some letters before this time. but for two reasons. The first is that for several days I had such sore eyes that I could not read or write. could hardly distinguish a letter. and the second is that I have not had the time. George wrote to Julia that I had been ordered to report to the Head for duty. so that you know why I am here. I can not stop to tell all the little particulars now. I came up on the Boston leaving St. Augustine at one o"clock last Friday and reaching here the next morning. I am mistaken it was Thursday I started and reached here on Friday. I have been trying to do what little I could in the office since but it has not amounted to much. Dexter ment to do me a favor when he got me ordered here for duty and I am very greatly obliged to him for his kindness. It is pleasant in the office. and I should like to stay. but think it very doubtfull whether I do. The truth is that it requires good. and rapid writers here. and I am neither the one or the other. and then I have been out of practice so long that I have lost what little skill I ever did possess. Therefore I think doubtfull whether I shall be able to give satisfaction. I would much rather not have left the Co at all. than to come, make the trial, and be returned. But I am equally gratefull to Dexter for his kindness. Besides Dexter there are three or four very pleasant young men in the office and as I said before I should be glad to remain. but shall not be disappointed if returned to the Co any day. But of this you need not speak to any one. I am talking to one who has a right to know all my cares and troubles, and who is therefore my confidant. Do not say more than that I think I may not stay here long. I am in the Asst Adjt Genls Office with Dexter. Mr Landon is in another office. that of the Commissary of Subsistence. We all mess together and I need not tell you that it is pleasant to be with him and Dexter. and be employed once more in something like my former occupation. I have much to say to you. but for want of time must defer it. I received your letter of June 21st today. Many thanks. Also two papers. I learn that a boat is going to St. Augustine tomorrow and I must write a line to George tonight. I wish he had been sent here instead of me. Have patience with me. my _love_. a little while longer until I either get settled and acquainted with my duties here. or else am returned to the regt. and then I promise that you shall have your usual allowance of letters. You know that I am loving you all the same. and am not forgetting you. Please continue to write and direct as heretofore for the present. and I will run the risk of getting the letters when I can.

With love. darling. Ever your own
Ned

No. 118 Rec'd July 21st/63.

Hilton Head S. C. July 5th 1863

Dear Sara.

It is Sunday. I have been at work most of the forenoon. have just returned to the office. none of the other clerks have come in and none of the Officers are in and as I have not yet been assigned to any special duty. I have nothing to do. The clerks are obliged to be in the office a part of the day on Sunday but if there is not much to do they get through earlier than on weekdays. Yesterday. the 4th. Maj. Smith dismissed us about the middle of the forenoon. I spent part of the day in Dexters' tent reading and the remainder of the time in strolling about the place. It was a very quiet day here. no demonstrations except the fireing of the national salutes. All the men were relieved from duty as much as possible. but on account of the movement of some troops. and the sailing of the Arago. the dock was a scene of noise and bustle all day. I left the office about eleven o"clock and reached the end of the dock just in time to see the Arago leave. It was a fine sight. She is you know one of the largest Ocean Steamers. She runs regularly between New York and here and carries for her protection four guns. As she left the dock her deck was crowded on the side toward land. Among the crowde were civilians. officers. and discharged soldiers with quite a number of ladies. A few of the latter were using their handkechiefs pretty freely as though they were leaveing somebody behind. After she had turned about and got fairly started on her journey. She fired a national salute. It was quite inspiring. the noble ship moving off so grand and gracefully amid the smoke and thunder of her own guns. It brought back vividly to my mind the scene in this same harbor on the 7th of Nov 1861. I believe that was the finest sight I ever saw. What a change in the place since then. Today I am seated in a comfortable office located on precisely the same ground that I marched over late that night. I can almost look from the window and see the very place in the Fort where I lay that night. It is not thirty rods distant. but the parapet of the Fort is too high to allow me to see it. About the same distance to my left is the place where we landed. jumping into the water pell mell in order to be as near the first one ashore as possible. but the Generals Headquaters are close by and shut off that view also. A little before noon (by our time) the navy fired a salute. at noon the heavy guns on the Fort thundered forth a salute which fairly shook the island. A salute was also fired from the Fort at Bay Point. opposite. All the vessels in the harbor. and there are a good many here now displaying their flags which set them off gaily. At night the Post Band came down on the end of the wharf and favored us with some good music. This band is quite an aristocratic affair. it has ninteen members. I suppose there are but few better bands in the country. in fact it is too good for me. I can"t appreciate it. I should enjoy hearing some of our one horse country bands much more. They play splendidly no doubt. but opera music is not to my taste. I prefer something that has less music and more melody in its composition. Last night. it being a national annaversary the band could not help playing some national airs so Hail Columbia and Yankee Doodle were played. but in a manner that seemed to imply that they considered they were doing Miss Columbia.

and Sir Yankee. a great favor. Fiddlesticks. I had rather heare an old negro murder the tunes on a cracked banjo. In the evening quite a number of rockets were fired. and colored lights burnt both from the signal station on top of the old plantation house. and from the shipping. To our usual dinner we added some chickens. and pies. washing down the whole with several bottles of Ale and wine. a present from Lieut Sealy. and Maj. Smith. Late in the evening I assisted in drinking a pitcher full of lemonade in Dexters tent. and then retired to my own quaters. So much for my 4th of July/63. You must acknowledge that I did not spree it very extensively. I must go up to the General Hospital before night and see if a man from our Co who has been sick a long while is still there. I can do him no good if he is. but it is a duty to see him. and one which I have no disposition to escape. I like it very much here. that is in comparison with company duty. Of course it cannot be as pleasant as my situation at the Falls. Nothing in or about Military can ever be so agreeable to me as buisness in civil life. The respect that must be shown. and the perfect obedience rendered to everyone who happens to be superior in rank is very galling to me. Then too. the constant restraint one is under. The most trifling favors can only be obtained at the expense of much ceremony. I do not speak now in a complaining manner at all. All this is doubtless necessary to preserve discipline. and without discipline an army is nothing but an armed mob. I was only giving reasons why soldiering was so distastefull to me. The longer I remain in the army. the more I become reconciled to strict discipline. and the enforcement of an implicit obediance of orders. You would doubtless like to know what my duties are here. I would tell you if I could. but as I said before I have not yet been assigned to any particular duty. I have done nothing of any amount yet any way. I have helped the other clerks a little. but have generally had to beg the job. for it is hard work to remain idle in an office with half a dozen buisy ones around. Office hours are nominally from half past eight until half past four. No intermission at noon. the clerks takeing turns about going to dinner. One must be on hand all the time. but it is impossible to be very regular. and the clerks usuly remain until the buisness of the day is done. When a mail is to leave for the North. it often happens that some one of the clerks is detained in the office until late at night. As I told you before I think it very doubtfull about my remaining although I have heard nothing said or intimated to the contrary. I enjoy myself very well and shall be very glad to remain. Shall try to make myself as usefull as I can. and if returned to the regiment. shall go with a good grace. There are fifteen or sixteen of us clerks and ordilies that mess together. We draw our rations for ten days. and have a man detailed to cook for us. So that we are at no trouble about our food only to walk down to the mess room and eat it. There are six clerks in the office where I am. Mr Landon is in the Office of the Commissary of Musters. We mess together. I do not think of more to write you about myself now. I think prehaps you had better alter the direction on my letters to that which I will give you. If I go back to the regiment. Newt. can send my letters to me. and I shall not be so long in getting them as I am now. when they have to go to St Augustine and be sent back to me from there. This is a lovely Sabbeth afternoon. I am with you in heart love. though the distance that separates us is long. Good by dearest.

With love I am ever your own
Ned

No. 119 Rec'd July 21st 1863.

Hilton Head S.C. July 15th. 1863

My Dear Sara.

You will complain by and by I fear if I do not write more. I often write the most when I have really the least spare time. It is soldiers bed time now but we are premitted to burn our lights as long as we please. so that I can take my time and finish my letter if I choose. There is but very little doing in the office just now. For several days past I have done scarcely anything. it is very tedious and decidedly unpleasant to be mopeing about in an office biteing ones nails and waiting for something to do. Yesterday (Sunday) I spent the forenoon in the office. and most of the afternoon sitting on the piazza in front of the building which we use for quaters. Its a small building close by the side of <u>Genl Gillmores</u> Head Quaters. not over four or five rods from the beach. It was very pleasant yesterday to sit there in the shade with the cool sea breeze blowing. and look out upon the bay. and watch the shipping as they swayed to and fro. rocked by the long steady ground swell. Both steamers. and sailing vessels wer coming in and going out. The buisy. bustling lively little steam tugs were raceing about. from the dock to the vessels reminding one of a little important wire puller on an election day. At the end of the wharf lies the Fulton. mate to the Arago. looking down apparently with silent contempt at the noisy. splashing. pigmies around her. She is being unloaded and prepareing for her return trip. A little further off. lies <u>one of the Moniters</u>. looking the most harmless thing imaginable. At the distance its deck appears to be level with the water. and in fact it is only eighteen inches above it. its turret looks like a very large hogshead with a smaller though longer one a short distance from it which is the smoke stack. A white canvas awning is spread over her whole length and the crew are seen enjoying themselves as best they may beneath it.

I tried or pretended to try to read while looking out on this pleasant scene but made but little progress. I spend a good deal of time walking on the beach and upon the wharf. watching the vessels. not a very intellectual enjoyment perhaps you think. well I do not know as it is. but I wish to get the benefit of the sea breeze. and the noise and bustle of the wharf pleases me after the quiet of the whole day in the office. Posted out of harms way behind one of the big <u>piles</u> which project up above the dock for the purpos of fastening the vessels in their places by the dock. I love to watch the work going on around. In one place a little engine is blowing and sputtering as it moves the machinery that hoists three or four barrells of beef at once out of the depths of the <u>Fultons capacious hold</u>. a little ways off a darky is driveing back and forth a pair of horses or mules. bringing up at each journey a barrell of provisions. a bale of hay. or a couple of huge shells from the hold of a ship or schooner. the cars rumble back and forth carrying to the store houses the goods just landed or perhaps bringing down a load of commissary stores. which a fatigue party of soldiers. white or black. are loading upon some of the smaller steamers to be taken to some of the distant posts. If while I am watching. one of the vessels starts out. its an extra treat. If convenient I like to have a hand in and help the dockman loosen. and throw off the "lines" that have held the ship in her place. I used to stumble at the idea of calling a rope which you could scarcly reach around with both hands a "line; and it often amuses me now as I sometimes tug to lift the big loop up. and off from the pile to which it has been attached. I am seldom on the dock in the buisiest time. for my time to stroll about is not until after tea. sometimes five. oftener six o"clock. This morning as I was going down to breakfast. I saw the hospital ship. the Cosmopolitan lying at the wharf. and the ambulances buisy conveying the wounded to the hospital. I went down knowing that our four companies had suffered severely in the attack on <u>Ft. Wagner</u> and not doubting but that I should find some of the boys that I knew. The most of the wounded had been taken away when I got to the boat. but I found a couple of our men. one of them wounded severly. the other not so bad. I could learn nothing with regard to our loss. Matters are kept very close. no tales are told out of school. We were driven back at the assault on <u>Ft. Wagner</u> but we shall have it yet. Genl Gillmore is after it. and it is only a question of time. Maj Smith, the Asst Adjt Genl told Dexter that "<u>the 7th Conn</u> covered themselves with glory." Had they been supported by the regiments ordered to do so they would have gone into the fort. would have captured it. As it was their support failed them. leaving them exposed to a deadly cross fire. Of the eleven officers in the detachment. six are wounded or missing. <u>Lt.Col Rodman</u>. is severly wounded. he has been taken up to Beaufort.

Among the wounded were about forty rebels who were wounded and captured in the first days fight when our forces drove the rebels from their first. and second lines of entrenchments. capturing a number of prisoners and several pieces of artillery. One hundred and fifty rebel prisoners were brought down here today. I have great confidance that Genl Gillmore will succeed. the great trouble is that (I think) his force is to small. but still if he is allowed to go on I have but little doubt but that he will succeed. The Fulton goes North tomorrow (the 15th) I shall mail my letter. but I have heard that no mail is to be taken from the Head except of course that which is special. I have also heard that none was taken by the Arago so if you are a <u>long long</u> time without letters it will not be entirely my fault. The last letter I received from you was dated June 21st. I think I received some before I left <u>St Augustine</u> that I have not acknowledged the receipt of. I must look them up and see. Its decidedly provoking. I"ve doubtless got letters in the mail for the 7th now lying in the office here. but I can"t get them until they have been down to St Augustine and back. I have got Georges' cane which he wants to send home and I must be sure and take it to the Express office in the morning. We have received N.Y. papers of the 9th. Hope that <u>Gen Gillmore</u> will soon have good news to send North. Good by my love. Accept many kisses and much love from.

Ned

CHAPTER 7

THE SECOND CHARGE ON WAGNER

From the book:

Notwithstanding the failure of the first assault, General Gillmore hoped with the combined fire of land batteries and gunboats, the principal guns in Fort Wagner might be disabled, and the enemy be driven away; or at least the way opened for a successful assault. Accordingly four sand batteries were erected within an average distance of about 1,600 yards from the fort. These mounted fourteen mortars and twenty-seven rifled guns.

Such expedition was used that on the 18th of July they were ready to open fire. It was designed to attack on the 16th, but heavy rain storms submerged nearly all of the batteries and destroyed much powder. This compelled a delay of two days, during which only sufficient firing was done to obtain the range of the mortars.

In the meantime the Confederates were improving their opportunity to strengthen the fort. The magazine was thickly covered, the embrasures were stopped with sand bags, even covering up many of the lighter guns on the land side so as to preserve them from injury until they should be wanted.

Soon after midday all our batteries opened, and the navy, which had been awaiting their completion, closed in opposite the fort and took an active and effective part in the engagement.

Late in the afternoon, General Gillmore signalled to Rear Admiral Dahlgren that the assault would be made at twilight. This signal was read by a Confederate officer, who by a Ruse de Guerre had managed to get the key of our signals from a Union prisoner.

Consequently as the head of the column marched out into open ground from the first parallel, the guns in Wagner, Gregg, Sumter, and on James and Sullivan's Island opened upon it rapidly and simultaneously. This fire was severe, and when our troops approached so near the fort that the fire from our guns and the navy had to be suspended, the garrison, which, while our fire was going on, had been safely ensconced in the bomb-proof, ushered and added to the cannonade a destructive musketry fire.

The leading regiment, the Fifty-fourth Massachusetts, went forward on the double quick until they reached the moat. There the fire was so hot that they were temporarily checked, but being rallied by Colonel Shaw, made their way up the slope against the opposing bayonets of the enemy to the top of the parapet, driving the enemy from most of their guns. "It was here, on the crest of the parapet, that <u>Colonel Shaw fell</u>; here fell Captains Russell and Simkins; here also were many of the officers wounded."

During this fight the few men of the Seventh Connecticut who escaped from the first charge, amounting to about seventy, acted as provost guard under the command of Lieut. I. E. Hicks, Provost Marshall. General Strong directed them to act as rear guard and "let no man pass to the rear unless he was dead."

During the first hour or two they could obey orders, but when the final order to retreat was given, with General Strong, Colonel Chatfield, Colonel Shaw and many of their other officers killed or seriously wounded and the converging fire of more than forty cannon raking their ranks, an orderly retreat would have been as unwise as it was impossible. It was every man for himself, and the fleeing columns came on like a whirlwind, not stopping until they reached their camp. There they pulled themselves together and many of the officers commenced drill for the sake of steadying their nerves, and making them ready to repel a sally from the fort should one be attempted.

The chief loss fell on the <u>Fifty-fourth Massachusetts, colored regiment</u>. It had been recruited from the best colored men of Boston and vicinity. Governor Andrew requested Colonel Shaw to take the command. He had a choice between this and an exceedingly desirable commission in a white regiment, but accepted the command of the colored regiment, because of a chivalrous desire to help a despised race lift themselves up to the respect and honor of their fellow men.

No. 120 Rec's July 27th 1863.

Hilton Head S.C. July 19th. 1863

Dear Sara.

It's a dull rainy Sunday. I have nothing to do this morning. so improve the time in writing to you. The Arago came in last night. but I can get no letters until the return of the boat which started for St. Augustine early this morning. That will not be under a week. and then I shall have to wait to have it (the mail) sent to Morris Island where I expect to be then. The office of the Asst Adjt General is to be moved to Morris Island. We have had everything packed up ready to go for two days now. We may go today. but probably not until tomorrow. It will be a little more exciteing to be so near the scene of action. and our curiosity can be gratified by knowing all that is going on. which at present is not the case. All the doings up there being apparently suppressed as much as possible. this is all right and as it should be. I do wish our force was greater. I fear that since Lee has been permitted to make good his retreat into Virginia. and the rebels have been so severely beaten in the South West that Charleston will be largely reinforced so that it will be utterly impracticable for us to more than hold our own here. I have great respect for our soldiers on the Potomac. but very little for our commanding Generals there. I earnestly wish that ten thousand of those veteran troops there might be sent down here and placed under Genl Gillmores' command. and at his entire disposal for a couple of months. I think they would soon show the world what American soldiers can do in the way of fighting. when led and commanded by real live Generals who understand their buisness. I should feel perfectly confidant of the down fall of Charleston and Savannah. and such other places in this vicinity as are of importance. within six weeks. I mailed you a copy of the New South yesterday. if it reaches you. you will find an account of the attack upon the batteries on Morris Island. The four companies of our regiment engaged done themselves. and the regiment great credit. Their praise is upon the lips of all. Major Smith. Asst. Adjt. General told Dexter. that "the 7th covered themselves with glory." Had they been properly supported they would have possessed themselves of Fort Wagner. without a doubt. and that to with less loss of life than was incurred by the failure of their promised support. Our boys did all that was required. or expected of them. Did it nobly and well. The account contained in the paper I can add nothing to. for I was ignorant of the precise plan of attack until I saw it in the paper. I have seen and talked with our wounded boys in the hospital. but they even did not know about anything but just what was required of them. The attack being made before daylight. the boys did not even know their supporting regiments. There seems to be no excuse for the stampede of the <u>76th Pa</u> but the lack of pluck. They have hitherto been considered one of the finest regiments in the department. There are twenty two of our boys in the hospital here. I think none of them have dangerous wounds. though many of them are exceedingly painful. None of the wounded but such as were able to help themselves got off the fatal field. It was impossible to bring off a wounded man. much less a dead. hence we are ignorant of who are killed and who are dangerously wounded. I shall hope to find out of the hundred and three missing. that many of them are unhurt though prisoners. On account of the darkness it was impossible for those who escaped to tell whether their comrads whom they saw fall. were killed or not. there was no time to stop to determine. The fire through which they passed on the retreat is described as being terriable. The peninsula over which their way lay was completely raked by grape. cannister. and musket balls. solid shot. and bomb shells. One of the boys told me it was utterly useless to attempt to dodge. for they were as likely to be hit while dodging as while going straight on. The companies engage(d) were. A. B. I. & K. Out of the eleven officers accompanying the battallion into the fight. but four came out unhurt. One was brought off the field (Lt. Col. Rodman) and six are missing. It is impossible to describe the feelings of our boys when after bresting the storm of death that greeted their approach. they had succeded in mounting the parapet and were driveing the rebels from their guns. to see those who were to support them fleeing like a flock of frightened sheep and leaving them. a mere handfull. to their fate. The men all speak in terms of the highest praise of General Strong. They find no fault with the plan of attack. or the part they wer called upon to perform.

    The Arago brought us papers of the 15th. with accounts of riots in New York. A more disgracefull scene was never enacted in our country. New York owes it not only to herself. but to the State. and the country. to utterly destroy every member of that infernal mob that can be met with. and that too without the benefit of judge or jury. But while that was being done. by no means should its instigators be premitted to go free. I consider <u>Gov. Seymour</u> as vile a traitor as lives and he should be hung at the same lamp post from which the body of the gallant Col. O'Brien was swung. The acts of the rioters have filled me with less anger and disgust than has his miserable. temporizeing. traitorous. course. He is as black hearted a scoundrel as ever bore the name of Seymour. If there is any one thing that I am particularly gratefull for just now. it is that my name is not Seymour. Our government is not wholly without blame in this matter. It has borne? to long with the rebels at home. It has not maintained its dignity in a proper manner. The disgracefull outbrake in

N.Y. is but the legitimate fruit of the teachings of such men as the Woods, Seymours, Vallandigham. Eaton. Gallegar. and of such papers as the N.Y. News. and Day Book. the New Haven Register. Hartford Times. N.Y. World and others. Government has been to lenient with these villains. and now we see some of the fruits of it. Talk of arbitrary arrests: where there has been one. there ought to have been fifty.

Evening. Directly after laying down my pen this afternoon I started down the beach towards the wharf. I saw the Cosmopoliton (which is used expressly as a hospital boat) coming in to the dock and knowing that she came from Morris Island went down on the end of the wharf hoping to get some definate news with regard to what was doing up there. particularly as I am slightly acquainted with the clerk of the boat from whom I hoped to get information. I was not prepared for the news I learned. and the sight which greeted my eyes as she neared the dock. Another desperate fight for the possession of Ft. Wagner took place last night. The works were again stormed by our forces. and as I learn the rebels driven out of them. but such a destructive fire was opened upon them by Fort Sumpter. and battery Bee. on Cummings Point. that it was found impossible to remain in them. and after holding possession for an hour and a half or two hours. we were obliged to leave them. spikeing the guns. Our loss when compared with the number engaged has been terriable. The Cosmopolitan. and the Mary Benton. had in all six hundred and sixty or seventy wounded on board. Fifty one wounded. and three dead. wer taken off the Mary Benton here. the remainder are to be taken to Beaufort. The Cosmopoliton proceeded immediately to Beaufort with all her load. On board of a third boat I am told there are about one hundred and thirty more wounded. makeing the loss in wounded alone Eight hundred. that we have taken off the field. The whole loss then cannot fall below twelve hundred. The numbers of the wounded I received of the clerk of the surgeon in charge here. who oversaw the removel of the wounded. and are doubtless correct within a score. either way. The dead taken off here were some that had died on the passage down. No correct, or authentic account of the battle can be had. as nearly as I can learn the attack was made late in the evening. The regiments engaged suffered severely. The 54th Mass. (colored) lost it is said nearly six hundred men. they are said to have fought bravely. their Colonel was killed. His name was R. G. Shaw. his father is I believe chief justice of the State of Mass. Colonel Putnam of the 7th N.H. was killed. Colonel Chatfield of the 6th Conn is wounded but not seriously. Lt. Col. Speidel. of the same Regt. is either killed or a prisoner. Col. Barton of the 48th N.Y. is wounded. Generals Strong and Seymour are both wounded but I believe neither of them seriously. Of the Capts. and Lieuts. lost. I could gain no idea but their number was large. Some think we have achieved a partial success. but I fear nothing commensurate with our loss. I fear it would have shocked the nerves of a great many good people at home who never see soldiers except in holiday dress. and who know nothing of war except what they read in the papers. and who have read so much about men being killed. and terribly wounded on the battlefield that they have come to think that it was not so great an affair after all. that the soldiers become used to it as the old lady's eels did to skinning: to have stood on the dock tonight and seen the wounded men taken from the boat and placed in the ambulances. I thought that I had become. in part at least. used to it. but my heart sickened and my head grew dizzy with the sight. It is just possible that what renders the sight more unpleasant to a soldier is the consciousness that his own turn may come next and prehaps very soon. These men were all right from the battlefield. their faces begrimed with the smoke and dust of the fight. their cloths torn. wet. and dirty. glued to them by the drying of their own blood. with which they were colored. So great is the number that only a few of the worst cases could be attended to. and those but partialy. It is better you know to do a little for all than to do all that is necessary in each case at once while a half dozen others might die for the lack of a very little attention promptly bestowed. But bad as the sight was at the wharf. I knew. and could not help thinking. that it was nothing in comparison to the field of battle itself. It is utterly useless for me to attempt to describe one. you have read the descriptions of hundreds. But it is a thing that must be seen in order to be realized. We now expect to go to Morris Island tomorrow. but we may not. I have one of my unpleasant colds in my head. and have felt particularly unamiable all day. and have made more blunders in writing than were at all necessary. One thing I am a little puzzled about. I received the Tribune of the 6th that you sent directed to me here. but no letter. true you may not have sent me any. but it seemed almost certain that you had. I am going to leave my letter till morning. wish I could be with you this evening dearest. Good night.

Monday afternoon. We have been employed most of the forenoon in prepareing to leave. We shall doubtless get off this evening. I have heard this noon that our forces had suceeded in retakeing and holding Fort Wagner and that it is now in our possession. I hope it is true. but am not certain of it. The heat is almost insupportable today. what weather for fighting. when its as much as one wishes to do to breathe. I will write to you again the first opportunity. With love. I am as ever

Your Own
Ned

No. 121 Rec'd Aug 9th 1863.

Hilton Head S.C. July 28th 1863

Dear Sara.

The Fulton came in yesterday bringing a mail by it. I received your letter of the 21st. By the Aragos' mail I received yours of June 28 & 30th. Mr Landon went to St. Augustine on buisness and he brought them up to me. I also received a letter from George and he said he was told that letters came for me in a preceeding mail. they have not been sent to me. George was going to try to look them up. I did not go to Morris Island as I wrote you. I expected to on the afternoon that I finished my last letter to you. Just before night the Lieut. sent word to the office that I was to remain behind to take care of things. and would be sent for if wanted. I did not exactly like the idea for I dislike being left behind but thus far it has been very fortunate for me. I have had a very hard cold for the last fortnight and for several days after the boys left had considerable fever. have not been ill enough any of the time so that I should have felt justified in staying out of the office. but it would have been very unpleasant to have continued on duty. As it is I had had nothing at all to do. with the exception of a few little errands. It is now over a week since the office was moved and I have not yet been sent for. Of course it is much easier to remain here and have nothing to do. and so long as that is my orders. it is not my fault. but if I am to remain long in the office I am anxious to be makeing myself acquainted its duties for which I have no more opportunity now than as if I were in the regiment. In fact since the first week I have had scarce anything at all to do any way. as there has been but little buisness done in the office since Genl Gillmore went to Folly on the 8th. I can give you no very definate ideas of a clerks duties here any way. it is differant from any other kind of buisness. it is writing principally of course. Returns of various kinds have to be made out at stated periods of all the forces in the Department. Orders and letters to be written and copied. Discharges to be made out and I don't know what all. I know that there is a vast amount of usless work done. a world of flumery and circumlocution to be gone through with before the simplest thing can be accomplished. Everything is done in what appears to me to be a style a hundred years behind the age. I suppose it is military but it looks to me very foolish. I have not succeeded in getting my pay yet. should have had it but for the removal of the office. Clerks in this office get 40 cents per day in addition to their regular pay. this is drawn monthly. its a little addition and helps to pay mess bills.

Evening. I have been quite buisy this afternoon running about. Have succeeded in getting my four months pay though am not quite satisfied that it was done by the proper Paymaster. Have also been helping to get the mail up to our wounded boys in the hospital. Some of the very badly wounded are dying quite rapidly. Four died last night. I think they are having as good care as is possible under the circumstances. The Cosmopoliton went North yesterday with a load of wounded including the officers. None of the wounded men belonging to our regiment in this hospital have died yet. One poor fellow. who was carrying his arm about on a pillow. the other day when I saw him. in such pain that it was impossible for him to sit still. has been obliged to have it taken off between the elbow and the shoulder. He was very thin and pale today. has suffered much and I am afraid will not live. The rebels preformed a good many amputations upon the wounded that they took prisoners. A good many more than it is thought to have been necessary. their plan being to prevent just as many as they could from ever entering the field again. A plan worthy of the devils incarnate who proposed. and practiced it. By agreement we took up the rebel wounded in our posession the other day. and exchanged with them for our wounded in their possession. I thought I did tell you what four Cos. were up at Morris Island. but must have forgotten it. they are A. B. I. & K. Capt Chamberlin of A. is a prisoner. unhurt. Capt. Tourtellotte of K. is wounded and a prisoner. Lt. Phillips of K. is a prisoner unhurt. Capt. Burdick. and Lt. Wilson of Co. B are both dead. Lt. Col. Rodman as you are aware was brought off wounded. he went North on the Arago. That Sergt. Dexter that you saw should have been Sergt. Decker of Co. I. he was slightly wounded in the head. We have no news of importance from Morris since the last unsuccessfull attack over a week since. Genl. Gillmore is planting heavy guns and getting ready for buisness. every one is confidant that Ft. Wagner will have to succumb before long. and then the rebels may bid good by to Ft. Sumpter. It would be boys play for Genl. Gillmore to knock it down with his heavy guns mounted on Ft. Wagner Brick and Stone. stand but little sight in this war. nothing but thick embankments of sand. and ironclad batteries can arrest our progress. This afternoon I received your letter of the 9th and some papers. also a letter from Sara B. I think on the whole that you have enjoyed yourself pretty well during your brief reign as mistress of the mansion. I judge so from your letters. Am glad that it has passed so pleasantly. You do not seem to get over the fear that I am not interested in the recital of the events of your daily life. Once again let me assure you that I am both interested and concerned in whatever pertains to you and yours. The most interesting letters that we soldiers receive are those that relate the petty incidents of every day life connected with our loved ones at home. I cannot force myself to imagine any open disturbance attending the Draft in Salisbury. If there is any. the <u>leading men</u> of the <u>Democratic party</u> (so called) who <u>voted the Seymour ticket last Spring will be responsible for it</u>. Our friend Waters is afraid of his shadow. is a great talker. and hears as many and as large stories. as any one. and the worst of it is believes them all. at least the foolish ones. I am really glad your Uncle Harrison is on the right side. he ought to watch the man. or men. who threatened to burn his buildings. and if he can catch them in the least treasonable act put them through for it. if he can by good luck catch one of them fireing a building <u>put a ball through him instantly</u>. You may think me harsh. but it is the only medicine that will cure the disease. Soothing Syrups have been tried altogether to long now. There never was any virtue in them to arrest the progress of <u>Copperheadism</u>. and all excuses for their further use are now. since the N.Y. riots. effectualy silenced. I am glad to hear that I am likely to get the boots soon. I am almost barefoot now. my shoes seeming to take particular delight in failing me as quickly as possible since I wrote for the boots. they looked good for six weeks service when I wrote the order. but now a good piece of one foot is on the ground. I shall look for the boots by the next steamer. With regard to my health. it has been very good thus far until this wicked cold attacked me. it seems determined not to leave me. I have not yet felt any of the difficulty that troubled me when in the Bunk. True I have done little or nothing in the office as yet. but I think I can remain without being troubled with any of the ill effects feared. The confinement is nothing to that of the Bunk. Under ordinary circumstances the clerks get through by half past five or six. frequently before. When a mail is about to leave it often becomes necessary to remain late at night. and prehaps to be in earlier than usual in the morning. I am still very uncertain whether I shall be kept long or not. am very indifferant about it. but of the two would prefer to stay. more because things are so unpleasant in our company than for any other reason though there are others. Had the matter have been left with me at the begining I would not have come. but I do not want Dexter to ever know that I have thought such a thing. for he was very kind and did me a favor that nine out of every ten men in the regiment would jump at with egarness. I don"t think I shall ask many of my correspondants to direct to me here yet awhile.

    We have good news from all quaters by the lasts mail. I think that a vigorous. and thorough prosecution of the war for the next three months. would end the war. Follow up promptly the advantages gained. and the end would quickly come. But unfortunately for us that has never yet been done. therefore I doubt if it be now. The rebels owe Genl. Meade a vote of thanks and a gold medal for his kindness to them in allowing Lee to recross the Potomac undisturbed with all the vast amount of plunder he had gathered. I will not attempt to tell what <u>our</u> government owe him. While I rejoice over. and am thankfull for our late successes. I am by no means sanguine of a speedy end of the war. We have gained a partial possession of the South West. have captured a few guns and some ammunition. all the prisoners have been let loose to fight us again. Exchanged or not every man of them will be in active service again in less than one month. Let those doubt it who are fools enough to trust Southern rebels' <u>honor</u>. As if they possessed such a virtue. If our people will make up their minds. and go to work in earnest. and capture Richmond and some of these southern cities. then and not till then. shall I begin to expect that the much talked of "back-bone" will begin to crack. I see by a New Haven paper that George F. Gardner. former Lt. Col. of our regiment. and a man named Keeler formaly a 2nd.Lt. in our regiment. are both drafted. It is not at all likely that either of them will come to the war. but I heartily wish Gardner might be obliged to come as a private. and be put into our regiment. He is a rank pro slavery Copperhead.

<u>Lakeville</u> must be getting to be quite an aristocratic place. suppose I should feel quite like a stranger there if ever so fortunate as to get home. Should think the Centre folks would be a little jealous of you. Sara B. rather comes down on Miss Martha. and I don"t wonder. Martha has succeeded in dupeing us all a little. I was a fool to ever trust her at all with the knowledge I had of the Richardson tribe. I ought to have known that there could no "good come out of them." but I am slow to believe ill of a lady and fealt least I should do Martha injustice through my dislike of the family. But the end proves my first impression true. she is strongly tainted with the family complaint. dishonesty. Have not yet answered her letter and would not if I could avoid it with decency. must do it soon and have the matter off my hands. I must say that her disregard for truth is only equalled by her impudance in following you so presistantly. I think I shall yet have reason to be glad that you have left Lime Rock. Well the strawberries. and the cherries are all gone the second year and I have not had one. Sara B. boasts of her ability to climb cherry trees. I guess she will have to help pick the apples this year. for Pitt won"t be home. I haven"t heard anything about the prospect of the apple crop this year. You see we are a little interested about that because it is possible to get them. A Charleston lady was very kind to one of our wounded boys belonging to Co. A. while he was in the rebel hospital and brought him many little dellicaces among them fruits of various kinds. She also took his soiled and bloody clothes and returned them to him nicely washed. when he came away she gave him her photograph. Kind wasn"t she! The rebels tried to prevent her seeing our soldiers when they found she sympathized with them. With love. dearest. I remain as ever your own

Ned

No. 122 Rec'd Aug 10th 1863.

Hilton Head S. C. July 30th 1863

Dear Sara.

I mailed you a letter yesterday but as I have today received a piece of important information. and the <u>Fulton</u> is advertised to start at eight o"clock tomorrow morning. I hasten to communicate it. The remainder of our regiment is to be relieved from duty at St. Augustine and sent to Morris Island. Their place is to be taken by one of the regiments that has suffered so severely in the late actions on Morris Island. It is to be hoped that no more such disasterous assualts will have to be made but operations are only airly begun and chances are plenty for hard fights yet. Our regiment stands high with the General commanding. and if there is difficult and dangerous work to do it will likely be called upon. I am sorry of course that we are obliged to leave such safe and pleasant quaters but it is no more than fair. We have fared compareitavly well thus far and can claim no exemption from the dangers of the battle field. We will hope for the best. I think it likely that I may be ordered to rejoin my Co when the regt. arrives at Morris. Col. Hawley will get back all his detailed men that he can. for he has always been averse to having them detailed out of it. This request to have them returned will be more likely to be granted under present circumstances than under any other. If the regiments that left here for Folly nearly all the detailed men out of them were returned. It is no more than right that I should go with the boys and share their dangers and hardships. If I do not and we should be fortunate enough to reach home I could not claim my share of the praise and gratitude that I know would await us. This is but a selfish reason. there are also higher and nobler reasons. I have nothing to add to what I have said in former letters in relation to rejoining my Co. I only ask that under whatever circumstances I may be placed. I may be able to act well my part. so that you may never have reason to be ashamed of me. I cannot help wishing that George was in my place. or that in some way he was well out of it. I fear for him as much from disease as from any other danger. He is very well at present I take it for he writes about going on guard. The exchange is to be made immediately. but it will be three or four days yet before they (our regt) reach Morris. Of course I will do all in my power to keep you acquainted with whatever befalls us. Dexter wrote me yesterday that the Moniters and the new <u>Ironsides</u> had commenced a vigorous fire upon Ft. Wagner. that we had heavy guns mounted within 600 yds of Wagner and about 3,000 yds from Sumpter and thinks that within a month at least we will have Charleston. It is utterly useless to calculate we can only work. and hope. friends at home must <u>wait</u> and hope and be prepared to hear evil as well as good news. You need not mention to Julia what I said about fearing for Georges' health. she will worry enough without having any help about it. Now my love once more a "good night" it is late and a friend who is stopping with me wishes to write a line and there happens to be but one serviceable pen at my quaters.

With love. I remain as ever

Your own   Ned

CHAPTER 8

SIEGE OF WAGNER

From the book:

The Forty-eighth New York was sent to St. Augustine relieving the Seventh Connecticut, which reached Folly Island August 4th, and from there proceeded to Morris Island joining the other four companies. From that time onward the regiment was constantly engaged in digging sand, carrying siege material, or serving artillery.

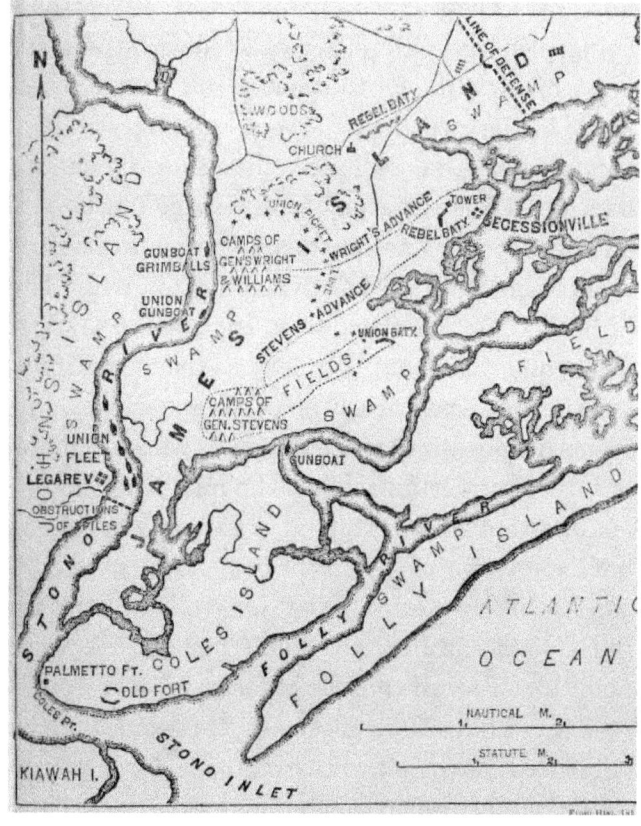

FOLLY ISLAND AREA MAP

323

No. 123 Rec'd Aug 15th 1863

Hilton Head S.C. Aug 6th. 1863

Dear Sara.

    Your long and interesting letter of the 24th I have just received and read with great pleasure. I expect to have leisure this morning to answer it. I am still here awaiting orders. <u>Capt. Sealy</u> of Gen. Gillmores Staff who haves charge of the office and clerks is here and has been for two or three days. I may return to Morris with him. though he has said nothing about it as yet. I am much better than when I wrote you my last letter. The Arago came in early yesterday morning brining papers of the 1st. I was most agreeably surprised yesterday afternoon to receive a visit from Tom Norton and Lee Wells. They spent a few hours with us (Mr Landon and I) and we went down and saw them on board the boat which was to take them to Morris. The boat that took our boys from St. Augustine to Morris did not stop here on its way up but kept directly on. there were a few sick ones in the hospital that were to go to Beaufort. So after the regiment was landed the boat returned here. Tom and Lee being in charge of the sick remained with them until they were delivered at Beaufort. and while waiting for the boat to return to Morris. made us the visit. As I wrote you would be the case. the boys were very angry at being relieved by the 48th. about which they cared a great deal more than they did about going to Morris. I was in hopes to have got a line yesterday from George but he probably was to buisy to write. Tom and Lee were in fine spirits. they are not certain but that they will now be returned to duty in the Co. I hardly think Tom will. The Arago brought down over three hundred men who have been engaged to take the places of the soldiers who have been employed about the Head as teamsters. blacksmiths. carpenters. and laborers drawing extra pay. nearly all of whome will be at once returned to their regiments. It seems hardly fair that these fellows who will not enlist should be premitted to throw us out of comfortable jobs. The men who came yesterday are an ill-favored sett. I am told that they are a part of the crew engaged in the riot in New York. that they have come away to avoid the draft is very evident. and many of them certainly look capable of any piece of villainy. Genl. Gillmore needs every man in the Department in the field and I cannot blame him for takeing this method of recovering a great many if he could do it in no other way. but it would have been more palatable if decent men could have been obtained. No decided advantage that I am aware of has been gained on Morris since I wrote last. I suppose we are every day strengthning our position. and prepareing for aggressive operations. A few reinforcements have been sent Gen. Gillmore. I do not know exactly how many. Another <u>colored regiment.</u>

the 1st. North Carolina. arrived last week. The colored men are very usefull here now. they can stand the heat better than we can and they do a sight of work. The Southern negroes from habit work very slow. it is extremely difficult to hurry them. Its as much as a wide awake stirring Yanke can do to keep quiet and see them work when there is any cause for haste. I have no doubt that ordinarily one northern white laborer will do more in a day than any three negroes. while in event of a necessity he would easily do more than half a dozen of them. Quiet reigns here at the Head. The only buisy places are in the Quater Masters Department. the Ordnance yard. and the Dock. Fatigue parties are at work on the dock from early in the morning until midnight. It would astonish you to see the quantity of shot and shell that has been sent away from the Ordnance yard here. and I suppose but a small part of what have been used have been sent from here. None of those used by the navy go from this yard. they have a separate Department elsewhere. One of our 7th boys who was wounded in the attack of the 11th died at the hospital here on Sunday last. he was wounded in the hand and though it was very painfull. yet no danger was at first apprehended that he would even loose his arm. but finaly it grew so bad that the arm was amputated near the shoulder but to late to save him. the ball was poisened doubtless. One would suppose that those who claimed to be the most enlightened and chivalric people on the earth would disdain to stoop to such a barbarous method of riding themselves of their enemies. Poisond arrows will do very well for savages. but white men ought not to copy their example. But <u>poisend bullets</u> are a favorate missel with the Charlestonians. Our officers were obliged to leave their ladies at St. Augustine. I presume they will soon return home. it will probably be long before our regiment will be again in a situation where they can rejoin their husbands.

Your last letter was very interesting indeede. I am greatly pleased to hear all about yourself and how you pass away the time. I am glad that you can enjoy yourself so much. time does not seem so long when it is passed pleasantly. Mr. and Mrs. Merwin are getting quite fond of you I should think. Well I don"t wonder. I am to blame dearest for not having been more particular to answer your letters correctly. one reason I have not done so is that I have often written to you when I had not your letters by me and it was not convenient to get them. I remember that you have related to me your various conversations with Martha. and I thank you for doing so for I like to know all about the matter. I approve of all that you have said in reply to her. I cannot see how she can expect to be considered. and treated as an intimate friend by you after all that has passed. I must say that I am astonished at her brazen obstinacy (I can call it by no other name) in presuming to claim your love and friendship after all that has passed. The truth is your friendship is valuable to her because she gains credit by it. She knows very well that she does not stand "A." No 1. with the very best class of people in Salisbury. and because you have to much principle to openly quarrel with her she still hopes to be able to use you for some purpos. I don't think she will be able to get much the start of Sara B. though she can annoy her by her pertinacity. I must answer Libbies' letter soon. am sorry if she is allowing her love for Charlie to increase. I was in hopes she would be able to forget him for a while at least. but there is no dictating a womens love. and it would only make matters worse I fear to attempt to argue with her. for my own part I believe he is forgetting her. I would not willingly do him injustice. but you know he is quite young. is fond of society especially that of the ladies. of which he is doubtless favored with a good deal. his mother we know will do her best to attract him elswhere. and I think it more than likely that before this he has come to look upon his affair with Libbie as being ended. an agreeable little love affair with a pretty girl who really loved him. on the whole a very start for a young ladies man. Now Sara. don"t you let Lizzie know that I think so. or see what I have written. My. she'd my eyes out if she could get a chance. for when did ever a lady like to be made a cats-paw of for the diversion of any gentleman. or like others to think that such a thing had been done with her. No. no. I can't afford to lose Lizzies friendship. not to say love. just yet. I do believe that if the two old ladies had not made so much fuss about it. it would have ever have made half the trouble. Am sorry that Harvy's health is so poor. hope he will be benefited by his visit to New Haven. But if he really has got a lung difficulty he never can get over it. The very best. and the first <u>thing</u> that he ought to do would be to pack up his traps and start for St. Augustine. he can live longer there than anywhere else in the United States. If I were in his circumstances I should go there immediately. I think. supposing the war to end as we all hope it will. that if alive and well I should try to get into some buisness by which I could support myself and wife in that ancient town. I do like the place beyond all others it has been my luck to see South. I think it a paradise for a lazy man. he can live so cheaply and with so little exertion. I am not surprised at Ed Hubbards promotion. such things have become to common to excite in me any other emotion than that of disgust. You say "he may be worthy of it." very possibly but you must remember that the wrong is done to men in the 19th over whose heads he has been promoted out of another regiment. he may be worthy of it. I <u>know</u> he is not <u>fit for it</u>. he may not lack in bravery possibly not in military knowledge. but such a wild rattled brained boy is not a fit person to hold a commission and to have almost unlimited control over men old enough to be his grandfather many of them.

His brother Jim is no more fit for his position than I am to take Mr Reids place in the pulpit during the Rev. Dr's vacation. (Editor note: Rev Adam Reid will marry Sara and Edwin 5 October 1865) But money. influence. and impudance. carry the day in these matters. Occasionaly a man so signaly distinguishes himself. that they are obliged to take notice of. and promote him. but it is done with such a flourish of trumpets and loud sounding proclamations of the justice and equity of the military authorities. that if the man possesses the least spark of modesty. he is made to feel that he is very insignificant and that insted of simple justice having been done him. a great favor has been conferred. which the poor services he can render Uncle Sam during the remainder of his mortal existance will not be sufficient to repay. John Reid has had many narrow escapes if all that he has related is true. There is nothing improbable in any of his stories that I have heard. I am very glad on his fathers account that John is succeeding so well. If he can be of any good to his country now it will be some palliation for his past uselessness. Well. Reid still continues the even tenor of his way in the Co. he is the company clerk and as such is excused from all guard and fatigue duty. Am sorry to learn of R. McArthurs death and am a little surprised for if any one could stand an attack of fever I should surely supposed he could. for I remember him to be a very healthy. active. vigorous. young man. I have not learned whether any other of the Salisbury men (except Mr. Nott) were injured or lost in the fight. I suppose Charlie Landon will not return with the Co. as I learn he has accepted a commission as Major in one of the colored regiments. I received quite a long letter from Mr Randall this morning. it came by the Fulton but has been to St Augustine and back. he is very hopfull of our final triumph and thinks the Northern rebels will not be premitted to have things all their own way just yet. Thanks my love for your prayers that I may be preserved in safty through the year to come. In no other way can friends at home so benefit us as by offering fervent. ernest. prayer for our protection. guidance. and safty. We know that the Almighty ruler of the Universe is the hearer of prayer and that the "fervent effectual prayer of the righteous availeth much." Who can tell dear Sara how much of harm and danger your ernest solicitations to the throne of grace have warded off from me. This is a pleasant thought to me. that prehaps I owe my present existance. through God. to you. it strengthens and encourages me. So you are out to Emmas' again. Well you are having many pleasant hours now. I am glad of it. I could not bear to think of your being unpleasantly situated. I fear you would miss very many of your enjoyments from the day you became a poor mans wife. You know George always told you that you would be better off not to get married for you were able to take care of yourself.

I must confess that my better judgement tells me there is much truth in his remark but I am altogether to selfish to advise you to follow his advice. I think I shall not write more today. I may add something more before the Arago sails which will not be yet under a day or two.

Evening of the 7th. And rather late too. Have not much to add. Received the Boots this morning safe and sound and am rather tired after wadeing around in the sand a considerable with them today. I like them very much and they are just right for me. Also found a Dozen Pen Knives that I sent for some time ago in the box. Georges Box is here also. I have written to him this afternoon. Have been guite buisy today writing. and this afternoon packed up a box of Stationary and sent it to the office on Morris. The <u>176 Pa Regt</u>. Drafted nine months men are to go <u>home on the Arago</u> tomorrow. they have embarked this afternoon. I could not help wondering when I saw them go by. if I should ever see the happy time when we should be "homeward bound."

Good by dearest. Love from
Ned.

No. 124 Rec'd Aug 31st 1863.

<u>Morris Island S.C.</u> Aug. 12" 1863

Dear Sara.

After several false starts I am at last near the scene of action. While I am writing the sound of the bombardment reaches my ears. while by stepping outside the tent I can see the sky light up almost momentarily by the flashes from the heavy guns and can trace the course of the <u>morter shells through the air by the streak of light made by the burning fuse.</u> There is much more fireing tonight than there was last or during the day. On Monday morning I received my orders to pack up. and come by the first boat. Mr Landon also had similar orders. the office in which he is employed having been ordered to move to this place. Along with my order came various little errands to be done for the boys. The day was intensly hot. with scarce a breath of air stirring. I was obliged to be out running around the most of the day. and when at last I got on board of the boat with my things. nearly eight o"clock in the evening. I was completly exausted. The boat started very soon after. and putting on my overcoat I lay down on the deck with my knapsack for a pillow and slept well till morning. Awoke at daybreak to find the boat nearing the landing. Quite a brisk bombardment was going on and had been pretty much all night. it ceased soon after sun rise. Some delay occurred about our getting off the boat. but I at last accomplished it and found my way up to our mess tent and made sure of a good breakfast. All the forenoon. which was very hot was occupied in getting our things off the boat and up to our quaters. The afternoon I have spent at work in the office. This evening I have been up to the regiment. which by the way is but a <u>few rods distant</u>. and was heartily glad to meet the boys again. It has been but a little while since I left them but I felt as though it had been a long time. Was glad to find the boys all well. George is looking finely. better than I have seen him before in over a year. he is well. and rejoices in feeling like himself once more. There is no spare time for the men now. Every moment when not actually employed on duty is necessary for repose. George will probably be detailed as a sharpshooter. he has been practiceing a little at mark with a number of others.

<u>Aug 14th 1863. Evening.</u> Have been very buisy for the past two days. Our offices is in three large tents pitched a few rods apart. it is very inconvenient of course on many accounts. but we hope to have better accommodations <u>when</u> we get into Charleston. I would like to write you a little of what is going on around us. but there has been an order issued strictly forbidding officers. privates. citizens. or any government employees from writing or giving any account of what is being done. The order is very definate. and so plain that there can be no evasion of its meaning. This is as it should be. for although probably not one in a thousand of the letters that may or would be written. would contain any information that would ever benefit the rebels. yet it is right that even the slightest chance of enlightning them should be defeated. This island beats all that I have seen in the South yet. I have not been to the upper end where our batteries are. but so far as I can see the surface is the same. The island is long and narrow. I should think not over a mile or mile and a half in width. and here at the lower end much narrower. It is a collection of small sand hills varying in size from a good sized hen.coop to bluffs thirty and forty feet high. On one side is the ocean. on the other a marsh and Folly Island river which is nothing but a small creek. There is no room for anything like a regular camp for one regiment and the tents are tumbled about in all sorts of shape. standing in beautifull disorder. Our regiment is camped with a number of others on a narrow strip of sand between the ocean and the sand hills. in reality they are right on the beach for at high tide the water comes within a few feet of their tents. The sand is of the finest kind and it is the most tedious and tiresom work to walk in it that I ever attempted. The highest sand bluffs form an irregular line of hills running lengthwise of the island. On these the rebels had most of their guns mounted so as to bear on both <u>Folly and James islands</u>. they still remain in their original position. The bathing is splendid. the finest I have seen since we left Tybee. The surf rolls beautifully. The men are not slow to improve such a fine chance for pleasure and comfort combined and morning and evening the surf is filled with the bathers. Bathing is not allowed after Eight in the morning or before Five at night. From the bluffs and the landing. Secessionville can be plainly seen. I had a good long look at it through a glass the morning I came. I could not see the battery which proved the death of so many of our brave men last year because from here we look at the place from nearly the rear of it. but the lookout and buildings all looked familliar and I could tell just where the battery was located. I could not help a feeling of sadness when I remembered our gallant Hitchcock and our disasterous repuls. which under better Generalship might have been avoided. The situation of Ft. Johnson is plainly seen. Ft. Sumter is in plain sight. Ft. Wagner. Gregg. Moultre. and Battery B. are not to be seen. the bluffs rising higher towards the north end of the island in their direction. Our fleet lie just abrest of us prehaps a couple of miles from shore. At low tide the turrets of the lost "Keokuk" are visable between us and the fleet. Why in the world Admiral DuPont should allow the rebels to take off her guns unless he was perfectly willing and wanted to do the rebels a favor. I cannot see. The wreck lies so far out that it seems as though a single wooden gunboat with the ordinary amount of pluck in those who manned her could have prevented the removal of her guns. The spires of Charleston can be seen over the tops of intervening trees. It is said that it is three miles distant. and also that General Beauregard has once or twice sent word to Genl. Gillmore that unless he left the island at a certain time he would shell his camps. to which Genl. Gillmore replied that if he attempted it. he would burn Charleston. I do not know that the report is true. I suppose that Genl. Gillmore could damage the city very much from here. if he would not burn it entirely. Lighthouse Inlet which divides the two islands. Morris & Folly. is a narrow creek. not over twice the width of the Housatonic at the Falls. It is deep enough to admit our light draft gunboats and transports. A small floating dock serves for a landing place. The old Planter is used as a ferry boat crossing every half hour nearly during the day. The wreck of the blockade runner "Ruby" which was run on shore near the latter end of June lies high and dry at low water on a sand bar a few rods from the north end of Folly Island. You may remember about the squabble that there was over her cargo. both our people and the rebels got some of it. first one party would board her and get a lot of good. and then the other would drive them off. I guess the rebels got the greater part for they had batteries on the lower end of Morris Island and were therefor more than a match for our boys rifles. I am told that our boys found large quantities of the goods in the rebel tents on this island after the forcing the rebels to take French leave. I have been quite buisy since I have been up here. two new clerks have been detailed. but what they are wanted for I cannot see. but suspect that some of us may be relieved shortly. Think I have written you all of interest I am premitted to at present. With love I remain as ever your own

Ned

CHAPTER 9

FORT SUMTER

From the book:

August 16th General Gillmore issued an order directing that the breaching batteries should open on Sumter at daybreak on the morning of the 17th. The batteries served by the Seventh Connecticut were as follows:

Battery Hearncy, First Lieut. S. S. Atwell, Seventh Connecticut Volunteer Infantry Commanding, comprising three thirty-pounder Parrott rifles and three Coehorn mortars.

The guns to operate against Battery Gregg with shot and shell unless otherwise directed, and the mortars against Fort Wagner, exploding the shell just over the forts.

Battery Ward, Capt. B. F. Skinner, Seventh Connecticut Volunteers Commanding, comprising five ten-inch siege mortars, to fire against Fort Wagner, exploding the shells just before the striking.

Battery Strong, Capt. S. H. Gray, Seventh Connecticut Volunteers commanding, one ten-inch Parrott rifle against the gorge wall of Fort Sumter, firing shot and percussion shell, commencing with the former.

"The breaching guns were served from day to day with great care and deliberation. The firing from the batteries in the second parallel was seriously interfered with, and, at times, partially suspended, by the galling fire from Fort Wagner to which the cannoneers were exposed. The combined fire of our mortars and light pieces, aided by gun-boats and ironclads, failed to subdue this annoyance entirely, and we were obliged to turn some of our breaching guns upon the work. There was imminent danger, indeed, that our most efficient, because most advanced, batteries would be hopelessly disabled before the work should be accomplished. Nothing of the kind, however, happened. A heavy northeasterly storm set in on the 18th, and raged for two days, very materially diminishing the accuracy and effect of our fire.

Soon after midnight on the night of August 21st, the Marsh Battery opened on the city of Charleston, firing only a few shots.

Brigadier General Terry was ordered, on the 26th of August, to carry the ridge at the point of the bayonet, and hold it. This was accomplished, and the fifth parallel established there on the evening of the same day. This brought us to within 240 yards of Fort Wagner. The intervening space comprised the narrowest and shallowest part of Morris Island. It was simply a flat ridge of sand, scarcely twenty-five yards in width, over which the sea, in rough weather, swept entirely across to the marsh on our left.

Final operations against Fort Wagner were actively inaugurated at break of day on the morning of September 5th.

TEN-INCH PARROTT RIFLE SERVED BY THE SEVENTH AGAINST FORT SUMTER.

General Gillmore says in his report:

"It was repaired at the suggestion and under the supervision of Captain Gray, Seventh Connecticut, the battery commander, who was a skillful mechanic, by chipping off the bands for a distance beyond the fractures and enlarging the diameter of the bore this distance, from an eighth to a quarter of an inch. This left a band of iron, as it were, around the muzzle. The gun was fired three hundred and seventy times after this, without any difference in the range or accuracy being noticed."

No. 125 Rec'd Aug 31st 1863

Morris Island.

**Headquarters, Department of the South,
Morris Island, S. C.,** Aug 21st **1863.**

My Dear Sara.

    I would like to write you a good long letter tonight but suppose I shall not be able to more than make a beginning before the tent will be half full of boys talking and laughing and it is impossible for me to write under such difficulties. I am not very pleasantly located just now. The boys had all secured tents for themselves before I came and so I am left out. none of them seeming disposed to share their tent with men and no more can be obtained for our use. so I have to have my bunk in the office tent. I do not mention this as a hardship. not by any means. but is not pleasant because there is no chance for retirement. for as long as I burn a light the boys feel privaledged to come in. not stopping to think or care. whether it is agreeable to me or not. I succeded after quite a struggle in finishing a letter to Lizzie last night. Today I received your favor of the 11th. a good long interesting letter. to which I wish I could pen a worthy reply. Now you doubtless think that I could. if I would. tell you all about the bombardment of the rebel Forts in Charleston harbor. but the truth is my dear that I know scarcly anything at all about it. The Head Quaters is between two and three miles from our batteries. none of which are in sight from here. A ridge of sand hills runs through the centre of the island and completly shuts off all view of the rebel batteries. No one is allowed upon these hills. a guard being posted upon each knoll. I cannot get premission to go up to the front to see what is going on. for no one not employed is premitted to do so. I might steal away around the hills. dodge the guard. and thus get there. but I never yet attempted anything of that kind but that I got caught. Should I try it. the least I could expect would be to be caught by the Provost Guard and spend a few days under their guardianship. along with turbulant negroes and all grades of refractory soldiers. have the pleasure of being marched up to the front to work in the trenches under guard. lose my place in the office. and be returned to my Co. So I think I'll not run the guard. though I do assure you that it is hard work to remain guietly here while there is such a splendid sight so near by. The report of the guns. an occasional jar of the ground. little clouds of white smoke telling where a shell has burst high in the air. is all of the bombardment that I can obtain. Nor am I any better off as regards learning the result of operations. I cannot hear anything. or rather I hear everything and so much that I can depend on nothing. So that though within two miles of the scene of operations and at the Head Quaters of the Department. I shall have to wait and get my news concerning operations from the N.Y. papers. Dexter and Mr Landon managed to steal away somewhere this morning where aided by a good glass they were favored with a good view of Sumpter and had the good fortune to see the rebel flag shot away while they were watching the Fort. This has occurred several times today. but it is quickly replaced. Fort Sumpter but seldom returns our fire. in fact many of her guns have been removed and mounted at other places where they can be used to better advantage. The walls of the fort are said to present a sorry appearance. they have been breeched in several places. Five of the arches are distinctly visable. There is no doubt I think but that we can utterly demolish the Fort from our batteries on this island. With the exception of the forenoon of the first day of the bombardment. the iron clads have not taken a very active part. though I believe the Ironsides has been pretty buisy today. The wooden gun boats have barked awy spitefully at Fort Wagner. I suppose. but with what effect have no means of knowing. Some of them carry quite heavy guns and ought to do some execution. but it appears to me. that they are trying to fight at to long range to accomplish much. It is useless to try to fight and have the injuries all on one side. nothing can be accomplished so. Doubtless there is good policy in not exposing the navy more. There is no doubt of our ability to knock down fort Sumpter with our land batteries. after that is done the rebels still have powerfull batteries which we must depend upon the navy to subdue. therefore it is not advisable to risk their being crippled when it can be avoided. We have had quite a gale of wind for two days this week which prevented the navy from doing much any way had they been desirous of it. The wind blew the waters into the bay so that the tide rose much higher than ordinarily. Our office tent stood very close to the bank which being of fine loose sand. rapidly gave way before the action of waves. and almost before we knew it our tent was down about our heads. this was in the evening. there was nothing for it but to pack up the desks. move them into an adjoining tent and let the tent lie. All the next day the wind blew so furiously that it was useless to attempt to put it up. So all day we lay idle. The second day was clear and still. and we had the tent put up in a new place. where the waves will not be likely to reach it. Admiral Dahlgren has a tent on shore close by the Generals and our tent is almost between them. close by them. The Admiral does not spend much time on shore. but he has a large tent on shore for his accommodation. It is very provoking that one can know so little of what is being done. or at least accomplished right under his very nose. and it is blamed provoking that I can't get half an hour of quiet in the evening to write to you. Now here I am trying the second evening to finish this letter. four of the boys are playing cards. two are talking and smoking. and so it will be until they get to tired to stay any longer. by that time I shall be glad to tumble into my bunk and go to sleep myself. Well my love you need have no fears for my safty while I remain in the office. Our other boys are continualy in danger. George is detached from the C<u>o</u> as a sharpshooter and has gone into camp further up nearer the front so that I do not always see him when I go up to the Co. It really does not seem right for me to remain here out of harms way. and in comparitively comfortable circumstamces when the other boys are roughing it so severly and exposed to danger hourly. I wish I had grit enough to ask permission to return to duty in the Co. If I were not so lazy and did not love my ease so well I certainly would do so. I feel ashamed of myself when I read how anxious you are for my safty and think that there is no cause for anxiety. that the other boys are doing well and nobly while I am sitting quietly in the tent. If I had come out with the intention of getting some "soft thing" as the boys term the clerks life and other easy positions. I would not write so many words about it. but as I did not. and have kept my end up for nearly two years with the boys. I cannot feel quite at ease about it. If I answer your letter at all it will have to be in my next. With love. good by

Ever your own
Ned

No. 126 Rec'd Sept 14th 1863.

Morris Island.

Headquarters, Department of the South,
Morris Island S. C., Aug 28th **1863**

My Dear Sara.

Am behind agan as usual. Wonder what you are doing just now! I have just returned from supper. It is raining hard. Dexter and another clerk are engaged at a game of chess. two other clerks are smoking and talking. a fifth and the most troublesome one of the lot has just come in. I say the most troublesom. because he is one of those kind of men who think that whenever <u>they</u> open their mouths to speak. every one is interested of course. and that whatever <u>they</u> have to say is more interesting than any other subject the audience may be engaged about. He is a citizen and confidential clerk to Col. Turner who is Chief of General Gillmores staff. He is a very pleasant fellow and has no other intention than to be agreeable. but with so much self esteem he is apt to overdo the matter. Now i'll tell you what I had for supper. Tea. with sugar. and milk. buiscits and butter. fried cakes. real nice ones. good fresh gingersnaps just from the oven. the remains of our dinner which was a good fresh beef stew. and pickled onions. There now isn"t that a pretty good bill of fare for a soldier! I think so decidedly. I have not lived so well before since we came out. We have a good cook. he makes capital buiscit. nut cakes. and ginger snaps. and does things up nice generally. At present there are seventeen in all in the mess. I think I have told you before that we pay two Dollars apiece each month which goes to pay for "Extras." One member is selected to take charge of the fund and see to its proper expendature. While at the Head we used to sell portions of our rations which we did not want. and buy different kinds of food. but as I said before we are living better now than then. and the reason is we have a grand good cook. So much for our Commissary Department. I have not been very buisy today especially in the afternoon. They have been moving Capt Sealys tent (the other office tent) this afternoon and buisness was in a measure suspended in consequence. I told you what a time we had with our tent. how the high tides washed away the bank and let it down upon us. Well. the tides now are higher than then and have washed away the bank nearly to the other tent and would have had it down before morning if they had not moved it. Now you would like to have me tell you some news. But I cant. We haven"t captured Charleston yet. but we calculate to. and are at work at it all the time. The almost impregnable sand forts of the rebels give us much trouble. Fort Sumpter is considered as laid upon the shelf. From what I can learn I judge that Fort Wagner is our greatest trouble now. Ft. Johnson troubles us some but can doubtless be taken care of when we get to it. All in good time. we shall have the rebel sand heap yet that has been the death of so many of our brave boys. A nice little thing was done two nights ago. One of the rebel rifle pits annoyed us exceedingly and it was determined if possible to capture it. The attack was well planned and neatly executed. it took place just after sundown. A detachment of the 24th Mass. did the buisness. A part of the attacking party carried besides their arms two shovels apiece and no sooner was the pit in our possission than the shovels were put instantly at work and in a short time we were intrenched in the pit and able to hold it. Of the eighty rebels who were in the pit we captured Sixty seven including two Lieutenants. Two or three were killed. Our loss was one Lt. and two privates killed. one man missing. and five wounded. The whole thing was executed with surprising rapidity and was all over with before any but the party engaged knew anything of it. We are losing some men in killed and wounded every day. Yesterday just as I came out from supper seeing an ambulance drive up to the hospital. I walked down to it. It contained two men who had just been wounded by the bursting of a shell. In a few moments another man was brought along. wounded by the same shell. The men were on picket at the time. two were killed outright. and eight wounded by that one shell. These pieces of shell make frightfull wounds. one of these men had three severe wounds. his left arm was completely smashed in two near the shoulder and hung by the pieces of torn and mangled flesh. At the hospital. one tent is used as an operating tent and is furnished with two tables of proper height. the wounded are taken there and their wounds dressed and then taken to the other tents where they are as well cared for as could be expected. Well. last night there were one on each table. with the Docters with coats off. sleeves rolled up. and instruments in hand. two more men on strechers. lay at the door of the tent waiting their turn. and in the mean time they were regaled with the sight of the operations being performed within. I often think that I will go down and witness an operation. but cannot quite get my courage up to the sticking point. for I always think I <u>may be obliged</u> to witness one sometime. I have not seen George since I wrote my last letter. but believe he is still all right. Our regiment are now all employed in the batteries. manning the guns. Therefor we have lost no one except the two men killed by the sharp shooters. but I fear every day that our turn will come. The rebels have filled the ground with torpedoes. between fort Wagner and our works. The Engineers have dug out several in excavating the approaches to the Fort. I am told that to one of them a dead negro was attached by a cord in such a manner that in the attempt to move the negro the torpedo would be exploded. fortunately the infernal machine was found by the workmen before the negro was discovered. and the hellish design of the chivalric inventor frustrated. This story was told me by one of the Engineers. I have not heard it contradicted. and it is generally believed. If it be true it is certainly one of the most fiendish designs that I ever heard of in modern times. We have had high winds for several days past and the nights are frequently very stormy. The prisoners captured the other night are still on the island in charge of the Provost Marshall. They belong to the <u>61st North Carolina</u>. Many of them do not seem to care much about being captured. profess to be rather glad of it, but some are regular secesh. full of fire and fight. A day or two since. by Major Smiths order. Dexter done up and sent to the Governor of our State a battle flag captured from the rebels when we landed on this island. It belonged to the 21st S.C. and bore the inscription "Pocotaligo Oct. 21st 1862". The 21st S.C. is an artillery regiment and fought us at Pocotaligo. The flag was captured by a private in the 6th Conn. who shot the color bearer through the head. he fell forward upon the flag which is stained with his blood. Today two other flags that were taken by two privates in the 9th Me were sent to the Govenor of Maine. The health of the troops notwithstanding their labors and exposures remains good. and if things are not progressing as rapidly as we could wish. we think they are sure. and that it is but a question of time. I cannot always get the "New Souths." now. but succeeded in getting a copy of last weeks issue which I send you. N. Y. papers of the 22nd have been received here. but I have not seen them. The rebs are represented as being rather "scary" at their prospects. I hope we may soon succeede here. and then another by no means light feather would (be) laid upon the already weak backed Southern mule. A few more of the same sort would break its traitorous back. I earnestly hope that Charleston will not be allowed to escape this time. I hear that somme of the "Elected" in Salisbury are procuring substitutes. Dr. Gregory who used to be with Dr. Welch. is Asst. Surgeon of the 17th Conn. now on the island. I was slightly acquainted with him in Lakeville. but doubtless he has forgotten me. Will Reid went up and saw him the other evening. I have kept very close to the office. going no where but up to our own regiment. have not yet seen Ed. Hubbard.

With love my darling. I remain
Ever your own
Ned

P.S. The head clerk in the office has just informed me that the torpedo story is true. he has just had a conversation with the Captain who saw the negro removed. A nice operation. I would respectfully present it for the consideration (if such creatures are possessed of any) of Northern Copperheads.

E.J.B.

No. 127 missing

Editor's Note: This next letter is written following the Union capture of Fort Wagner and Fort Gregg.

No. 128 Rec'd Sept 23rd 1863.

Headquarters, Department of the South
Morris Island, S. C., Sept 11th **1863.**

Dearest Sara.

Your welcome favor of the 30th was received yesterday. Thus I am favored with another proof of your love. and long suffering towards me. For you still continue to write faithfully and confidantly though you get no return. I have been striving for an opportunity to write for several days. am almost discouraged. I owe so many letters. all of which I wish to answer. Now just for once i'll pour out my little budget for your inspection. and see if you can find any that you think I can well aford to pass long unnoticed. <u>Firstly</u>. I am indebted to Mr. Randalls people three letters. one to Mrs. R. one to Carrie. and one that I wish to write Mr R. partly on buisness. <u>Secondly</u>. One to my friend Miss Northrop. Mrs. R.s' sister. <u>Thirdly</u>. One to my brother in New Preston. <u>Fourthly</u>. One to my friends in So. Canaan. <u>Fifthly</u>. One to my sister Sara B. in reply to her "pencil sketch" "taken on the spot." received with your last. <u>Sixthly</u>. One to a lady friend whom I "cousin." Miss Elliott the same whom you may recollect taught school one season in Lakeville. and lastly to your own dear self I owe not only letters without number. but love. and devotion. more than I can express. Now Sara examine the lot and see if candidly. I am not owing more than I am worth! I feel that your decision must be aganst me. and fear that you will be led to hesitate about casting in your lot for life with such a bankrupt. Now it is past bed time. I have not been at work this evening. and your question probably is why is not the letter nearly finished instead of just begun! Well. I went up to the Co just after supper on an errand. coaxed Wells to come down and spend the evening with <u>us</u>. McNeil came too. and we deliberated on various matters untill roll call. when they were obliged to retire. Last evening slipped away from me in the same manner. I went down to Mr. Landons to see if he received letters from home. found he had not. and read to him and <u>Saml Wolcott</u> your account of Ebens' funeral. Mr. Landon has not received any letters in several weeks and Saml has heard of his brothers death only from others. Once chatting with Mr. L. time past unheeded and it was quite late when I left him to return to our own office. Samuel is at present in the same office with Mr. L. as clerk. I am glad that he has been so fortunate and hope he may be retained there. It has been very quiet for the two days past. No fireing of any account has been done. Three days ago the navy fairly got their foot into it and were obliged to fight. They were making some demonstration up the harbor when one of them got aground under the fire of the enemy. I think it was the <u>Weehawken</u>. The rebels were not slow to discover her predicament and opened upon her with a will. The other ironclads thus forced to fight in order to save their consort opened in earnest. and all day the air was filled with the thunder of their heavy guns. The engagement was with <u>Fort Moultre</u> and the batteries on Sullivans island. In the forenoon a magazine exploded in the rebel works. The explosen must have been terrific. it jarred the ground cleare down here nearly or quite ten miles distant. with two or three miles of water intervening. I think our fire must have been very damageing. it is said that several guns in Moultre were dismounted by it. <u>Moultreville</u> was fired about noon. it is supposed by our shell. it burned all the afternoon. and far into the night. I know not whether any damage was done to any of the navy. the grounded moniter got off at the next high tide. and since "quiet has reigned in Warsaw." The soldiers really took courage for the day. and really thought and hoped that the Rip Van Winkle who commands the navy had got waked up. and was going to demonstrate that our iron clads were really what we had so proudly boasted. and fondly hoped they were terrible engines of destruction and able to cope with anything that now obstructs the passage to Charleston. But we were doomed to be disappointed again. their consort once more afloat they retired to their old anchorage and not even a growl has since escaped them. I learn that in the late N.Y. papers. the correspondants have lashed the navy for its inactivity. most unmercifully. I hope justice has been done. and that the credit for what we have thus far accomplished has been awarded where it belongs to the army. Matters are all dark now as regards future movements. We have accomplished all that can be done from this island. we are in possession of every fort of it. and no further offensive operations can be carried on here. We (the uninitiated ones) have all the while fondly hoped that if we could only reduce Forts Wagner. Gregg. and Sumpter our iron clad fleet would do the rest. our work was done here and that Charleston or at least its site would be in our possession. But all such calculations have proved incorrect. The navy either cannot or will not perform what we expected of it. The whole concern turns out to be only a big "Bug a boo." If. therefore. we are still bent on capturing the city. I see no way but to remove our force on to James island and fight our way up nearly its entire length. That operation would be equivelant to several Morris Island campaigns. and to do it we must have very large reinforcements. Therefore I am at loss to think what will be the move if indeed any is attempted at present. It seems a great pity to suspend operations now for a single day even. for the exasperated rebels feel that they are indeed in the long talked of "last ditch." and are employing every moment in strengthning their already almost impregnable defences. I am at loss what idea to form of our next probable move. An order has just been issued granting furloughs not exceeding thirty days "to enlisted men who have been distinguished for gallantry or good conduct in the present campaign." so the order reads. The number granted is not to exceed two per cent of the force present for duty. That would look as though there was to be a cessation from active operations for the present. They seem to have got the matter in a fog concerning the furloughs or else I do not understand the order rightly myself which is very probable. for it is not very definate. In our Co eighteen men have been employed in the batteries. those eighteen drew lots yesterday afternoon to see who should go home. Barnes was the lucky man. Now George. Brinton. and one other were detailed as sharp shooters before the men went on duty in the batteries and have been doing the most exposed. difficult. and dangerous duty ever since. A number were also detailed for duty in the magazine and have been on duty there ever since our regiment came up here filling cartridges and shell. and yet none of these men had a chance to draw. It may be that they are to have a chance outside of the Co. but if they do not. the grossest injustice will have been done them. I almost felt certain that either George. or Brinton. would be one of the fortunates. and could not restrain a feeling of vexation when I learned who the fortunate one was. An attempt to take possession of Fort Sumpter. a few nights since resulted in a failure and the loss to us of about a hundred marines and sailors taken prisoners. I have been unable to learn definately about the plan of the attack. I think there was a misunderstanding between the land and naval forces. if indeed there was to have been any concert of action between them. I talked with some of the soldiers who were in the boats. they say that they rowed around nearly all night. did not attempt to land on Sumpter. lost their way and were glad to get back as best they could. And so the rebels still keep their detested rag flying from its sheltered walls. and fire their morning and evening gun from some little piece that has escaped the pursuit of our shot and shell. I would have paid liberally for an opportunity to visit Fort Wagner. but I cannot get the desired privilege. The wildest stories are told by those who have visited it. which I always reduce one half before I attempt to believe them. I have been both amused. and disgusted in reading over the list of the drafted men from Salisbury and Canaan and seeing upon what contemptable excuses they have avoided their duty. They must be proud of themselves. their friends must honor them. and highly the ladies must admire them. particularly some of the young men. Had I escaped in the manner many of them have. I should never expect my presence to be tolerated in any decent society after it. But I suppose that whatever lurking feeling of contempt the <u>fair</u> may at first be disposed to entertain for them. it will soon pass away and they be received with the old favor. acting upon the principle that "A live coward is better than a dead hero." Am finishing my letter in the morning. Ed Hubbard has just been in to get a Pass to go to the Head and from thence North. I do not think he can get permission to go home. even from New York. he has mounted his shoulder straps already. though he is not an officer in the N.& S. service until he has been mustered in by the Mustering officer of the Department to which his Co is attached. I have known of instances where men have run around so for months with their commissions in their pockets unable to get mustered into the service. But it is not likely to be so in Hubbards case. only I don"t like to see anyone so fast about it. let them wait until they are fairly entitled to them before they mount the gilt patches. I don't see but what you will have to give up ever hoping to get any answers to your letters my dear. How long think you. will your patience hold out and you permit matters to continue in this way! Bear with me a little longer dearest.

With love I am as ever
Your own
Ned

CHAPTER 10

FORT SUMTER

From the book:
    General Gilmore issues a congratulatory order:
Headquarters Department of the South.
In the Field, Morris Island, S. C.
September 15. 1863. General Orders.
    "It is with no ordinary feeling of gratification and pride that the brigadier-general commanding is enabled to congratulate this army upon the signal success which has crowned the enterprise in which it has been engaged. Fort Sumter is destroyed. The scene where our country's flag suffered its first dishonor you have made the theater of one of its proudest triumphs.
    The fort has been in the possession of the enemy for more than two years, has been his pride and boast, has been strengthened by every appliance known to military science, and has defied the assaults of the most powerful and gallant fleet the world ever saw. But it has yielded to your courage and patient labor. Its walls are now crumbled to ruins, its formidable batteries are silenced, and though a hostile flag still floats over it. the fort is a harmless and helpless wreck."

No. 129 Rec'd Oct 7th 1863
Morris Island

Hospital of the 7th C.V.
Morris Island S.C. Sept. 25th 1863

Dear Sara.
    I gladly improve the first opportunity when I have felt able to write in several

days. to drop a few lines to you. I am the more anxious to do so because I know how much you prefer hearing of my welfare from myself direct to that of learning of it from others. I am on the gain now dearest. and hope by the time this reaches you to be about my accustomed duties. Still my recovery may be somewhat prolonged beyond what I hope. A severe cold caught a few nights since is now keeping me back. I caught it in this way. Our hospital consists of three large tents set close together so that the only chance for ventilation is to keep the two ends open as much as possible. My bunk is one of the first at one end and of course while I had the full benefit of the fresh air. I was also more exposed to takeing cold. I suppose I must have thrown the cloths off from me in my sleep as I have endeavored to be very carefull when awake. My disease is not entirely checked yet but hope it soon will be. I have now been in the hospital twelve days. Had I done as I ought and given up a fortnight before I did. doubtless all would have been well now. Do not worry for me dearest. for I assure you it is unnecessary. I have all the care I need if not all that I could wish. The nurses are very kind and obligeing. You know Norton and Wells are among them. they have done all for me that they could have done for a brother. Tom was taken sick himself a day or two after I came in and has not yet returned to duty. but Wells has been unremitting in his kindness and care of me. Our surgeon is a good natured jolly sort of a man who laughs and jokes with the patients while going his rounds. listens to all complaints and grants cheerfully all reasonable requests. He understands his buisness too I think. So my dear. being in the hospital is not always the worst thing that can happen to one after all. The hospital is full now. twenty in all. I think. and a few have had to go back to their companies rather earlier than the Dr. wished them to I think in order to make room for those more in need of care. But now it is to have quite a cleaning out. A dozen or more are to be sent to the Genl. Hospitals either at Hilton Head or Beaufort. A surgeon came in the hospital this morning with Dr. Jarvis and picked out those who are to be sent away with the same coolness and as little ceremony as a farmer would select the chickens he would take to market on the morrow. Two who are to go are not able to be moved. and ought not on any account to be stirred at present. but there is no help for them. go they must and may God be mercifull to them. The Dr. missed me in his easy carless examination of the patients. though he would not have been apt to have taken me if he had seen me. Two days ago a member of our Co was brought in here fearfully. probably mortaly wounded. He was on fatigue at the time at Ft. Gregg. was struck by a solid shot or large piece of shell on the left hip. and cracking through bone and muscle carried away far more than its Shylock pound of flesh. I do not suppose there is one chance in a thousand for him to live. Yet I think if anyone can live through such a frightfull wound he will do it. He is young. not over twenty I think and has always enjoyed the best of health and spirits since we came out. has never been sick a day. I have not talked with him any in relation to his wound. but he is cheerfull and I guess thinks he is going to recover. He cannot see the extent of his wound else I think he could have little hope. I did not mean to tell you a wrong story when I wrote you that Barnes was coming home. Wolcott will probably explain the matter. I was in hopes to have sent you a lot of letters and a few other trifles by Sml but my things are all down at Headquaters and I could not get them in time and make up the parcel. I felt very bad the day he left for home and was as sick then as at any time I have been here. Am heartily glad he was permitted to go. it will do his parents so much good. Mr Landon was so much better as to be able to leave the hospital yesterday though I hardly think he will return to duty just at present. Dexter calls up to see me often. if I get well shortly I shall return to my old situation. but if detained too long by my illness will lose it of course. as there must be some one to do the work which cannot be delayed. Yesterday there was a grand review of all the troops on this island by Genl. Gillmore. who has been promoted to a Major General. Dexter told me that it had been pretty much decided to move the Head Quaters from here to Folly Island. It has been talked about ever since we obtained possession of Wagner and Gregg. Want of room here is the principle reason I believe for the move. Dexter was to have sent my things up today if they moved. but as it is now afternoon and they have not come I conclude they have not started yet. George has called in to see me twice. and has probably written home about me. it is very difficult for him to come down here and as there is no occasion for him to take so much trouble he will not be likely to be down again soon unless on other buisness. I received your letter of the 9th. also some papers on the 22nd. Your favor of the 7th was also received the 14th. the day I came in the hospital. You will pardon me I know if I do not attempt to answer them now. and doubtless before I am better able to do so. I shall be in the receipt of others from you that demand attention. and they will be put away like dozens of others you have written without ever being answered. Dearest will you not at last grow weary of writing letters that are never answered! Now I have told you how well off I am here in the hospital so that you know I am not suffering for anything I must open my heart a little to you and tell you how lonesome. dreary. and homesick I am at times. Everything will look dark and dreary do what I will to prevent it. I feel that it is wicked for me to complain and murmer. me. who am so much better off than a dozen of my fellow patients. But. O the unsatisfied yearning towards the dear ones far away. the longing for the old familiar scenes. which we may never look upon more. the craveing after the loving words and tender offices of those we know would be so happy to wait upon us. A man must either have no conception of what the word. <u>home</u>. means or never have known what it was to be loved and cared for by any human being. who will not feel lonely and desolate when sick in the army. For several days past the weather has been cool and cloudy and the wind has blown most of the time hard and cold. particularly is this the case today when there has not been a ray of sunshine. and I have sat while writing with a big blanket wrapped around my shoulders. There is considerable heavy fireing today but I (am) told it is principly from the rebel works on Sullivans island. The Navy has not made a move I believe since the afternoon when one of them got aground and their fire had such an effect on the rebel works. It is rumored that there is to be a change of Admirals. I hope it may be true and that some one will be sent who will be able to save the fast failing credit of our navy in these waters. Now darling having as I think written considerable of a letter for an <u>invalid</u>. I must give you my love and bid you good by.

Ever your own
Ned

No. 130 Rec'd Oct 7th 1863.
Morris Island.

Hospital of the 7th C.V.
Morris Island S. C. Sept 30th 1863

My Dear Sara.

Still in the Hospital you see. but I am happy to add still on the gain. Am very week and gain strength very slowly. Have not got along as fast as I hoped to when I wrote you last. I then thought I should be about my business by this time. My disease is not entirely cured yet. and my cold still hangs on though slowly giving way. I take no medicine and have not had any for the last ten days. They got out of the kind I needed. and have not yet been able to get a supply. fortunately my case has not been a bad one and I have not suffered for the want of it. only it is taking me much longer to get well. I have only been out doors once. that was yesterday. Our hospital has been moved from where it stood on the beach. back a few rods on to the sand hill and I walked up when the patients were being removed. I asked the Dr. this morning if he was not going to let me go out doors today if I would be carefull. and he said. <u>yes</u>. so I have permission you see. and mean to avail myself of it by and by. Am sorry it is not pleasanter for I shall not dare remain out long enough to take any comfort. Now I suppose you think perhaps that I have written any quantity of letters during the past week. and are a little inclined to be half vexed because this is not my third instead of my second letter to you. But my dear this is the second letter I have written in the hospital. I have not felt able to write much as I long to attend to my correspondance. I have spent the time mostly in reading. reading anything that came in my way. be it theology or fiction. though the latter has been largely in the excess. I have read Dickens "Great Expectations" and am at present employed on "Bleak House" with the "Antiquary" by Sir Walter Scot in reserve when Bleak House shall have been disposed of. I rather like Dickens' works when as at present I have plenty of time to read them. I find his descriptions of characters very entertaining while the manner in which he handles and ridicules the sons and fathers of society is exceedingly interesting and amusing. Last Sunday night eleven of our brother patients were removed to Beaufort. We still had seven left and have been reinforced by the arrival of four more from the companies. one man has been returned to his Co today. so that we now have ten in all. but I guess the hospital will soon be full again for the sick list of the regiment is very large. The men are much worn down and exhausted with constant duty and hard fare. and do not so readily get over the complaints that attack them. There has been a great deal of hope and strong expectation among the most of our regiment that we should be sent back to St. Augustine. as it seems that the 48th N.Y. are doing very badly there. but all our hopes are pretty effectually blasted now for the 24th Mass has been ordered there and are buisy packing up and prepareing for the move. The 97th Pa has also been ordered to Fernandina. so that the 7th Conn may just as well give up all hopes of luxurating in fair Florida any more. and make up their minds that their lot henceforth is cast among the rough places of the Department of the South. However the boys take a grim delight in the rather poor consolation that the 48th at least are not to be allowed to remain in the coveted place. The man of whom I wrote you as having been so dangerously wounded a day or two before. is still alive and the wound doing as well as could be expected. he is still in a very precarious condition and but little hopes are entertained of his recovery. His case excites a great deal of interest on account of the severity of his wound. and the manner in which he bears up under it. I do not think it possible for him to live. and do not believe the Drs' think he can either. though they talk very bravely to him about his recovery. The Head Quarters have at last been removed to Folly Island. so I shall have to wait a few days for my letter from you. I think Dexter will send it over but he may keep it until I join them. I do not mean to be impatient or to complain but I am anxious to be able to return to the office as quickly as possible. With myself there are now seven clerks in the office. the full complement. One of the clerks who was taken sick before I went into the office so far recovered that he was granted a furlough for thirty days. which furlough has nearly expired. as he has never been relieved from duty and is the friend of the head clerk to whose Co and Regt he belongs. I think it more than likely that if he should return while I am absent sick he would take the place and thus I should be left out in the cold. I have been quite unfortunate first in having been left behind at the Head and secondly in being sick. and thus kept out of the office so long. It has given other clerks who came after me the opportunities for getting the start of me in becoming acquainted with the buisness and thus keeping me at the foot of the class as it were. As I have no penmanship to boast of I had hoped to make myself so thoroughly acquainted with the buisness in all its details that I should be found usefull. and by steady application perhaps make myself liked and tolerated. But after so many drawbacks I am almost discouraged and if our Regt were only somewhere where the duty was not so hard. I should hardly ask to go back to the office. Wolcott lost his place by going home. but of course no one under the circumstances would have hesitated a moment about that. He may get some other situation equally as good when he returns. Well. my dear. I have eaten dinner and muffled up in my overcoat have been out and down to our Co where I spent over an hour talking with the boys. and now I am trying to finish my letter which I do not wish to lay aside for another day. Perhaps you would like to know what we have to eat here in the hospital. My fare is principally toast. and crackers. three times a day. with tea morning and night. Fresh meat soups are a frequent dish but of them. I dare not eat at present. Hominy. Rice. Farina. and Corn Starch we have most every day. but I heartily detest all such kind of food especialy when I am sick. Last night for supper we had some very nice fresh tomatoes for which we were indebted to the Sanitary Commission. they have also furnished us with other creature comforts quite as agreeable. So the second Fair has been held without being graced by our presence. Well. we must hope with all our might that we may be more fortunate the third time. I can"t see as we are missed much after all and really begin to doubt whether by another year we should not be considered incumberances. as being in the way. not wanted. no place for us. and people be wondering why we will presist in remaining around in other peoples' way. and why we don"t go back to the army where we belong. the only fit place for us. rough. half civilized fellows. I know my love that there will be a place for me in your heart and that you at least will not wish <u>all</u> of the soldiers back in the army. but I think some of us would be sadly in the way of some of the miserable shrimps who have sold themselves out so cheaply to avoid doing their duty to their country. I ought to look over some of your last letters and pay a little attention to answering them. but as I am getting a little tired and it is most night. shall have to put it off once more. No news here. The rebels try to worry us all they can. and occasionly open their batteries upon us at a tremendous rate. but seldom do any damage. Meanwhile our working parties keep steadily about their buisness strengthening the works and mounting Parrot guns on Gregg with which I should think Charleston could be easily shelled. as it is about a mile and an eighth nearer the city than is the "Swamp Angel" with which the other shells were thrown in. We have rumors of Rosencrans' defeat. I hope it may prove not to be so bad as reported.

Oct. 1st. Supper was brought in last night just as I was about closing my letter so it was laid aside. while eating supper your three letters of the 13. 20. & 23rd. were received and before they were scarcly read it was to dark to write more. We do not attempt to read or write in the evening for we are supposed to go to bed like the chickens at dark. However two or three of us who are getting better get together in one anothers beds and keep up a quiet chat as long as we can conveniently in order that we may get as tired and sleepy as possible before going to bed. for the nights are awfull long and as we do nothing all day. and are lying down most of the time it is difficult to sleep but a small portion of the night. So my letter was not finished last night and as there is no need of haste in sending it under a day or two I will try to answer your three dear good letters. But in the first place my love. I must tell you that though your last letters have not been answered it does not follow that I have taken no note of their contents. that I have not read them. and been mindfull of. and gratefull to you for all the items of news they contained. both such as related to yourself and also matters of general importance. I do assure you dearest that I cannot sufficiently express my gratitude for your goodness in being so kind as to continue writing your dear interesting letters while I was so remiss and recreant in my duty. Dear Sara. I am not worthy of you any way. I know it. feel it. and realize it more and more each day. I am too selfish. too careless of the comfort and welfare of others ever to be a good lover. a good husband. or a good man. My conscience often reproaches me for having ever sought your love and pressed you to be my own when I know my unworthiness. and how incapable I shall always be of making you as happy as I wish you to be. and as perhaps another might have made you. But I strive to quiet my mind by the reflection that I love you as dearly as another possibly could and by assuring myself that I shall certainly do all in my power to make you happy. I must confess my dear. that my failure to contribute to your happiness in the small matter of writing to you frequently. is rather a damper to whatever consolation I might other wise receive for the last declaration. but I hope to be more faithfull hereafter. I remember Mr Brant of whose death you write me. it is truly a sad case. a hard case. I do not think that there was anything at all peculiar about the 28th. and can see no reason why they should have been kept any where to recruit. the better way and the proper way was to do as was done. send them home as quickly and as expeditiously as possible after their time was out. So much is said on all sides about their being so worn down by fatigue that I cannot help wondering what they would say and what would become of them if they had the fatigue. and exposure that falls to the lot of our regiment. to encounter. If they choose to over eat why they must take the consequences. You may think me unfeeling. but I do not think I am. I only wonder what would have become of 28th Regt of braves. if instead of nine months. they had had three years to remain in service. To my mind the upshot of the whole matter is. that they have never taken any care of themselves thinking that it was only a short time any way and they could live it out. and that nobody has ever taken any care of them. If Landon and Bostwick. and Meatson. were a fair sample of their officers. I am more than convinced that I am right in my supposition. I suppose of course Bert will consider himself very much ill used by having been drafted and that he has now good reason to growl and complain and find fault with the present Administration. pardon me but I cannot help wishing that he could not be exempt. Am glad that you remembered me to Selina. Yes. dearest. I hope we are going to visit her one day. but when that <u>day</u> will be I think will tax even your strong womens' faith. and hope. to determine. I <u>have</u> been living pretty high for a soldier and now I am getting my pay for it. It was doubtless eating heartily and exerciseing so little that was the principle cause of my sickness. I know it had the same effect on another clerk who was taken sick the same day I was. but he having a stronger constitution. stood it out a few days and got over it. Am aware its rather a humiliating confession to make. but the truth is one cannot judge so well here. as at home. how much he can endure. and its not always a proof that a man has been gormandizeing because what he has eaten has made him sick. I regret to hear of Fred Harrisons illness and hope it may not terminate fataly. I can well imagine how much his friends must be alarmed for it always appeared to me that bleeding at the lungs was a sure precourser of sudden death. I think things must have changed some since we came away if Drs. Knight and Bissell have got so as to docter. or rather consult. together. When I came away I know that <u>Dr. Bissell</u> would on no account allow an opportunity to slip of venting a low. petty. spite against Dr. Knight. It was the only thing I ever had against Bissell.

who in all things else minded his own buisness in a most exemplary manner. But manys the time I've ached to knock him down for his contemptable treatment of Dr. Knight behind his back. Of course I am glad if all differances are amicably adjusted. and the <u>lion</u> and the <u>lamb</u> can lie down together. The disasterous attack on Ft. Sumpter was. I think if the truth could be known. about as discreditable as disasterous. Its not likely that the real truth about the planning and execution of the attack will ever be known. One thing is pretty certain. that is that there was no cooperation between the land and the naval forces engaged. had there been and the naval force promptly and ably seconded? by the land force the result might have been differant. The soldiers would have done their part and supported the naval force if they had given them an opportunity. but the navy wanted all the glory. and they have got it. I have very little hope of Charlestons' being ours in several months to come if ever. I am in hopes that our people will show no more leniency. but promptly burn the place to the ground if it can be done from our present position. and if not at the first opportunity. You have of course found out all about the "White flag." on Moultrie. The present rebel flag is so nearly white that it appears like a flag of truce at a distance. Am heartily glad you have learned to distrust those lying N.Y. papers. Don't never believe any war news you read in them until it has been confirmed in time by other papers. I know of somebody who would have loved dearly to have been a party to that ride you so much enjoyed. Or what would be better still. be your single companion to a quiet moonlight ride all to themselves during some of the beautifull evenings just passed. Annie Deane is certainly favored in hearing so often from her husband. and is still further favored in having such a good kind husband. But <u>Adjutant Deanes</u> facilities for writing exceede mine as greatly as do his ability. George is exceedingly faithfull in writing. probably more so than one <u>husband</u> (not to say lover) in a hundred. As he improves every opportunity whenever he has anything to write. he seldom fails of having a letter ready for every mail. Many. many thanks my love for what you style your "long yarn." it is very interesting and I have had the requisite "time" to read and appreciate it. Unworthy as I am. I still hope to receive many more such. I regret that Lizzies' visit should have been terminated so suddenly. and unpleasantly. it is really quite unfortunate. and too bad. She wouldn"t make a very good soldier. would she? getting sick every time she went from home. or to a new place. I am decidedly of the opinion that if you saw Aloah Bushnell. and did not "shake hands" that most certainly "distance lent enchantment" to the view. else how otherwise did you escape! I learn that he has been afflicted with the "<u>shakes</u>" pretty severly lately. it may have a good effect and help to cure his "shaking" propinsity. Well I ought not to talk so. none of us are without our little weeknesses. and peculiarities. and really I have a great respect for Aloah. who is certainly a most exemplary young man. By the way does he exert his influence still for the benefit of his country? I learned of the death of Mrs Gale a few days since from her cousin. I need not tell you that I was greatly schocked and surprised. It did not seem possible that one who I left in the full enjoyment of youth and health was now lying in the silent grave. I know I am telling the old story over again. but after all it is ever news. We all realize that though it is old as the hills. still it is ever new when we lose a young friend so unexpectedly. I used to be very intimate with her. and her family. and saw her a great deal both at her home where I was a frequent caller. and at Mr Randalls where she often visited and called. All who knew her were her friends. for they could not well be otherwise. Before I left I had grown to be very familiar as I was well aware of her engagement with Mr. Gale. there was no harm in it and of course my calls were either made in company. or when the family were present. She was a fine singer and a good musician. and the last three or four months I spent at the Falls I shook off many a fit of blues while listening to her music. She has gone home very early. but she has been prepared for several years. she will be sadly missed by her friends for she was always light hearted and gay. and made all pleasant around her. My dear. I do remember that we are on the last year now. and that it is now time to count the months that remain for my weary servitude. I rejoice. but very quietly and with the rememberance that even on the last week on the read home. even the destroyer may overtake me. I try to be patient and hope and above all to remember that "not even a sparrow falleth to the ground" without the knowledge of our Father which art in Heaven. After all dearest it is sad to be in a hurry to have our time which is short enough at the best hurry away scarcly caring what becomes of it. I have very. very. much to repent of in the two years that have passed and. well. I suppose I shall continue going on. doing wrong. and repenting. and never really improving so long as my life is spared. I cannot say my dear that I do always enjoy those evening chats with my friends. on the contrary I must say that I very often wish them in Flanders. and am thoroughly vexed at their presisting in staying and talking. for if they would go about their buisness I should be left to myself and my correspondance. Again the conversation is not often such as one need ever mourn to have lost. Did I tell you how vexed I was one night. when they came near burning the office tents up? Two or three of them were talking while I was vainly trying to write. disgusted at last. I left leaving the candle which was nearly burned down still burning. thinking that when it had burned quite out they would clear out. We had no candlesticks and the candle was stuck on a little block. which sat on my desk on which was lying a number of papers that I had been to work at for several days at odd spells. I had them nearly completed and calculated to finish them in the morning. Well. what do the dunces do but go off and leave the candle nearly burned down and still burning. it burned down completly. burned up my papers and in a few moments more would have burned up the whole office. doing a large amount of irreparable damage. Fortunately one of the printers happened to look in and was just in time to prevent further loss. As it was it cost me a day and a halfs labor. over which I felt anything but amicable. for the work was none the pleasantest.

Oct. 2nd. Takes me three days to write a letter now you see. Yesterday I received another letter and some papers from Mrs Randall. making two from her since I have written. I think you'l agree that my duty is plain in that quater. Today received a short letter from Dexter and some New Souths which I will send along. He does not like the new quaters very well. all the tormenting insects are very thick and the prospect is not pleasant generally. Capt Sealy is going North on a furlough. his wife came down to see him a few days since on the last steamer but one I think and he is probably going to "see her home." I am real vexed to think I can"t be in the office now. for the boys will have jolly times while the Captain is away. I learn that our batteries have been fireing on the ruins of Ft. Sumpter some for a few days past. it is thought the rebels were at work in it making it into a battery. which was the reason for again opening fire upon it. Now my love I think you will give me credit for a tolerable lengthy letter. whether it be interesting or not. As I am getting better. I must lay aside my reading and endeavor to do my duty to my correspondants. I gain strength very slowly a short walk tires me. as does sitting up for a long time. We have no easy chairs here you know. and one must either sit bolt upright without any support. or else lie down. You have not told me whether Julia is boarding at home. or at Mr Tuppers. George told me the circumstances about he(r) leaving Mr Dodges'. but when I saw him last she had not decided at which place to board. I have not seen George since writing my last letter. presume he is usually well. Adjutant Moore visits the hospital frequently and I have had more conversation with him here than before since his promotion. but we have never been quite so unreserved and intimate as before that time. his conduct towards us is so far as I can see the same as of old. I do not feel very anxious to have you hard at work again teaching though no doubt you would prefer a pleasant situation to remaining at home. Do not be in such haste to secure a situation as to make a bad bargain. and have all winter to repent of it in. Praying that your health and happiness may be continued to you. and that in

Gods' own good time. we may meet again. I remain with much love

Your own
Ned

No. 131 Rec'd Oct 15th 1863

Head Quaters Dept. South
Folly Island S.C. Oct. 6th.1863

Dear Sara.

You(r) letter of the 24th has just been received. and if you will accept of a short letter a reply I will try to answer it this evening. But in the first place a few words about myself and present whereabouts. Great. I. is the principle subject of my letters at all times. I get fairly ashamed of myself some times to think I have so little else to write about. Last Sunday morning the Dr. told me that I might return to my duty. I was glad of it but really felt quite ill that morning. having spent a feverish restless night owing I suppose to having either eaten a little to heartily or to exerciseing too much the preceeding day. However I said nothing to the Dr. but shortly after went down to Genl Terrys' and got my pass. but by the time I got back found myself tired out. so gave up going that day. lay down and slept all the afternoon. Monday morning felt much better. but pretty weak. Found on inquiry that the Hd Qts. was about four miles from the landing at <u>Light House inlet</u>. Did not fancy the walk much especially as I had all my things to carry. Finaly concluded to leave my knapsack and most of my things and return for them when I got stronger. So I rolled up my blankets and swung them together with my overcoat. and haversack containing a few necessary things. over my shoulder. bid the boys good morning and started. By the way one of the nurses brought my traps down to the Ferry for me. and I shouldered them after crossing. The morning was cool. clear. and lovely. and the walking on the smooth hard beach. splendid. Another time I should have enjoyed it exceedingly. but though I started to walk very slow yet after going about half a mile I was too much fatigued to enjoy anything but rest. I found I had greatly overestimated my strength. And now my good fortune overtook me. Two men rode by me on horseback. I did not notice them particularly but thought from their dress. which was about half uniform and half citizen that they were either Sutters or Citizen clerks. They had passed me a few rods when I noticed them stop, turn round and ride back. they rode to me and one of them whome I then saw to be an officer asked me if I was sick. I replied that I had been. had just left the hospital. and was returning to duty. he then asked what Regt. I belonged. and where I was going. upon answering him. he said that their Brigade hospital wagon was just behind. would overtake me in a moment and if I had a scrap of paper about me he would give me an order to the driver to let me ride and if I wished it the wagon should take me to the Hd. Qts. though it was about a mile further then their camp. I thanked him. and said that I could not think of putting them to so much trouble as I could easily walk the remaining mile. I handed him my pass and on the back of it he ordered the driver to let me ride and signed his name to it. Bidding me good morning they turned their horses and again rode off. The wagon came up in a few moments and I had a nice ride for about two and a half miles. and then walked the remaining mile very comfortably. arriving here a little after noon. Now you will probably think that it was a very small matter for the surgeon to give me a ride. and that it was nothing more than what he ought to do. Granted. But such actions are very. <u>very</u> uncommon. and therefore the more praise to the kind hearted surgeon. I ought to have told you before that all teamsters and wagoners have strict orders not to allow any one to ride in their wagons. so that twenty empty wagons might have passed me and I not have been able to have persuaded a single teamster to disobey orders and let me ride. The name of the officer. was H. W. Carpenter. Asst. Surgeon of the <u>117th. N.Y. Vols</u>. Well. I was pretty tired when I got here. but have got pretty nearly rested now. I am far from being well yet. but hope to gain fast. Have been in the office at work most of the day but have not accomplished much and don"t mean to until I have more fully recovered. Our new situation is an improvement on the old one on Morris. It is on the sand bluff that rises abruptly from the sea shore. The offices are not over ten rods from the beach. The place is thickly wooded with pine and palmetto. with a few live oaks sprinkled among them. The most of the trees have been cut away and the ground cleared up for twenty or twenty five rods back. The office tents form two sides of a hollow square. the Generals quaters the third side furtherest from the beach and are situated on a little knoll shaded by live oaks and pines. The fourth side is open with a guard pacing backward and forward who permits no officer whatever his rank to ride a horse within the enclosure. The open space is perhaps twenty rods square. Our tents are out a little to one side but very conveniant. Our mess tent is some little distance to the rear of all the rest out in the woods. We do not fare as well as we used to on Morris. The mess is larger than there. which makes it worse. for to live well and get along nicely. eight or ten are all that ought to mess together. We now have twenty-two or three. We have changed cooks too. and made a bad bargin in that respect. We are going to be awfully tormented with insects here I see at once. Tired as I was last night the infernal fleas kept me awake three fourths of the night. Mosquetoes. sand flies. and gnats are also abundant. We have plenty of music here. There are three or four good bands. and the boys tell me they play in front of the Gen<u>ls</u> quaters nearly every night. The General and part of his Staff went to the Head today. but will not probably remain there long. Dexter went to the Head this morning. there is still quite a number of articles of office furnuture down there. and he goes to bring them up. I can easily imagine what an exciting time there was at Mr Wolcotts when Samuel appeared to them so suddenly. Thats just the way I should like to visit home. entirely unexpected. only I should not want any one present but just my loved ones so that we could have just as much of a scene as we chose all to ourselves. I dislike exceedingly. a public exhibition of feeling. though of course it cannot always be avoided and then I respect it though I would much rather not witness it. On the whole I believe I could put up with a tolerable "large" scene after all if I could only meet you once more. But I suppose like a good. dutifull girl. you have done as I have told you. and given up all idea of seeing me for another eleven months to come. If we survive those. we shall doubtless meet. not till then. I wish I could get hold of some of those nice grapes. Wonder if Sam wont think to bring down a box full! Packed in sawdust they will keep two or three months nicely. and bear transportation most any where. Now Sara though I am very selfish with regard to your kisses. I think you ought to have given <u>Sam one</u>. hes' a brother soldier you know. and I believe he has earned <u>one</u>. However you need not lie awake about it only I would not have scolded if you had kissed him just the least bit in the world. Well I can remember when you wouldn"t kiss me any more than you would an African. and would run as if for dear life if I attempted to coax acquaintance with your lips. But things have changed since then. Wonder how they came to. and why! Well I find that this letter is pretty much all about Barden after all. but as you are supposed to be slightly interested in said personage perhaps you"ll not care much. If I am fortunate in recovering my health quickly. I <u>will try</u> to be more punctual about writing. but while I am feeling semi miserable most of my writing must be in the office. With much love my darling. I am ever your own.

Ned.

No. 132 Rec'd Oct 19th 1863.
Folly Island.

Head Quaters Dept. South
Folly Island S.C. Oct 11th. 1863

My Dear Sara.

I can only be with you in spirit tonight love. and I can give you no more convincing proof that you are in my thoughts than to write to you. It is Sunday evening you see. I have been at work all day as buisy as I could be. It has not been much of a Sabbeth to me I can assure you. I have read three. or four. chapters in the Bible since supper but that cannot go far towards paying for the broken Sabbeth. Some how it most alway(s) happens that there will be an unusual amount of work to be done on Sundays. We cannot arrange to prevent it. whenever the work comes in it must be attended to. Generaly. I do not have so much of the Sabbeth in the office as I did in the Co. but it would make but little differance now. for all the boys are on duty nearly all the time. Our head clerk has been sick for the last two days and unable to be in the office and of course that makes more work for the rest of us. I think he is not going to be able to attend to his duties regularly again for some time. When Capt Sealy gets back which will be in a few days. he will probably go North on a furlough. We are having our usual evening treat of music. The band now playing belongs to one of the N.Y. Regts I believe. they are Dutchmen I should think by their looks. and I should think they were playing Dutch music. for its' the querist combination of sounds that ever I heard. They have been playing over an hour. and I have not heard a single note I believe has ever been played before. We have music every evening when the General is here. We are having magnificent weather now. the sky clear and cloudless. and the air just the right temperature to be pleasant. I thought surly that band had left for they have been silent some time. but they have commenced again. I do wish you could hear them. they are playing another of the oddest tunes. (if it is a tune) that ever you heard or dreamt of. its rather pretty too quite lively. and played very soft. Yesterday I went up to Morris. to get my knapsack. I was not fit to go. but could not get along without my cloths and so was obliged to go. I walked up there a distance I should think of about four miles though some try to call it five. Was pretty much used up when I got to the hospital. I did not succeede in doing all my errands satisfactorily and shall have to go again. or send. I had a good dinner with the boys at the hospital of codfish and potatoes. bread and butter. and afterwards had a good glass of blackberry cordial which did me good. Went into the hospital and saw the sick boys. Some of those I left there. have got well and returned to their Cos. and others had taken their places. All were doing well. I found Barnes there. he was up. dressed. and writing. so concluded he is just comfortably sick. I found the young man who was so badly wounded doing nicely and bidding fair to get well. Another man from our Co is in the hospital severely wounded in the ankle. it was thought at first he could have to lose his foot. but at present the wound is doing well and he will probably escape with a stiff joint. He had the piece of shell that hit him brought down. and keeps it beside his bed. its an ugly looking piece. and I should think would weigh six. or seven pounds. Another man belonging to Co B. wounded the same day. was still less fortunate. A piece of shell shattered his leg so badly that he was obliged to have it amputated. Its a wonder he escaped so well. or rather at all. for the shell exploded so close to him as to burn off his eye brows and whiskers. I went down to the Co. and saw the boys. had but a short time to stay and was sorry for I found that George had got back. and I wanted a good long talk with him. but could only have a few minutes. Wells brought my knapsack down to the dock so as to save me all the fatigue possible. On the ferry boat I was fortunate to find one of the Head Quater orderlies. who had come up with a team to get a large cage that had been made for the Genls eagle. and I rode clear down here with the teamster. The cage occupies about the centre of the square in front of the Genls quaters. and there "Mr eagle" holds his court with all the dignity of a "King of birds. I got some medicine at the hospital and now hope with a little care. that I shall soon be well again. I have no news at all to write. Dexter has dropped in and we have been having a chat. and a smoke. that is he did the smoking. and I the talking. He recommends to me. to learn to smoke. thinks it would be beneficial to my health. What do you think of it? Do you want a husband that uses tobacco? I have often thought myself that smoking a little would do me good. but as I have no liking for it. have never tried very hard to become accustomed to it. I know that the Drs. never caution the men against the use of tobacco. but rather encourage it. As I do not get my rations of "whiskey" now. perhaps I need some other stimulant. What think you? The Arago has gone by on her return trip to N.Y. She ran within half a dozen miles of the shore. when she went by. and was boarded off Morris Island and the dispatches put on board. All the vessels passing between the Head and Morris keep near the shore. and we can scarcly look seaward without several vessels being in sight. I could not help thinking tonight as I sat on the bank and watched the Arago. of the time when we first came down here. and my first sight of the Southern shores. And then I wondered if I should live to return perhaps in some noble ship like the one passing. if so. how gladly I would bid farewell to these low. sandy. pine covered shores. and how eagerly and impatiently I should strain my eyes to catch a glimps of our more rugged Northern coast. I almost begin to hope sometimes when I remember that we are on the last year. and that as the boys say. we shall soon be "nine months men." The time will pass very quickly if I remain in the Office. I shall have to confess to you that I am getting to be desirous to remain in it and am a little sorry there-at. for I wanted to feel all the time perfectly indifferant as to whether I left or remained. Well my love I have only just candle enough left to close and direct this by. and I am to lazy to go to the office and get more. especially as its about ten o'clock. So with love. my dear. Ever your own.

Ned

No. 133 Oct 22nd 1863
Folly Island.

Head Quaters Dept South
Folly Island S.C. Oct. 14th 1863

Dear Sara.

It is rather late to think of writing much of a letter. I have but just come to my tent. from the office where I have been buisy all day and all the evening. I think it is about nine o'clock. am rather tired and not feeling very well. but I received your letter of the 4th. today. together with the papers. and as it is possible that I may go to the Head in the course of a day or two. thought it better to write if only a short letter tonight. for if I put it off. it might be several days before I should have an opportunity again to write. I think I mentioned in my last letter that our head clerk was unwell. He still remains quite sick. and it is important that he should be in a hospital somewhere where he can have good care. At present he is staying in his tent and the boys are waiting on him as best they can. There is no comfortable hospital on this island. He wishes to go to Beaufort. If he does I shall probably go with him to take care of him on the way. We want some things in the office which were left behind at the Head. and I should stop to get them. On the whole t"would probably be rather a pleasant trip. and I think it would do me good. The most objection I have to it. is that I have been out of the office so much. that I do not like to be gone a day from it. until I have become more thoroughly acquainted with all the details of the buisness. It appears almost usless to write. that "there is nothing new" here. It seems as though you must know it of course. But really I dont think that we need be ashamed yet awhile to have it said that "all is quiet in the Dept of the South." All things fairly considered. Genl Gillmore has actually accomplished more with his little force in three months. than has the whole vast army of the Potomac. since the War commenced. That army has been a stupendous humbug. a vast moving slaughter house. It has. with the most men. and means. accomplished the least of any army in the field. With all its vastness. it has. at times had much ado to protect Washington and has only succeeded in doing it by the aid of large bodies of State Militia or minute men. who have rushed to their assistance. On the whole it has done little or nothing towards ending the war. and never will. In this Dept at present. it requires every man and the utmost vigilance to hold possession of what we have. The rebels are in great numbers. and strongly intrenched. on the adjoining islands. from which we are separated only by narrow creeks. easily crossed. A bright lookout. and constant readiness to repel an attack are therefore essential to our safty. We cannot afford to lose what has been so hardly won. and is of so much importance. I can easily imagine how beautiful our loved New England is looking in this delightful Oct. weather. Every season has its beauties and pleasures. but I think the Fall is the "cream" of the whole year. I am right glad we three "precious" invalids are all better. if we had only managed to have hung on a little longer. perhaps we should have had the pleasure of seeing a deputation of our friends from Salisbury. down here on a kind of home brewed Sanitary Commission. That wouldn"t have been bad. and much as I detest what the soldiers call "playing off". (pretending to be sick when nothing ails one. that he may get rid of doing duty) I fear I should have been sorly tempted to have hid my book and been in a great deal of pain when I saw the Dr coming. if I could have been sure of scaring some of you down here. Should think from Mrs Nortons' eagerness for news. that she did not hear often from Tom. I know he writes often. He was fretting. and fuming. at a great rate while I was in the hospital because they did not write to him from home. I have written to my uncles people. (Mr. Jones). Did it the first thing after writing my second letter to you. I have endeavored to write very frequently to them without any regard to their answering my letters. since the death of my couisin last April. They wished to hear from me often. and had but little time or opportunity for writing themselves. and I am glad and willing to do the little in my power to cheer them in their affliction. I had only written to them once after coming to Morris Island and I suppose they feared that I was very ill. I have had a hearty laugh. and many a quiet "chuckel" over Eddies' reply to my brother. It was decidedly rich. True as the book. my brother probably supposed that I wrote frequently to the <u>older</u> members of the family. and quite likely they had heard from me since he had. I correspond quite regularly with him. about once a month I write. If they do not hear from me at the expected time they think something is wrong. Of course I am behind with them as well as everybody else. but i'll catch up yet. Am going to try to write a letter a night until I get even. Not much of a chore to do that I suppose you think. It would"nt be. if I did not have to work till bed time frequently. and then after one has been writing all day they often feel as though they did not wish to see a pen again for a week. I wish you were acquainted with my brother and his wife. she would almost worship you. I know. She is a lively kind hearted woman.

not remarkably intelligent and never enjoyed many advantages. thinks the world of me. and seems to love me next to her husband and her kittens. My brother is a carpenter. was doing well at his trade. until attacked by a severe fit of sickness from which he has never fully recovered. This was while I was in the Store. I received a letter from Sara B. the other day. but she said nothing of going to Chicago. thought she should take a trip somewhere after closing her school. but did not intimate that she thought of spending the Winter away. You will doubtless miss her very much. I doubt whether I shall be able to answer her letter in time for her to receive it at S. If you will send me her address. I can mail my letter to her new home. So the L. R. school is to "be promoted" to a man teacher. Well Martha can go West now for all thats hindering her. Do you hear from that lucky lover of hers now? Is Robert Mitchell in Chicago now. and do you know what he is doing?

My eyes wont keep open much longer love. I"ll warrent you are asleep before this time. May you have pleasant dreams my darling. Good night.Ever your own.Ned

CHAPTER 11

THE DINGIE PLAN

From the book:

On the 16th of October the Seventh left Morris Island by steamer and landed the same day on St. Helena Island, where they went into camp, tents floored and bunks built.

They were provided with forty little flat boats, each capable of carrying fourteen men and an officer. Twelve men were to use paddles and two to steer, one at each end of the boat. The boys called the boats "Dinkies." They were made to order in New York at short notice, of inch pine boards, hastily put together and primed over with lead and oil. They seemed so frail that the boys said a blow from the butt of a rifle, or a heavy step would send them to the bottom.

A novel drill with these boats was commenced. The first consisted in handling the boats, launching them, then landing them again. Then as the boats lay on the sand the men withdrew to a short distance, and at the word of command approached the boats on a run, launched them, embarked, paddled back and forth in an array which someone called the drill of the "horse marines;" then at a signal paddled to the shore, landed, hauled up the boats and scrambled up the bank ready to meet a hypothetical enemy. Of course the boys could only guess at the object of the "Dingie Drill," but they easily guessed right.

The plan was to make an attack by night on demolished Fort Sumter by means of these boats, hoping to surprise and capture the Confederate garrison which still held the ruins.

Lieutenant Colonel Rodman rejoined the regiment October 24th, but on crutches, unable to resume command. <u>Maj. O.S. Sanford</u>. the next in rank, commanded the regiment.

No. 134 Rec'd Nov 1st 1863.
Folly Island

Head Quaters Dept South
Folly Island S.C. Oct. 21st 1863

My Dear Sara.

    Have been trying to get at writing all the evening. and have at last succeeded. but it is so late I can do but little more than to make a begining on a letter.

    Your favor of Oct. 9th was received last evening.- that is I found it here on my return from the trip to Beaufort and the Head which I mentioned in my last that it was probable I should take. As the little incidents of the trip. though of no importance. and not worth the telling. may interest you a little. and will at least serve to lengthen out a letter. I will give you an account of my five days travels. I met with an accident to begin with. We were obliged to start from here at daylight in order to be at the Inlet in time for the boat. which we were told was to leave at 7 o'clock. but did not until 9. Mr Foster the citizen clerk of whome I have before made mention was to go to the Head with us. on his way home. being through with his duties here. He and the Sergeant (who was sick) occupied the same tent. or tents for they had two officers tents placed close together. the back one being used to sleep in. and the front one having served Foster as an office. I was awakend before daylight by hearing Foster (the tents are all close together) talking and thrashing around. jumped up and dressed. and started to go in and fix up the Sergeants things ready to be off. As I was about entering the tent. I saw there was no light in it and heard no one stirring. was just singing out to know the reason when I was struck by some object. thrown with violence out of the tent. The blow nearly knocked me down. and I cannot see why it did not quite. for it was heavy enough to have felled a horse. <u>nearly.</u> As it was it staggered me back several paces. whirling me around like a top until the pain. and faintness. obliged me to sit down. Foster was pretty well scared and hastend to relight his candle and came to my assistance. Found that I had received quite a severe cut just over the inner corner of my right eye. which was also slightly injured. Then came the explanation. It seems Foster in coming out of one tent into the other had put out his light and while in serch of a match had stumbled against a heavy pine stool. which in his vexation he had thrown with all his might out doors. not dreaming that any one else about was up. much less coming near his tent. The corner of the stool had struck me and I have reason to be thankful that it did not hit me an inch lower down in which case it could hardly have failed to have put my eye out. The same blow on the temple no doubt would have been sufficient to have killed me. Well. I was in a pretty fix. all the arrangements made for me to go. Passes made out and signed and the ambulance waiting to take us to the boat. and there was I. a three cornered hole cut in my head from which the blood was running. and half crazed from the effects of the blow. I could not well back out. though Dexter insisted that I should. Fortunately I had a few strips of adhesive plaster. with which Dexter did up the cut. and tying a handkerchief round my head helped my patient into the ambulance and was soon "en route." First though Mr Foster and myself went up to the Generals mess tent and had a good cup of coffee. some cold ham. and bread and butter which the cook had got up and prepared for us. It was a beautiful morning and I enjoyed the ride up the smooth. hard beach. exceedingly. It "was by all odds the most pleasant ride I have had in over two years."

Arriving at the Inlet we crossed in a small boat rather than wait for the ferry boat. and soon had the Sergeant aboard of the "Escort" and snugly stowed away in bed. and feeling quite comfortable. The "Canonicus" was also going to the Head. and I found that our Regiment (which I knew had been ordered to St. Helena. but supposed had already gone) was going down on the two boats. Six Co<u>s</u> on the Escort. and Four on the Canonicus. Our Co. was on the Escort. and I had a short chat with George who was not feeling very well. We took breakfast together on board of the boat and I intended to have talked with him more but what with a slight touch of my old friend. "Sea-sickness." and attending to my patient. did not get to it. The Regiment has gone to St. Helena Island. which is about eight miles distant from the Head. across. and up the Bay. for the purpos of drilling. to act as "Boat infantry." T"is said they are to be armed with a new kind of rifle. the best yet made "Spencers breech loading rifle." I have seen one of them. and they are certainly the most perfect thing of the kind I have yet seen. Well. as I am told the boats are small. to hold eighteen men. ten of whome are armed. and the remainder are oarsmen. This at present is all I can learn about it. I think the boys will stand a good chance to see hot. and dangerous work if ever called to act in the proposed capacity. They will probably remain at St. Helena a month or six weeks drilling unless their services should be sooner required. I believe St. Helena is a comfortable place to be stationed. The boys will be able to supply themselves with fresh fish. oysters. and sweet potatoes from the negroes who inhabit the island. On reaching the dock at the Head. where the boat was obliged to stop. I got my patient at once on board of the "Genl Hunter."

which boat runs between the Head and Beaufort. and at dark we landed at the latter place. I had a letter from Col Smith to the Surgeon in charge of the Hospitals. requesting him to admit the Sergeant to some one of them for treatment. The Sergeant insisted upon accompayning me to the Surgeons office. which though quit near he was in no condition to walk to. but he was obstinate and went. but we did not see the Surgeon who was out. However his clerk did the buisness for us. and calling an ambulance we were soon on our way to a hospital which was quickly reached. the patient admitted. put to bed. given his supper. visited by the Dr. and made comfortable for the night. before eight o"clock. It was much better luck than I had expected. for I had feared that we should be obliged to remain at the Head. over night. which would have been more than unpleasant. After my unlucky adventure in the morning. everything went very nicely with me. I expected to have gone to the hotel to have stayed. but the Hospital Steward told me it was quite unnecessary. that a bed in the same room with the Sergeant (by the way the room was small and there but two beds in it) which was unoccupied. was quite at my service while I remained. So as soon as the Sergeant dropped into a doze. I blew out the light and "dropped" into bed. Yes into bed. A real <u>bona fide</u> bed. with matress. clean sheets. and a pillow. It was the first time I had enjoyed such a luxury since I left home. Perhaps you think I lay and thought over my good fortune. and speculated as to the probable time when I should again be so fortunate. But I didn"t. I just comfortably settled myself among the bed cloths and went right to sleep. Had a pretty good nights rest in spite of the "luxury." was obliged to get up once or twice to attend to my patient who in spite of a dose of twelve <u>grains of Dovers powder</u>. did not sleep near as soundly as I wished him to. I forgot to tell that we started from here on Friday morning. Saturday I was to spend in Beaufort as my patient was in want of some articles of clothing which I had not time to purchase before the boat left for the Head. The day was mostly spent at the stores. for I had many little commissions. to execute for the boys. I however hunted up two men from our Co who were sent to Beaufort from our hospital while I was an inmate of it. Found them much better than I expected. and getting along finely. At night took my lodging again at the hospital. felt quite unwell and very tired at night. and got to bed as quickly as possible. and into a sleep from which the Sergeant could only awaken me once though he tried more than that. he had been much worse during the day. and had a high fever all night. Sunday I expected to come down to the Head. The time for the boat to leave is half-past eight. but that morning with little or no notice given. she went off at Seven o"clock. The object was to reach the Head before the sailing of the "<u>Fulton</u>"

which was to leave in the forenoon. I had plenty of company in my "misery" for quite a number were like myself left behind. some of whome were to have gone North on the "Fulton." Among the latter was <u>Major Woods a Pay Master</u>. I divided the time in the forenoon between the Sergeant whose fever continued unabated. and watching the wharf thinking another boat would surely go down. as the Hunter had left so early. But not even a <u>canoe</u> left the dock after seven o"clock. In the afternoon I went out to the camp of the 1st S.C. Vols. (cold) and had quite a visit with one of the Asst Surgeons. who was formaly a private in our Co. The day was beautiful and though I fretted a little at first. soon gave it up and enjoyed myself as well as I could. In looking out <u>after</u> (litterallay) the boat. missed my breakfast at the hospital and so took it at the "Stevens House" the only hotel in the place. and a decidedly one horse affair. My patients fever gradually gave way towards night. so that he was able to get to sleep. and rested very well all night. as did also your most "<u>obedient</u>." But I woke up with a high fever in the morning and felt too unwell to rise. but was obliged to. and after doing what more I could for the Sergeants comfort. who was much better. bade him good by and made sure of being aboard of the boat in season. I was really quite ill. every bone in my body was acheing. and shivering with fever. and it seemed utterly impossible for me to keep on my feet and do the days work. which I knew was before me at the Head. I lay on the cuishoned settee all the way to the Head. and crawled off the steamer feeling half dead. Here I had a lot of Clothing to draw from the Quater Masters. for the boys. A quantity of Books to pack up which had been left in the office. several other <u>Official</u> errands to do. and every <u>Sutters' shop on Robbers Row</u> to visit on some little commission. I need not tell you that I spent a most miserable day. The fever continued unabated "till some time in the evening. I had a good dinner at our old mess room. for though so sick I had a hearty appetite. A little before sunset I got my work done. and myself and things on board the "Alice C. Price." I would have given any price almost for a State room. so that I could have gone to bed. but it was a small boat and there were none for the Officers that were also passengers. The best thing I could do was to go into the saloon and lie down on the cuishoned seats. which was the best accommodations the "Shoulder Straps" could get. and on the seats. sofas. and floor. a dozen of us passed the night. I barrowed a blanket of the Steward and getting into a sweat. the fever left me. and I got a sleep between seven and eight. Slept about an hour and was awakened by the entrance of <u>Crane</u>. the artist. and <u>Sawyer</u> the Herald reporter. and two or three officers all as happy and jovial as whiskey could make them. What with their smoking. swearing. singing. whistleing. and loud talking. sleep was out of the question. They remained an hour or more. drinking. and enjoying themselves. and then left for the "Nantasket" which lay along side and was also going to the Inlet. On that boat they had secured state rooms. so I was most gratefully relieved of their presence. Got to sleep again about midnight and slept until daylight when I was wakend by the boats getting under weigh. After a passable breakfast I was agreeably surprised to find myself feeling very nicely indeed. The day was clear and bright. the sea as still and smooth as a mill pond. and I enjoyed the trip very much. We reached the Inlet at noon. but the "Nantasket" which left the Head before we did got in to the dock first. so we were obliged to come to anchor. We were much heavier loaded than the other boat. but we gave her a close rub. with all her advantage. When we came up to the buoy. she was barely half her length ahead. and our boat was obliged to slacken her speed to avoid a collision. I went on shore after awhile in a small boat. and went up and took the second table at Mr. Landons' mess. However I made out a hearty dinner. and was much gratified on coming out to see that the Price was coming to the Dock. which she reached before I did and my box which was pretty heavy was landed before I got there. "Borrowing" three or four negroes at work on the Dock from the Sergeant in charge. I had it carried around to the ferry boat Dock. and Mr. L. and I sat down on it had a good chat while waiting for the boat to come across. Mr. L. I am sorry to say is in very poor health. and does not seem to recover at all. At last the old "Croton" came across but she had on a load of logs which must be unloaded by a lot of South Carolina negroes. and if ever patience is required. it is to wait for a Southern negro to do. anything. no matter what. It seems utterly impossible for them to move. except at the most moderate of snail paces. But all things must have an end. and the last log was found finally. We crossed at once. and getting a man with a horse and cart to take my box down we were soon on the road. and at dusk arrived safely at "these Hd Quaters."

 Now if I haven"t spun a sufficiently long yarn over a simple trip to Beaufort tell me. and I will try to add something more. I find that a Commission has come for my patient while I was gone. and the Sergeant is now a 2nd Lieut. No news here at present. I learned while on Morris yesterday that large quantities of ammunition. shot. and shell were being taken up to our batteries at the front. Also that another three hundred pounder Parrot gun had been mounted in an advantageous position. The first gun of that size was rendered usless by bursting some time since. Genl Gillmore has gone to the Head tonight. is to be absent two or three days. General Seymour is in command during his abscence. About the sermon. I have a copy. there were several sent on. If you have sent the one in question I can give it to some one who will be glad to read it. It is a good sermon. as are all of Dr. Reids. but I have heard him preach much smarter ones. which no one ever thought of having printed. I do not think it a fair criterion by which to judge of his abilities. I have sat up late and written this letter very hurridly. so beg you will excuse its appearance. With love ever your own.

Ned

P.S.

I return your wrapper as it is convenient to do so. It has had quite a journey. Love to Lizzie and Sara B. if she is not already out of reach. and to any other of my "sisters" that you know. and kind regards to enquireing friends

Think Julia is liveing rather "high" for a soldiers wife. Boarding at "first class" hotels &c. Good night.

Ned

No. 135 Rec'd Nov 1st 1863.
Folly Island

Head Quaters Dept. South
Folly Island S.C. Oct 24th 1863

Dear Sara.

Your letter of the 14th. together with a paper and the sermon. came to hand today. Have just come in from the Office where I have been at work until after nine. and have left a good pile of work in my desk for the morrow if it is the Sabbeth. I expect to be very buisy all day. and shall be hardly able to get up town to hear Mr. Reid preach. And what is worse I shall not get through in time to call at Mr. Jones' to see his daughter. and get a glass of ale. Its bad to have so much buisness on hand. I expect to be able to <u>retire</u> in about ten months. when I shall not be obliged to labor on the Sabbeth. and shall have leisure to call upon my friends. I shall not probably get the letter you sent by Wolcott under a week or so. and shall not see him at present. If the Regt. was on Morris Island. I should try to run away some day and go up and see him. but its impossible now. I suppose he has had very fine time of it. and been made quite a <u>lion</u> of. I am glad of it. Its rough enough in the army. and when a man does get a chance to enjoy a taste of civilized life it ought to be made as agreeable as possible for him. But my love. you must not expect if we all live to get home that the "receptions" will be either so very many. or very grand. All coming home together will be very different from what it is when by rare good luck one man is permitted to run home for a week or so. Each house will have a "lion(!) of its own. which it can <u>fete</u>. and honor to its satisfaction. I regret exceedingly dearest that you were so grievously disappointed. and I feel annoyed when I remember that you had such good cause for feeling so. and that I am to blame for it. I find that I have written you four letters. since leaving the hospital Oct. 5th. a little over one a week and nearly a week of the intervening time I have been away. during which I could not write very conveniently. You are very lenient to me my love. Do not think I do not appreciate your clemancy. I have been very grateful for your kind favors. which have come to me so regularly. whether answered or not. That fact itself telling me plainly that they were dictated by a loving. patient. womens' heart. and sent fourth on their mission of cheer. and comfort. to the object of her love. who though he sadly tries her patience now. she yet trusts will one day repay her for all her love and sacrifices. And <u>I</u> would do it. if it lay in my power. but it does not. Your letters <u>do</u> "do me good" and you <u>must</u> not speak so lightly of what I prize so highly. It seems that apples are quite cheap after all this year. that is in comparison to what they were a few years ago. I had supposed that they were very scarce and high. We pay five cents apiece to the sutters for all that we get here. and can not get them all of the time for that. Our wounded friend of whom you enquire. was able to be moved when the Regt went to St. Helena. Tom came over to the Head the day I was there. to draw rations for the hospital. he told me that all the patients stood the move finely. and that all wer doing well. This of course would include our case. and be satisfactory. but I asked him in particular about Parmalee who he assured me was getting well. You see I have not lost my situation yet. but I wish you to bear in mind what I have before told you--that you must not be surprised to hear of me. as back to my Co any day. The clerk who was home on a Furlough has returned and is now in the office. another has also been detailed. so that there are now eight clerks in the office besides the one whom I waited on to Beaufort. Tis not likely he will ever be in the Office any more. I think your Father has done wisely in deciding to go into buisness again at the old stand. With his past experience. he will undoubtedly conduct his buisness somewhat differently than he used to and thus avoid a great deal that used to be not only very unpleasant. and annoying. but also very unprofitable. Tell him for me that if the old "<u>suckers</u>" insist upon establishing the old order of things. he must tell them flatly that he has "effected a change of base" and that it is impossible. You had not told me about Mr Williams' houskeeping arrangements. I supposed that he. and his family had gone to California some time ago. as the last that I heard of his movements was that he had come home after them and that Mrs. W. had decided to go. Mr W. will never amount to anything any way. his friends will have to support him and the nearer he keeps to Salisbury and Stockbridge. the better it will be for him. I think there has been a great deal of sympathy waisted upon him. perhaps though I should say that it has been mis-directed. for none of us are free from faults that call for both sympathy. and forbarence from our fellow men. I have no news to write you as yet. Genl. Gillmore returned tonight from a short trip to the Head. The rebels have been noisy as usual for the last few days. and with probably the same result. I scarcly get time to look at a paper but from the little I can learn we are not gaining any ground any where. if we are holding our own. The Grand Moving Slaughter House of the Potomac. seems to be trying to dispose of a few more lives. at the least possable advantage--its usual terms. It strikes me that Foreign matters are getting decidedly "mixed." though perhaps the "mixture" is more in the heads of the newspaper makers. than any where else. I am much pleased at the result of the late elections. it is very gratifying.

Tis quite late. and I am getting awful sleepy. so beg you will excuse a short letter this time. With love I remain

as ever. Your Own.
Ned.

No. 136 Rec'd Nov 5th /63
Folly Island

Head Quaters Dept. South
Folly Island S.C. Oct. 28th 1863

My Own Dear Sara.

Your letter sent by Samuel reached me yesterday. as I have already answered one of the same date. written a few hours later. there is but little <u>answering</u> left to be done for this one. I have just finished a hasty scrawl to one of my sisters in reply to her letter about the middle of August. and believe I can safely say that I have improved the first opportunity to answer it. I don"t expect to do more than to commence a letter to you this evening for besides its being late as usual. it is cold as the mischief. The Arago went by bound North yesterday. and I felt a satisfaction in thinking that if the mail carriers had done their duty faithfully she had a letter on board for you. I do hope that ere this you have received some of the letters I wrote directly after leaving the hospital. I am sure that since that time I have written enough so that you ought to receive <u>one</u> a week from me. No news since I wrote last. The rebels seem to be getting more and more uneasy. and have paid us more than the accustomed amount of iron compliments. During the last day or two. I believe we have replied.

partially. I do not learn that their increased activity has done us any harm. Quite likely a man or two has been sent to his long home. and perhaps a score or so maimed for life. but when thousands are falling all around. what are a few? Everything. you reply. to themselves and their bereaved. and distressed friends. Very true I grant it. but our people have become so accustomed to the grand slaughters. of the Army of the Potomac and other large bodies that they refuse to notice so small a sacrafice to the shrine of Liberty as the lives of two or three men and the maiming of a dozen. It is a bright moonlight night. and the rebels appear to be very buisy. the sullen roar of their heavy guns reaches me very distinctly as I am writing. It is probably a good dozen of miles from here to the rebel batteries that are fireing. Under favorable circumstances. the report of cannon on Morris Island. and the rebel works on James and Sullivan Islands have been heard at Hilton Head. a distance of Sixty miles. I had quite set my heart on seeing Wolcott on his return. but it is now impossible for reasons that I have before given. and I really feel quite disappointed. Samuel is one of those who can give a good interesting. and correct account of what he sees. and hears. and I had hoped to feel myself transferred several degrees nearer home. while listening to his account of matters and things. in the old "Iron" town. He is the only "<u>real</u>" Salisbury boy that has been home. and returned to us so that he is as much a "<u>curiosity</u>" to us as he has been to you. and will probably have as much talking to do. to satisfy all the enquiries of the boys. as he did to gratify the solicitude of the friends at home. Samuel is a very discreet young man. and possessed of good sense. As you intimate. he is a good soldier. brave. prompt. energetic. and in earnest. I have never heard him growl. and complain. as much as I have done. either about the officers. or the unpleasantness of the service. Whatever may be his real feelings. he gives as little vent to them here. as you indicate that he has at home. He is always cheerful and seems disposed to make the best of everything with as little complaint as possible. He stands head and shoulders above ninteen twentieths of his Co. which though favored by our balance for various reasons is considered by the men as about the <u>scaliest Co.</u> in the Regiment. This is between us two. Doubtless Samuel would stand up for his Co. and resent any insinuation that implied they were not at least <u>par</u>. On the whole while I should have been much pleased to have had a member of our own Co. gone home to S-. as that could not be I am exceedingly gratified at Samls good fortune. and while it is not in my nature to forbare envying him that parting kiss. I do not grudge it to him. "twill be long enough doubtless before he gets another one. Have I told you that our camp in the wilderness is graced by the presence of a lady? a real bona-fide "woman! Well it is. She is the mother of <u>Capt.</u>

Bragg one of the Generals Aide De Camps. She stays in the Generals quaters. has been here about a week. of course we only see her when passing. I have been feeling much better for several days past. better in fact than I have before. since the fore part of July. am in hopes that I shall continue to feel well. for this trying to keep about when one is feeling miserable. is a hard way of dragging out ones existance. My "sore head" is very nearly well. though I still wear quite a suspicious looking patch between my eyes. Guess it will leave me a scar to remember Foster by. for life. Have not heard from the regiment since I came back from the Head. Mean to write a line to George in a day or so. and ask him to give me an account of their doings. if he is not to buisy. Suppose that ere this the old Store is again filled with "New Goods." and that there are a plenty of "lookers" if not buyers examining them. I remember well every nook. corner and cranny. in the Store and fancy that I could go into it in the dark after the Goods are arranged and put my hand right on any kind or quality that I might be told was in the stock. I could never make a successful merchant. I cannot talk glibly enough for a salesman. If there were nothing to do but to cut off. and do up. and take the pay I should like it first rate. but to force a piece of goods onto one customer which they do not want. to haggle half an hour over a shilling. more or less on the price of another article. with a customer who really wants the goods. to take all sorts of abuse from every body. and not dare to get angry at them in return least a few dimes should be lost. are things that I cannot do and dont like to do. so as I said before I can never earn my "Salt" in that vocation. I trust that you are well. and happy tonight. dearest. May you always be thus blest. With love I remain. Ever your own

Ned

No. 137 Rec'd Nov 9th 1863
Folly Island

Headquarters, Department of the South
Folly Island, S.C., Nov. 1st 1863

Dear Sara.

Yesterday I received your letter of the 20th and today the one of the 15th thus. as very often happens. the last coming first. It's Sunday evening again. have been in the office all day but not driven with work. have managed to finish a letter commenced last evening to Sara Bosworth. I feel real vexed about it. for it is such a poor apology for an answer to her letter. but I have not the time to tear it up and try again. though much I wish to. Yesterday afternoon much to my surprise. I learned that our Regiment was on the upper end of this island. had just landed. and had been hurried up under circumstances that justified the conclusion that some <u>work</u> was contemplated. Soon as I had my supper I started to find them. I knew about where they were and that it was some ways. but was hardly prepared to find the walk so long. it must be nearly or quite two and a half miles. and as it was most night and I had no pass. was obliged to make quick time travelling the distance in order to get back before the Countersign should be used. Found the boys bivuacked on a little wooden knoll on the bank of Folly Island River or creek. and nearly up to Light House Inlet. They were without tents. in light marching order with two days provisions with them. and all wondering what they were there for. It is surmised that another attack is to be made on Sumpter. and the fact. that for the last few days we have been bombarding it very spiritidly would seem to justify such a conclusion. I could give them no information for I am in as total ignorance of their destination as they. and it was by the merest accident that I learned of their presence. I found them well and in good spirits. Even their short sojourn in a comfortable camp has evidently had a very beneficial effect upon them. Twas dusk when I reached them. and some had already unrolled their blankets and lain themselves down for the night, while others were still seated around little fires. doing a little primitive cooking. Had a good livly chat with them while I staid. Found George much better than when I saw him on the boat. They are all much pleasured with their quaters on SC Helena. they have evidently lived in <u>clover</u>. during the few days of their stay. Sweet Potatoes. Oysters. Fish. Oranges. Eggs and fresh Milk. are some of the items in the bill of fare which they told over to me. I hope they may <u>all</u> safely return to their land of plenty. They brought their boats with them. the new rifles. if they are to have them. of which they are not certain yet. have not arrived. Col Hawley who went to Washington to see about them returned on the Fulton reaching the Head just in time to join the Regiment on their way here. I some expected to have gone up today and made them a longer call. I wish to very much. but thought I could not get away very well. and besides. found that I did not feel much like repeating my last nights trip. Have heard from them by one of the other clerks here who also belonged to our Co. who has been up to see them tonight. They are as I left them. waiting. watching. and wondering. I consider that they are liable to be employed in very dangerous service. and await the termination of whatever plan may be on foot with a good deal of anxiety. We have had our usual allowance of music tonight. I should like to have you hear the Band which I wrote you about not long since. as playing much queer music. it has played several times since then. and far surpasses any thing of the kind I have ever heard. Our Post Band which plays every night. and is gotten up in fine style with German Silver instruments. and all in shape. and are really a fine Band. and make excellent music. cannot leave off where the Dutch Band commenced. I never supposed such music could be got out of old brass. Yesterday morning we were mustered in for our two months pay. thats one of the advantages of being here. that we are very regular about getting paid off. we shall doubtless be paid in a few days. Our Extra can. and should be drawn every month. I will try to "heed your request" to be very careful of <u>my</u> health though I must say that it seems odd to me that any one should be so very solicitous about my health. I know that I have no buisness to say so. but somehow I have come to look upon myself as a kind of unimportant piece of drift wood. about which no one was particularly concerned. Doubtless. long absence from friends has done much to favor the idea. if so it may be called. though I feel confident that I have some kind of friends who really feel and interest in my welfare. Of course there thoughts do not include <u>you</u>. for you are not reckoned as a <u>friend</u> in spite of <u>your</u> declaration once made that "nearer than friends we can never be." I suppose you"ll be vexed at my allusion to that letter. which you have often teased me to destroy. but I cant help it. You must remember that that letter caused me trouble at the time. and not wonder that I now have to plague you just a little bit about being such a poor prophetess. We are a thousand times more than <u>friends</u> now. though a wise Providence may not permit to be all that we hope to. to each other. Am glad that you have given up all idea of seeing me until my term of service expires. for its decidely a rational proceeding. I will tell you now that if I had been a little wiser when in the hospital. I suppose I should have but little difficulty about getting a furlough. The only difficulty would have been to have got a Surgeon's Certificate of Disability. and I think I could have pretty easily got that. However am glad I did not try now. though I may be sorry before the remaining ten months are up. With regard to writing as much about "Barden" and telling you any more about him. I will only say that I would not object to do so if I could only tell you <u>some good of him</u>. I have never had the least doubt in the world. <u>dearest</u> but that you would be a most excellent housekeeper. my only fear. and trouble has been least I never should be able to furnish you with any house to keep. You regret your broken Sabbeth. what do you think of mine? I am getting to be a perfect heathen hardly know when Sunday comes. My Bible has been untouched today and as it is very late. fear it will remain so. True I am not to be expected to be able to observe the day with much outward respect. but my conscience tells me that I do not have the Sabbeth in my heart as I should. I do not doubt my dear that you do not find enough to do. and that but little if any of your time is wasted. If it is pleasant for you at home. am glad you are stopping there this Winter. Of course you would have no buisness to attempt a district school. and go trudging about in the snow and water begging your bread from door to door. When Spring comes there will be a call for "School Marms" some where I"ll warrent. I think the new milliner uptown must be my old friend Mrs. Manley. Now that I think of it I have an indistct recallection of some ones telling me that she was at work for Mr Olin. She is an excellent woman. but I have an idea that she is not fitted for that buisness. and shall be mistaken if she gives good satisfaction. but as there is no telling what a woman is capable of until she has made a trial she may succeede first rate and really I don"t know as I have any sufficient reason to doubt but that she will do so. I have never forgotten the last words of her husband to me when I called to bid him good bye. he was then nothing but a walking skeliton. and had been unable to speak loud for a couple of months. "Good by Barden" said he as he shook my hand. "if I was only able I should be with you." his wifes family are all the rankest kind of "Peace forever." and I have no doubt that they <u>now go</u> the Copperhead ticket its whole length. But that did not deter one of the sons from accepting a lucrutive Office in one of the Mass. regiments as Asst. Surgeon. Of course Mrs. and Mr. is not at all responsible for all this. and it has nothing to do with our subject. but while thinking of the matter I have also kept writing. When I left. Mrs. Mr. had a beautiful little. blue eyed. flaxen haired girl. about two years old. in a few months after her husband died. it was also taken away from her by death. The Presidents called for three hundred thousand more troops. looks as though the War were about to commence again. it having been ended several times by the news papers. But unfortunately for the Country. their kind of endings do not answer the contract. I do not know how the President will succeede in getting the new recquisition for men supplied but suppose they will scatter along. a few at a time first enough to keep the slaughtering buisness good. but not fast enough to do anything towards ending the War. I think I have some idea of the <u>very very</u> very much which you speak of in relation to seeing me. but I regret to tell you that you cant. cant. <u>cant</u> do it for ten months to come. I havent the least idea that the "charm" will have departed when I get home. unless you see fit to take yourself off to parts unknown. then indeede it will have literally departed. But I have no fears that a charm that has proved so powerful for so many years will lose its efficiency in the short face of ten months. I like the letter "about Sara" first rate. write me another. please. I havent commenced the use of tobacco in earnest yet. though I did <u>smoke</u> a whole pipe full of it the other night after supper. But then I had been making a pig of myself and eaten to heartily so took the smoke as medicine. Wont let me kiss you if I chew it! How do other ladies get along? most all the nice young men I know of practice the <u>manly</u> accomplishment. and I never could see but what they got their full share of kisses. However you neede never fear of my becoming addicted to that habit. I could hardly do it to save life. I have a very vivid recollection of the first and only chew of tobacco. that I ever put in my mouth. It was hardly more in amount than would cover your thumb nail. and I was not five years old at the time. but it has lasted me ever since. and is fair to. as long as I live. The typhoid fever seems to be very fatal in town this year. I remember the young Englishman. whose death you mentioned. By the way I saw Tom last night who told me that our wounded friend is still doing nicely and gaining rapidly. and unless somthing unforeseen. and unexpected occurs will soon be well again. Now my love I must go to sleep. and see if I can be so highly favored as to dream of you. It really seems to me that it would be but a simple act of justice to us pour unfortunate. to premit us to dream every <u>Sunday night</u>. of the "girl we have left behind us." Good night. With love. every your own

Ned.

No. 138 Rec'd Nov. 18th 1863
Folly Island

**Headquarters, Department of the South,
Folly Island, S.C.**, Nov. 6th **1863.**

Dear Sara.

I am just going to date. and commence a letter to you. but shall not promise to write much for tis late. and I have just returned from Hilton Head. and am quite tired. I found your favor of the 23rd awaiting me. and am desirous of making sure of its being answered in proper time. so write a few lines this evening. Yesterday about ten o"clock Dexter came in and asked me if I would go to the Head. and return as quickly as possible. Some of the old records of the office were wanted for reference immediately. I started in a few moments. walked to Stono Landing. which is at the lower end of Folly Island. and a little over two miles from Head Quaters. It is always splendid walking on the beach when the tide is out. as was the case that morning. and if I had not been in a hurry to catch the boat. would have enjoyed it much. Found when I reached the Landing that one boat had just gone. felt a little disappointed for it was the <u>Ben Deford</u>. a boat we all like. Another boat was to go in an hour or so. and I took a look around. trying to get an idea of the situation. I could see <u>Legarsville</u> a few miles up <u>Stono River</u>. the place where we stopped over night at the end of our Johns Island tramp. over seventeen months ago. I fancied I could see the very house we stayed in. and think with a glass I should have had no trouble in identifying it. We have been nearly all over the Department since then. and now we are within less than a dozen miles of the scene of our disasterous James Island campaign. with a strong probability that some part of our force will have to go over the same ground again. or at least operate against the same positions of the enemy. but it is to be hoped in a manner less calculated to ensure our defeat. About 1 o'clock we left the dock on board the steamer "Monohunsett." the "Ben Deford" was now well on her way. nearly out of sight. but owing to having two schooners in tow was not making very good time. Once over the bar we were after her at a rate which promised to soon pass her. and we did in about a couple of hours after starting. Reached the Head at 1/2 past 5. having been five hours on the way. should have made the distance (About fifty miles) in less time but there was quite a sea on (as the sailors term it) and a head wind blowing. The boat did not come up to the dock but anchored out in the stream. Went on shore directly in a small boat. by this time it was dark and to late to do anything but get my supper and to bed. I lodged at the Port Royal House paying half a Dollar for the privilege of sleeping in a little room about ten. by twelve. feet. in size. in which there were four single beds. but very little better than my own bunk. I shared the room with two other men. and the bed with any quantity of insects. some of which I found. to my great vexation. had accompanied me home. Got up early this morning and. by sharp hurrying got ready to return on the "Escort" which left. for Light House Inlet at about ten. O'C. reached the latter place at four. and landing. chartered a horse cart to bring me and my box to Hd. Qts. which I reached just at dark. Surprising the boys a little. who thought I had made a quick trip. and receiving the promise that I should have all such buisness to do. which offer I think it decidedly doubtful about my accepting. On the way down from the Inlet. I came past where our boys are encamped. but as it was late. had no time to stop and see them. They are still waiting for something to "turn up." Our batteries. and I believe a Moniter or two. are pounding away at Fort Sumpter yet. Some of the regiments have been partially reinforced by the arrival of conscripts. Our regt. has not received any. the 6th. Conn. I hear has received a few. I suppose Mr. Platt considers his duty to his Country done. and so has taken him a wife and intends to settle down and rest upon his "Laurals." Nine months is a long time enough to remain in the army for pleasure. I'll agree. but must say that those who get off thus cheaply. are fortunate. On the whole I dont know but Platt is right. It is rather discourageing to see how little effort is being made by the Northern people to end the War and. thus release us from our unpleasant position. If we could only get the politicians and government contractors into the ranks and make them face both the "music." and the enemy. I think the soldiers would willing remain a while longer to "put them through." Think I will close and send this apology for a letter rather than trust it until another evening. least it should get neglected. It will at least serve to show you that I am thinking of. and loving you. Good night.

Your own
Ned

No. 139 Rec'd Nov 26th/63
Folly Island

Headquarters, Department of the South
Folly Island, S.C., Nov. 11th 1863

Dear Sara.

Your letter of the 30th was received two days ago. and I intended to have answered it in time for the return of the Arago. which leaves in the morning. but have been unable to do so. We have been having a few days of very cold weather. and as we have no stoves up in the Office we have hardly been able to do the necessary work. and letter writing has been out of the question. It is pretty cold tonight. and I am anything but comfortable while writing. but must delay no longer for the weather may be more severe tomorrow than it now is. My life now is less eventful than yours. and the record of one day is the record of many. Last Sunday I worked hard all day. and during the evening. Monday. received your letter. but the day was very cold and we worked but little. in the evening we built a large fire near our tents and managed to keep warm. and enjoyed ourselves as well as we could. Yesterday was another cold day. we left the office early. and most of the boys gathered round the fires again. directly after supper I started to visit the boys in the Regiment. As yet they have not been called on for any duty. They removed a short distance from where I first saw them. and are now partially sheltered by some large Quater Masters sheds. though most of the protection they receive from the weather is in little huts that they make of bushes. and their rubber blankets. They are pretty well fed. and all seem to feel pretty nice and enjoy themselves as well as the uncertainty of their position will admit. Last Saturday they all went back to the St. Helena to have an opportunity to clean up. and get a change of clothing. and returned Monday morning. George says he feels first rate. and his health is good. It being late when I started. it was after dark when I found the boys. and there was but little time to stay. and I spent the most of it talking with George. He sayes that Barnes is very sick and that it is not probable that he can live. He was getting well when the regt first came up here. but was taken worse very suddenly and is now very low. unable to sit up at all. George does not think we shall see him again. It is not likely we shall. for I never knew a person with his disease get well when it had reduced them so low. It is much to be regretted that he had not been allowed by the Captain to go home when we thought he was so sure of his Furlough. It would have saved his life undoubtedly. The Furlough by rights belonged to him. He had served faithfully with the regiment in all the dangers of the <u>Morris Island campaign</u>. had done his duty in the batteries serving the guns that helped to breach Sumpter. and had in a fair way. drawn his chance to go. and when the order reducing the number permitted to go home. came there was no question which of the two men should have been allowed to go. and a still greater reason why Barnes should have gone. was that he was then unwell. Suffering from disease the result of exposure and severe duty. But on the contrary the man whom the Captain permitted to go had been Mr. Moores clerk all the time and not at all exposed as the rest of the men were. but he was a New Haven man. and Barnes an upcountry fellow. Only a little specimen of Military justice. thats all. I must answer your letter and that will be all I can find to write about this time. and it is so very. <u>very</u> dull. here. One bit of news. the same number of men as were allowed before. are to be permitted to go home on furloughs. One can go from our Co. I am a little curious to see who it will be. the men of our regiment had not heard of it until I told them last night. I suppose they do wish any to go from it. until after the buisness now in hand is accomplished. I wish George might be the lucky one. but there is little hope of it.   <u>Sunday Eve</u> - Nov 15th.   Over a week since I have sent you a letter. too bad. but I have been at work very hard. and have not been able to finish even this one. I have worked very hard every moment of this day. and never left the office till a few minutes ago. It is now nine O"clock. I am tired. sleepy. and ill natured. and am just going to close this letter and go to bed. I now expect to be returned to the company in a few days. Capt. Sealy told Dexter the other evening. that the General told him (Sealy) he must reduce the number of his clerks. So some of us have got to go. and it will either be me. or another man from our Co. (There are three from it in the Office) possibly both of us. The Capt has gone to the Head today he will bring us our last two months pay. and then probably some one or two of us will have to travel. The <u>evil</u> day may be put off a short time longer. as the head clerk has been granted a Furlough. and will go home on the return trip of the "Fulton". in a few days. we may all be retained until his return. Dexter thinks I shall not be discharged. and I should not be if it lay with him. but the head clerk is perhaps as favorably disposed towards me. as he is to the other man. and perhaps not. if. as is likely two of the clerks leave. it will make the work pretty hard for those who remain and I cannot say if I care very much to stay if obliged to work all the time as I have for the last few days. I have not seen any of our boys since commencing this letter. but suppose they are still up in their bivuack wondering what will all their fuss is going to amount too. There has been a good deal of heavy fireing today. up to the front. down here. all is quiet and the Band played tonight as usual. Your last letter contains some things which I wish to answer. but cannot tonight. I have already missed two mails. and do not mean to another just yet. I will try to write again <u>very</u> soon. if possible in time to go in the same mail with this. but may be disappointed. Beauty of my dearest I am loving you as fondly as ever. and wishing to be with you. let me once more assure you that I do not forget you because I fail to write frequently. I have done my best. to be faithful since I left the hospital. and will continue to do so. Excuse. haste. Errors. and a short letter. and believe as ever your own

Ned.

No. 140 Rec'd Dec. 3rd 1863
Folly Island.

**Headquarters, Department of the South,
Folly Island, S. C., Nov 22nd 1863.**

Dear Sara.

Have just remarked to Dexter that Mr Reid was abut through his days labors by this time as it is a little after 2 o'clock. Am writing in the office. having a few moments of leisure. but am not at liberty to leave as Cpt. Sealy may come in at any time. and I might be wanted. I have taken a step upwards within the last three days. My position heretofore has been in the back office "at the foot of the ladder." The head clerk has gone home on a furlough. and I have been directed to come in the front office and take one of the books to keep. not to take the head clerks you understand. for I cannot do that. and I do not believe that the Capt will keep me on the book long. as I am by odds the poorest writer in office. I was surprised when ordered to take it. as it is a part of the work with which I have had nothing to do. and consequently knew nothing about. while others of the boys have. I should really love to succeed with it and please the Capt but have little hopes of doing so. Thus far I have got along very well as there has been scarcely anything to do. but when a hurrying time comes. as there will. and does every few days. I do not believe I shall be equal to the emergency. I am real glad you have been reconciled to the probability that I may rejoin the Company. that is as I wish you to be. Now one step further if you please and make up your mind that perhaps it would not be the worst thing that might befall me. A portion of the Invalid Corps has been sent to this Department. and I learn that it is proposed to make all details for clerks and orderlies in the various Staff Departments from among them. thus relieveing all the men able to do duty in the ranks. from their snug positions. I can find no fault with the idea though it take off my head with the rest. Dexter told me the other day that Capt Sealy was going to reduce the number of his clerks. that the General had told him he must do so. I expected then to have been relieved before this time but presume that none of us will be dismissed until the return of the other clerk. I received your letters of the 3d & 8th by our last mail. and commenced a reply to have gone back in the Fultons' mail but was cheated out of completeing it. You are often disappointed about receiving letters from me now. <u>love</u>.

 I am sorry but do the best I can. For the past two weeks I have only written to you. with the exception of one letter to Mr Randall requesting him to send me some things. I think if I were in the Company I should find time to write you oftener. There are many spare moments of time in the office. enough if I could take advantage of them to enable me to attend well to my correspondance. but it is not pleasant to be obliged to leave a half finished letter among ones papers to be overhauled by some one of half a dozen young fellows. who might perhaps be mean enough to read it. Our regiment has gone back to St. Helena as I presume you have already learned through Julia. They are to have the new rifles. so I calculate that if any fight occurs they will have the post of honor. which is the post of danger. I was vexed at not being able to see George again before they left. but had to work night and day almost just at that time and could scarcely get time to breathe. I must write to him though I have no particular business. The weather for several days past has been splendid. and the nights magnificent. today it is cloudy. and cooler. We have no stove up in the office yet. The longer I stay here the more I like it. and the less disposed to do a soldiers' duty. I had at one time even begun to think I might remain a long time here. possibly the remainder of my time. The chances for it look very slim at present. Thanksgiving comes this week. my third one in the army. it is also probably my last. but whether I shall spend the next one with my friends time only can tell. I received a letter from Carrie the other day. She is at New Haven at school this Fall. and writes that she enjoys it exceedingly. She is a little inclined to be a flirt. about which she tells me in her letter. that her mother gave her a long lecture when she was last home. and adds. that she does not see as it has done her a bit of good. She knows very well what I think about that class. for I always expressed myself pretty freely on that point. but seems to like to touch me up about it. It does not seem possible that she is seventeen. a little girl when I went there to live that I used both to tease. and be tormented by. a young lady now. whom it would not answer for me to be kissing very often now. even though she were disposed to allow it. Have received a note from George since commencing this. He is well as are also all the boys except Barnes whom he sayes he has not heard from. by which I conclude that Barnes has been sent to the General Hospital at the Head or Beaufort. for if he were on St. Helena George would know about him of course. The Regiment has received 109 conscripts and Substitutes. nearly all of the latter class. Our Co. has received but 3 which is a great plenty of the kind. Doubtless the four companies that suffered so severely at the attack on Wagner get the larger proportion of them. The expected new rifles have been received. but as there was but 500 of them Col Hawley will not give them out because there are not enough to arm the whole regiment. As I told you I thought it would be. a New Haven man is the lucky one permitted to go home on a furlough from our Co. Perhaps the man drew lots for the chance. if so of course it is all fair and the Capt not responsible. At all events its a more just selection than the other. for the man is one of three sharpshooters that our Co. furnished. the other two. were George. and Henry Brinton.

 Another poor letter my dear. but remember I do not forget you and that when I can I will write you some more long letters. With love. ever your own

Ned

No. 141 Rec'd Dec 3rd 1863
Folly Island

Headquarters, Department of the South
Folly Island, S.C., Nov. 27th 1863

Dear Sara

 In one of your late letters you speake quite exultingly of receiving frequent letters from me. and seem to think them the foreshadowing of good things to come. by this time you will have found that you took courage all together to soon. for my letters to you have been very few indeed for the last three weeks. I have just come from the office where I have been very buisy all day. and have left half a days work undone. it was not absalutely necessary that it should be done tonight. but would much better have been. I guess its about nine o'clock. but the Arago goes North tomorrow and Sara would expect a letter by her mail. so though Ned's eyes ache a little he will not relieve them from duty until he has written a few lines more to the girl he left behind him. Received two letters today. one from Sara B. and one from Mr. Randall. Yesterday, recieved a letter from one of the boys in our Co. enclosing me a photograph of our esteemed friend and patron. James M Townsend Esq. One of the few really good men. who are granted to us in this age of evil and wickedness. I have been favored with one or two short letters from him. short note merely but the genuine Christian spirit prevades them. The photograph is an excellent one. his signature is on the back. It is needless to say that I prize it highly. I think I will send it to you for safe keeping. as photographs fade out badly when subjected to our soldier keeping. Two or three days since I received a letter from Carrie Randall and one from an uncle in Flushing N.Y. with whom I correspond as regularly as I can. He is Methodist Minister always about his "masters business." his letters are very earnest. full of advice. and instruction. Now all these letters. with some of much older dates ought. to be. and must be answered. When and how is not so plainly seen. Mr Randalls' letter is dated Nov 13th. he sayes. "I am going to send you a box of things next week." I hope that he has been delayed about it. so that my letter to him written the 19th will reach him before he starts it. as I sent for a number of articles that I need and I would not wish to cause him any more trouble than is necessary. Dexter has had such bad luck with a barrel of things sent him from home that. I fear I may be equally unfortunate. His barrel was started from home some six weeks since. and has not yet been heard from. Dexter thinks it lost. he values it at about $75. Wonder if George has received either of the boxes. you mentioned that Julia was sending him. I forgot to ask him when I wrote. I learn that Barnes is improving. hope it is true and that he may yet be able to weather the storm. I must write to George again right off. for if Mr R. sends me the box soon as he expected to. or before he should receive my letter. he could direct it in the Regiment. and I want George to look out for it. Well Thanksgiving has come and is passed. A more. lean. cheerless. miserable one I never passed. our Mess made no provision for any extras. and as it happened we had a much more miserable dinner than common. Dexter and I made a desparate attempt to fill ourselves by an appeal to the Sutters. but only succeeded in getting a small can of tomatoes and another of chicken. for a good deal of money. and were forced to the conclusion that we had made a very bad bargain. even though in reference to the occasion we were disposed to be liberal and not particular about trifles. The weather had been for several days quite warm. but night before last a sudden change took place it came on very cold. and yesterday was about as blustering. <u>raw</u> unpleasant weather as ever graces a thanksgiving day at home. I spent the day in the office. buisy most of the time. several of the clerks. got passes and went away. some to Morris Island and some to their regiments on this Island. Liquer was not plenty except with the officers who of course never lack. Some of the clerks and orderlies made an ineffectual attempt to get some whiskey in the afternoon but Capt. Sealy would not sign the order. so they wer forced to go dry. Several companies of the Engineers. camped just above us have put up quite a large building. covered with tarpaulins and old tents which they use for meetings and social gatherings. Well the various newspaper correspondents. of which there are about a dozen hanging about Head Quaters here. gave the Officers a supper in it last night. and I suppose they had a gay time. I regret to learn of the sickness in the family. and hope it may be neither long or fatal. Once more my dear accept much love. though in a very brief letter. and believe me most truly.

Your own.
Ned.

No. 142 Rec'd Dec 14th 1863
Folly Island.

Headquarters, Department of the South,
Folly Island, S.C., December 5th. 1863

Dear Sara.

Have just left the office. should have staid and written my letter there only I wanted my overcoat on and as it is as warm in my tent as in the office it wont pay to go back. For the past week the weather has been very cold. we have no stove up in the office yet and it has been with great difficulty that we have got along with our work. letter writing has been simply impossible. The Arago goes North tomorrow. and I must have at least a few lines ready for you. to go by her. Your letters of Nov 20 & 25" were received by our last mail. I hope that Clara has recovered ere this. and that your fatigueing watches. are at an end. I trust that you will be spared the affliction that the loss of the little one would cause. Miss Libbie is really getting to be quite a traveller. I think she is exceedingly fortunate. and no doubt will have an agreeable. pleasant time. Your uncle H. may have something more than a "cousinly" interest in making his visit. Selina and her family have hardly been at the West long enough to make a merely family visit necessary. I suppose the present relationship is not near enough to be an objection to one of a closer kind. I think whoever is fortunate enough to get Selina. will have no cause to repent their choice. In one of your late letters you mentioned your relatives at the West. I have no recollections of ever hearing you speak of them to me. and think that I never had been told of them before. I hope my dear the time may come when I can have the pleasure of visiting with you all your relations wherever they may be. but the prospect is a dismal one at present. I often think that if I live to get out of the army I shall be obliged to seek my fortune at the West. But the time is yet to uncertain and distant. to warrant the laying of plans. or building of air castles.

We have had quite a time here this afternoon. a fine flag pole has been raised in front of the Generals quaters in the open square which I think I have before described to you. it is also just in front of our office tent. About four o"clock everything being in readiness the General (Gillmore) raised the Flag. himself. The german band. which I have often mentioned was in attendance and played "Vive el America" as the flag arose to the breeze. Capt Hamiltons Light battery. of six pieces was in position on the beach a few rods distant. and fired a national salute. at the end of which three rousing cheers for the flag were given. A large number of officers were present. among whom were Generals. Terry. Vodges. Gordon. Turner. and Foster. with plenty of Colonels, Lt. Colonels, and Majors. and Captains and Lieutenants. as plenty as blackberries. a short speech was made by a Mr. Shay. from N.Y. an old friend of Gen Gillmores. now on a visit to him. The whole thing was well managed. and passed off beautifully. Frank Leslie's artist Mr. Crane was on hand and as I observed buisy with his pencil and paper. so no doubt you will be favored with a good representation of the scene as Mr. C. is a good artist. The pole is in two pieces so arranged that the upper half can be lowered. or raised at pleasure. three days ago while attempting to raise the top part to its place. it got loose when nearly to its place and came to the ground with a crash. breaking it into several pieces. so the flag raising was necessarily postponed until a new pole could be made. You said that you expected to hear the next thing. that I had caught a cold. Well you will not be mistaken. I find I have done just that thing today. Received a short letter from George yesterday. sayes he feels first rate. I have been feeling very well along back. and but for the cold should have written to you frequently. and at considerable length. But until I can find a place to write where my fingers will not be as numb as sticks I can write but very little to anybody. even to you. my loved one. Good night. love fromNed

No. 143 Rec'd Dec 19th 1863
Folly Island

Department of the South, Headquarters in the field,
Folly Island, S.C., Dec. 12th 1863

Dear Sara.

Your letters of the 27th Nov. and Dec 4th were received yesterday. I am pretty much turned out of house and home in the office or I should have written you a longer letter. Am writing in my tent this evening which I always prefer to do when the weather will permit. It has been a tremendous stormy day. having rained furiously all the afternoon while the wind has blown a gale. But it is a warm storm. though it will doubtless clear off "Nipping" cold. You remember my trip to Beaufort a few weeks since in charge of our head clerk. then sick. well he returned four days since and resumed his duties in the office. The next day the clerk who has been home on a furlough. and whose place I wrote you I was partially filling returned. and took his old place. All the clerks are now present for duty: nine in number. There is not so many needed. two, at least will doubtless be discharged. and it is right that they should be. It is decided who one of them is to be. but he will not return to his regiment for Capt Sealy told him to remain about the office a day or two as he should probably turn him over to one of the other offices. Who the second "unfortunate" is to be, is not so certain; it probably lies between "your obedient servant" and another member of our Company named Smith. I think the chances are about equal. and was in hopes the matter would be decided today, but it has been so very stormy that there has been very little done any way. Smith took my place in the back office when I left it. and I have not retaken it. as there is evidently some change to be made. and I have no wish to make more commotion than is necessary. Yesterday and today I "hung" around the office. doing a little copying. but nothing of any amount. and heartily wishing myself either settled at work. or if I must go. well on my way to the regiment. If there is anything on earth I dislike it is vying about a place of business in such an unsettled state. Mr. Landon has moved over. and settled within a few rods of us. within the week. Have spent part of one evening with him. I believe his health is very good at present. Had a letter from George a few days since. he writes in the best of spirits. and I think that for soldiers. the boys are enjoying themselves first rate. I do not find myself dreading a return to the company very much after all. and some times doubt whether I care much about it any way. Of course you have heard all about the sinking of the Weehawken. I have not been able to learn yet whether two. twenty. or all of her crew were lost. Shall know when we get the next N.Y. papers. Everything joys quietly along here as usual. no news and no probability of any. As usual, the rebels are "pretty much used up." "nearly cleaned out." "in the last ditch" &c &c. (Editor's Note again: &c was the accepted abbreviation for etc.) and yet I think the war as likely to last a year or two longer as any way. I fail to see any prospects of a speedy termination of it. Thanks for your notice of lecture. and all the interesting particulars. Is the Mr. Cowdry you mentioned as waiting upon Miss. Grant. the same who took my place (<u>again</u>) in the Bank. Do you see that "again". Well two or three of the boys with an Accordian are in an ajoining tent. and for the last fifteen moments have been steadily hammering away at "Home again." No wonder I come near getting their song in my letter. They have not minded the matter much now, having commenced "ever of thee" I'm fondly dreaming Am glad your little sister still continues to improve. and that your weary watchings are nearly over. I begin to look for my box from Mr Randall. Have but little hopes of getting it. Dexter received his barrel after a long while. most of the "goodies" spoiled of course but the valuable part all safe. I will write again. love. as soon as this question of "go or stay" is settled. With love. Ever your own

Ned.

No. 144 Rec'd Jan. 4th 1864
Folly Island.

Headquaters, Department of the South,
Folly Island, S.C. December 27" 1863.

Dear Sara.

    I have two letters from you unanswered. Yours of the 13th, received on Christmas day. and of the 20" received this morning. I have left the office with its noisy crew and come over to my tent to write. It is quite comfortable without a fire. the weather having moderated very much last night. For the last fortnight it has been clear. cold. and windy. day after day. an exact counterpart of the weather usually experianced at the North. at the present time of year. Our Christmas passed very quietly. Nearly all the officers at Head quaters went to Hilton Head on the 22d and did not return until the evening of the 24th. They had some kind of a ball. or party at the old Headquaters building there. on the evening of the 23d. I know but little about the preformance. not being particularly interested. Doubtless a good deal of Liquor was disposed of. and a great many foolish things said and done. I know that we took things easy. for during the two days of their absence not the least thing was done in the office. Christmas day we did but very little so that we have had three holidays the past week. The weather was too cold to admit of our enjoying them in any other way than by sitting around the fire reading and talking. I spent Christmas eve in quite a homelike. and sensable manner. The Camp of the 169th N.Y. Vols. is a few rods below ours. They have been employed for some time in building a little church. of such materials as could be obtained and I venture to say that a more neat. tastefull. and appropriate. place of worship was never constructed when all things are considered. It was completed and beautifully trimmed. and service held in it on Christmas eve. I had the good fortune to attend. The chaplain of that Regt is an Episcopalian. and of course the church. and service. are episcopel also. I wish I could give you a description of the little church. but know that I should not succeed in giving you any thing like a definate idea of it. It is small of course. will prehaps seat a hundred persons comfortably. (if they wear no crinoline) it is built. or rather made of a light frame work covered with palmettoe leaves which are woven in. The roof is canvas. supplied by old tents. The ends are covered with a kind of narrow. rough. boards. split out of pine. and the joints are battened with the stalks of the palmettoe leaf which are nailed over them and being of a bright green color form a pleasing contrast to the light color of the siding. The chancel is built on the end opposite of the entrance. and is complete in all the requisite appointments. being supplied with a reading desk. pulpit. alter. alter railing. &c. almost mineature in size. but sufficiently large to answer the purpos. The pulpit and alter are particularly neat. and pretty. being formed principally of grape vines twisted into a variety of shapes. and fastened in the desired form. The whole was as prettily trimed for Christmas as is possible for man to do any work of that kind unaided by ladies. Over the reading desk. the Apostals creed is plainly lettered on a white ground. and bordered by a broad wreath of the bright green leaves of the Magnolia. Directly over it is a large bright Star encircled by a wreath of green with bright scarlet holly berries interwoven. Of course it was well illuminated on Christmas eve. and was a pretty sight. The chaplain did not preach. but after the service was read. a Mr Taylor of the Christian Commission. and the Chaplain of the 112th N.Y. made short addresses. On Christmas morning at ten o'clock I again attended service there. accompanied by Mr. Landon. The chaplain gave a very interesting sermon. and at the close of the service administered the communion. to I think about fifteen. I was mistaken in supposing that none but those who were confirmed churchmen were admitted to their communion. and so did not partake with them. After the service I found out my error. and have to regret that my ignorance prevented my enjoyment of a great privelage. I went down to church again this morning. but being hindered by some work in the office. only got there in time to hear a very good sermon from Malachi. 3 & 7. And now hoping to be permitted to attend upon any an all of its services. I will bid good by to the church for the present. The afternoon of Christmas day was spent very pleasantly but not very profitably. Shall have to plead guilty to having imbibed three glasses of hot whiskey punch. and one or two of Sherry wine during the day and evening. went to bed at nine o'clock perfectly sober. and slept in spite of the racket made by half a dozen drunken fellows of our mess in an adjoining tent. As usual on such occasions our cooks were both fully drunk by noon and we had little or no dinner. I made a requisition on a Sutters shop and purchased the requisites for a passable supper. No New England goodies were on hand to appease the appetite. and I could only hope that the good things had been received by the boys in the regiment and that they were enjoying them. Thus passed Christmas 1863. May there be no cause for any ones passing the one for 1864 in the army. I have not yet heard of the arrival of the donation to the boys. am waiting rather impatiently to do so. for I want my things and want an opportuunity to run down and see the boys. Hope to hear from George tomorrow. Mr. Randall wrote me a note of the 9". saying that he had started the box that day. and giving me an inventory of its contents. He said there was another small box in the Depot directed to me. but I am at loss to imagine who its from. I must have more friends somewhere than I have taken credit for. George wrote me that he thought of reenlisting. and I am a little curious. to say the least. to know how he decided the question. I am not positive. but think. that I wrote you my views upon the subject in a late letter. I answered Georges letter at once. and though I did not advise him (as he did not ask me to) at all about the matter. I told what I thought of it. I would not advise any of my friends to reenlist now. neither would I dissuade them from it. I do not believe there will be any personal benefit derived from it whatever. for I think that just as great, and the chances are greater. pecuniary inducements will be offered. to volunteers next Fall as are now being held out. The true patriot can enlist as well at one time as another. I do not believe your fathers idea "That if the boys reenlist. their danger will be over by the time that their nine months have passed." I have not the least shadow of faith in any such speedy termination of the war. and see no reason why the rebels are not as good for another years struggle. as they were for the one just passed. I shall believe the war ended (when) I see it. and not when Greeley. Raymond. Bennett & Co. say it is. Newspaper prophesies have long ago play(ed) out effectually and soldiers pay no more attention to their speculations about peace than they would to the twittering of a bird. I dont think I shall reenlist until my present term has expired. but hope that everybody else will. I suppose you have doubtless heard from George the particulars of Barnes' death. I am still ignorant of them.

Dec. 30" Morning. The mail leaves this morning and I have only time to close my letter without writing half I intended too. have carried it in my pocket the last two days. vainly trying to find time to finish it. and must now send off. without even time to look it over and correct mistakes. Have not heard from George this week. am a little impatient to. for I want to know who are reenlisting in the Regt. I hear from various sources that a large portion of it are doing so. Expect to go down and see the boys in a few days. In haste. Good By.

Love fromNed

No. 145 Dec 26th/63 (Note: received date)
Folly Island

Headquarters, Department of the South,
Folly Island, S.C., 1863.

I improve a few spare moments this evening in commencing a letter to you. with but little hope of making much progress. with it. for their cold evenings the office stove is surrounded by a lot of idlers who are not as quiet by any means as they might be. I am feeling very uncomfortable from the effects of a severe cold. Nearly all the boys are complaining in the same way. We are having Southern Winter weather now. cold and stormy most of the time with occasionly a day or two of clear comfortable weather. I shall probably remain in the office for a short time longer at any rate. Since the return of the two clerks mentioned in my last. I have had but very little to do. No regular duty perscribed. but have assisted the head clerk when I have done anything. This is of all duty. what I least like. I want something to do. during office hours. want to do it and have my days work done. Have seriously thought of asking permission to return to my Co. but hoping that something better may yet turn up. have pretty much concluded to await the course of events. Only one clerk leaves the office at present. and he is provided with a good berth as clerk in one of the other offices of the Dept. received a short note from Mr Randall the other day stating that he should send the things for which I sent. by a vessel which was to leave Bridgeport on the 14th. freighted by the "Good Samaritan's" at home with "creature comforts" destined for the Connecticut soldiers. to enable them to enjoy a "Merry Christmas" in good earnest. I sincerely hope it may arrive in time. though of course. I shall not be permitted to enjoy the good things the vessel may be bring. Am sorry he sent my box by that vessel. for I have very little hopes of getting it. Would prefered that he had sent it by Express as I directed. The only thing that has occurred to relieve the usual monotony of affairs here. was the execution of a man belonging to the 3rd N.H. Regt. for the crime of desertion. He was a substitute. had only been in the service about a fortnight. I will tell you what I have learned of the case. He joined the Regt. on the 15th of November. deserted on the night of the 28th. on the morning of the 1st of Dec. he came to the camp of the 9th Me. Vols. on Black Island. claiming to be a deserter <u>from Fort Johnson</u>. his intention being to deceive the officers. get sent North as a rebel deserter. and thus escape the service. He was taken to the Guard house (as are all deserters from the rebels) where he was at once recognized. his name was John Kendall. but he claimed when he came back that it was Thomas. he was identified beyond a doubt. A Court Martial was at once convened. he was found Guilty of the charge and sentenced to be shot. The proceedings of the court were approved by Genl. Gillmore who ordered the execution to take place within forty eight hours after the receipt of his approval by Genl Terry who commands on Morris Island. The execution of the sentence was finally deferred one day longer. when it was carried into effect in the presence of every man on Morris Island who could be spared from duty. He died protesting his innocence. and claiming the statement he made as to his being a deserter from the rebels to be true. He was undoubtedly a hardened villain. as there was no doubt as to his being the John Kendall who deserted on the night of the 28th Nov. Two other deserters have also been caught. and were obliged to be present and witness the execution. I believe they have not yet been tried. Well I took your advice, and tried to improve my spare moments during the day in writing. and this letter is the result of said labors. I only send it because I wish you to see how that plan works. I will not try to tell how many times I have been obliged to fold it hastily and cram it into my pocket. until it is so dirty that I doubt your ability to read it. We have had, and are still having very cold weather. We think. though it would hardly pass as such in Salisbury. My cold is still troublesome, and besides I have had a sore mouth for several days past. which is very unpleasant. but my general health is first rate. A large number of the soldiers in the Department are reenlisting as veteran Volunteers. for another three years. In an Indiana Regt two hundred have already reenlisted. George writes me that the numbers 325 have put their names down to reenlist in our regiment. I would not have believed that fifty could be induced to reenlist. out of the whole regt. Doubtless many would back out when it came to the point. and the papers were presented for them to sign. but no doubt a large number will reenlist. I think I shall be content to serve out my present term, before I bind myself for a longer period. though if needed no doubt but that I should be ready to enter the service again whenever it became necessary. All who reenlist now get a furlough for thirty days. I should love dearly to get hold of some of the good things you have been making. and have sent on for Christmas.

I wont write another word. Do you see that big blot? With love, ever your own

Ned

Editor's Note: Here appears a large script signature E. J. Barden

P.S. Dec 27th

I learned this morning from Capt. Mills that Barnes died on the 17th. the mail is closing unexpectedly. and I have no time to write.

Ned.

LETTERS FROM EDWIN JANES BARDEN TO HIS FIANCE
AND LATER HIS WIFE, SARAH MARIA JONES,
WRITTEN DURING THE CIVIL WAR

SECTION 4 – 1864

No. 146 Rec'd Jan 19th 1864
Folly Island

Headquarters, Department of the South,
Folly Island, S. C., Jany 10th **1864**

Dear Sara.

 I regret that another mail has gone North without a letter for you from me. but think you will be satisfied with my explination. I started to go down to St. Helena. to get my box. on New Year's day. Went to Pawne Landing to take the boat. found there was none going. the weather was cold and windy. and rather unpleasant. The next day I went to Stono in the afternoon and went to the Head on the Ben. Deford. General Gillmore and several of his staff officers were on board. making a trip to the Head and Beaufort. The boat reached the Head and came to anchor about eleven o'clock. a tug boat came along side. and I went on shore in her and went directly to the Port Royal House and was in a short time snug in bed. I got over to the regiment about eleven o'clock and had a buisy time from then "till late at night talking and visiting. The boat for the Head usualy stops at St. Helena about half past nine. and I was ready waiting for her quite a while. before I found that one had passed that morning about daylight. A small boat was going over. but as there were a number to go. and it would be difficult to take my box for want of room. concluded not to go in it. but take my chance at getting over some other way. I had calculated on having a good visit with George that day. for I had had but little chance to talk with him. he had obtained permission to go over to the Head with me. and we were quite disappointed. It being very difficult for him to obtain permission to visit the Head, which he was desirous of doing. having some business there. I persuaded him to go over in the small boat. it was the last I saw of him. for I did not finally succeed in getting over until four in the afternoon. and then we passed each other on the way. I had considerable business to do at the Head. in the way of errands for the boys getting a quantity of Blanks. and some office furnature. from the old office and our last month's extra pay from the Quartermaster. It was too late to do more than make a small beginning that night. and impossible to get ready to leave for Folly on the boat that was to start early in the morning. Spent the night at the Port Royal. and the next forenoon finished my business and was ready to return. but no boat was to leave until daylight the next morning. The day was foggy with a light rain and quite uncomfortable. About six o'clock I got a horse and cart and took my goods down onto the wharf and waited with what little patience I had left. for the boat to come up for her loading. The night was very dark. and the fog so dense as to render it impossible to see anything a rod distant. At ten. orders came to put the stuff for the steamer on board of a tug and take it out to her. as it was too dark for her to find the dock. The luggage was quickly loaded and we set out to find the Steamer but after over a two hours search. were obliged to give it up. and return to the dock where we lay until daylight enabled us to find the object of our midnight search. It was about Seven. in the morning when I got on board of the "Collins." which was to take us to Folly. We had a schooner to tow up. and did not get started until after nine o'clock. and then the fog was so thick that we could not come up the inside channel. which is much the nearest. but were obliged to go "outside" as it is called. where we found quite a high sea running and a head wind. All day we toiled on. making but little progress and at five o'clock came to anchor abrest of North Edistoe. still twenty miles from Stono Landing. The boat. which is a propeller. and had no loading to keep her steady. rolled and pitched about in a most uncomfortable manner. particularly for me. who was as thoroughly sea-sick as was ever a landsman taking his first trip. I had stood it pretty well until about the middle of the afternoon. ate a hearty breakfast. and dinner though it took some forcing to get down the latter meal. and was determined if possible to brave it out. But I had to yield. and before supper time was snugly stowed away on a coil of rope amidships. I secured a comfortable berth and "turned in." while they were at supper but soon turned out again to pay tribute to old Neptune. The berth being clear aft. the motion was much worse than amidships and I began to think I should have to leave the comfortable berth. for my coil of rope. but at last succeeded in getting to sleep and managed to pass the night quietly. With daylight. my sickness returned again. and I was obliged to shift my quarters to where the motion was less. and the air more fresh. About nine. the anchor was hove up and an attempt made to take the schooner in tow again. but after laboring in vain for over an hour and a half it was found impossible to get the line on board of her. they were obliged to give it up. she hoisted sail and put back to Port Royal. and we started on towards Folly. The wind had risen a considerable during the night and blew "dead ahead" as the sailors say. raising quite an ugly sea. We were until almost three o'clock in running the twenty miles to Stono Inlet. Once more on shore I could "navigate" finely. and being furnished by the Quartermaster with a horse and cart and an Irish driver was soon on my way to Head Quarters which I reached about dark. and ended the most unpleasant trip I have attempted since coming in the office. The first two days the weather was cold. raw. and most decidedly uncomfortable. the remainder of the time it was wet. foggy. and disagreeable in the extreme. I left on Saturday. and returned. or rather reached here again on Thursday night. After setting. the Saturday on which I left. the sun never shone again until after my return. Was about thirty six hours in making a trip usually made in six. at the most. eight hours. I had no idea of being gone over four days. but was absent six. and at the end of the trip found myself out of spirits. and worse yet out of pocket. I shall be in no hurry to try the trip over again. The boys are all looking finely. and appear to be enjoying themselves quite well. All express themselves much pleased with their rifles. and indeed they are pretty pieces. the finest things I ever saw in the gun line. It is not near the trouble to keep them clean that it was the old ones. and as there is no bayonets to them. there is not as many articles of accouterments to be carried about. and taken care of. The boys think that they may be the means of getting them into some hot place yet. but that would only be the fortune of war. Some little excitement was still observable in the regiment on the re-enlistment question. and a few re-enlisted while I was there. but the buisness was pretty much over with. I did not feel any increase of desire to re-enlist myself while there. and shall not do so yet awhile. the opinion of Mr Dodge. and other wiseacres as to the wisdom of the proceeding. notwithstanding. Over three hundred and twenty-five have re-enlisted out of the Regt. They are all to have their furloughs at the same time, and go home together. I am heartily glad so many have re-enlisted. and wish that every old soldier would do the same--if they wish too. Would do all in my power to induce them to do so except to set the example. Considering the number. less re-enlisted in our Co. than in any other one in the Regt. Great efforts were made by the officers to induce the Co. to do better. but. without avail. their talking only made the matter worse. I belive the Co. furnishes only fourteen veterans. None of the Salisbury boys are among the number. Doubtless many will think that <u>our</u> patriotism has oozed out somewhere. but I do not think it is so. and believe that if our services are needed. at the expiration of our present term of enlistment. they will be again promptly and freely offered. The Salisbury boys did not enlist for money. neither will they fight for it. I think they are all true. and ready to do their duty. of which I think they are as well qualified to judge as any one. and doubt very much that they will be grateful for advice upon the subject from the "Stay at home rangers." It will be plenty of time for the members of that Society. to question our patriotism when <u>they</u> have spent twenty-eight months in the service. Well I have my things all safe and sound and in good order. probably shall have no occasion to send for any mor things at least for <u>necessary</u> articles. Today is the Sabbeth. I have been down to Church. and heard a good sermon. the little building was full. and not the least interesting. and attractive member of the congregation. was a good looking. well dressed lady. Shoulder straps. though they may be all potent at home. are of but little account. and fade from view here in the army. when a lady appears. Have been at work most of the day. and have witten a hurried letter to George this evening. Have not written to Mrs. Randalls people. acknowledging the receipt of the box yet. though Mr. R. requested me to do so at once when it was received. Have a letter commenced to her. but could not content myself to sit down an finish it until I had written to you. Mr. Smith. one of the clerks here. and a member of our Co. has obtained a furlough. and I found had gone. when I returned. He is one of the recruits that came out to us. about fifteen months ago. It is not pleasant to see those fellows favored in the way they are. while we who have stood the wear and tear so much longer than they and have not received half as much remuneration must still continue to bear the brunt. We have good reason to think that Smith will obtain a commission in a colored regiment. and I do not expect to see him back here again. Our chaplain has resigned and his resignation has been accepted. so that we shall lose him. It will be a great loss to the regiment. and one that will not be likely to be replaced. or made good. I can hardly blame him for leaving us. Government has reduced the pay of chaplains and in event of wounds. or dying in the service. they get no pensions. neither they nor their heirs. Mr Wayland is a brave man. and on the battlefield exposes himself as much as any soldier. The boys will be sorry to lose him. I have a great deal which I ought to say to you. for your last letters have scarcely been alluded to. I hope to get an opportunity to write you another letter to go in this same mail. but felt as though I ought to finish the one commenced to Mrs. R. Shall be glad when it gets warm enough to write in my tent. after the days work is done. It is too noisey in the office to make much headway. We have had a deal of cold weather this winter. more I think. and more severe. than in either of the two last winters. How is it at the North? Are you having much sleighing? War news seems to be pretty scarce now adays with a fair prospect that the derth will continue for some time yet. To judge from the paper. one would think that all hands had concluded to let the rebels alone for a while. and settled down to a job of president making. for the next term.

    Begging pardon for my remissness in duty. and loving you fondly as ever.

I remain truly.
Your Own.
Ned

No. 147 Rec'd Jan 27th/64
Hilton Head.

Head Quarters Dept. South
Hilton Head S. C. Jan 21" 1864

Dear Sara.

We are back again you see in our old office. reached here on the 16". are still all upside down. not having had time to get settled. The buildings are not large enough for office use. and an addition is being made. Meantime we cannot be idle and wait until all is ready for occupation. but get out our Stationary in the most convenient corners we can find. and are very busy. surrounded by unopened boxes. chairs. tables. desks. and other office furnature. waiting until the room is ready for their occupation. It will be. it is. a great deal more comfortable. and pleasant. here than it was while we were in the field. We had a pleasant day to move in. and as comfortable a time during the operation as could be expected. All the Staff officers. with their offices moved with us. or will soon follow us. Mr. Landon did not come with us. but doubtless the office in which he is will be moved here shortly. We came down on the "Ben Deford." left Pawne Landing on Folly Island about three o'clock and came to anchor here in the harbor about nine in the evening. We remained on board during the night. and came on shore the next morning. were very busy all that day in seeing our things removed from the boat to the office. We spent two nights on the office floor. but the third day our quarters were prepared for us and we took possession. We are provided with wall tents. and excepting mine there are but two persons in each. Through a little piece of meanness on the part of some of the clerks I was shoved into a tent with two of the orderlies. It will not make much difference as I do not expect or wish to spend much time in my quarters. Things are not as agreeable with us now as they might be. perhaps they will improve. If not I do not wish to stay. The location of our quarters is as pleasant a one as I think could be found. close by the beach. where it will be cool in the summer. and away from the quarters of the civilian laborers. They are on ground that was within our camp lines. when we were camped here directly after landing in/61. and I have many a time marched over it while posting my relief when on guard. At present my work is sadly behind hand, and I am much pressed for time. I hope soon to be more at leasure. and think there is a reasonable prospect that I shall be. Your letter of the 11th was received by the "Fulton" mail. Your last four letters. about which I have done nothing but to acknowledge the receipt. are still kept by me in hopes that the wished for time to answer them will soon appear. George has surprised me much by re-enlisting. I thought he had settled it, not to do so, but it seems that he has re-considered his decision. I respect. and honor his patriotism. and the motives which have induced him to take the step. but do not honestly think that it was his duty to do as he has in the matter. I have great fears that his health. though good at present, is not robust enough to stand the severe test of another three years campaign. I am desireous of seeing him for though I have had three or four letters from him lately. he has not said anything upon the subject. Shall try to go over and spend the night with him after we get settled. So you have commenced school again on a small scale. I presume that the Mr. Phelps you mention, is the same one that used to visit at Mr. B.'s. Well death has been very busy of late. in your midst. I sympathize with you in the affliction which has fallen upon the family. The stroke was not only sudden but fell where least expected. The frail. feeble. child. was spared. and the strong. man. in the prime of life & usefullness taken. You will all miss him very much. as I judge from the frequent mention of his name in your letters to me during my absence. that of the two he has been even more attentive to you than he used to be when I was with you. I trust he is at peace in the better land. "Not lost but gone before." Mrs McNeil also rests from her labors. I had always a great respect for her. coupled with a feeling of almost pity for it seemed to me that either from necessity or choice she exerted herself beyond her ability to endure. Besides I cannot help thinking that anyone that is obliged to live with Mr. M. their life time. is an object of commiseration. I hope that the ravage of the disease that has proved fatal to so many of our townsmen has ere this been stayed. Thus far it seems to have attacked only those who were at least in comfortable circumstances. Should it break out to any extent among the poor. a great amount of suffering must unavoidably ensue. I regret my dear. that you should take my feeble attempt at dissipation so much to heart. I will not attempt to argue that it was right for me to indulge. even to the limited extent that I did. I ought to have remembered that you would be displeased to know of it. but not thinking anything of it myself. I foolishly mentioned it while speaking of the manner in which I passed the day. You speak of the indulgence. as being a habit. I cannot but take exception to that term. and deny that I am in the "habit." of drinking. If I were. my better judgement would tell me that you were not only perfectly right. but wise in requireing me to break it off at once. under the penalty of forfeiting your respect. your love. <u>yourself</u>. if I persisted in it. Why my dear. if I were in the "habit" of drinking even the little that I did that day. as often as once a week. or once a month even. I should consider myself a drinking man. and by no means a fit person to be trusted with the love and happiness of any woman. I assure you that. though a little surprised. I am not at all displeased at your remarks. or rather expression of feeling. I am perfectly willing to hear whatever you may have more to say upon the subject. but trust that my weak <u>defense</u> will satisfy you that I am not yet a confirmed toper. neither in quite so dangerous condition as you seem to fear. I have received a note from George this afternoon. he thinks he shall go home on the next trip of the Arago. the Fulton leaves for N.Y. tomorrow. it will be another week before the Arago will leave. Am finishing my letter on the evening of the 23rd. as I got to sleepy to complete it last night. You must question George well. and. closly. he will tell you the truth. whether it be favorable to me or otherwise. Good night my love. believe me ever. and truly your own.

Ned

No. 148 Rec'd Feb 5th 1864
Hilton Head

(Printed letterhead)
HEADQUARTERS, DEPARTMENT OF THE SOUTH
Asst. Adjutant General's Office
Hilton Head, S.C., Jany 30th **1864.**

(Note: January 29 was Ned's 29th birthday)

Your letters of the 15th & 17th were received two days since. Many thanks my dear for taking so much trouble to write to me. when you are weary and disturbed. You get poorly paid. for all your love. and affection. so freely bestowed upon me. I often think lately. that it would be much better for both your present and future happiness if instead of continuing constant in your love for me. you could learn to forget me and bestow your affections on a more worthy person. some one better fitted to be your constant companion through life. and who could minister to your happiness as I shall never be able to do. And I sometimes almost wish that it might be so <u>for your sake</u>. but my selfish nature will get the better of me. and the same feeling master me which led me to pursue you so persistantly with my suit. until you promised to be mine. And while I regret my disability to make your happiness all that I could wish it to be. I am proud and happy in the consciousness of being the object of love. and interest. of one so fine and good. Do not accuse me my dear of love sick sentimentality. I do assure you that I often think of these things. and much more of late than heretofore. and never without regret that I am not more worthy of you. and with anxiety in regard to your future happiness. Should it be my fortune to have it confided to my care. I ought to write you a long letter tonight. for I have plenty of material ready at my hand. but have still some work to do. things that must be got ready for this mail. I find it much more pleasant here than on Folly. and like it better than I did last summer. but presume. should I be here when the hot weather comes on. with its swarms of fleas. to torment ones life out. that I should be glad to escape to the desolate. barren. but breezy sand hills of Morris Island. Business is quite lively here. which is of itself a relief from the dreary. jog trot. treadmill. state of affairs on Folly Island. So many civilian labors have been brought here. that one now sees more civilians than soldiers about. The camps are all at a little distance from the village and as the men are not permitted to go outside them without permission but few are seen without visiting their camps. Quite a number of Storehouses have been erected since we left here last summer. and a good many little improvements made. But the place is still the same old bed of sand. into which one sinks anckle deep at every step. and which every little breeze drives in clouds before it. fairly turning what would otherwise be the greatest blessing into an almost curse. "<u>Robbers Row</u>" the center of business. not military. seems to be thriving. New shops are continually being put up. and there is now quite an assortment of most all kinds of goods to be found by a little patient. persevering search. A "Theatre" is in process of erection. quite a good sized building considering. and no doubt it will prove a very paying speculation. Amusements of any kind are so few. that almost any apology for a play. will be sure to be patronized. Not the least of the improvements to be noted is the frequency with which one meets <u>ladies</u>. I actually had the good furtune to meet <u>three</u> at once. tonight while walking up the Row. and before I reached the office. saw three others riding on horseback. The shoulder-strapped. gentry devote one evening each week to dancing. I happened to stumble upon the place where they meet for that purpose the other evening. and stood at the door for some time. enjoying the fine music and the sight of beautiful ladies. It was a glimps of civilization we soldiers are not often favored with. and to be enjoyed all the more because of their rareity. Re-Enlisting still continues extensively. A large proportion of the old soldiers in the Dep't will become Veterans. Probably nearly all have re-enlisted in our regiment who will do so. Confidentially Sara. I am half vexed at George for the step he has taken. Under existing circumstances. I do not think it was required of him. The present prospect is that none of the rest of our party will enlist. at least. now. That of course has nothing to do with influencing him. but still if our lives are spared. it would be pleasant to all go home together. Though George is in good health now. it looks to me very much like tempting Providence. to risk it for another three years. He cut his foot severely last Sabbeth and is at present laid up in the hospital. I learn that the wound is not serious. and he will doubtless soon be out again. I am trying to get him a place here in the office. but things do not look very promising at present. I do not know whether he has mentioned the fact to Julia or not. Do not say anything to her. or any one. of it unless she should speak about it first. Mr. Landon has got Wolcott in the office with him. a good place. and where he is likely to remain the remainder of his term. I wished it very much for George. but Mr L. knowing that he had re-enlisted. did not suppose he could. or would. delay his furlough. on any account. and had given in Wolcotts name for detail before I knew of it. and told him to the contrary. Tom Norton is now acting as a kind of <u>Post</u>.. Post Master. and haves charge of the mail for the forces on <u>St. Helena Island</u>. He is over every day or two now. and I see him frequently. McNeil has managed to worm himself into the Quartermaster's Department over there. and I learn manages to spend the best part of his time in speculating. Wells remains in the hospital. Pitt is fat and hearty. full of fun as ever. The other boys are well. Quite a little excitement exists at present among the civilians here. who are shaking in their shoes for fear of a Draft which is about to be made. The rascals sneaked off here. to avoid the Draft at home. and are now in a strait to get home again to avoid it here. Our office has been fairly besieged by the scamps this afternoon. all anxious to get papers. to go North on the steamer tomorrow. Yesterday I was Twenty Nine years old. I intended to have written to you last evening. but was to sleepy. and weary. I fear that the best part of my life has gone. and I have literally nothing to show for it. I have accomplished nothing of what in boyhood I have often purposed to. though I do not see that my boyish plans. and aims. where too high for accomplishment with a moderate amount of energy. and faithful application. I have but little disposition to hope. and take courage. for the future. for I have too often deceived myself. with promises. that have not been fulfilled. Well Dear Sara. I am not answering your letters. but please be assured that they were read with interest. and I am very grateful for your continued "long suffering." with me for my tardiness. I suppose Dexter will not remain with us much longer. He has received permission from Washington. to appear there. before the Board of Officers. for the examination of applicants for positions as officers in the Signal Corps. He will go well recommended from this office. and if he passes the examination. which he doubtless will. will receive a commission. It will be. or rather the Signal service is. the most desireable branch. for many reasons. of the service. I think he will go North on the next boat. or the next but one. though something may get occur to alter his arrangements. so it may not be worth while to mention it just yet. How do all my "Sisters" do nowadays? Tell Sara B. that I do not forget her. nor the fact that I owe her a letter. Give my love and a brothers kiss to Lizzie.

Please consider yourself embraced. kissed. and bid goodnight. by your own

Ned.

No. 149 Rec'd Feb. 15th 1864.
Hilton Head.

HEADQUARTERS, DEPARTMENT OF THE SOUTH,
Asst. Adjutant General's Office,
Hilton Head, S. C., Febuary 5th 1864.

Dear Sara.

Your letter of the 24." was received yesterday. I will try to write you a few lines tonight fearing that I may not have another opportunity before the sailing of the Fulton. We are <u>very</u> busy indeed. and the prospect of having any leisure for some time to come is very poor. Dexter expects to go North by the Fulton. Our head clerk is discharged from the service and will leave the office in a day or two. We expect two new clerks. but it will take some time for them to get acquainted with the business. The business of the office seems to increase all the time. I have worked very steadily ever since we came down here. Get into the office about half past eight in the morning and seldom leave it before ten. or eleven at night. My health is good. I am feeling first rate. and should enjoy it pretty well. if I only had a little more leisure. I regret my love. that you should so often be disappointed in not receiving letters. You have just grounds for complaint aganst a lover who cannot afford to give you one letter a week. But my dear I write to no one but you. with the exception of occasionally a hurried note to George. I assure you that I will care for you first if I ever do find a breathing spell.

Besides the two clerks just mentioned. who are soon to leave. a third one. Mr. Smith. is not expected to remain long. He returned last night from a furlough. while North he went to Washington and was examined for a position in the colored troops now being raised. was accepted as 2nd Lieutenant and now awaits his commission, and orders, which may come by any mail. but may be delayed for some time. I have been disappointed. in trying to get George in the office. Capt. Sealy said he must have men who have had some experience in the business and I suppose he has got them. as they are now quite plenty. I am sorry but while I remain I shall keep a good lookout. and do my best to get a position for him somewhere. Last Sabbeth afternoon I went over to St. Helena and spent the night with the boys. My object in going was to see George though I had no particular business with him. He is quite well. and the wound on his foot is healing rapidly. Spent a very pleasant night. Tom Norton is Post Master for the Regt. and has an officers tent to himself. staid with him. Called on our Captain. and Adjutant Moore. in the evening. The Captain's wife is living with him. in his tent. Col. Hawley's wife is also with him. George does not intend to go home at present. and I think he is wise in defering his visit as he thinks of doing until the hot weather comes on. I think my love I have received all of your letters. and among them the "one you refer to in your last. which is. I take it. the one containing the temperance lecture. You have doubtless ere this received my reply to it. I will not allude to it further until I hear from you again. I regret that I have not something to send you by Mr Dexter. but my oppertunities for obtaining trophies are very few. and I have none at all. Am very sorry that I had not collected some shells while we were on Morris and Folly. and should have done so had I the least idea of having a chance to send them to you. I may send a package of letters which I have by me. My time is drawing to a close and as the probabilities of getting home again increase I am desirous of saving all of our correspondence that I can. You will find that some of the letters are soiled. the effects of several days transportation in my pockets.

It is nearly eleven o'clock and I must close

With love. ever your own
Ned.

No. 150 Rec'd Feb 27th 1864.
Hilton Head.

HEADQUARTERS, DEPARTMENT OF THE SOUTH,
Asst. Adjutant General's Office,
Hilton Head, S. C., Febuary 12" **1864.**

Dear Sara.
Your letter of the 31." Jany. was received today. and not being very busy tonight. take the opportunity to commence an answer. Have just returned from the Hospital where I have been to see George. Our regiment having gone on the expedition to

Jacksonville (or Florida I should have said) the hospital on St. Helena was broken up. and all the inmates sent over here to the General Hospital. George is in first rate health. and spirits. and I venture to say. more flesh than you ever saw him. Some how or other the wound on his foot does not heal up as readily as he thought it would. Still it seems to be doing well. does not pain him any. He cannot walk on it yet. and haves to get around on crutches. He is enjoying himself quite well. as he is in the convalescent camp where the patients do just as they are a mind to. and are not under any unpleasant restrictions. Dexter did not go on the last boat North as I wrote you. but now intends to go on the return trip of the Atlantic which arrived today. she having taken the place of <u>the Arago</u>. for the last two trips. He could have gone last week but chose to wait because a friend of his. who is also going to Washington to be examined for a position in the Signal Corps. could not get away until this week. We get but little news as yet from the expedition which went down to Florida last week. As there is but little rebel force in the whole State. I do not expect to hear any stirring news. for I do not think there will be any fighting. I am at loss to account for the taking of so large a force down there. for there can be but little doubt that. one half as strong a force could march all over Florida. with but little or no molestation. unless troops should be sent to reinforce the rebels. There is but little to be feared on an expedition through Florida, but its snakes. and poisonous insects. The weather has been quite rough for several days past. but is moderating a little now. Things in the office continue about as usual. varied only during the last week by one of our clerks getting drunk last Sunday night keeping it up on Monday. and requiring Tuesday and Wednesday to recover from the effects of it. Captain Sealy was pretty angry. though he said nothing. and another such a drunk will surely be the cause of his dismissal from the office. I was obliged to leave my own work for a day or two and do what I could of his. You have probably learned ere this that you are not to expect George home just at present. His being lame just at present is quite unfortunate. for I had a good chance to get him a place in an office. you must say nothing about our attempts to get a situation for him for he does not wish it known. Nothing new here at the Head. The officers enjoy themselves one night each week in dancing. an upper room in one of the commissary buildings furnishing the hall. and a fine band. the music. a moderate number of "fair" ones supply the necessary partners. I do not know but that I have told you before that a <u>theatre</u> is being built. it is quite a good sized building. and was rapidly approaching completion until the expedition sailed. The <u>48" New York</u>. are the originaters of the enterprise. they have gone with the expedition. and work on the building is pretty much stopped. General

Gillmore made quick work shutting up one of the Sutler shops here the other day. He sent his negro servant out to buy some oranges. I do not remember what the man charged. but the price was exobitant. the general at once wrote a note to the Provost Marshal. who sent a file of men and arrested him at the same time closing his store. The goods were confiscated and sold within three days. among them were a large quantity of apples which were turned over to the hospital. George told me they had all they could eat. I have not yet changed my mind about re-enlisting. though. I sometimes think that perhaps I ought to do so. I know you love me dearly Sara. and you would no doubt be sadly disappointed. if after waiting so patiently and faithfully for three long years. you should be called to wait for <u>three</u> more. it would be most discourageing. But though I love you very dearly Sara. I am often forced to the conclusion that it would be for your future happiness and welfare. if you could forget me. and bestow your affections on one more worthy. and then I reproach myself with having done wrong in seeking to win you for myself. I shudder at the thought. that it is possible that our union might be fatal to your earthly happiness. I could bear to be unhappy myself. but it would make me miserable to think that I had ruined the welfare of one I love so fondly. Loving you as I do. l know that I shall be blest with a priceless treasure if ever permitted to make you my wife. I <u>cannot</u> but regret that the advantage of the union would be all on my side. I fear that I can never make you happy as I wish you to be. My dear. you have <u>too much</u> confidence in me. except when I tell you that I am not good. am not what I ought to be. am not what you think me to be. then you <u>will not</u> have confidence in me. and believe that I am telling you the truth. I do not wish you to take back any thing that you wrote in regard to the Christmas day affair. You would not have been just either to yourself. or to me. if you had said less. <u>Only</u> I did not wish you to think that I was in the <u>habit</u> of drinking at all. much less to excess. I have not "resolved to not write anything more about myself." and trust that I shall always be ready to account to you for the manner in which I conduct myself. How can I ever think otherwise than "kindly" and lovingly of you dear Sara. I feel assured that let what will come. I shall always love you. and be a better man for having been the object of your love. and favored with so much of your society. God grant that whatever <u>our</u> future may be. <u>yours</u> may be a happy one. and your path through life a pleasant one.

Febuary 13." This day has passed like most others. rather more quiet than some. I have been moderately busy. have had a plenty to do. without being hurried. Received a welcome. and quite a long letter from Mr. Edgar. I had forgotten how our correspondance stood. but was under the impression that I was owing him a letter. and was therefore very agreeably surprised to receive his favor. acknowledging the receipt of a letter from me some time ago. and excusing himself for delay in answering it. After supper tonight I strolled down onto the wharf. as I frequently do nights before returning to the office. and was most agreeably surprised to meet Mrs Hawley and Mrs. Mills. the wives of our Colonel. and Captain. I only saw them a few moments as the boat was just leaveing the dock. They are on their way to <u>St. Augustine</u>. our regiment is on the expedition and as it is doubtful whether they will return to St. Helena very soon. if at all. I suppose the ladies are going to St. Augustine to await for them to get settled somewhere. They met me very cordially. and I need not assure you that it was a great pleasure to me. to again press a ladies hand. and be favored with her smile. Such favors are to rare with a soldier. not to be highly prized. So that on the whole. I count this a bright day in my calendar. The late mail brought but very little news. nothing as I can see of importance. I suppose that it will be quiet for a couple of months yet until the roads have become settled. then <u>the Army of the Potomac</u> will begin summer marches to defeat and death. My hopes of success. one now fixed upon the <u>army of Gen'l Grant</u>. If he is properly supplied with men. and let alone by the President makers. I think he will accomplish great results in the coming campaign. But I have learned to not pin my faith to strongly on any one man. and while I hope. I do not forget how liable I am to be disappointed. I suppose of course the Unionists at home "intend to re-elect the soldiers friend." <u>Mr. Buckingham</u>. again the coming Spring election. I have heard nothing as yet in relation to the subject so presume there will not be as violent an opposition to him as there was last Spring. I commenced a letter to Mr Randall the other evening. and I must try to finish it. to have it ready for this mail. I have managed to write just one letter a week for several weeks past. and have got so that I know better than to expect letters from any one but you. I am sure of at least one from you each mail. I wish you were equally as sure of one from me. and as seldom disappointed. With love.

Ever Your Own.
Ned

CIVIL WAR LETTERS 1861-1865

Editors note: This letter refers to the bloody Battle of Olustee 20 February 1864

No. 151 Rec'd Feb 27th 1864.
Hilton Head.

HEADQUARTERS, DEPARTMENT OF THE SOUTH,
Asst. Adjutant General's Office,
Hilton Head. S. C., Febuary 23rd **1864.**

Dear Sara.

Your letter of the 7th. was received by the Fulton's mail. which arrived Sunday afternoon.
Thank you my love. for such a dear good long letter. I do not know how to answer it as it deserves. You love me to well Sara. and are not willing to believe ill of me. I am more than fortunate. and cannot be enough grateful to the good God who has blessed me with the love of one so pure. and good. It would be impossible for me to forget you. and I could never love another. as I do you. All my hopes of earthly enjoyment are centered upon you. and I know that my choice is a good one. and will never fail me. My fears are for you. I have reason to distrust myself more and more every day. and to become more and more convinced that I am not. can never be. a fit companion for you. I cannot disguise the fact if I would that I am not the same as when I parted from you long ago. I feel confidant that I am the same to you. and look forward to meeting you again. as the happiest period of my life. I shall never be able to do for you all that I wish. for I would like to make your future free from trouble or care. while much I fear. that by marrying me you will only add to your life-troubles. Perhaps you wonder why I should tell you this now. why I did not think of it while pressing my suit so earnestly. before I had prevailed on you to commit yourself. My dear. I have done wrong in more ways than one. and not the least of my errors has been that of putting off the thinking and reflection. until now. that should have been done years ago. It was easier. and pleasanter. then. to push them aside. Now they will be considered. will no longer be put off till a more convenient season. Doubtless the probability that I may soon be permitted to return home. has led to. and kept up. the train of thought that has pursued me. almost unremittingly. lately. Pardon me for troubling you thus when I should be writing something more becoming a lover. but I deem it right to warn you. though I fear that if I found you attempting to profit by them (the warnings). and save yourself while yet there was time. all my courage would fail me. and I could not. would not let you go. Dear Sara. we will hope for a bright. and happy future. and that all will be better than my trouble

borrowing spirit fears.

    You will here ere this reaches you. of our late disaster in Florida. As yet nothing very definate has become genarlly known in relation to it. Evidantly. we have met with a severe defeat. Nothing official is known respecting the extent of our loss. but from what I gather from differant sources. think it must be over a thousand. The loss of our regiment is said to be sixteen in all. from which I judge that they had but a small share in the engagement. Lt. Dempsey of Co. E is reported killed. but shall hope to see him turn up all right as have Col's. Henry .&. Sammons. Capt's. Hamilton & Langdon of the Light Artillery together with several Lieutenants. all of which were at first reported killed. or badly wounded. but have all turned out to be alive. and some of them not even wounded. Florida is of much more importance to the rebels than I had supposed it to be. What I wrote in my last letter respecting the State and the rebel force to be found there. was no doubt quite near the truth. as things were a year ago. But there has been a great change since. and Florida it seems. has become one of the chief Commissary Departments of the rebel army. No wonder then. that they kept a watchful eye upon it. and while no more men than were absolutely necessary. were kept there. a large force was held in readiness to hasten forward at a moments warning. I regret exceedingly our defeat. It seems as though nothing but disaster attended our expeditions in this Department. True we have accomplished something. but we do seem to be exceedingly unfortunate in these incursions. which if we were successful in would be of great advantage to us. and damage to the enemy. It is reported that the gunboat Housatonic was blown up a few days since. by a torpedo. as I have not heard contradicted suppose it is true. There can not be the least doubt I think. but that the rebels are determined to fight to the bitter end. they are growing doubly desperate but that only makes them fight the better. I trust that the policy of our Generals will be to hold all we have gained. through the coming summer. and lose as few men as possible. then after the election is over with in the Fall. I shall hope and expect. to see an earnest. vigorous. and successful campaign immediately commenced. which will in a short time end the war. I may be unwise. perhaps foolish. but I cannot help believeing that in some way. the war is being kept up for political capital. and that there is no hope of its being ended until after the Presidential election. I must apologize for the appearance of this letter. it was after ten o'clock I think when it was commenced and I have written so hurridly that I have made bad work. The Fulton leaves in the morning and my letter must be finished tonight. Work done in a hurry is generally ill done. and my letter is certainly pretty good proof of the fact. Some questions in your letter I will answer next time.

> Good night love.
> Ever Your Own.
> Ned.

No. 152 Rec'd March 8th 1864.
Hilton Head.

HEADQUARTERS, DEPARTMENT OF THE SOUTH,
Asst. Adjutant General's Office,
Hilton Head, S. C., Febuary 29" **1864.**

Dear Sara.

I sent you a half finished letter three days since, when I laid it aside the night before. I thought there would be plenty of time to finish it before a mail left. else I should have completed it before retireing. An unexpected opportunity offered, and I thought best to send it without delay. Very sad news I was obliged to write you, I pray that there may not occur the necessity for the like again. Returns from our defeated forces are very meager. and unsatisfactory. Evidantly efforts are being made to hush up. and keep the matter as quiet as possible. Our regiment at present is at Jacksonville. Our forces still hold Camp Finnegan, a distance of six or eight miles from the city. All quiet at present so far as I know. An attack from the rebels has been expected. but it is doubtful if they make it. as under the circumstances, it would be differant from their usual plan of operations. The Veterans of the 6" & 7" returned last Saturday (the 27") I went over to St. Helena and spent the night with them. Our camp equipage was left at St. Helena when the Regt. went to Florida. and a few men left in charge of it. Most of the Hospital furnature and stores were left. also. Wells was one of those left behind. The Veterans will proceed at once to join the Regt. taking with them all the Regimental property. Reid has been detailed as clerk to Capt. Gould. who is in command of the Invalid Camp. and present Commander of the Post at St. Helena. and so will not at present rejoin the Regiment. Our sick and crippled. in the Hospital here. will remain. I could not but regret to see how bad a use some of our boys had made of their short liberty. and the money they will now have to pay for so dearly (I look upon all the bounty money paid down. as merely wages advanced. which has got to be earned in the hardest kind of a way). A large number of them have spent not only the whole amount received on re-enlisting. nearly $500. but a good deal more besides. which they had laid away. or borrowed from their more prudent comrads. I learn that $15. & $20. were paid for a pint of whiskey on board of the steamer. on their way home. It is difficult to conceive how men can be so foolish. but it is so. Quite a number got married. which though foolish enough in all conscience. perhaps kept them from worse follies. They were all in good spirits. and seemed ready to set about their new three years with as good grace as could be expected. Well Sara Dear. I do not feel like re-enlisting yet. neither do I feel like shunning any duty. or danger. that as a soldier. may lie in my path. during the coming six months. The remark made by one of our regiment. now on extra duty. that "he never wanted to see the D----d regiment again." meets with no favor from me. I <u>do</u> want to see it again, I am perfectly willing to join it any day and share its hardships. and dangers. I would hardly ask permission to remain in the office. were it necessary in order to do so. Neither am I anxious to leave the office. I shall be glad to remain. and do my duty here. as well and faithfully as I can. but do not feel that I am hardly entitled to the credit of being a soldier while thus engaged. I am in no hurry to die. and shall do all that I consistantly can to preserve life and health. but if it should be my fate while in the army. I could wish to fall as did Pitt. doing my duty faithfully. bravely. and in the front of battle. Our little party who have so often talked of the pleasure we should have when we returned home. has been sadly broken up and disorganized. First, Johnnie Sweet was discharged for disability. next poor Brinton went to his long home amid the Southern sands. Then Mr Moore was promoted which added another year to his term of service. next. Barnes was called to his last account. George has voluntarialy increased his bondage for three years longer. Dexter is in a fair way for promotion. which will remove him from us and attach him to the service for an indeffinate period. lastly Pitt has fallen. We know not what the coming six months may have in store for us. we will borrow no trouble. for surely "He" who doeth all things well. hath us in charge. and will do for us as seemith him good. But we lay no more plans. and as the hoped for return grows to be more and more an individual matter. less is said about it. though doubtless it is thought of all the more. When a man is promoted. to be a commissioned officer. he is sworn in to the service for three years from the date of his mustering in to his new rank. Mr. Moore was mustered into the service as 1 Lt. Adjutant some time in Febuary/63. I think. That binds him to the service for a year and a half nearly. over the time of his original enlistment. Should he be again promoted. he would, when mustered in to his new rank, be sworn in for an additional three years. and so on at each new promotion. Officers in the regular army. hold their commissions from the President, and are not sworn in for any definate time.

Tuesday Evening. Mch 1." Am very tired tonight. having been hard at work since eight o"clock this morning. and all the afternoon have had an exceedingly vexatious piece of work on hand. Must finish my letter though. as it will not do to put it off longer. I think you asked me in one of your last letters to have my photograph taken. I intend to do so. the first opportunity. which means when I can get time. There are two or three establishments for taking them here. but think that they all do very poor work. I have not looked into a paper since the last mail arrived. but suppose if there had been any stirring news I should have heard it. Met Mrs. Hawley in the street today and had a few moments conversation with her. She thinks she will be obliged to go North. as there is an unwillingness at Head Quarters to have the wives of officers remain here. She does not wish to go. she seems very much attached to our regiment. and takes a lively interest in all that pertains to it. takes a deal of pains to visit the hospitals where our sick and wounded are. and does all that she can to cheer and comfort them. She does not wish to leave. and I hope may be permitted to remain. I believe she is universally loved and respected by the members of our regiment. I spent Sunday evening with George. He is not likely to leave the hospital just yet. and I am in no hurry to have him. for I like to have him here to talk to. How much longer are you going to stay at Emma's? If you enjoy yourself there. of course I am willing you should stay. but I am half a mind to be vexed to think that others can be so freely favored with your company which is denied me for even one short hour. Please tell me if you think that Lizzie cares for Charlie now. and if all their letters. and love tokens have been returned. You would know if she kept his picture yet. Tell me if you know. what has become of him. Perhaps you think me over curious. but I have quite a desire to know how their love affair will terminate if indeed it has not done so already. Good Night.

Love from
Ned

No. 153 Rec'd March 24th/64

<u>HEADQUARTERS, DEPARTMENT OF THE SOUTH</u>
Asst. Adjutant General's Office,
Hilton Head, S. C., March 14th 1864

My Own Dear Sara.

You have been disappointed again in not hearing from me as you ought to have done. I am more than sorry for it, but believe me my dear. I could not help it. Several mails have left here for the North since I wrote last. and I cannot tell you how vexed

I have felt at not having been able to send you a word of love and rememberance by them. I have been very very busy for the past fortnight. working until eleven and twelve o'clock some nights and not getting through with my duties then. But if I have not had the time to write. I have not ceased to think of you. and all the more perhaps because I knew I was doing you wrong in withholding from you the letters which you so much desire. Dearest, my term of service is drawing rapidly to a close. but I think the remaining six months appear longer than did the whole time a year ago. I have never until quite lately felt very confident of seeing you again. but now so far as we can judge there seems to be a reasonable prospect of it. and my hopes. and impatience increase as the prospect brightens. If I could only meet you as I wish to. confidant that I was such a man as you believe me to be. as I <u>ought</u> to be. and that with the blessing of Providence I should be able to make your life happier by a union with me than it otherwise could be. I should be very joyous in the anticipation of the meeting which we hope will soon take place. I distrust myself. and have good reason to do so. I have perfect faith in you and <u>know</u> that you are <u>all</u> that I believe you to be. all that I could wish you to be. Do not think that I am merely using idle phrases to fill out a letter. I am sincere in what I say now. if I have never been before. I beg your pardon for still writing upon a subject which you wished me to drop. but it is not easy to remain quiet about that which is so much upon my mind. I received a letter tonight from Mrs. Randall. still kind and interested in my welfare as ever. She is glad I have not re-enlisted. I begin to doubt myself whether it is really my duty to do so. I have been inclined to think all the while that it was. but do not know but that my friends at home are as good judges of the matter as myself. Mrs. R. though a very dear friend. has not the same interest in me that you have. and being very patriotic and plain spoken. would no doubt tell me if she thought that I had ought to remain. If I were sure of remaining in the office. it would only be making sure of a tolerably comfortable situation to re-enlist. and of course I should take no credit to myself on the score of patriotism for so doing. and perhaps I could not do better. as I have no idea of what I shall be good for. or able to get a living at when discharged from the army. But there can be no certainty about staying here. How would it suit you my dear. if I should act upon Sara B's proposition. re-enlist. come home on furlough, marry you. and return to the army. or office which would be the same thing? I think I could save enough to partially support you. If the project pleases you I will think seriously of it. I may not be able to marry you in a long while unless I should do so. You said in your last letter that I need not marry you until I got ready. What an idea. As if I was wishing to postpone that much wished for event and it was you that were urging it! Had my ability been equal to my desires. you would have been my wife years ago.

March 15" 1864. Have just re-read your letter of Febuary 29". had I done so last evening before I commenced writing I should not have said so much upon the one topic of which as you say my last letters have been so full. I beg your pardon. I am but a sorry lover to be making you miserable with such dolorous. bug-a-boo stories. I found your letter of March 7" at my tent last night after leaving the office. today received a number of papers. and a magazine from Mrs. Randall. Should like to know. when. where. or how. I am to find time to read them. So you are enjoying yourself among the nobility are you! What a pretty figure I should cut at an evening party, would'nt I though! Should be as much out of place as a pig in a parlor. I have of course no objection to your sending my letter to Sara. but had I thought of your doing so.1 would have written a little differantly. and tried to have made it more presentable. It would give me great pleasure to write to Sara. I feel vexed. and angry almost. that I cannot find time to do so. I hope you will soon see her and be able to make her a good visit. Tell her please. that I do not forget her. that I shall write to her the very first opportunity. I learn that the preformances at the theatre here last evening were very creditable. they are to be repeated twice during the week. The Arago has been looked for all day. but has not yet arrived. Shall hardly expect a letter by her mail. but will try to write you again in time for her return mail. Good Night. a kiss and love from Ned

No. 154 Rec'd Apr. 1st 1864

HEADQUARTERS, DEPARTMENT OF THE SOUTH
Hilton Head, S. C., March 26th **1964**

Dear Sara.

The boys have all gone to the theatre tonight. there is noby in the office but a single orderlie and myself. I hope to be able to begin and finish a letter to you without interruption. Your letters of the 9" & 13." were received three days since. I cannot tell about the letters you mention that I have not acknowledged the receipt of. as the last few letters I have received are in another coat pocket. up at my tent. I think though they have all come safely to hand. Dexter arrived yesterday on the "Fulton." Well. and in good spirits. He does not yet know whether he was successful or not, at Washington, and guess that he does not care much about it either way. He resumes his place in the office. We have had a good long chat about home. and the subject is not yet exhausted. We have had a succession of equinoxial storms during the past week. and the weather is hardly settled yet. No news about the Head. or throughout the Department. I should have liked much to have heard Mr. Reid's sermon on the death of our boys. Wonder if they will not have it printed. It might not be of so much general interest. but it would be highly prized by the friends of the families. and by the comrads of the boys. Had I thought that my letter to you was to have been seen by other eyes than those that always regard so lightly all my faults. I should have given it more thought. and have written it more carefully. as it was I wrote carelessly merely intending to get the bad news to you as quickly as possible. and then when I felt more like it. writing at more length. I am most happy to be able to testify to my friends many virtues. and good qualities. and if ought in those few hurried lines could help to comfort. or lighten the heavy load of grief. so crushingly cast upon my friends. I am thankful. Well I have been disappointed in my writing as usual. had hardly got nicely commenced before along comes Col. Smith with a job that has occupied me all the evening. and to no purpos. for he required information which after hours of diligent serch I am unable to furnish. and which will have to be referred to the head clerk after all. It is very annoying. I might better have gone to the theatre with the boys. I should have accomplished just as much and saved myself much vexation. I went one night last week. did not go until late. and had been there but a short time when there was an alarm of fire. which very shortly emptied the building of the spectators. The engine house (or one of them. for there are two or three in the place) stands close by the side of the theatre. we made a rush for it. got out the engine. and after tugging it a good distance through the sand to where the fire was said to be found there was none there, not even so much as a smoke. It seems that by some accident a fire had broken out in one of the upper rooms of the old plantation house. but had been promptly extinguished. Not a little disgusted to think that I had worked so hard for nothing. I steered for the theatre again. determined not to be cheated out of all of my evening's entertainmnet. A fire here would be no joke I can assure you. All of the buildings are constructed slightly, of light, inflamable materials. which would ignite from a spark. and the building would be consumed. before a fire in an ordinary building would be fairly kindled. Millions of dollars worth of property are stored (in) them. while our ordnance yard would be found to be most uncomfortably near in the event of a general conflagration. Dexter tells me there was much indignation expressed at <u>Genl. Seymours</u> proceeding throughout the North. I have no hesitency in pronouncing it righteous. I think that he. and Benham. of James Island memory. should be laid away together. Thank you for enquiring. and reporting. concerning Lizzie and Charlie. I think she had better forget him. if possible. at least give up all ideas of a renewal of their love affairs. Of course you will not tell her I said so. but I really do think that it will be for her future happiness to do so. You can kiss her for me some time when you meet her now. and I hope she will be able to write me the letter which she mentioned. Hasn't she a plenty of photographs. so that she could spare me one? You, and Carrie. are the only ladies of my acquaintance that have favored me with their shadows. but I am favored a thousand fold thus and am content. I have not had any photographs taken yet. Am waiting for a new establishment to commence operations. which is expected to open soon. I have seen some pictures taken by the same firm in Beaufort. and think they are the best that I have seen taken here at the South. There are now five establishments of the kind in operation here. As there are but few troops here at present. should think that they could not all get rich. Stores and Restaurants. are being opened every few days. At the present rate of increase they will soon have to live upon one another. One of our orderlies has only six weeks of his time remaining to serve. He is congratulated on all sides. and has many offers from others less fortunate to exchange places. But to no avail. He thinks he shall be content to stay at home hereafter. and he has good reason to think so. for he has a young wife awaiting his return. no doubt as anxiously as he is expecting it. These young wives are terriable attractions. and one must be careful about trusting themselves in their power. But I shall be just rash enough to venture it. if fortune ever favors me with an opportunity;

With love. ever your own
Ned

No. 155 Rec'd Apr 9th 1864
Hilton Head.

HEADQUARTERS, DEPARTMENT OF THE SOUTH,
Asst. Adjutant General's Office,
Hilton Head, S. C., April 2nd. 1864

Dear Sara,

Your letters of the 20th & 25th have been received within a few days. many thanks for them and all your multiplied favors to my underserving self. I could not believe at first that so long a time had passed without my writing to you but upon looking in my diary I can find no mention of having written to you during the period you were without hearing from me. I shall make but poor work of writing tonight. as you see I have made a miserable beginning. My eyes are troubling me just at present. and I can scarcely see the lines on the paper. besides one of the clerks is going home on furlough tomorrow. and there is all sorts of jabbering going on so that it is almost impossible to keep track of ones thoughts. I suppose that there is no doubt but what I could have had a furlough if I had chosen to ask for it. but I do not think it advisable. My time is nearly out. and when I <u>do</u> come home it will be pleasanter to feel that it is for good, that I shall not be <u>obliged</u> to hurry back into the army at the expiration of thirty days. Circumstances may lead me to regret my decision. but at present I do not think it advisable to ask a furlough. I cannot answer your two letters tonight dear Sara. please excuse me. Be assured however that I appreciate them and prize them highly. I do not wonder that you are inclined to think sometimes that I do not care for your letters, so many of them remain unanswered. I beg you to believe me. I do care for them. <u>all</u> of them. and I regret that it is not in my power to do justice by them by answering them as they deserve. I am glad you had such pleasant times. both at home and at Mr Bosworth's. By the way, is the "Eva Fay" you mention a sister of Mrs Sellect's. and is she as pretty as Minnie used to be? So you are going to teaching again are you? I think it would be much pleasanter if you could obtain some such situation as the one at Mr Barnum's. but such good places are not always to be had. mores the pity. What has become of Miss Richardson, sympathizing with rebellion somewhere I suppose? I mean to write to Sara B. within a week if it is a possible thing. and I think it will be if my eyes will only behave themselves. I did not send you the music my dear. I do not know whether George did or not. Shall be most happy to hear it if I ever get home. Have often thought of sending you some pieces of music. but feared that it would not reach you. But I received a present yesterday that I know came from you. a pocket knife. both neat, useful, and much wanted. What put it into your head my dear. to make me this present? You ladies are so much more thoughtful than us men. No doubt I stumble past many little things every day that would be acceptable presents to my dear wife. that, is to be, if I only was sharp enough to see them. I have had my name engraved on the knife, also the date I received it. so that I hope I shall not lose it. There is a splendid engraver at work here in the office now. engraving medals. Have you noticed my new seal? <u>General Gillmore</u> is having a party in his quarters this evening. the band is there. and no doubt they are all having a gay time. I did go up and go to meeting with George the evening he expected me. and have done so once since. I took him up three letters. a few nights since. am very glad to beg any service to him. though it may be ever so small. He has ever been kind to me and laid me under lasting obligations to him which I can never forget. the rememberance of the many pleasant interviews I have enjoyed with you at his house. will never be effaced from my memory. I have taken to another evil habit lately. that of smoking. Dont know but I shall become a confirmed smoker before I get home. but think I shall be able to break it off if you wish it. Mrs Randall would give me an awful scolding if she knew it. she does so detect it. I cannot blame her for the smell of tobacco makes her sick. Dear Sara please pardon this poor return for your excellent letters, and believe me. as ever.

Your Own
Ned.

No. 156 Rec'd April 19th 1864
Hilton Head

Headquarters, Department of the South,
Hilton Head, S. C., April 9" 1864

Dear Sara,

Your letters of the 29" and Apl 1st were received by the Fulton's mail yesterday. Two came for George which I carried up to him last evening. found him doing nicely. he has gone into the cooking department of the hospital. he still finds it difficult to use the wounded foot much. So you have some new teeth! You had not told me that you thought of getting them. though you did tell me about having several of your old ones extracted.

Well you have beaten me in more ways than one. it seems as though I could almost stand it to have a <u>sound</u> tooth drawn if only some fair lady's arm would encircle my neck during the operation. I well know that a dentist's professional embrace is not very desireable. but there's no telling what a poor banished fellow would not submit to. for the pleasure of once again having a pair of plump. white. arms thrown around his neck. Only think of it Sara. thirty-one long. dreary months. away from civilization. and the girl one loves. I tell you its tough. The nearest approach to social intercourse with a lady. that I have enjoyed in all that weary time. was the privelage of touching her hand (gloved at that) and holding a few moments conversation with her. But I promise my self that at the expiration of five months more I shall be repaid for all my <u>sufferings</u>. by the welcom greeting that awaits me from my loved one at home. I don't half like this idea of watching. and counting the time so closely. but what's a poor fellow to do to avoid it! One can"t prevent the birds from flying over his head. though he may keep them from building nests in his hair. Thoughts of home. and the amount of time intervening. before I can again see it will come to me. though I do try not to allow them to occupy too much of my time. I remember the last time I saw poor Pitt, full of life and health, and of hopes of soon seeing his beloved home. he told me how many days more we had to stay in the army. it was over to St. Helena. not more than a couple of weeks before the expedition started for Florida. Together with Wells, and Norton, we were talking about our return home. and rejoicing in the rapid approach of the joyful day. But one. at least. of the merry. hopeful party was not to again see the much loved spot. and receive the greetings of the dear ones there. I hear nothing of James Deane. of late. and so suppose that all is well with him. The 19th seems to be a very safe regiment. at least it has been thus far. Their good fortune in being permitted to keep out of harms way is often spoken of. I learn that my brother has enlisted in the regiment, was much surprised to hear of it, as I had not dreamed that he thought of such a thing, neither did I think that he would ever be accepted, as a soldier, as his general health for the last few years has been very poor. Well my dear. you and Julia have doubtless had a good laugh at me about my present of a knife. but I doubt if you have enjoyed the joke any better than George and I have. Of course the laugh is all against me. I can only wonder at my stupidity in not thinking that it might be for George. Luckily he does not want the knife. and I have settled the matter with him. I wrote to George Burrall by the last mail to send me another dozen of knives such as I had sent me last summer, they are wanted in the office. I directed them sent by mail. there is some risk attending that way of forwarding them. but the loss will not be great. If you get the little papers I send you, you will have seen that the theatre going temptation has been removed from me at least for the present. It is probable that the building will be again opened for amusements of some kind ere long. I believe I was quite moderate in my attendence. going but three times. and the number would have been still less but for the solicitations of friends. It is getting to be quite a busy place here. There can not be less than twenty-five stores. and sutlers shops. on Robbers Row. of various kinds. and some of them are establishments of considerable importance and doing a large business. I count seven restaurants. in active operation and an eighth. the largest of the lot nearly completed. Our two newspapers. though small in size are very profitable affairs. the editor of the "New South" I am told having cleared eight thousand dollars on his paper the last year. There are four photographic establishments in operation. and a fifth in process of erection. Two barbers shops, two shoemaker shops. and a tailor's shop are among the other evidences of civilization and progress. The "Palmetto Herald" is at present printed in a couple of officers tents joined together. but the proprietors have a building in process of erection on a desireable "corner lot." it is of two stories. not as large as your fathers store. they intend to occupy the lower front room as a store. one of the firm said tonight that they had been offered three thousand dollars. for the rent of that one room. for one year. by a responsible party. That is rather ahead of Salisbury rate of rents. Did I tell you in my last that I had bought me a watch? You probably remember the fate of the one I used to carry at home. since losing that I have been without a watch until I bought this one. which I think is a very good one. Its a silver watch with hunting case. Cost me twenty Dollars. and a watch seller on Robbers Row told me tonight that it was worth twenty-eight or thirty. so guess I have lost nothing by my bargin. unless indeed. I should lose the watch. which I will look out for. It is not as elegant an affair as your grandfather's but then you know I had not his long purse to buy it with. I suppose Miss Winchell's death has prevented the old gentleman from committing a piece of grey haired folly. which would have caused his children some uneasiness. I assure you my dear that your looks needed no improving. at least for me. and I flatter myself that I am as much interested in them as any one. I hope you will find your teeth not only ornamental. but useful. My teeth have decayed sadly. and now I have scarcely a single sound one. I assure you upon my honor. that <u>my looks</u> have <u>not</u> improved in my absence from you. and furthermore. if you are looking for the return of a handsom young gallant. from the wars. you are doomed to disappointment and had better be setting your cap without delay for better game. I intend to have some photographs taken in a few days. may be when you get one. you will be convinced of what I tell you. I have got an ugly scar between my eyes. that looks for all the world like a souvenir of some bar room brawl. and which has not even the merit of being an honorable scar to recommend it to your favorable notice. If you dont get weary of me before I have been home a year. I shall be a more fortunate man than I have ever dared to hope. Don't talk about the shadow of my ugly phiz's "gracing a page in your album." it would put a good looking picture out of countinance. and if possessing any of the spirit of its original. would find fault with them for being better looking. We are anxious to hear the result of the Spring election. shall know by the next steamer. which will be plenty soon enough if <u>Seymour is elected Governor</u>. No active operations to chronicle for the past week. and a general dearth of news of any kind

With love
Ever Your Own
Ned.

No. 157 Rec'd April 23rd 1864.

**Headquaters, Department of the South,
Hilton Head, S. C.,** April 16." **1864.**

Dear Sara.

In an awful hurry as usual. and have only time to apologize for not writing a decent letter. We are about to move again. and this time to that vast slaughter pen for soldiers. Virginia. I think one is excusible for not feeling highly gratified at the change. In the South and South-west operations are commenced, and prosecuted with energy, vigor, and a determination to succeed. but in Virginia, it does not seem to be so. Judging from the past, one can hardly be expected to entertain very confident hopes of the success of any aggressive movements upon its blood-stained fields. No matter, its the fortune of war. Our regiment has already gone. left yesterday morning on the "<u>S. R. Spaulding</u>" the boat which brought them up from Jacksonville. They did not land. though they remained in the harbor about twenty-four hours. Tom came on shore and I saw him a few moments. left him in the office while I went out to bid good by to George, and when I came back he was gone. They took all from the Hospital that were able to go. though they were unfit for duty. George will not be fit for duty in a long while. yet I should judge. he cannot march at any rate. if they want him to fight they will have to carry him into action. Our Regimental Band came on shore and played several tunes in front of General Gillmores quarters. I heard them, but was to busy to go out and see them, afterwards I saw, and heard them, as they were playing in front of the Port Royal House. They make excellent music, and present a fine appearance. They are a credit to the regiment. The officers commanding our regiment make strenuous efforts to get us boys at Headquarters relieved from duty and returned to the regiment. for which I can assure you I do not thank them. I have more than half expected that I should not be wanted and should therefore be returned to the Regt. If such had been the case I should have had nothing to say. It would be perfectly right. and what I should expect. But it does vex me to be hunted down like a runaway negro and I earnestly hope that our officers may not be gratified by succeeding in their persistent attempts to get us back. If my services are not required here. I am ready to do my duty to the best of my ability in the ranks. It's being doged after in such a manner that I object to. Well the regiment has gone. and I am still in the office where I hope to remain so long as myself and my work give satisfaction. I expect the <u>Headquarters of the 10" Army Corps</u> will go North within a week or ten days. though I know nothing definate about it. This movement upsets my fine prospects of passing a pleasant summer here. I will not deny that I thought I should have a compareitavely easy time of it for the remainder of my term of service. but I now expect hard work, and rough fare, but if I am to remain in the office. I shall be much better off than if in the ranks. I received a letter from you by the last mail. also a paper which I think is from Sara B. When shall I ever get an opportunity to write to her! I do not feel at liberty to sit down and scribble her a few lines. I feel that I owe her to much in the way of a letter. to admit of it. and so I begin to fear I shall not make out to write at all. There is to be an execution here tomorrow. Three men are to be shot for desertion. It seems hard at first thought. that men should die for attempting to escape from service that is so disagreeable as soldiering. But it is not so. If the penalty of death was not at times rigorously executed. there would be no such thing as keeping our army together. In the present cases, there is not a single mitigating circumstance, in the favor of the prisoners. that I know of. They are substitutes. who have received large bounties to do the work from which they have tried to escape. They have made <u>three</u> desperate. and well nigh effectual. efforts to escape to the enemy. and now they must suffer the conseuences of their bad faith. and treachery. We are all hard at work now. closing up all our work and getting the office affairs in condition to turn over to our successors. and also arranging and packing things for our own removal. There will be but little peace. and less leisure, for the next two or three weeks, and I guess I might say for the remainder of my time. I hardly know where to have you direct to me, and think perhaps you had better not write at all until you hear from me again. It is late, and I very tired so please excuse.

Believe me, my darling. ever your
Own, Ned

No. 158   Rec'd May 3rd 1864

HEADQUARTERS DEPARTMENT OF THE SOUTH,
Asst. Adjutant General's Office,
Hilton Head, S. C., April 25" 1864.

Dear Sara.

I commenced a letter to you yesterday. and had nearly filled one of these sheets. but as I could not write to suit me. tore it up. and concluded not to write until I felt in better humor. The fact is, I am suffering from the effects of a severe cold. which has troubled me for the past week. Yesterday being a very stormy day. I felt more than usually uncomfortable. and out of sorts. This morning is bright and beautiful, and I am some better. though my head still feels very badly. We are about ready to start. and as a couple of new clerks have come I manage to turn over the most of my work to them. So I think I shall succeed in finishing a letter to you during the day. The time for our departure was some days ago. fixed for tomorrow. and I have not heard for certain that we shall not start then, but things hardly look like it to me now. but we shall undoubtedly go within three. or four. days. I have not yet fully arranged my personal effects for a "tramp." but have them so that I can do so at a short notice. We have got to bid good by to the many little comforts. and conveniences. with which we have gradually surrounded ourselves. during our long period of quiet, and inactivity. Its almost as much of a change. as from civil life to soldiering. We can take nothing with us but what we can pack in our knapsacks. and carry on our backs. Dexter is going to send home a box of extra clothing and blankets. and I shall put in a few things of my own which I wish to save. they can remain at Mr. D's until I return. I have put a number of your letters which I have saved. in a small cigar box. together with a few other little trifles. sealed it up and directed it to you. shall put it in with the other things. and have Newt ask his folks to deliver it to you. I fear you will not like to be bothered with so much of my old truck. and I hesitated long about presuming so far upon your kindness. as to send this lot to you. I should not have done so. but for the letters. which I do not wish to destroy and which it will be very inconvenient for me to keep with me. There are a very few shells, which came from Folly Island. those are for you, I had just commenced picking up. and selecting a lot to send to you. as we were ordered away. Like all the rest of my good resolutions. and fine projects. the idea came too late. The light colored pen holder. was made from a piece of the old house in St. Augustine, the oldest in America. It was made. and given me by Pitt. and being the only keepsake I have of him. is of no light value to me. The dark colored pen holder is made from a piece of the old Spanish "Treasury Chest" which in olden time when Spain held dominion over Florida. was used to hold the State Treasure. George whittled it into its present shape if I live to get home I intend to have it neatly turned, mounted with a gold pen holder. and present it to my lady love. I have taken quite a fancy to collect all the signatures I can of officers of note. You will find quite a lot of them. I intend to turn them up. and fix them in a book. Some one has relieved me of quite a number. that I once had. and I was much disappointed upon looking over my collection to find them missing. You will find my diary. such as it is. for about a year and a half I think. It was kept for you. I am sorry that it is not in a more readable shape. and doubt very much if you find yourself repaid for the trouble of deciphering it. You will see that it was dropped at the time of my coming into the office. just at the time when it should have been kept up. and more pains taken with it. I have valid excuse for not keeping it up. but perhaps you will not regret it. after reading what I send you. I had some photographs taken the other day and enclose you one. I believe it is a pretty correct picture of your humble servant. but its a miserable poor piece of work. We soldiers cannot get good work of the kind done because the artists will not take any pains for us. They will take only one negative and we are obliged to take just what pictures they happen to print from it. be they good. bad. or indifferent. there is no redress as they require pay in advance. I certainly should never have submitted to such usage if I had known how it was to be beforehand. It already begins to look dull around here. The last of the 10th Corps will leave by tomorrow. So many troops leaving will make a deal of differance in the business of the place. The wants of the colored troops who to a great extent will take the place of the troops leaving, will not be so great, nor so varied, and I think that "Robber's Row" will decline somewhat from its present flourishing condition. I will write to you again soon as practicable, with love,

Ever Your Own
Ned

P.S. Would you have any objections to my using printed envelopes directed to you. I can have them very neatly printed. as well as not. but am not perfectly sure that you would like me to use them in our correspondence. Am going to have some printed to use in writing to my friends. Capt. Sealy uses them in addressing letters to his wife. love from

Ned

P.S. No. 2. Feeling rather desperate. am going to smoke a cigar. and. having a number of N.Y. papers just received. shall then read myself to sleep.

Ned

No. 159 Rec'd May 14th 1864
Va.

Head Quarters 10th Army Corps
In the Field, May 10" 1864

My Own Dear Sara,

I must write you a few lines tonight at any rate. for I am real hungry for a letter from you. and I remember that I told you you had better not write until I sent you word where to direct. Well here we are in Virginia. within about fifteen miles of Richmond. and trying hard to get nearer. We landed on the night of the 5th at what is called "Bermuda Hundred" a short distance above City Point. the place where the flag of truce boat has met the rebels for the past year. It is 75 miles above Fort Monroe. and six or eight above Harrisons' Landing of Mc.Clellan notariety. The morning of the 6th the advance was commenced. at present, Corps Hd Qtrs is established at a small plantation owned by one. Hatchen, or Hatchel. about 5 miles from the landing place. The troops have penetrated from 8 to 10 miles further in crossing the Petersburgh and Richmond Rail Road which they have destroyed for some distance. There has been considerable fighting with as near as I can judge rather indifferent success. The loss in killed and wounded is not large. but the men suffer terribly from the heat. and also for want of food but the latter want is now provided for. and in fact there has been no negglagence. or needless delay in supplying it. but always in such moves we are necessarially short of rations for the first day or two. The heat is very oppressive. no cool sea breeze tempers it. There has been no rain for several days. and the clouds of dust raised by every passing team are almost stifleing. **Evening**. The troops are returning from their days work. they seem to be in good spirits. and cheering lustily. The bands are playing, teamsters yelling, wagons rattleing and there is a noisy time generally. The Rail Road above mentioned has been destroyed for several miles. an attempt of the rebels to get in our rear by flooding Gen. Terry's Division on its left was handsomely repulsed. The 4" N.J. Light Battery lost two of its pieces for a short time. but re-captured them. their loss in horses was severe. thirty-five I think. Our forces have retired behind their entrenchments bringing off all our own. and rebel wounded. We were masters of the field. but have retired for some good reason. Our Regt. has been engaged today. I can learn nothing. with regard to its loss. but am told that it was very slight. I saw all the boys the day we marched up. George was with them. well. and hearty. I was with them but a few moments. they are some distance in advance of our Hd. Qtrs. and I do not expect to see them often. My eyes trouble me so that I can scarcely see to write. and I shall have to make my letter short. Dexter and myself have been detailed for duty at Corps Hd Qtrs. There are now five clerks here. I have material on hand for a long letter. but fear you will not get it very soon. We had a pleasant voyage up from Hilton Head. were but two days making it. were on the Arago. had plenty of room. and fine weather. The sail up the James river on the 5th was delightful. I was busy most of the time coming up. but found time to keep my eyes open to the beauties of the scene. I hope I may some day be permitted to sit down quietly and tell you all about it. We have good news from Grant. and hope he may continue to be successful. You may direct to me as "Clerk at Head Quarters 10th Army Corps." Fortress Monroe. Va. Please write soon as convenient. Pardon my delay in writing. Be assured of my constant love. and believe me ever. am truly your own.

Ned.

CIVIL WAR LETTERS 1861-1865

Editor's Note: The following letter refers to the Battle of Drewery's Bluff

No. 160 Rec'd May 20th 1864.
Va.

**Headquarters.10TH ARMY CORPS,
In the Field, near "Hatcher's" May 16", 1864**

Dear Sara,

Your letter of the 9th was received three days since. and one of a previous date. just after closing my last letter to you. I cannot tell you how welcome they were. it had begun to seem an age since I had heard from you. I ought to have written to you before as I have had but little to do for several days. but have waited hoping to be able to write good news of our successes in this quarter. but though there has been some fighting and we have gained some advantage. we have not yet met with complete success. Last Friday and Saturday, were hard days for our boys. they were fighting nearly all the time. and succeeded in driving the enemy from a line of intrenchments and rifle pits over a mile in length. which formed a part of the defenses of Fort Darling. or rather commanded its approaches in the rear. Yesterday (Sunday) there was but little if any fighting. We held a position so near as to be under the fire of the Fort. It is extremely difficult to obtain any definate information with regard to the situation of our forces. and their prospects of success. The little we learn is obtained from the couriers who carry dispatches. and each one tells a differant story. This morning there has been a severe fight. commencing I should judge about daylight and lasting about three hours. When I awoke there was a heavy cannonading going on. and soon the sharp rattle of musketry was added to its roar. The morning was clear and bright. and though the battle ground must be from five to seven miles distant, it seemed not over two. At times the sharp hissing sound made by the shells. was plainly heard. General Gillmore and all his Staff are. and have been for the last four days, up to the front. Nothing certain is known as to the result of the fight this morning. all has been quiet for the last two hours, with the exception of the occasional boom of a single cannon.

Evening. Today has been a sad day for us. The rebels largely reinforced. fell upon us at daylight, in such overwhelming numbers. as to drive us from the hard earned vantage ground we had taken from them on Friday and Saturday. Their attempts to crush us out. by hurling upon us a vastly superior force, has failed. Desperately contesting every foot of ground we have retired from the bloody field. bringing off our wounded preventing the broken. and shattered fragments of our regiments from falling into the enemies hands. and not allowing the defeat to become a rout. We are this evening behind our intrenchments, and are doubtless secure. I have no doubt of our ability to hold our present position. The rebels took advantage of a dense fog this morning to approach within easy range of our forces. and daylight found them strongly posted. with artillery all in position all along our front. They at once charged and as I learn were twice repulsed with great slaughter. but some regiments on our flank gave way. the rebels opened a cross fire upon us. and flesh and blood could stand it no longer. We were obliged to retreat. The <u>7th Regt.</u> was in the center. or nearly so. as I learn. they held their ground as long as it was possible for men to do so until. their ammunition expended. and overwhelmed by superior numbers they were obliged to retreat. They have suffered severly. but it is utterly impossible to form any estimate of their loss. Many are probably taken prisoners. Our little party has been called upon to furnish another sufferer. Reid has lost his right hand. It was taken off by a round shot as he was coming off the field. Amputation was performed just above the wrist. I saw him this afternoon. soon after his arrival in camp. He is in good spirits and I think his chance for recovery very favorable. He is to remain in camp tonight, where we did the little that we could to make him comfortable. He could have been sent to the landing today, but would have been obliged to ride in our army wagon. and we thought it better to wait until tomorrow, when a more comfortable conveyance can be most likely obtained, in an ambulance!

    The rains of the past few days have rendered the roads almost impassable. and the continual passing of heavy teams, has reduced them to a continuous line of mud in which the wheels sink clear to their hubs. George stood it very well. much better than I thought it possible for him to do. but the exertion of the first two days was too much for him, and he was obliged to remain in camp when the Regt. went out three days since. so that he was not in the engagement this morning. Yesterday, orders wer sent to the camp. for every man, except some wounded. to rejoin the Regt. at the front. accordingly this morning they started. but instead of joining the Regt. I learn they were ordered to remain in the intrenchments. so I think there is no doubt but that he is safe and unhurt yet. McNeil is with George, Brinton. and Olin were unhurt when last seen by the wounded boys who have returned. but that is no guarantee of their safety. I am anxious to hear from them. I saw Wells. and Norton. both of whom are on duty in the Hospital Dept. and are well, though of course much fatigued. and well used up. The camps of two Brigades are near our Hd Quarters here. The remnants of the regiments have come in this evening. and are anything but dispirited. judging by their actions. they have been cheering like mad. There can be but little doubt I think but what the rebel loss in killed and wounded. equals. if not exceeds our own. They have doubtless taken many more prisoners than we. though we have taken a good number. I cannot stop to write you more tonight. as I must write a line to Dr. Reid. Tell Julia that I left George in good spirits and feeling very well, last night. I have no fears of safty now. and she will probably hear from him soon. Will write again soon.

Love from
Ned

P.S.

Have not time to look over. and correct my letter. please excuse errors.

<u>Lt. Charles Wood</u> was killed in action at the Battle of Drewry's Bluff on May 14, 1864

No. 161 Rec'd June 4th 1864
Va.

HEADQUARTERS.10TH ARMY CORPS,
<u>In the Field</u> near Hatchers Va May 30" **1864**

It is late, and I am very weary and sleepy, but it has now been two weeks since I have written you a word, and I can delay no longer. though it be but a few lines in the hope of an apology for a letter. that I now send you. I have received several letters from you, some of which had been sent to Hilton Head. and were of course rather out of date when they reached me. Yours of the 22nd was received about the 25th. I have not the courage to look at the mail bag when it is brought in. for I know that I have forfeited all right to expect anything from its contents. and the sight of it brings to mind unpleasant recollections of unanswered letters, and duties that I owe to correspondants. The two past weeks have been extremely busy ones. We have all worked hard. and been much broken of our rest. Never getting through our work, Seldom quitting it until eleven or twelve o'clock. and sometimes working till one. o'c. and then perhaps being turned out before morning. either to do some writing or else by some attack of the enemy which would bring us all but for an hour or two. Up in the morning again at sunrise. and at work, never finding an idle moment that could be improved in writing to ones friends. I do not speak complainingly. for I well know that the brave boys in the ranks are having it twice as hard besides being constantly exposed to danger. and necessarialy very indifferantly fed. but mention it to show that I have some reason for my long delay in writing. For several days past there has been no fighting, seldom a gun fired, it was becoming quite monotonous, the rebels sought to relieve it this afternoon by the exchange of a few shots. A lively fire was kept up for about an hour. mostly by the artillery. I think there was but little gained or lost by either side. About the time our little affair was over. we began to hear the heavy reports of artillery, miles away to the northward, which we suppose came from Grant's forces. who are approaching in that direction. A large portion of the force from this Dept. has been sent to Grant, we suppose. We are sufficiently strong to resist any force the rebels are likely to bring against us. Secure behind our intrenchments we can defy five times our number. let them attack us as they please. Our Corps Hd Qrs has not been moved since we first pitched our tents. We may remain for two. or three weeks perhaps. and are quite as likely to pack up and start tomorrow. such is the uncertainty of everything in the army. If I could only have the privilage of writing to you daily. I could probably manage to write tolerably interesting letters by giving an account of the doings of the day. but as it is. there is no such thing as beginning at the commencement and telling the whole story. In fact I have been so hurried that I have had no time to note down half the items of interest. either in my memory. or in my diary.

I have not heard from Will Reid since he left. should be glad to. but have but little fear for his safety. as he was doing well every way when he left. Think Mr. Reid did perfectly right in going at once to Fortress Monroe to see him. I wanted to advise him to do so when I wrote. but did not feel at liberty to do so. It is a great misfortune to be so crippled. but if Will recovers he may be thankfull that he has escaped thus. Capt. Sealy has just sent in for us to go to bed. I suppose he does not like to have us around looking to sleepy. during the day. It is near twelve. now. Good night. believe me. your own loving, Ned.

Will write again the first opportunity.

Ned.

CHAPTER 12

AT BERMUDA HUNDRED

From the book:

Up to half an hour after sunrise on the 2nd there was perfect quiet along our entire front, although towards midnight a furious cannonade had been kept up for an hour over our heads by the rebel batteries. At that time an attack was commenced along our line by the enemy's pickets advancing from their posts as a line of skirmishers, strongly re-enforced, and two or three feet apart. In the woods on the left this attack was extremely rapid and sudden. A few steps placed the enemy in our pits, in a position, which, favored by the direction of part of the line, enabled them to cut off and capture a large part of Company B.   Such part of Captain Mills' command as was not captured, with the exception stated below, fell back slowly, contesting the ground, to a position nearer the works, which they held until later in the morning they were re-enforced and reoccupied and held their first position. In the open field the advance of the enemy began a few moments after firing and had been heard on the left. The enemy moved toward us in good line, but slowly and hesitatingly. I opened fire along the whole line, and in two minutes they had all dropped to the ground, where they lay firing from such cover as they could get for a few minutes longer, when the entire line rose and ran to the shelter of their rifle-pits at full speed, followed by our cheers and bullets. From this cover they never ventured again, contenting themselves with a dropping fire from it until we abandoned nearly our entire line. On the right the movement of the enemy was by a dash across that part of the line which ran along the edge of the woods, nearly at right angles with the general direction of the line. This movement, of which at the time I had no information, cut off nearly the whole of the two companies posted there, together with the major commanding the regiment. Word had already been passed to me repeatedly along the line that "our left was turned," "was cut off," "had fallen back," and at last that the enemy were occupying our rifle-pits on the left. I refused to believe these statements, having great confidence in the strength of that position until I saw our skirmishers falling back across the open held toward the works; but I passed the word to Major Sanford on the right. No communication had yet reached me from that officer. I had seen a body of thirty or forty rebels dash from their pits into the woods in a direction that placed them in the rear of Companies C and H. Groups of our men now began to be led to the rear of the rebel lines under guard; straggling skirmishers were seen falling back toward the works on our right; the enemy's fire began to enfilade our lines from the woods on our right; the position was critical. I sent at last the question to the right, "Where is Major Sanford?" The answer came back "He is cut off." Up to this time my duty had been simply to hold my position and await orders. It now became necessary to act. The choice was plainly between capture and a perilous retreat across the open fields to our works. I therefore gave the order, with a reluctance which I never felt before in performing a military duty, to fall back. This order was executed after almost all the rest of the division line of picket had given way, under a severe fire from the enemy, and across an unprotected field, but with little loss; and painful as it always is to order a retreat, I had the satisfaction of knowing that the order saved a hundred men and rifles to the service, and of receiving the unqualified approval of my brigade, division and corps commanders, for the course adopted. On the extreme left a part of Company E, in a favorable position, did not leave their ground at all. The men who fell back to the works were reorganized and at once pushed forward to the picket line for the most part on the left. The entire line, except at the former post of Companies C and H, was gradually re-established and held.

  The conduct of officers and men throughout the affair was admirable, but I may be permitted to speak especially of the extraordinary coolness and <u>courage of Capt. Charles C. Mills</u>, of Company G, who received early in the fight a wound which it is greatly feared may be mortal.

No. 162 Rec'd June 7th 1864.
Va.

HEADQUARTERS. 10TH ARMY CORPS,
In the Field near Hatchers Va. June 2nd. 1864.

I steal a few moments in which to drop you a line. but must be brief as usual lately. Our regiment has been on picket for the last twenty-four hours. have not yet returned. This morning the rebel(s) made an attack on our picket line where our regiment was posted. and succeeded in driving them out of their rifle pits with much loss. Have not been able to learn the amount of our loss. as the Regt. is still at the front and will not be in until late this evening. Have just seen Tom, who was on his way to the landing after the mail. So far as I can learn George. Olin. &. McNeil are all right. Wells it is probable is taken prisoner along with the other field Hospital attendants, <u>Capt Mills of our Co. is mortally wounded</u>. shot through the right breast. cannot live, Sergeant Hawthorn is wounded. in the side. it is thought not seriously. one Lieut was killed. our Major Sanford, and Cap. Dennis taken prisoners. The regiment suffered severely. I heard the Asst. Adjt. Genl. say that "we (meaning our forces) had lost about two hundred men. and a number of officers." As I do not know, or hear that any other regiment was engaged. I suppose the loss must have all been ours, I believe the enemy are not in possession of the rifle pits from which they drove us. I do not think I ought to have written at all, until I could have done so with more certainty, but am so much hurried lately that I must write when I can. Norton will write in the morning. and will then be able to state correctly our loss and will know with regard to the boys. Our regiment is encamped under fire of the enemy. and whenever they open on us, are obliged to hasten to the entrenchments for cover. Many of the little shelter tents are torn by the pieces of shell, and bullets which have passed through them where they stand. Three days since. a sergeant of our Co was mortally wounded by a fragment of shell, while in camp. he died the next day. Two days since our Second Lieutenant. and a private of our Co were killed by a shell while going to the entrenchments. I have not been up to where the Regt is camped yet. should have done so tonight, but they will not be in until it will be later than I can be so far from my quarters. Wells called on us yesterday afternoon. I did not suppose he was going to visit Richmond so. soon. The regiment must be pretty nearly used up. Our company, if there are enough left to be so called. are now without any commissioned officers. We had no 1st Lieut. our Second Lieut. & Capt. are now both gone. but Mills was a good officer in a fight. brave. cool. and self possessed. He was married about the time he took command of our Co. and has had his wife with him most of the time until we left the Dept of the South. Probably she will never see him move. There has been a good deal of artillery fireing for the past two or three days. Not a steady fire, but a furious and rapid one for a short time, with hours of quiet, intervening. The enemy have acted as though they thought we were withdrawing our forces and are probably suspicious that the whole of this command would be taken away. to reinforce Grant. We are very strongly intrenched. and can hold our works against twice or three times our number. There is no probability of our being driven away from here. unless. Lee should succeed in defeating Grant. and then suddenly throw a large force upon. and overwhelm us. We shall get N. Y. Papers of the 1st to-night. I hope we have good news. We have been so roughly used by the rebels. of late. that I have lost all confidence. and await the news with much anxiety. While writing I can hear the dull, sullen roar of distant guns. which must come from where Grant is fighting. We have heard the same. at intervals for the last three days. You must excuse me if I am rather down hearted this afternoon. Our recent reverses do not give to affairs a very promising look, and I do not see how any one who has any interest in the speedy. and successful termination of the war. can feel very cheerful, at the present prospects. I think the present time, the most critical of any since the war commenced. Everything seems to <u>hang upon Grant's success</u>. I have much confidence in his ability, and patriotism, but he has fearful, and tremendous odds to contend with. Under existing circumstances, he should have three men to the rebels one. to be even with them. The rebels are fighting with an energy and determination. which. had it been shown by our own armies, would long ago have swept the last rebel out of the Confedarcy. **Evening**. I ran away after writing the above. and have been out to where our boys are stationed, getting thoroughly wet through as it has been raining hard. but that does not matter. as it is quite warm. George is all right. he was not out on picket. as he was a little unwell, Hawthorn is the only one of "our" boys hurt. The regiment had not all come in. and there is yet no return of the casualties. There seems to be no doubt but what <u>Wells is a prisoner.</u> I think most of if not all, that we cannot account for, are prisoners. During the firing this morning. a solid shot, and a bullet passed through the tent of Brinton. The boys all have holes dug just back of their tents into which they dive like rats. when the rebels open fire. that is when they do not have to go to the intrenchments. Cannot write more now. Please excuse haste. Pardon delays in writing, and believe me ever your own

Ned.

No. 163 Rec'd June 9th 1864.
Va.

HEADQUARTERS 10TH ARMY CORPS,
In the Field, near Hatchers Va June 5. **1864**

Dear Sara,

It is a rany Sunday morning. with every prospect of being an unpleasant day. By some unaccountable mistake I got up much earlier than common, and while waiting for breakfast, thought it a good time to commence a letter. The past two days have been very quiet ones with our forces here. not over half a dozen shots have been fired by either side. In the office we have been busy as usual. On Friday, heavy firing was heard at intervals during the day, from across the river in the direction of Grant. Just at night it became more and more distinct, and for about two hours there was almost one continuous roar of cannon. I never heard anything like it. It must have been a terrible fight. We are probably twenty-five miles from the scene of conflict, and the reports resembled volleys of musketry heard at a distance of two or three miles. Saturday morning a few shots were heard and then it was still for the remainder of the day. To-day we have heard nothing from that direction. I thought surely we should hear something from Grant to-day. but no news has reached us, not even a plausible rumor. I await the news with much anxiety. for I surely think that if our army had gained a victory we should have heard something of it by tonight. The perfect silence of the last two days. after such severe fighting. adds to my fears. As I have before told you, I am by no means confident of success. for I can realize something of the tremendious odds Grant has to encounter. I hope for the best, and believe that the present commander of our army will do all in his power to ensure the success of our arms.

Evening. The mail has come. but brought me no letter. I thought surely I should get one to-night and feel a little disappointed though I know I have no right to expect one. I could bear my punishment tolerably well, were it not that I fear you may be sick and unable to write. Have seen George to-day, he is at work in the Q.M. Department, and is feeling pretty well. About an hour before sunset, the rebels opened fire upon our entrenchments. and shelled away vigorously for about half an hour, have not heard what it amounted to, or whether there were any casualties. It was very pleasant at the time. the clouds having cleared away. the sun shining, bright. and clear, the air still. and I was just thinking how like it was. to many Sabbeth evenings. I have seen in New England. We are but a little out of range of the enemy guns. say three-fourths of a mile perhaps. safe enough however. It is said that the rebels are mounting a two hundred pounder on one of their batteries. to annoy our fleet in the river. if they do. they can easily throw shells to where we are. But we anticipate no trouble from that source. George tells me that Col. Hawley is trying to get all the men of the regiment. now detailed on detached service. back into the ranks. which are very much thinned out by our late losses. Do not know whether he will try to get us from Hd Qrs again or not. Should think he would be satisfied with the efforts he made to attain the same end at Hilton Head before we left. But he may think he has a better excuse now, and try it again. We are still very busy in the office, and get but very little time to sleep. We have no tents and have to sleep in the one used for the office. The officers are always around more or less until late at night. and some of them are sure to be prowling around after something early in the morning. One morning we had the honor of being awakened by Gen. Gillmore in person who wanted a clerk to do some writing for him. You can be assured we did not stop to take another nap after his summons. I do not think that Hawthorn of our Co was seriously wounded the other day. He has been taken to Fort Monroe. and I do not expect to hear from him except by the way of home. The last I heard of Cap. Mills he was alive. A bullet hole through the lung is not always fatal, but is exceedingly dangerous. I hope the Captain may recover. He will stand as good a chance as almost any one, being in good health, and with a fine constitution, uninjured and unimpaired. The whole loss of our regiment in the last affair was ninty. of whom seventy-five are reported missing. three or four killed. and the remainder are wounded. I think that Wells has suffered no further harm than to be taken prisoner, which is bad enough of course. You mention in one of your last letters, that Sara B. asked you for my address. I wrote to her before we left the Head. and am sure enclosed my address. Have been more than half expecting to hear from her lately. Please ask her if she ever received it.

 I had a fine lot of cherries tonight. all that I could eat. with the fun of sitting up in the tree and picking. and eating at my leisure; presume they are hardly ripe yet up home.

 Must close now. Hope to hear from you soon, and how you are progressing with the school. With much love I remain,

 Ever Your Own
 Ned

No. 164 June 13th 1864
Va.

HEADQUARTERS 10TH ARMY CORPS
In the Field, near Hatchers Va June 7th. **1864.**

Your letter of May 29th was received last evening, and that of the 3d insts. tonight. I trust you have found out ere this. the reason you were so long without hearing from me. and are satisfied therewith. Doubtless you will be inclined to think that if I had time to visit George "frequently." I might also have found time to write. But you must please bear in mind that George was nearly all the time exposed to imminent danger. that I was therefore very anxious to be assured of his safety. at the earliest possible moment. therefore I would run away from the office long enough to pay him a "flying visit." I had begun to feel anxious lest something was amiss with you. that I did not hear from you sooner, and am happy to learn by your last two letters that you are well. Have not been quite so busy in the office today. We are getting the upper hands of the work a little. I have got an awful cold tonight which is anything but pleasant. All remains quiet along our lines here. There has been no demonstrations by either side. since last week when the rebels gobbled up our regiment. Today we have heard some firing from across the river in the direction of Grant. We have also learned the cause of the heavy firing heard in the same direction last Friday night. In the papers the credit of the fighting was given to the <u>18th Corps</u>, in reality it was done by two divisions of the <u>10th Army Corps now serving under Gen. Smith</u>. Several well known officers who never belonged to the 18th Corps were lost. It is very plain that no credit is to be given to the tenth Corps if the old rascal who now commands this Department can possibly avoid it. Our men gained all the advantage that we ever obtained here. and then when it was lost through the incompetency of Butler and his clique. they have, as far as it lay in their power, thrown the odium of the defeat upon us. <u>Ben Butler</u> is the most unmitigated humbug in the military line I know of. He is a good mate for Seymour. and one ought to be a private in some company of stay-at-home rangers in which the other was corporal. He will do very well as military governor of some captured province, but is utterly unfit to command in the field. Some of the biggest lies it was ever my fortune to hear have been told in regard to his operations, and successes, in this campaign. I will defy rebeldom itself to beat them. General Gillmore has more military talent in his little finger, than ever existed in the whole Butler family. With the single exception of Seymore, there is no man holding a commission in our army for whom I feel a greater contempt than I do for him. I have yet to learn of one single success that we have ever obtained that we owe to his generalship. From his first entrance in the military arena at Big Bethel. down to the present time. his operations in the field have been one continual series of disastrous blunders.

Our present staffing place is at the house of a farmer by the name of Hatcher, as bitter a rebel as ever went unhung. On our approach, he sent off his cattle, horses, and most valuable negroes. retaining only four old dilapidated specimens of the latter kind of property. For certain reasons he was unable to move his family, consisting of his wife and two little children. and so he remained with them. The first two days of our stay here, he was permitted to go about at his pleasure, but on the second day he absented himself for the greater part of the day. and when questioned as to his wheareabouts failed to give a very satisfactory answer. Since then he has been kept under guard, a soldier attending him with a drawn sword. or loaded gun, wherever he goes. He owns a large farm, mostly of excellent land I should judge. and is doubtless wealthy. He had most of his crops planted when we came here. and they were just springing up. Our troops camped in his corn. and wheat fields and in a few days, scarcely a vestage of their recent cultivation remained. There is scarcely a rail left on the place. all used up for fire wood. Hatcher claims that he has been damaged to the amount of twelve thousand Dollars. and had the impudence to ask Gen. Gillmore to pay him. the General referred him to <u>Lt. Col. Morgan, Chief Commissary of Subsistence</u>. who told him that when the war was ended, if he could prove to the proper authorities that he was at this time a <u>loyal citizen of the United States,</u> he could doubtless get his pay, as he refuses to take the oath of allegiance, he will have some difficulty in establishing that fact. The farm house is a small one story-and-an-attic affair. with a few little out buildings and negro houses standing near. The tents of Gen. Gillmore and his staff are pitched around. and within a few feet of the house. under the shade of the few trees that surround it. Hatcher was about building a new house for which he had the lumber all out. and in piles near by. Not even a shingle of it now remains. it has all been used for government purposes. Many houses have been burned by the straggleing soldiers. The fires that swept through the woods the few first days of our being here burned up nearly all the fences. the passage back an forth of troops and heavy teams and artillery. have beaten the ground as hard as a pavement. The movement in this direction I think has thus far failed of accomplishing any decided advantage. I cannot see as we are doing any good here now. We have a strong line of intrenchments. which we can hold against an assault by three times our number. but what is the use of holding the position? I cannot see any. at least not enough to justify the employment of so many men who might be doing more good some where else. We have not a sufficient force to advance with. and at the most I dont suppose there is one half our number of rebels in our front. but they are strongly intrenched as well as we. and we can not drive them away. or make any diversion in favor of Grant unless we could be largely re-enforced, which it is not likely we shall be. I do not think it likely that we shall be permitted to remain long in this state of comparative inactivity, but cannot guess what the next move will be. Give my love to Lizzie and say to her that I believe her assertion that she still cares for me, and that it gives me great pleasure. I congratulate her upon upon being so fortunate as to have a gentleman visitor, if as you say they are scarce, and of unfrequent occurrance. I suppose in that case they will be apt very shortly to become very <u>dear.</u> Please ask Sara B. if she received the letter I wrote her from Hilton Head. Not that I would urge her for a reply until it was convenient for her. as I am much in her debt on the score of correspondence. but if she has not received it. I wish to write to her with as little delay as possible. I have but very few friends like Sara. and cannot afford to lose them by neglect on my part. So you are both getting sick of school teaching! I do not blame you. I never fancied the occupation and wonder how any one can endure it. except perhaps as a penance for some great guilt. I learn that Will Reid has arrived at home. am glad of it. he has had better luck than I anticipated. now if he keeps temperate he will soon get well. He may have great reason to be thankful that he has escaped ever thus. We have but three months more to serve. but if I mistake not. there is more of danger in those three months than there has been in all the rest of our long three years. There is no certainty of any of us seeing home. until our feet are on the threshold. In our last fights here, I have known of men who were killed in action on the very day that their term of service expired. We can only do our duty and hope for the best. Mr. Landon and <u>Sam Wolcott</u> still remain at the Head nor do I hear of there being any prospect of their joining us. If they are wise they will remain where they are as long as possible, if it is until next Fall. Do not expect me home until you hear of me alive and well on the 7th of September next. then it will be reasonable to do so. In the mean time believe me with love and affection.

Ever Your Own
Ned

No. 165 Rec'd June 15th/64
Va.

HEADQUARTERS 10TH ARMY CORPS
In the Field, near Hatchers June 11th. **1864.**

Dear Sara,

Your letter of the 5th was received last evening. I think my last letter to you was forwarded without the usual address at its commencement. if so please excuse. I am in the habit of leaving the address until the last thing, as I am frequently interrupted, while writing, and obliged to lay aside my letter for hours, and in the mean time some one might get a peep at it. All remains quiet here. On the 9th. a demonstration was made on Petersburg. but the place was found to be strongly fortified. and the rebels getting notice of our approach. reenforced so rapidly. that an attack on the place with our small force was not deemed advisable. and we withdrew without giving battle. Col. Hawley's Brigade formed a part of our force. and our regiment was along. I have not seen George since writing last; he is not at work at the place where I last saw him. and I have not been out to the regiment for some time. Col. Hawley was in the office yesterday. I saw him looking towards me. as though he would like to drive me back to my company. I have no acquaintance with him; never spoke to him but once. and that in reply to a question when I was in the hospital last summer. News from Grant is not what one could wish. but on the whole as good as could be expected. I see that Mr. Lincoln is re-nominated for the presidency. am glad of it. but do hope that the matter may now be allowed to rest until the proper time for action arrives. and that all the energies of the country may be directed to the prosecution of the war. Have not had so much to do for the last few days; have got along quite easily. Found time to write to Mrs. Randall a day or two since. Day before yesterday in company with one of the clerks I took a walk in the afternoon over to the James river, about a mile distant at the place where we went, but in another, it runs within half that distance of us. Our visit was to the lookout station which is in the top of a tall pine, one hundred and twenty feet high. A platform four or five feet square is firmly secured to the stem and branches, and is boarded up. around to the height of between two and three feet. a slight frame work upholds a roof of green boughs to protect the look out from the sun. A bench formed a comfortable seat. and altogether it is a cosey place to sit and while away the hours of a pleasant day; but the day we visited it, the wind was blowing very fresh. and there was rather too much motion. especially when one thought of the possibility of a tumble. from the lofty perch. Access is had to it by means of a ladder fastened aganst the tree. Not being partial to lofty heights when there is any room to doubt their stability. I was more than half inclined not to go up. but not liking to back out. I did. and was well paid for doing so. and there really was no danger about it. We had a fine view of the country for many miles around. Two church spires in Richmond could be seen over the intervening trees. distant in a straight line. I suppose about sixteen or twenty miles. <u>Chaffin's Bluff</u> with the rebel works there, commanding the river for four or five miles. lay before us. and by the aid of the glasses we could get a very good idea of them. and see the rebel flag flying from the commandants tent. A rebel tug boat with a flag of truce flying came down the river to within about half a mile of our fleet. when she was warned to stop by firing a blank cartridge from our foremost vessel. The tug then came to anchor and a small boat put off from the vessel that fired the shot. to communicate with her. I should think that about a couple of hours passed before her mission was finished. and she started on her return. I have not learned what the business was that brought her down. While the conferance was going on under the flag of truce. the rebel fleet came down this side of Chaffin's Bluff. and commenced firing. fired in all perhaps twenty-five or thirty shots. We were much puzzled to know what it meant. as being in their limits there was no enemy for them to fire at. At first we could only see their flags. the banks of the river lined with trees. hiding them from view. but after a while they came down past a bend in the river. and through the glasses. we got a good view of them. I saw four boats, there were more but just then a staff officer came up to reconnoiter and we were obliged to come down. Our own fleet lay just at our feet. four monitors. and several small gun boats. One of the monitors is a large two turreted one, the others have but one turret. Supply schooners are anchored a few rods lower down the stream. The position of the fleet is just above Dutch Gap. and opposite <u>Farris Island</u>. The river here is but a little wider than the Housatonic at the Falls. but is about twenty feet in depth. The water is much the color of the brook at Mr. Sage's when they are washing ore.

Evening. I have done a very uncommon thing today. The Staff officers have been away most of the day. and as there has been but little to do. I went into my tent and had a couple of hours good sleep. I am still a good deal behind on sleep. and it will take me some time yet to catch up. I shall have more to do tomorrow. the mail tonight having brought me a plenty of work. George came down from his work to see us in the afternoon. he was with the regiment the other day. in the advance towards Petersburg. is well with the exception of a hard cold. As I have work for the rest of the evening I must close by assuring you of my constant love. and bidding you good night.

Ever Your Own,
Ned

No. 166 Rec'd June 18th/64

HEADQUARTERS 10TH ARMY CORPS
In the Field, near Hatchers Va/ June 14th 1864.

Dear Sara,

Your favor of the 9th was received last evening. Be assured dearest. that I appreciate your goodness in sacrificing your own comfort. that you may please. and gratify me. I fear that I do not always make as much effort as I might to write you. I am dreadful selfish. Sara. and too fond of my own ease. to do all that I ought for the comfort of those around me. even for those I love. Your letter contained quite a little budget of items. So they still continue to "marry and give in marriage" in town. There can be but little romance about either of the bridals you mention. as the parties have passed the age for day dreams. and air castles. and are doubtless looking upon the married state with mature and sober views. Dexter is quite unwell. he was taken night before last with severe pains in his back. and head. he is rather better today. and I am in hopes that he will recover without being obliged to go to the hospital for treatment. he is sweating nicely now which I think will brake up his fever. I went up to our camp on Sunday afternoon to see the boys. The regiment was out on picket. but I saw nearly all the boys that I cared to. Met Mr. Moore and had a long talk with him. have had no chance to see him long enough to have any conversation with him before. since we left the Head. He is not looking quite as portly as formerly. but is quite well. he had a very narrow escape during the action of the 16th. a shell exploded directly in front of him. and within a few feet, stunning him severely. The boys are quite jubulent over the prospect of a speedy release from the service. they are now counting the days. I do not keep account of them myself. but am pretty sure to be told how many more we have to stay. whenever I meet with any of the boys. All agree that our time will be up on the 10th of September, that being the day that the last company was mustered into the service, which according to late orders from the War Department fixes the time for the discharge of the original members of the regiment. Probably no one in the regiment. or service. is more anxious to be free once more. than I am. but I do not think it well to let ones mind dwell upon it too much. Reenforcements of the one hundred days men are arriving here. The left of

Grants army reached the James river yesterday; and today Grant himself is reported to be at Bermuda Landing. Perhaps this quarter may yet be the scene of important operations. It has been very quiet for the past week. only an occasional shot being fired. to remind us that we are still in the face of the enemy.

I received a short letter from Mr. Landon the other day. He and Wolcott were well. and I judge are having a fine time of it. he wanted to know when we were going to send for him. A request that he be relieved from duty. and returned to his company has already gone forward. and probably he will join us in the course of two or three weeks. It is not likely that he will be obliged to do duty in the company as his services will be required in some of the mustering offices. and his health is not good enough to permit of his doing duty in the company. Accept a short letter this time. and much love. from

Ned.

No. 167 Rec'd June 30th 1864.

Hd Qrs 10th Army Corps
In the Field June 25th 1864.

Dear Sara.

Your favor of the 14th ins"t. was received last night. I have also received two other letters I think since writing you last. but forget their dates. I have been unwell for several days past, which is the reason you have not heard from me. For three days I did not do scarcely anything; am better today and hope to get along without further trouble. Dexter has not returned from the hospital yet; guess that he is enjoying himself too well to be in any hurry about coming back; he was quite sick for a day or two. but was getting well when sent to the hospital. I ought to have gone down myself and staid a few days. but hardly dared to on account of Dexter being away. Several of the boys about the office are unwell. one of the orderlies has been sent to the hospital today. and another of the clerks is sick this afternoon with a fair prospect of having an attack of fever. I saw George, both yesterday and day before, and had a good long chat with him each time; he sayes that a large portion of the regiment is on the sick list. It is difficult getting excused from duty if a man can possibly walk. and when there is any alarm. every man is obliged to drag himself to the entrenchments and be ready to assist in repelling an attack. From what I can learn it seems that our regiment is getting sadly demoralized and is by no means the 7th Conn. of old. Want of proper officers is the greatest trouble. Nearly all the old officers are gone. those now with the regiment are very incompetant and most of them unfit to hold any position in the army whatever. Col. Hawleys system of putting in men as officers who were ready and willing to do his dirty work instead of seeking out the really meritorious men and promoting <u>them</u>. is ruining one of the finest regiments in the service. The Col. at present commands a Brigade. and has done so for nearly six months past. Cap. Bacon at present commands the regiment. He is a very fine man and no doubt means to do just exactly right. but he <u>cant</u> be a military man. and would do much better as President of a ladies tea party than as commander of a regiment. <u>Gen. Gillmore</u> has left us. to the sorrow of every officer and man in the 10th Corps. The papers I see are filled with lying reports in regard to him. There is not a word of truth in the reports that are circulated intended to damage him. I cannot tell you all about him now. if I ever live to get home I can explain. One thing I want to assure you of. that is that all these newspaper stories intended to injure Gillmore. and the vile slander that appeared in the Tribune about two weeks since, in particular, are the vilest kind of lies. as you love me. dont believe a single word of them. Gen. Gillmore is one of the best. bravest. and most able generals that we have. He is true as steel. void of political ambition. and only desirous of wining fame in the legitimate way that every soldier should. in the line of his profession. We felt bad to lose him. I can assure you. Although he has hardly spoken a dozen words to me since I have been in the office. and does not know my name. more than that of a hottentot in the wilds of Africa. yet it seemed as though I had parted with a dear friend. He came into the office before leaving. shook hands with each of us. and bade us good by. His departure is regretted by every member of the Corps. We are confidant that he will vindicate himself against the villainous charges made against him. and be given a command of more importance than any he has yet held. You ask me "where we are." A little less than five miles from Bermuda Hundreds Landing. a little over a mile in rear of the line of entrenchments which we have thrown up across the neck of land. or peninsula, bounded on one side by the James. and on the other by the Appomattox river. We are not over a quarter of a mile from the James river where our gun boats lie at a place called Dutch Gap. Our Headquarters are in tents pitched in the door yard of a rebel by the name of Hatcher. who with his family remained when we approached. but who took good care to send off all his most valuable stock including all his best negroes. I do not know the exact length of our line of entrenchments. but think they are about four miles. Our Corps at present are holding this line.

 I have not time to write more tonight. hope to write again in a day or so.

With love. Ever Your Own
Ned

CIVIL WAR LETTERS 1861-1865

No. 168   Rec'c July 11th 1864

<u>U.S. Hospital, Point of Rocks, Va.</u>
July 5". 1864

Dear Sara.

Here I am in hospital once more; not very sick. and don't mean to be if I can help it. Since I wrote my last letter to you, I have had an attack of fever which the Doctor broke up without difficulty. but I am troubled with the same disease that took me to the hospital last summer. I came down here yesterday forenoon, the distance from the Corps Headquarters is about three miles. I have been wanting to come here for the last week. and finely asked Col. Smith to send me which he did without delay. Dexter is still here. not very sick. able to be about and does some writing for the Doctor. This is a fine place for a hospital. The patients are all in tents. most of which are situated in the shade. the ground is high and clear of woods so that we get the benefit of any little breeze that blows. A small farm house with barnes and out buildings stands close upon the bank of the river (the Appomattox) and are used as quarters for the Doctors. cook houses. medical Purveyors office and by the Sanitary Commission. The pontoon bridge crosses the river just above the hospital landing. The banks on both sides are high. rising in some places abruptly to the height of seventy-five or eighty feet above the level of the river. just in front of the hospital and a few rods apart. are two ledges of rocks which jut out from the woody banks and seem to hang over the waters of the river which wash their base. These I suppose give the name to the place. It must have been both a picturesque and beautifull place before war set its destroying mark upon it. Gun boats and a few schooners lie in repose in the bend of the river. and at the landing. just below. and on the opposite side. is another landing which is much used just at present for landing supplies for the troops near Petersburg. On the opposite side of the river. directly in front of us is Spring Hill. where we have a small redoubt mounting a few guns. The left of our line of entrenchments between the two rivers. rests upon the Appomattox a short distance from here. The rebels are but a little distance from our works. and from a battery near, threw several shot around our hospital the other day. We have a tall lookout built on the rising ground near by. it is a hundred feet high I should think. One would think that a view of nearly all creation could be obtained from it; but Dexter. who climbed to the top the other day. tells me that but very little can be seen. the country as far as the eye can reach being so thickly wooded, and nearly level as to present only the appearance of a vast forest interspersed with occasional clearings. and residences. The vast amount of fine tilliable land grown up to woods. is a matter of much wonder to us New Englanders. who are only accustomed to see woods growing where the nature of the ground percludes the possibility of cultivation. but here thousands. upon thousands of acres of as fine land as ever repaid the husbandmans toil. are covered with a dense growth of the everlasting Southern pitch pine. To me, it seems as though Virginia was not more than half settled. As I have before intimated. both the <u>Sanitary, and Christian Commissions</u> have agencies here, dispensing their welcom good things to the sick and wounded soldiers. I have no means of judging of the compareative merits of the two institutions. both are doing much good. Soldiers are a good example of the truth of the old proverb. that "the hog looketh not up to him who thresheth down the acorns." they make sure of all the good things that come in their way without stopping to ask what "Commission" furnishes them. I do not mean to intimate that they are ungrateful. or insenseable to the benefits which they receive through the agencies above mentioned. by no means. but carelessness becomes second nature to most soldiers. and they do not always take pains to ascertain to whom they are indebted. I think the patients are very well cared for here. of course we cannot have all the conveniences. nor all the varieties of food that are supplied to such post. or I should say General hospitals. as were those we used to have at Hilton Head and Beaufort. The worst cases. those that are likely to be long ones and all wounded. are sent every few days to <u>Fort Monroe</u>. Such a clearing out took place the day before I came. so that it is comparetively "cleared out" here now. I suppose there is nearly three hundred here now. Yesterday noon for dinner, we had roast beef, potatoes, squash, bread, corn starch, horse radish, all topped off with a piece of very good cake; in the afternoon a pail of iced lemonade was distributed through the ward. for tea we had bread and butter and preserved blackberries. of course we do not get the same profusion of the nice things that we do at home. each one has a small allowance put on his plate just for a relish. This morning our breakfast consisted of bread and butter and an egg for each man. with coffee as usual. our dinner was a fresh meat soup. and bread. with a little chopped cabbage for a relish. tonight for tea. bread and butter. crackers. and stewed dried apples. I think we fare very well for a hospital in the field. The surgeon in charge was formally Asst. surgeon of our regiment. he is now surgeon of the 10th C.V. There are two ladies here I think furnished by the Christian Commission. they are generally seen about the cook house. and tables at meal times dispensing the goodies if there be any. One of them is the Miss. J. B. Morse, who writes frequent articles for the papers. over the signature of J.B.M. She is quite young I should think though I have not been able to see her except with her "shaker" on. and I defy one to tell whether a woman is fifteen or fifty when their face is hid in one of those things. The other lady is Miss Barton. and is older. not exceeding thirty though I should think. I have been in Dexters room this afternoon helping compare the muster rolls of the hospital for the last two months. Dr. Porter (the surgeon in charge) was there. and as he was in a very jolly humor. the afternoon passed very pleasantly. We had two brimming full pitchers of excellent ale with ice in it. I tell you it was capitol. it seemed to do me more good than anything I have had in a long. long while. it was the first decent ale I have had since I left home. if I could only have a glass of it each day it would do me more good than aught else I could get. Our boys at Head qrs. have nearly all been more or less unwell for some time past. Owing to the heat and drinking the water. The day I came away Col. Smith had their rations of whiskey drawn for ten days. and mixing quinine with it gave it to them. it is just what is needed to counteract the evil effects of the water and keep off the fever. One of the clerks and our cook. must needs make fools of themselves and get drunk. and the cook in his drunken bout chopped his toe off. and was brought here to the hospital yesterday afternoon. I fear you will find my letter rather tedious and that I have been too particular in recounting little matters. I heard a piece of news yesterday afternoon that gave me great satisfaction. and that was that Gen. Gillmore's appointment as a Major General has been confirmed. it is but a simple act of justice both to him and the country. I have no doubt but that he will soon be assigned to some important command. it is not likely that he will take the command of 10" Corps again. at least while it remains under Butler. he will be very careful how he ventures himself within that mans power again. Let what will come, Gillmore is sure of the love and confidence of every member of the 10" Corps. We are without any reliable information of the doings of Grants army. for several days past. There has been but little firing in the direction of Petersburg. and all remains quiet along our line of entrenchments. I learn today that reenforcements are arriving at Bermuda Landing. Grant's Hd. Qrs. are still at City Point, which is about three miles from here. I am glad you are enjoying a short vacation. I had like to have forgotten to acknowledge the receipt of your letter of the 26th. I hope our friend Miss Foster has made a good choice of a husband. I think she will do her best to make his home happy. It appears to me that one must have good courage or a long purse to dare attempt matrimony during the existance of the present high prices of the necessaries of life.

July 6". Went in to help Dexter. and have been busy helping him most of the day. so that my letter is still unfinished. It is after sundown now and I cannot write much for I wish to send this by the mail that will leave early in the morning. I received your letter of the 30" this afternoon. Many thanks. Direct to me as herefofore. With much love.

Ever Your Own.
Ned.

No. 169 Rec'd July 14th 1864
Va.

Depot Hospital, Point of Rocks, Va.
July 9". 1864

Dear Sara.

I have no material for a letter on hand. but I have the time. and am able to write and feel that I ought to do so. but fear that you will find my letter something like a washing days dinner. "a picked up affair". I am feeling pretty well this morning have been steadily improving since I came here. and shall be able in a very few days to return to duty. If my work were here I think I should have no difficulty doing a fair days work today. but feel that I would rather be a little more sure that I am able to hold out before I return to the office. All that I require now is a little more strength which thanks to the good fare we get here I am rapidly gaining. The weather for the past two days has been extremely hot. and we have had but very little breeze to temper and render it bearable. There has been no rain for a long time and it has in consequence become exceedingly dry and dusty. This morning it is cool and cloudy. with a fair promise of rain before night. Most of the patients here are improving. there has been four deaths since I have been here. All cases that appear doubtful or likely to be of long duration are sent to the <u>General Hospitals at Fort Monroe</u>. One boat load of such was sent off three days since. I have just had my breakfast of bread and butter. coffee. and apple sauce. We have meat but once a day. always at noon. Yesterday we had quite a farmers dinner. fresh beef. new potatoes. and cabbage. Some of the men pretend to find reason to complain of our fare. but I am inclined to think that they never lived as well at home as they are now doing. I have noticed. and indeed it is an accepted fact among soldiers. that those who find the most fault with their rations are men who were accustomed to the poorest living at home. I have not wandered about much since being here. the weather has not been favorable for it. and in fact there is no inviting place for a stroll. I spent two or three hours the other evening down on the rocks. watching the boats passing. thinking of the olden time when this was the domain and favorite camping ground of the red men. Within a few rods of the principle ledge of rocks. stands a large noble. and thrifty oak. whose stately trunk and wide spreading powerful limbs have defied the storms of many a winter. the legend is that it was planted at the time. and to commemorate the event of the marriage of the Indian princess Pocahantes. it certainly is large and old enough to justify the story. its a splendid tree and I regret to see that the soldiers with their usual passion for destroying everything that comes in their way. have made a deep cut in one side of the trunk. it is not sufficient to kill the tree if not further injured. and I hope it will not be. There were some distinguished visitors here yesterday. senators Wilson and Sprague. I was not favored with a view of the "lions." It is all very nice for these Washington gentry to take a pleasure trip to the seat of war. look on at a safe distance. drink toddy with the officers. listen to their bragget speeches and return to Washington with the conceited idea that they know all about the matter. cause they have been. and seen for themselves. but I think that it would look better for them to stay at home and attend to their business or else if they want to see fighting shoulder their guns and take their places in the ranks where they can "see" to some purpose. Perhaps you think I am ill natured. but I can't help it. I am thoroughly disgusted with the officious meddling of these Congressional gentlemen. I see that the rebels are making another northern raid. I suppose they will be allowed to proceed. load themselves with powder. and retire at their leisure. I do hope that not a single man will be sent from this quarter to oppose them. If there is not pluck enough in the thousands of able bodied men in the state of Pennsylvania to defend their own soil against fifteen or fifty thousand marauding rebels, then I say let the rebels rob them of their last dollar. and burn the last roof over their heads. I was perfectly disgusted with the conduct of the cowardly Pennsylvania Dutchmen when Lee made his invasion of the state. I should hate to have any evil befall Philadelphia. it is so intensely loyal. and gloriously patriotic. every soldier who has shared her generous hospitality (and their name is legion) loves her and would most willingly do battle in her defense. but the rest of the state might take care of itself. I hear that the pirate Alabama has at last been disposed of. I fear "uncle Gideon" will be very angry. and perhaps dismiss the commander of the vessel which wrought the ruin. from the service; he will be apt to be very angry especially if any one should be indiscreet enough to wake him up to tell him the news. What a miserable. contemptable. old dotard. Wells is to occupy such an important position. and what a sin it is in the Administration to permit him to hold it. Our navy is fast becoming a by word. and a reproach, the soldiers used to look upon it as invincible. now they regard it with the utmost contempt. Some young talented and wide awake officer of the navy should be made its Secretary. and the old fossel who now occupies the place be removed to some conspicuous niche in Barnums Museum and labeled "One of the criminal errors of the Administration." Will that do for Mr. Wells!

Afternoon. Your letter of the 3d. was received this forenoon So you passed a rather of a dull fourth. did you? I have told you how I spent the day; probably my last one in the army. I had no idea of spending three anniversary days of our nations independence in the army when I entered it. I am quite gratified at the very general good taste which has prevailed throughout the North. in refraining from the usual demonstrations of joy and exultation on the day just past. To my mind. under existing circumstances. the usual 4th of July performances. would be as much out of taste. and as discordant as a cotillion party in the chamber of death. If God grants us success in putting down this rebellion. the day that seals its doom. will be a day to be celebrated throughout all times. then will be time enough to give expression to the joy and exultation which has been fettered. and kept in bondage. by the terrible strife of years. My impatience is sadly increasing as the day of my deliverance draws nigh. Until within a month past. I have not allowed my mind to dwell upon the matter much and thus the days and weeks passed by quite unnoticed. but of late I have been utterly unable to remain so unconcerned as formally and have I in spite of myself fallen to watching the time with feverish impatience. No doubt it is owing in a great measure to the state of my health. and to my not having been as busy about my work. I hope to have good health. and a plenty to do during the remainder of my time. I care not what the work be whether writing or fighting. so that I come out safe at the end. You are right in supposing that I have no definate plans for the future. I am not without apprehensions in regard to it. and have often of late tried to settle upon some plan with regard to it. but as soon as I attempt to do so. I find myself utterly at loss at once. I know that during my three years banishment. everything has changed. and I do not know anything how business is done now. or what the chances are. therefore I have no ground to speculate on. If my health was good. and I was able to perform any kind of labor. I would ask no odds of any one. but just there is the pinch. I can do nothing at present. and must wait until I get home before I can settle upon any course of action. With regard to marrying. Heaven knows that if the thing were possible I could make you my wife within an hour after my return. I earnestly hope that I may succeede without delay in securing such a position as will enable me to offer you a comfortable home. and a reasonable guaranty of its security. I shall never be able to give you an elegant home. or one entirely free from care. but it shall ever be my endeavor to render such an one as I can furnish. happy. I think now that I shall be back in the office before I write again.

With love. Ever Your Own.
Ned,

No. 170 Rec'd July 19th 1864

Head Quarters 10th Army Corps
In the Field July 15th 1864

I am back again as I told you I expected to be. Came back yesterday noon. Went right to work and have been very busy since. My work accumulated considerably while I was absent. and it will be a day or so yet before I get things straightened up. Am pretty tired tonight and have not felt quite as well during the day as I could wish. Dexter remains at the Hospital. he is not very sick. is about as well as ever. I think he will remain there for some time. as he has good accommodations. everything comfortable. and at present is not much needed in the office here. The Surgeon in charge of the hospital will keep him as long as he can. as he is quite valuable to him. I have had a very pleasant time at the hospital. and if I could have had something to do there would have been glad to have remained. It was rather dull after I began to feel well. to be around without any occupation. I received a letter from Sara B. yesterday. was glad to hear from her direct once more. I believe you school teachers can understand something. how little inclined one feels to set down to letter writing after having been busy with their pen. or poring over books all day. There is a great dearth of news here. We are in great ignorance of everything, except the rebel raid in Maryland and about that we care but very little. It is the universal opinion of the soldiers here that if the men at the North have not pluck. and patriotism. enough to defend the National Capitol. and their own fire sides. the rebels are welcom to both. People at the North have yet no conception of what war really is. and perhaps it is best for the country that they should have a good practical lesson on the subject.

July 15th. Have been down to camp this afternoon to see the boys. Was fortunate enough to find the Regt. in camp and so saw all the boys. and spent a pleasant afternoon with them. They have a very pleasant camp at present. and in a less dangerous situation than their former one; they have been obliged to go on picket every other night for some time past. but today they are most fortunate and are having two days of rest. George is not very tough. and is not looking quite as full in the face as when I saw him last. The all absorbing topic in camp is the nearness of the end of our term of service. All are full of hope that they may live to return home. and of course are anticipating much pleasure. My head is miserably barren of ideas tonight. and I am at loss for material to make out a letter.

Please excuse me for sending a partly filled sheet. and believe me as ever.

Your Own
Ned.

No. 171 Rec'd July 28th 1864.
Va.

<u>Head Quarters 10th Army Corps</u>
In the Field Va. July 21" 1864

I am a little behind in my letter this time. should have written last night. but went down to see George. so could not. Your letters of the 10th. 13th & 15th are all received. Have no particular excuse for not writing sooner. have not been able to bring my mind to it. Have been moderately busy. We have been favored with one real good old fashioned rainy days. commencing early in the morning. it continued with scarcely an intermission until nearly night. During the long dry spell we have had. the ground had become thoroughly baked to a considerable depth. I had occasion to dig a few holes in the ground. and found it impossible to make any impression with a spade. had to take a pick. and go at it like an Irishman in the ore bed. I am in hopes that George will come home with us. or I should say when his time is out. for it is not certain yet that we get home, you know. In fact there is but little fear of his being obliged to stay longer than the expiration of his original term. The men who reenlisted at the same time he did were mustered shortly afterwards but have not received their furloughs until within a few days. they left for home yesterday morning. I was in hopes that Col. Hawley would not learn that George had signed the enlistment papers, but by some means he became aware of it. however he is more reasonable than I had supposed he would be. and is willing that George shall be discharged provided it can be done without his incurring any responsibility to the War Department. The Col. fears that one copy of the enlistment papers has been forwarded to Washington. and that there may be some questions raised why the man was not mustered in. I do not think there is the least chance in the world that the matter will ever be noticed there. The enlistment papers if in existance at all are among the company papers at Norfolk. and there is no probability that they will be brought up here very soon if at all. The Col. wished George to write to the Adjutant General stating the facts and requesting to be relieved from any responsibility to serve longer, that he may have incurred by signing the enlistment papers. I do not think it advisable at all and hope that the Col. will not insist upon it. I am sure it is better to let the matter rest just as it is. Bap. Brooks the Coms. of Musters sayes so and he ought to know. I have but little doubt but what George will be mustered out of service with the rest of us. I told George that I should write to you just how the affair stands. and what I thought of it. but he is desireous that you should not say a word to Julia about it. He does not wish her to know anything about it. but what he tells her.

Sunday. July 24". I am making slow progress with my letter. Have not seen George for a day or two. for a wonder. Adjt. Moore came up to see me yesterday morning. we had a good chat. he thinks there is no doubt but that George will, or can, be discharged when his time expires. A new General has taken command of the 10th Corps. it is <u>Maj. Gen'l D. B. Birney</u>. he only came yesterday. so that I have had no opportunity to form an opinion of him. He is reported to be a good fighting man, and to have seen much service. I think too much of <u>General Gillmore</u>. and too much vexed at his removal from us. to fall in love very readily with any one appointed his successor. General Brooks was in command of the Corps only about a month. I did not like him at all. he has resigned and left the service. Do not know the reason he urged for his resignation but mistrust that disappointed ambition was the real cause. he has twice been appointed a Major General and both times the Senate has refused to confirm the appointment. and so they have run out. When in command here he was only a Brigadier. The confirmation of Gen. Gillmore as a Major General was as great a blow to his enimies at Butlers Hd Qrs. as it was cause of rejoicing to the members of the 10" Corps. I am not well tonight and do not feel like sitting up longer. With love

Ever Your Own
Ned.

CHAPTER 13

THE RICHMOND CAMPAIGN

From the book:

A new epoch had now been reached in General Grant's campaign. He had not accomplished what he hoped when he entered upon the wilderness battles, but forty-three days' fighting had shown him that General Lee would not take the offensive, and would fight no furious battles save behind intrenchments. The hope of engaging him in the open field and winning by the force of numbers had to be abandoned. He had crowded General Lee's army back to the strong intrenchments around Richmond; General Sheridan had put an end to General Early's attempts to invade the North through the Shenandoah valley.

General Sherman was evidently able to reach Atlanta. The Union Army was outside a circle 300 miles in diameter, and the Confederate Army had everywhere the inside track. A fall and winter campaign with artillery through Virginia mud was out of the question. Generals Sherman, Schofield, and A. J. Smith must take care of Generals Johnston, Hood and Price, until at length General Sherman could strike through to Savannah and Charleston, then up to Wilmington, coming between Generals Johnston and Lee, so that in the spring Generals Sherman and Grant could combine their forces and crush General Lee, leaving Generals Schofield and Thomas to perform the same operation upon Generals Johnston and Hood.

With this condition in mind we can better understand the movements in which the Tenth Corps took part during the remaining months of summer and autumn, 1864.

No. 172 Rec'd Aug 8th 1864
Va.

**Headquarters, TENTH ARMY CORPS,
Assistant Adjutant General's Office,
In the Field,** August 2d. **1864**

Dear Sara.

    I have got to be terribly negligent about writing. and almost expect a scolding every time I open one of your letters. I know that I may deserve one. but your patience still holds out. and I begin to be almost tempted to think that you have an inexhaustable supply of that article. We have had a few days of intense heat. it has been almost insupportable. tonight it is very sultry and close though the rain is steadily falling. We have made another ineffectual attempt to advance our lines in front of Petersburg. and have been bloodily repulsed. I understand the papers give a very glowing account of the affair. and strive to place it in the light of a success. so far as I can learn. we gained nothing at all in the end. and lost heavily. Gen. Sherman sends us a crumb of comfort in the way of news of the defeat of a rebel attack on his works on the 27" yet. No particular news about here. Mr. Landon and Wolcott arrived here last week. have met them both. and had a chat with them. Was down to the Regt. last night. found George usually well. My friend Brinton is quite unwell but I hope he will get along without any serious illness. My own health is quite comfortable at present though there is no day in which I feel much like stirring about or doing more than necessity compels me to. I have as many ups and downs as a confirmed invalid. George did not like it that Julia had gone out to Emma's again. he does not seem over pleased to have her there much. Of course you'l not mention to her that he has said as much to me. Went down to the Hospital at Point of Rocks Sunday. the day was one of the hottest of the year. but I was bound to go because I expected to meet my friend Moller (former Hd. clerk in this office) but was disappointed in that. however we spent the day as pleasantly as we could and as profitably as any of our Sabbeths are passed now-days. perhaps more so. as we attended a short service held in the Hospital. which is more than we are often priveledged of doing. I was fortunate enough to get a ride down in an ambulance. and. stopping at Gen. Terry's Headquarters on the way. where Mr. Landon is on duty. took him in with me. At night we walked back. taking it very leisurely. The Second Division of our Corps has been on duty with the 18." Corps for several weeks past. in the trenches before Petersburg where they have seen hard service. and lost many men. In one month. which included a part of both June and July. they lost 500. men in killed and wounded alone. without being in any open engagement. the loss was caused by sharp-shooters. and the enemy's shells. They were losing on an average twenty men a day by these causes alone. The Division participated in the fight of the 30<u>th</u>. lost over 300. men. a part of it went on picket that night. and on Sunday the 31". they were ordered to come over here and rejoin the Corps. The distance is about eight miles in marching which. thirty men died from sun stroke. and a large number suffering from the same cause were taken to the Hospital, which was in consequence filled to overflowing. so that they were obliged to pitch the shelter tents of the men in the yard. and let the least sick ones occupy them. Rumors that our Corps is to move before long are in circulation. but though I think it quite probable that it may. yet I know of no foundation for the present reports. Mr. Landon tells me of seeing you. and hearing you sing and play. thinks he never saw you looking more finely. and warming up with his subject. grew quite eloquent in your praise. and finely wound up by threatening dire vengance against me if I ever caused you trouble. or pain. or failed in aught of my duty to you. You have many friends, my dear, and I suppose a wholesom fear of them will shield you from harm. if love should ever fail in its duty. should you ever fall within my "<u>power</u>". Of course I was exceedingly pleased to hear such favorable accounts of your health. and so much said in your praise. but could not help wishing that I might set, and hear for myself. We hear that the rebels have burned Chambersburg, it seems to create no excitement or surprise, and the general feeling of the army. from highest to lowest. seems to be that of indifferance. I might almost say of satisfaction. Violent diseases require violent remedies; and nothing but a baptism of fire seems likely to arouse the North to its duty. I must close rather abruptly. for it is late and the boys have got to raising ned so that I cannot hear myself think. so good night and much love. from

No. 173 Rec'd Aug 13th 1864
Va.

Headquarters,   TENTH ARMY CORPS,
Assistant Adjutant General's Office,
In the Field, Va. August 7th. **1864**

My Dear Sara,

Your letter of the 1st inst. (Editor's note: This 'inst.' abbreviation for the Latin phrase *instante mense*, means "this month" and Ned begins using this in his letters now.) was duly received. Suppose you are about returning from church at this time. I have not been today. but am hoping that in a few Sabbeths more I can have the pleasure of attending church away from the strife and turmoil of war. amid the peacefull scenes of my New England home. The weather is extremely sultry. it has become very dry again. Virginia soil furnishes the finest dust. and worst mud of any that I ever saw. The fine dust lies on the ground to the depth of a couple of inches. teams are passing almost constantly and swerve to stir it up and the least breath of air is sufficient to send it flying in clouds that will almost smother any one unlucky enough to be caught in their track. The old story of "all quiet along the lines" holds true of our Command at present. three nights since a heavy cannonading was heard in the region of Petersburg. but I am uninformed of the results; the same evening our batteries here on the James opened fire. and threw a few shots in the direction of the Howlett house battery. I believe the rebels did not reply and probably no damage was done. We heare but little of operations around Petersburg now. I think that there is but little confidence of its being captured among our soldiers at present.

A very neat little affair took place here on Friday. **Brig. Gen. R. S. Foster,** one (of) our favorite generals, was presented with a beautifull sword, belt, sash, & shoulder straps. by the officers and men of his old brigade which he once commanded for a long time. The cost of the articles was something over eight hundred dollars; they were very elegant indeede. Gen. Foster is much liked by all who know him, both officers and men. a braver, truer, man does not hold a commission in our army.

Tuesday Evening. Its getting awfull hard work to write now. I cannot content myself to sit down to it, for I keep thinking all the time while writing how near at a close my time in the army is. and it makes the remaining time seem long. and tedious. I went down to the regiment last evening. did not see George. but spent most of the time with Tom. and Mr. Moore. sitting with the latter in his office until quite late. Of course the all absorbing topic was home. and our prospect of getting there. Two years ago when we used sometimes to talk about going home. we thought we would be perfectly satisfied to be discharged any time within a month after the expiration of our three years, now. when we are within a few weeks of that much desired time, we cannot think of willingly parting with a single hour more of our time to Uncle Sam, than we agreed to give him; and we are all jealously watching the progress of events and earnestly discussing the chances of being let out "on time." It is not certain that

Mr. Moore will be able to leave the service in September. indeed he does not think he will be allowed to. I think he could get out if he was determined to. and worked hard for it. He is not so very anxious to leave the service. if he could only get leave of absence to go home for a short time. but that he cannot do at present.

    A terriable accident occurred at City Point today. three barges loaded with powder and ammunition exploded simultaniously. they were mored along side of the dock. close to a large store house, in which was also a quantity of powder. shot. shell. field artillery. caissons. and quartermasters stores. this storehouse was also blown up. The explosion was terrific. City Point is about ten miles from us in a direct line. The explosion brought us all to our feet at once and all ran out to see if possable what had happened; we saw an immense volume of smoke rise high in the air. and continue to increase in quantity and blackness; an officer at once telegraphed to know what had occurred. and we received for an answer. "that a caisson had exploded." they only told us part of the truth. later in the day we learned the whole. A clerk in the quartermasters office here happened to be down there at the time. and was within a few hundred feet of the storehouse when it blew up. he escaped by a miracle. he sayes that the air was perfectly filled with missels of every kind from musket balls, up to 11 inch shell. huge beams. guns. wagons. horses. mules. and men. were all flying in the air at once and falling around in every direction. He remained there for an hour after the explosion when he came away it was thought that between four and five hundred men had been killed. Cannon balls, fragments of gun carriages, great bars of iron, and heavy pieces of timber. wer thrown to a great distance falling among the ajacent camps. and the vessels that crowded the river. killing and wounding a great many. Nothing is known at present of the cause of the accident. and it is not likely that the exact reason will ever be known. as all who were in the immediate vicinity were instantly killed. This afternoon we have heard heavy firing in the direction of Petersburg, and there are rumors that we have met with some success there. I hope my dear your patience with me will continue to hold out for another month and that you will still be able to forgive my short-comings in not writing. My health is pretty fair. so that I am able to keep about. With love. good night.

Ever Your Own
Ned.

Editor's Note: the following letter is about <u>the Second Battle of Deep Bottom, VA</u>

CHAPTER 14

BERMUDA HUNDRED ON THE JAMES RIVER

From the book:

August 12. In the evening the brigade received orders to prepare every available man to march at a moment's notice with knapsacks and two days' rations. Six hundred of the Sixteenth New York Heavy Artillery being engaged on the canal at Dutch Gap, the remainder of that battalion relieved the garrison of the Sixth Connecticut in Redoubt Carpenter.

August 13. In the evening the brigade, accompanying the First Brigade of this division, marched to Deep Bottom, crossed the James there, and before daylight took position near the picket-line of the Third Brigade, which had been holding Deep Bottom.

August 14. At daylight the division, under command of Brig. Gen. A. H. Terry, moved forward, the First and Third Brigades, driving in the enemy's skirmish line, driving them out of their rifle-pits, this brigade supporting. The Sixth Connecticut was detailed to assist the Third Brigade, and. supporting the One Hundredth New York, crossed Bailey's Creek to the right and took a battery of the enemy containing four 8-inch siege howitzers. A portion of the Seventh Connecticut, in throwing out skirmishers to protect two light batteries, lost five or six killed and wounded. The division remained on Kingsland Road until nine o'clock at night. It then moved by the right flank to the New Market road, and down to Strawberry Plains to a position in the rear of the Second Corps.

No. 174 Rec'd 18th Aug/64.

Hd. Qrs. 10th A.C. In the Field
August 14" 1864

Dear Sara

    I wish you could be here just about five minutes. at this present time. you would be able to witness some of the discomforts of a soldiers life. It is just noon. and one of the hottest days of the year. The sun pours down its fierce schorching rays with scarcely a breath of air stirring. to temper. and render them bearable. and then the dust. Oh dear I cannot give you any idea of it. The army is in motion. and teams, and horsemen are constantly passing, almost hid from view at times by the blinding clouds of dust. which is raised at the least movement. In the office all is closed up and in readiness for a start. Our travelling wagon stands ready all packed close by the tent. the desks are locked up. and since twelve o'clock last night we have been in readiness. to move. I arose early this morning. it being impossible to sleep after daylight on account of the flies. which are here more troublesome, and bite more ravenously than our northern mosquetoes; for an hour or so I was busy completing our preparations which being made in the night required some looking over after it was found that we had the time for it. I have passed the forenoon uncomfortably trying to sleep and expecting orders to start every moment. Gen. Birney and Staff left about midnight. We had not the slightest idea of a movement until about the middle of the forenoon yesterday when we were ordered to be in readiness by eight o"clock last evening. Nothing was known of our destination. and of course the usual surmises. and conjectures. were indulged in. Orders were issued to the troops of the Corps to be in readiness to march by eight o'clock in the evening with three days cook rations. I do not yet know how much of a movement it is to be. but think it doubtfull if the Hd. Qrs. of the Corps are removed at all: this trip. The object seems to be a demonstration against the enemy on the north side of the James at <u>Dutch Gap. Chaffins Bluff.</u> and possibly Malvern Hill. The whole of the 1st. Division, and two Brigades of the 2d Division together with a negro Brigade just arrived here from Dept. of the South under <u>Brig. Gen. Wm. Birney</u> (brother of the Maj. Genl. comdg.). crossed the James river last night at Deep Bottom. this morning we heard some firing in that direction and orderly who returned from thereabout the middle of the forenoon. reports that the <u>1st. Division under Ferry</u>, was hotly engaged in skirmishing when he left. A portion of our Corps, consisting of the 3d. Division. and one Brigade of the 2d. Division. with some light artillery. all under the command of <u>Genls. Ferry and Turner</u> are still here holding the line of intrenchments. I should hardly think it likely that our Hd. Quarters would be moved though they may be for a few days. I went down after dinner yesterday and saw George a few moments; the men were then making preparations for the move. I sincerely hope that no harm will come to any the boys. It is dreadfull to think of being obliged to fight such a day as this. I can hardly draw my breath sitting in the shade. We have a large working party employed some where in the vicinity of Dutch Gap. Yesterday morning early the rebels commenced shelling them. and our batteries were not slow in answering their fire; the gun boats also joined in the long range duel. and the usual quiet of our camp was disturbed for half the day by the "dogs of war." I do not know whether the enemy obliged our men to stop work or not. but understand that they were the first to stop firing. Several were killed and wounded in our batteries. but do not know how many. The rebels have a <u>100 pounder Parrot</u> mounted in their <u>Howlett House battery</u> with which they threw shells by though at a considerable distance <u>from</u> our Hd. Quarters. Did they but know it. a slight variation to the right in aiming their piece would at any time make this place to hot for us. I was most agreeably surprised one day last week by a visit from Mr. Edgar. I spent a couple of hours very pleasantly talking with him; talked as fast as I could. but had so much to say that I found when he left that we had hardly made a beginning. He is with the Christian Commission. stationed at City Point. Expects to remain in the field about three weeks longer. and then will return to FV again. He gave me some further account of the explosion at City Point of which I wrote in my last letter; it did not prove as destructive to life as was at first reported. I received your letter of the 7th two nights since. In a short time we may hope to throw aside our pens and paper. and be able to <u>talk</u> our matters over. You must prepare to grow weary of the sound of your own voice. for if I get home you will have to tell me <u>everything</u>. Only think of it! three years talking to be done. Will Sunday evenings be long enough to do all of it in? It is too noisy in the office to write. the boys have nothing to do and are amusing themselves in not the most quiet manner. Oh for one more good, glorious, quiet, New England Sabbeth.

With love. Ever your own
Ned

Editors note: The following letter refers to Sam Wolcott's death at the <u>Second Battle of Deep Bottom Virginia.</u>

No. 175 Rec'd Aug 24th 1864

Headquarters, TENTH ARMY CORPS,
Assistant Adjutant General's Office,
In the Field, Va. August 19th. **1864.**

Your favor of the 12th was received two days since. I regret to learn that you have been ill. and hope that you will not confine yourself so closely if you find it to be injuring your health. better let the school go. as I do not see starvation or the workhouse staring you in the face just at present. I have been extremely pained today to <u>learn of the death of S. Wolcott.</u> Mr. Landon tells me that he wrote yesterday. so you will doubtless have learned the sad news ere this reaches you. Mr. L. informed me of it this morning, but he could not help but hope that there was some mistake. as he heard of it in rather an indirect way; this afternoon. the official list of killed and wounded of the 1st. Division came into the office and I there found the confirmation of the evil tidings. for his name was among the killed. It is difficult to realize that he has passed from us. that we shall see his face no more. When I last met him he was full of life and health. and we were congratulating ourselves on the prospect of our speedy return home. the next I learn of him. he has gone to his eternal home. sent thither by the deadly bullett of a traitor to his country and God. I can only learn that he was killed instantly in the thickest of the fight and do not know whether his body found burial at the hand of friends or foes. Thus another costly sacrifice has been laid upon the altar of our country. We may not refuse this call for another of our dearest and best. for the good of our cause. but rebellious thoughts will arise within me. which I will not trust on paper. even to <u>you</u>. That good men like Wolcott are falling every day by hundreds to maintain the right. does not make his loss any the less. We cannot <u>know</u> the mass that fall. but this one we do know. we feel that this is <u>our</u> loss; we know the worth of our fallen hero. and can realize something what the salvation of our country is costing when the lives of such as Bosworth and Wolcott must be laid down to purchase it. Of Samuel's many virtues I need not speak. you knew him better than I did; but my acquaintance with him has taught me to respect and admire him for his many excellent qualities of mind and heart. Earnest in the cause in which he was engaged. he feared no danger, and shrank from no hardship or exposure. He has left a stainless record as a man. a soldier. and a christian. Of his loss to his family I cannot speak. for I have no words in which to express myself.

This fall has again lessened the number of our little band. and the pleasure with which we anticipate our return home is sadly embittered by mourning for the absent loved ones. who have given their lives for their country. We get but little news from the scene of the late operations. none of us clerks have been out to the front. and the little news we have learned has been mostly received from couriers. teamsters. and ambulance drivers. and is very contradictory. and unsatisfactory. I do not think the office will be moved at all. and shall not be surprised to see the Corps. back here in a few days. We have had nothing at all to do since last Sabbeth morning. Some work has been sent us tonight so that we shall be busy tomorrow.

Yesterday afternoon I went down to the old camp and saw George a few moments; today Mr. Landon and I went down to Point of Rocks. spent a few hours with Dexter. and had the pleasure of walking home in the rain. Was sorry to find Dexter quite unwell. and to learn that he has been so for several days past. he does not sit up all the time. I hope nothing serious will result from his indisposition. but he looks bad. and I cannot help feeling as though the 12th. of September could not come around any too quick for his welfare. He is with those who will see that he is well cared for; as a sick man in the army he could not be more favorably situated. I am very glad that George did not attempt to go with the regiment for he has been quite down while remaining quietly in camp. and must have certainly broken clear down had he tried the march. The weather has been cloudy with occasional showers for the last two days. today it has rained pretty steadily with a fair prospect of its continuing through the night. The dust is effectually laid for the present. and now I suppose we shall be troubled with the other great Virginia discomfort. mud. I had nearly forgotten to mention that Clark. the man who joined us with Wolcott. was wounded in the face. the ball entering one cheek. passing through the mouth and out through the other, probably not a serious wound. The loss of our regiment during the operations of the 14". 15". & 16th. was 6 killed, 28 wounded and 5 missing. there has been fighting since. but we have no accounts of our loss. I hope you are entirely recovered from your indisposition and that you will not confine yourself too closely this hot weather. With love.

Ever your Own
Ned.

CHAPTER 15

RICHMOND CAMPAIGN

From <u>the book</u>:
    August 24. Started with the First Brigade (the Third Brigade and the rest of the Tenth Corps to follow) to relieve the Eighteenth Corps before Petersburg. Took to hold about 700 yards of the line, are left resting on the Suffolk railroad. Here remained through the month, losing some daily.

No. 176 Rec'd Sept 3rd 1864.
Va.

Head Quarters 10th Army Corps
Before Petersburg Va. Aug. 28th 1864

Dear Sara
    We have effected a change of "base" you precieve by the heading of my letter. It was rumored at our old camp on the 23rd inst. that our Corps was coming over to relieve the <u>18th Corps before Petersburg.</u> on the morning of the 24th we were ordered to pack up everything and be ready to move by night. During the day we completed our preparations for a start and went to bed expecting to be called up to go before morning. We were not disturbed however until about four o'clock in the morning when all hands were routed out by a rebel attack upon our picket line. It was a small affair. resulting in a little loss to both sides. about an even thing I think as we did not lose any ground. the firing was limited to musketry until after daylight when a moderate fire was opened from one or two of the batteries on the line. and continued during the day. It was the first occurrence of the kind on the line since the latter part of May. and was probably caused by the knowledge that an exchange of

troops was being made by us; which information the rebels doubtless received from a deserter from the 47th N.Y. who skedadled just at night. and who knew that Gen Terry's Division had left the line during the afternoon.

The tents of Staff officers were all struck and together with their baggage was loaded and sent off by noon on the 25th. About two o'clock we clerks "took up our line of march." It was pretty hot. but we had nothing to carry but our haversacks and canteens. and after going a couple of miles we put the former on one of the wagons and jogged on with nothing but our canteens. We passed Gen. Ferry's quarters. and stopped a few moments to see Mr. Landon. I was determined to stop and see Dexter though it took me about half a mile out of my way and as the rest of the boys kept right on. I should have to finish my journey alone. I found Dexter in bed though he got up while I was there. he looks rather poorly. but guess he will come out all right. While resting a little. one of the boys came in and said they were taking off a rebel's arm just outside. I went out to see it done and becoming interested staid longer than I intended. so that it was five o'clock when I renewed my journey. It proved to be much further over here than I had supposed. and I did not arrive until about seven o'clock. hungry. weary. and a little foot sore. for I am not used to so walks. though I suppose the distance is not over eight miles. A cup of coffee and some crackers. with a piece of "salt horse" made my supper. and a large dry goods box turned down on its side formed a good shelter for the night. but what with excessive fatigue and a drove of large number one musketoes kept me awake. so that I do not think I slept half an hour all night. The 26." proved a very hot day. in the morning I walked out towards the front. going only about half a mile to a couple of fine houses where the commander of the former 18th Corps had his Headquarters until the rebels shelled him out. I had a good view of Petersburg, distant about two miles. with the intervening works of our own and the rebel lines; a battery close by was then throwing shell into the city. In the afternoon I was busy fixing me up a place to sleep in. A fatigue party were at work all day pitching the tents. our office tent was pitched just at night. and we got in our desks and nearly ready for business. Gen'l Birney and half a dozen of his staff who had remained at the old Hd. Qrs. arrived just at night. Yesterday we got things into good working order. and today we have been "keeping shop" just as quietly as though we knew nothing about moving. It is not quite as dull here as it had got to be over at our old "stand." About sunset or an hour or two before the morter practice commences on both sides and is kept up slowly until after daylight the following morning. There is also more or less artillery firing during the day. The musketry firing is incessant. not in volleys as in a fight. but a constant "spattering" of single pieces. I am told that it is not the pickets who fire. but the sharpshooters. and woe to the unlucky head that is exhibited above the breastworks. It is very fine to watch the firing in the night. to see the bright flashes of light when the pieces are discharged. and to follow the flight of the shells which is much like that of a falling star. only much less rapid. Our Headquarters are well out of harms way. but we have only to step out the tents to witness the "fireworks." There is one morter battery guite near us. and when in operation. it reminds me of old times <u>on Tybee</u>. I have been out to find the regiment today. it is in the trenches so I did not go to it. but I found George and spent an hour or so with him. Our men have been in the trenches since they came here. I suppose they will be obliged to stay in them nearly all of the time. with occasional short respites. It is dangerous going into the trenches in the day time, and whenever it is necessary to relieve the men in them. it is done in the night. The cooking is done some distance to the rear. and the food carried to the men by the cooks. I have received your two letters of Aug. 21" & 24." for which as ever, I am thankfull and would kiss you if I could. but for that you will be obliged to wait a few days longer. Tonight I received a bundle of papers from <u>Falls Village</u>. which shows that my good friends there have not forgotten me. Around camp nothing but home. and the prospect of soon being out of the service. is talked of. no other subject seems worthy of consideration. all are jealously counting the days. and it is hardly safe to intimate to any of the men. that it is possible that the exigences of the service may render it necessary to keep them a few days over their time. The time of nearly all of the detailed men about the Headquarters, expires within a month. a large portion of them belong to the <u>6." & 7." Conn</u>. I used to say that if I was discharged any time within a month of the proper date. I would be thankfull. and not complain. but at the long looked for period approaches, my impatience increases. and it all I can do to keep from grumbling because I must stay five days longer than I ought. The third Sabbeth from this I hope to be with you. The country about here is the pleasantest I have yet seen in the service; it has quite a familiar home look with its hills. valleys. orchards, meadows. and running brooks. but war has made sad havoc with the fair face of nature. in the orchards. the trees that have escaped the ax of the soldier, have been girdled by the horses who have been fastened in their shade. and have striped the bark clean off them wherever their gnawing teeth could reach it. The meadows are so crossed and cut up by roads. and dug so full of rifle pits, that there is scarcely a chance for the green grass to show itself. The woods are cut down and burned. nearly all the houses have been destroyed. leaving only their tall, gaunt chimneys. and a few foundation stones to mark their former site; not a vestage of fence can anywhere be seen. and nothing but little stunted lines of bushes show where what is now one large open common was once divided into tilled fields. The road to our Headquarters crosses the City Point and Petersburg rail road about a mile and a half. back. at the nearest point to our works that the cars are run. all the supplies for the armies before Petersburg are brought to that place. and taken from thence in wagons to their destination. This afternoon I caught sight of Gen. Meade as he was mounting his horse after an interview with Gen Birney. (Editor's note: **General David B. Birney died at Philadelphia, October 18, 1864**) We are now only about three quarters of a mile from his Headquarters. I have not seen Gen. Grant yet. suppose he does not often come around here. With love. and hoping soon to be with you. I am as ever.

Your Own
Ned.

(Down the right-hand side of page 3 Ned writes by an ink blotch, "Please excuse this blot. I had my letter nearly written to take another sheet.")

No. 177 Rec'd Sept 13th 1864
Va.

Headquarters, TENTH ARMY CORPS,
Assistant Adjutant General's Office,
In the Field, Before Petersburg, Sep. 8th **1864.**

I was exceedingly pleased last evening by the receipt of your letter of the 4th and I had thought it probable that you would not write again, thinking perhaps that I might be on the way home when the letter reached here. I must confess that I had not intended to write again. though I knew well that I ought to. for I am not "out of the woods" yet by any means. I have been pretty busy since we have been here. and then somewhat elated with the prospect of going home. so that it has been rather difficult to tie myself down to better writing. Dexter has beat me on getting home as he has on everything else. He was discharged from the Hospital on the 6th. and left early yesterday morning for home. so that he will doubtless reach Salisbury a day before my letter does. I went down to the hospital yesterday morning to see him off. but when I reached there about nine o'clock found that he had already been gone an hour. The distance from here to the hospital is about four, or four and a half miles, and I could not be too thankfull during my ride down there and back that I was well. and not either sick or wounded, for it is enough to half kill a well man to ride there over these horrid roads. a man must be rather more than human to be able to endure it when either diseased, or wounded. I missed seeing Dexter of course, but saw several other acquaintances. had a good dinner and then jolted back to the office. The head clerk's time expires four days after mine does. We have two new clerks in the office breaking in. and so I have nothing to do but to show my successor about the work. All the detailed men of the 6th & 7th whose time expires. have been relieved and ordered to their regiments in order to be present for Mustering Out. I have sent my Descriptive List to my Company commander in order that my Muster Out Roll may be perfected, and also a notice to him from Col. Smith, A.A.G. that I would be retained on duty here until required to be present to be mustered out with the company. I cannot tell you what day we shall get home. the question when we shall leave for home is undecided yet. The matter stands thus--It has been recommended by the Commissary of Musters that the men of the 6th & 7th C.V. whose term of service expires on the 11th & 12th of this month. be sent under the charge of one officer to the State in time to be mustered out on the 12th. those papers have gone to Department Head Quarters for the necessary approval. they will doubtless be approved. and should be back today. but may be delayed a day or two yet. Should they be returned approved today we might be able to leave here on the 10th. If they are returned disapproved we shall remain here until the 12th. in which case I could not hope to be home before the 16th or 17th. Col. Hawley has forwarded an application for permission to go home and recruit for the 7th Conn. it will I think be granted. and if so he will doubtless be the officer to take charge of the men on their way home. I cannot tell my dear, exactly how even after my arrival I shall come to you. but you may be sure I shall make haste to do so and that no unnecessary delay will be permitted in the matter. Our present Headquarters is being fixed up very nicely indeed but I cannot stop to describe them now. it looks as though the Corps might remain in the vicinity for some time. unless we should capture Petersburg of which there is no immediate prospect. The Petersburg and City Point Rail Road is being extended by a branch. so as to open communication thereby between our extreme left on the Weldon Rail Road and the base of supplies at City Point. The trains are now running; the road passes within a few yards of our office and it seems quite like old times to hear the whistle of the engine. and the rumbling of the cars. and to look up and see them so near. One almost thinks he is in a civilized country once more. Poor Sara B. I shall be so glad to see her and yet I shall dread to meet her for I cannot bring Pitt with me and I know how she loved her noble brother and with what almost crushing weight the full realization of her loss will be forced upon her when his companions return with out him.

Evening. Have been over to the regiment this afternoon. called on Mr. Landon on my way up and he went with me. We saw Norton. Moore. George and all of the boys that we cared to. Stopped at Mr. L's quarters on the way back and took supper with him. we had sweet potatoes. crackers. ginger cakes, cheese and tea. Got caught in the rain on my way back and got quite wet. The papers have not yet been returned from Dept. Hd. Qrts. There is no possibility of our getting off tomorrow. and probably not the next day. It is useless to think of getting home before the middle of the month.

With love. and a good night.
I remain, Ever Your Own
Ned.

CHAPTER 16

NED'S ENLISTMENT ENDS

From the book:

Colonel Hawley established his headquarters in a bomb-proof, there sheltering his horse as well as himself. The brigade was occupied in the heavy and trying duties of the siege. Many casualties were noted every day and no life was safe except under cover. The boys dug burrows or splinter proofs, into which they crept while off duty when the fire became hot. Those on duty had to dodge the shells the best way they could. This condition continued until September 4th. On that date private John Rowley of Company D, who as before stated shot his comrade at Olustee, was executed in accordance with the finding of a Court Martial.

On the 12th of September such members of the Seventh as had not re-enlisted were discharged by reason of expiration of their term of service. They numbered about 200 and were under command of Colonel Hawley. They proceeded to New Haven, bivouacked in the old state house on the green, where they were paid off and scattered to their homes.

Ned is released from his three-year service on 12 September 1864.

He returns home, marries Sara on October 5th, 1864 in Lakeville, Connecticut, with Rev. Adam Reid presiding.

He returns for civilian service again on October 8th for another year with the "Office, Chief Commissary of Subsistence, Armies operating against Richmond" as a civilian clerk.

He resumes writing, and Sara, now Ned's wife, begins to record this next series of letters with "No. 1"

CHAPTER 17

ON RETURN TO DUTY

From the book:

A resume of the movements from September 29th to October 10th is copied below from a letter sent home by an officer of the Seventh.

"We are now exactly on the ground where the enemy swept down upon us on the 7th of this month. I want you and everyone else to remember the doings of the Second Brigade, First Division, Tenth Army Corps, during the operations north of the James.

Behold our noble Seventh in its place in brigade line charging the rebel rifle pits on Newmarket Heights, on the same day prowling around the defenses of Richmond, peering into the windows of the rebel Capitol itself. Mark them on October 1st pushing through mud and rain toward the "seven hills" the seat of the Confederate government, extending as a skirmish line a mile long, dashing through field and wood, wading streams and climbing hills, never hesitating or wavering under the concentrated fire of the great guns of the enemy until they made out the grand inner line of the city's defenses. See the Second Brigade hurriedly occupy a dark pine wood and set itself as a wall across the path of the cowering foe. No breastworks, no artillery, only stout hearts and Spencer rifles. Forward go the skirmishers, and for some time hold the enemy in check. Our line lies down — the skirmishers are forced in, the butternuts following hard after with that well known inhuman yell.

Then at the word 'Fire by battalion, fire" up rose the brave boys and with a rousing union cheer gave them the contents of their rifles. Only a moment it lasted; no men could stand up and face such fire at a range of fifty or seventy-five yards. The attacking column melted away and was gone, only those killed, wounded, or scared to death remaining on the field. It was a furious, well delivered attack — it was a complete repulse.

We had lost some men; they had lost half as many men as we had on the field. We pursued them after bringing in their wounded and attending to our own. The next morning we buried their dead. They were from Alabama, Georgia, South Carolina, and Texas. The attack was by two full divisions, Field's and Hoke's."

On the 9th of October General Terry was assigned to the command of the Tenth Corps relieving General Birney, who had previously been in command of the corps while Terry commanded the First Division. On the 12th General Hawley returned from the North and resumed command of the Second Brigade. About this time the Connecticut soldiers in our command were permitted to vote at presidential election, commissioners appointed by Governor Buckingham having been sent to the field to receive their votes.

From Hist. 1st Lt. Batt'y.

CITY POINT.

CIVIL WAR LETTERS 1861-1865

Letter No. 1 Rec'd Oct 14th 1864
City Point.

OFFICE, CHIEF COMMISSARY OF SUBSISTENCE,
Armies operating against Richmond.,
Oct. 10th. 1864.

My Dear Wife,
    Believeing that you will be glad to hear from me as early as possible I hasten to improve my first opportunity to write. I suppose I was well on my way before you were up last Saturday morning. George left me at the Ferry Boat. on arriving at Jersey City I went at once on board of the cars. and in a short time we were on the move. I was disappointed in my ride that day and did not enjoy it all: how could I? when I was leaving my little wife behind me! Besides I was much perplexed as to my prospects of success in getting on. I did not travel the same road we did when the regiment went on in "61; the train being a through one. stopped but seldom and there were but few changes when it did. I was in no humor to read. and the seats would not be comfortable so that I could sleep; the country did not seem to be very interesting. or else I was not in a condition to appreciate it. so there I sat through the long dreary hours of the day: the most lonesome one certainly that I had ever spent. But the longest day must have its close. and the longest journey its end. so as five o'clock drew nigh the train entered the "Monumental City." I learned the name of the Quartermaster and the number. and street where his office is situated. from an officer. but not knowing anything of the city it did not look very certain that I should be able to see him and get off on the mail boat which I found was to leave almost immediately upon our arrival and our train was already three quarters of an hour behind time. I was at loss what to do. and did not at all relish the idea of staying in Baltimore twenty-four hours as I would be obliged to if I missed the boat. When the train stopped I looked in vain for any signs of the Steamer. or any one to enquire of; at last I espied a policeman and while he was directing me, a negro. on the watch for a quarter. offered to show me the way and carry my valise. I consented and he shouldered the valise and started off on a dog trot saying "he"d git dare afore de mail did." I followed at a not very dignified, or military pace, but it served to bring me to the boat just as the mail was being put on board. I had just time to take the oath of allegiance, get my pass and aboard the boat. and we were on the way. Had I taken the train the preceeding evening. I should have had plenty of time to have procured my transportation from the Q.M. and thus saved the Five Dollars that my fare cost me. but I need not tell <u>you</u> that I did not regret my stay. I felt so weary and

sleepy that I would have been glad to have gone to bed at once. but the state rooms were all occupied. I could have a berth in the cabin after supper. so I strolled up into the saloon and after lounging about a short time dropped into one of the large. cushioned chairs. and went to sleep directly. I was awakened by the pain caused by the cramped position I had sunk into; and found it to be about eight o'clock; went below and found the steward just clearing the supper tables; I made for a berth and stowed myself away at once. A mattress. pillow. and white bed spread constitute the bedding. Coat and boots off. and you are ready to turn in! My shawl and overcoat furnished abundant covering. I found no "little woman in white" in my berth. and I need not add that none came to me after I got in. but I thought of her. and ere I slept, breathed a prayer to the Great Protector that he would care for. preserve. and bless her. I did not get up until the boat touched the dock at Fort Monroe. at about seven o'clock the next morning. then I went on shore at once. got breakfast. visited the offices of the Provost Marshal and Quartermaster. at both of which I was successfull. and procured a pass. and transportation without trouble. Went on board the City Point boat. and about nine o'clock was once more on my way here. The weather was exceedingly raw and chilly and the wind piercing cold. and blowing quite strong. Every one kept in doors. and a couple of coal stoves in the saloon wer kept in full blast. There was quite a number of passengers. but there was plenty of room and the long saloon offered a fine promenade to those inclined to pedestrian exercise. I have no idea of the distance I travelled, but think it would be. estimated by miles. It was the most supremely miserable day I ever passed. The boat reached here about five o'clock. My trip from the time I left you was of thirty-four hours duration. It was better than I had hoped to do. and I do not know as I care much for the expense. on the whole. I would have liked to have got my passage free from Baltimore. as I undoubtedly could have done. if I had gone to the Quartermasters. but should have missed the boat had I done so. and probably it would have cost me as much as did my passage. to have remained there a day. I found my friend Carl. and Mr. Brown. where I left them. and received a hearty welcom. I have been busy today trying to get some insight into my new business. I find it more extensive than I had expected and my prospect of mastering it very uncertain. I shall do the best I can. and if I fail. it will not be from inattention to business. Carl is very good to me. and as he understands the business perfectly and will do all that he can to aid me. I am in hopes to get along. At present I am staying with Carl; he has assisted me today in getting into a mess, which he thinks is a good one. I got dinner and supper at it. and think it will do well. The expense for board will be about ten. or twelve. dollars per month, in addition to my ration. which amounts to about ten dollars. this is not extravagent. and is as cheap as one could expect to live out here. Today as I sat at work I heard the roar of the heavy guns in the direction of Petersburg. the roll of the drums. and the piercing notes of the fife and bugle near at hand, and it seemed quite homelike again. I felt at my ease again. and had to stop a moment to think. and note the change in my dress, to realize that I was not still a soldier, and that the events of the last three weeks were not all a dream. as I would to God that some of them were. I have thought of you much. these last three days. Sara. and have forced myself to realize in some degree. that you are my wife. I am gratefull for the blessing. and do not fail to thank the Bountifull Giver. for the rich gift. I wish that <u>you</u> had equal reason to be thankfull. It is late and I must close. though I have not said that which I meant to. Tell George I will keep my eyes open for him as soon as I can look around. Give my love to Julia, Libby, Sara and all the dear ones.

May our Mercifull Father have you ever in charge. spare your life. and preserve your health. May his love be yours, and may his peace be upon you. and so may you be happy here and hereafter. is the prayer of your loving, though most unworthy husband.

Edwin J. Barden

Direct to
E. J. Barden
Care Lt. E.H. Brown A.C.S.
<u>Gen. Grant's Headquarters</u>
<u>City Point, Va.</u>

No. 2 Rec'd 18th October 1864

Office, Chief Commissary of Subsistence,
Armies operating against Richmond.,
City Point VA Oct 14th 1864

My Dear Wife.

I am more than fortunate today. I have received both of your most welcome letters! the one not written in N.Y was handed me while at dinner. and the other while at supper. I should be tempted to sit down to a meal every hour if I could fare as well each time. I was most happily disappointed in receiveing your letters. for I remembered when it was too late. that I had not told you how to direct to me. and I was calculating how long it would be before I might reasonably expect a letter in answer to the one I sent you on the 10th inst. and regretting my stupidity in not leaving my address with you. when behold: the longed for and much coveted treasure appeared. Another proof of your goodness my dear. the slow progress of events to inform you of my whereabouts, but sent your message of love and cheer. after me at a venture, I regret that you did not enjoy yourself better in N.Y. after I left. I thought you would have a more pleasant time after I left, and sincerely hoped that it might be so. for it has not seemed to me that I ministered to your wants, and enjoyment as a husband should, and as you had a right to expect. I exceedingly regret that I am not more of a gentleman. and better fitted to move in the society to which you are accustomed. I was haunted with the idea while at home, that I was not much more than half civilized, and that nothing but the good nature of my friends enabled them to tolerate me. Since my return here I have still felt that there was much of truth in the thought. That I was a stupid bridegroom I am perfectly convinced! and felt keenly while with you my unfitness for my position. I would have done better if I could, and beg you to believe that wherein I failed to do all that I might, and should have done for your happiness. it was the fault of the head, and not the heart. it seemed to me that after I had left you, and was fairly on my way, you must feel relieved of the continual perplexity with which it worried you. So it seems there was a still more unpleasant place than N.Y. than No 83. Inter Hotel. I can appreciate your feelings of dislike to stopping at Chase's, the thought that aside from that it must be more pleasant than where we did put up. Am glad you succeeded in seeing a decent play before leaving the city. Libby will also be satisfied now that you have <u>been to "Walleck's"</u>. Having already written you about my journey, you are aware that the two days after leaving you were not the most pleasant ones that I have spent. I have been so constantly employed here that I have had no time to give way to feeling. I am to much accustomed to the mode of life to be very home sick! and whenever I have caught myself getting lonely, or dispirited, I have striven to apply myself more closely to business. for such feelings must not get the mastery of me now! I have too much at stake. I need not tell you that I should be happier if I could be with you, but perhaps it is but as it is. I often think that you love me better and think of me more highly than you possibly could if you knew me more intimately. I know that it ought never to be so, but I know too that there are times when reason, and cool judgement will have their way. and refuse to be controlled by passion or feeling, I expect to find my chief pleasure in striving to promote your happiness, and if it demands my banishment from your presence, so be it. I will try to be content. Mr. Randall's letter explains the reason of delay in receiveing Lieut. Brown's letter. I do not see how Mr Millspaugh could have been so stupid. I had told him distinctly two or three times, that if any letter came for me. to put it in Mr. R's box. I dont care much about it now. only it caused me a deal of vexation while in N.Y. Please keep the letter. I think I shall manage it so as to get back nearly all that I had paid out here. for my postage. I shall do all that I can to get a place for George, but I dont suppose I can do anything just yet, as I cannot get time for a few days yet. to run about much. I think the weather is colder here than at home. This is a magnificent moonlight evening, and very suggestive of love. Ladies, fast horses, buffalo ropes, buggies, oyster suppers &c. I cannot think what Mr Townsend could possibly want of me, am much puzzled over it! how should he know that I was in N.Y. and at the International! I am not so sanguine as you, and have no idea that he had another position to offer me. If George hears from him, please let me know at once. what it was he wanted. Hope its none of my evil deeds coming to life. Should have written to Mr T. tonight if you had not mentioned that George has already done so. I ought to have had the Photograph taken without your urging me to do so. after you had expressed a wish for them, I should certainly have gratified you, had you pressed the matter, but I felt a strange repugnance to have it done, for which I cannot account myself. I hope the girls will be pleased with their presents! its decidely shabby, and I am now exceedingly vexed about it. Did you get any more music, or any photographs?

A thousand thanks my dear wife, for all the love, and confidence, expressed in your last letter. It is most gratifying to me to know that in any way, I have conduced to your pleasure. I hope, and pray that you may not be disappointed and that I may be able to be to you all that you wish. You ask for forgiveness when I have nothing to forgive, and plead with me to believe you, when I have never thought of doubting you! You must not ask me but favors until you have some right to claim them. Believe me my love, that I am more than confidant that I can never be sufficiently greatful to the Good God who in giving you to me, has bestowed upon me such a priceless blessing. I cannot bear to have you talk to me so, as though the bargain was all on one side. <u>you</u> had drawn a prize, and poor I, a worthless piece of lumber. It is getting late and I cannot answer your letters as fully as I had intended to do. I trust you will excuse. Be assured that your kind admonitions are heeded and that I will endeavor to act upon them. I hope if my life is spared, that I may be a better man, and not always be the worthless piece of drift wood that I have hitherto been. My kind regards to all our friends, and love to all <u>our dear ones</u>. May God bless and preserve you my darling little wife. Good night. Edwin J Barden

MY DEAR SARA

No. 3 Rec'd 20th October 1864

> Office, Chief Commissary of Subsistence,
> Armies operating against Richmond.,
> City Point VA. Oct. 16th,1864

My Own Dear Wife,

    Highly favored again today, your letter of the 13th was handied me while at supper. I hope that by this time you are convinced that you are <u>not</u> to do all the writing. I presume that you will do more than your share, as you have ever done, but will endeavor to be faithfull to my duty in this respect, at least in a greater degree than I have been for the last six months. I could not well write the first night I arrived here, but did so the next evening. You were very nicely caught indeed, on your arrival home. Perhaps on the whole it was better to "face the music" thus, at home, than elsewhere at another time. I should have been happy to have met Mr and Mrs. Burrall, and many other of my friends, had the opportunity offered. but I should not have cared to have gone into company for that purpos. I regret exceedingly that I was obliged to leave the (to you) unpleasant business of providing you a home, to yourself! it should not have been so. and had I thought that there <u>could</u> be such an unpleasant scene as you describe, I would certainly have managed some way to have arranged the matter myself. I feel vexed that your father should have thought that I was begging for either you or myself. I thought he understood me, and being pressed for time, did not wait to enter into any details, or make a regular bargain. I told him distinctly that I should pay your board, and though he said it was no matter about it. I did not suppose that I had given him any grounds for the unpleasant remark that you quote. I regret that you agreed upon so small a price. for I fear that it will leave room for more of those <u>charity</u> remarks. If you wish to stay at home for the present, I would prefer to pay your Father <u>more</u> than would be charged elsewhere. I am poor enough heaven knows. and proud as I am poor, and so long as health is spared me intend to support both of us independent of the charity of friends. I cannot for the life of me see the necessity of all this fuss. If it is not perfectly convenient for your Father and Mrs. J. to have you there, they have only to say so!– if they are willing you should stay. let them say so and done with it. and I am ready to accept, and to acknowledge the favor, as I would from any one else. I shall have to leave the matter to you, but do not stay if it is unpleasant. As I have told you before, I think your Father when left to himself, well enough disposed towards both of us. Why Mrs. J. should insist upon makeing herself and all around uncomfortable. I cannot imagine. I am sure I have always used her well, at least as well as I knew how. I never considered it necessary for me to make love to you through her, and fear that if it had been so, I should have been without a wife, or a sweetheart, to this day. Why didn"t you tell me about the fuss that Lizzie had? I think it necessary that I should know about all that concerns us, so that I may know how to conduct under any circumstances. Libby is a dear good girl, and I love her much, but I cannot allow her to get into trouble, or jeopard her own interests, on my account. This has been a fine day, and if you are well, persume you have been at church. There has been quite a lot of notables here today. <u>Secretary Stanton, Quarter Master General Meigs</u>, and <u>Commissary General Eaton</u>, with a number of lesser lights. I have not seen them, not having taken the trouble to look for them. Would like to see them well enough, but don"t think it pays to put oneself much out of the way, to see public men, unless one is a great admirer of them. I had a visit from Gen. Mead the other day. he came into the Lieuts. tent where I was at work, to see how the chimney was fixed. I have been so busy the past week that I have not read the papers, and know of the result of the elections only by report. in our mess which is composed entirely of Commissary clerks, there are two or three McClellen men!– they are of necessity rather moderate in the expression of their sentiments, but it seems too bad that there should be such a thing as a McClellen man in the employ of the Government in any capacity. I think you need not knit me the smoking cap at present. perhaps I may want you to do something else for me. We have a chimney built for your tent, with a nice fireplace, so that what time we choose to spend in it, we can be as cosey as mice. You may be assured my love, that I not only think of you every day, but many times a day. I do not think that I realize yet that you are my wife, at least not fully. But I am trying hard to convince myself that I am really thus fortunate, and think that I shall succeed in time. At present however I fear that if the question were put to me unexpectedly, "married or single" I should, before I thought deny-not my master, but my wife. I think you should have received my letter of the 10th before you wrote your last one. I hope it has not miscarried. Do you wish to reclaim the "J." in your name? if so I will address you thus! Will not until I hear from you, as I do not know as you intend to sign yourself so only in private. I have a poor memory and have forgotten whether you told me you expected to be very busy this Fall, or not, I hope you have not much to do, but to enjoy yourself. If you wish to practice much I shall be happy to furnish you any facilities to do so, that may lie in my power. I hope you will <u>learn "John Anderson My Joe."</u> &c. for me, for I quite like it. I hope you got those other pieces of music that were mentioned, in which I was so stupid as to forget. I wish I might hear you sing tonight my dear. I have heard plenty of music this evening, but it does not satisfy me not being of the right kind. please be sure and tell me all your troubles if you have any, and what mortal has not? and if there is anything that I can do for you, do not fail to let me know it.

With love.
Your devoted husband
Edwin J. Barden

No. 4 Rec'd 22nd Oct. 1864

OFFICE, CHIEF COMMISSARY OF SUBSISTENCE,
Armies operating against Richmond.,
City Point Va. Oct. 18". **1864**

My Dear Wife.

I am not going to write much tonight, for its late and I am quite tired: tired of writing. for I have been at it all day, but feel as though I could take a good deal of exercise. and enjoy it. I have been very busy all day. but thoughts of my little wife would creep into my mind, elbowing, and crowding away figures, and calculations, very much to the detriment of the interests of my employer. I have worked very steadily since being here. and shall be very busy indeed all this month. Perhaps next month if I stay I may have more leisure. I find that I misunderstood Lt. Browne in regard to what were to be my duties. He expects to make a regular Commissary clerk of me while I understood that he only wanted me for a special purpos. Had I understood him fully, I doubt very much if I should have engaged with him, for I did not think I could succeed in the business. I should like exceedingly to get along and master the business. Can probably tell by the first of next month, whether I can or not. I find no fault on the score of salary. for there is no danger of my earning more than I am receiving. yet awhile. Lieut. Browne is a queer "Dick." very sociable and pleasant. and will do anything almost for those who please him. He professes himself well satisfied with me thus far. but the "tug" will come at the end of the month. when we come to make out the papers for Washington. Lt. B. has been in the business over three years. and has the whole routine of its operation in his head and at his tongue's and fingers ends. I fear he will think me slow, and stupid, as I must appear beside him. Will you ask George if he would like a situation as clerk in the Commissary. or Quarter Master Department. if he could get the same that I am having here. There is not the slightest doubt but that he could succeed admirably in either Dept. and I think it likely that he can get a situation in the Com's Dept. before long. I have not tried the Sutlers yet, for the reason that I have not had the time: but shall do so as quickly as possible. About every other shanty is a Sutlers shop. and there always seem to be plenty of men hanging around inside them, but some of them may want help for all that. I gave your Father a little pamphlet to read entitled "The Annual Conquest of Florida." I told him to let Mr. Whittlesey read it also. I wish if you can find it without trouble you would do it up snugly and send it to me. Do not take any trouble about it, for it is of no consequence if you do not find that one. I think you will find another copy among some papers that I handied you in an envelope. you can send me that one. Mr. Moller has just come in and finding me writing, sayes "remember me to her" and I answerd "of course." He expects to go North on a furlough shortly. and will visit Dexter whenever he does go. so you may have a chance to see him. he is very much of a gentleman. and I should think would be a favorite with the ladies. I believe I had another question to ask you, but cannot think of it now.

The weather has been very fine and mild for several days past and the evenings are delightfull. There are several fine bands of music about here. and some of them are playing until bed time every evening. It is very quiet up here where we are. all the bustle and business is down at the wharves and about the railroad depot. which are a few hundred yards away. If I get along nicely I think I shall like it very well here. I hope that you are well tonight. and happy and pray that you may always be so.

With love,
Your Devoted Husband,
Edwin J. Barden

No. 5 Rec'd 24th Oct. 1864

Office, Chief Commissary of Subsistence,
Armies operating against Richmond.,
City Point Va Oct 20th 1864.

My Very Dear Wife.

Your letter of the 16th was received today. I had not expected to answer it tonight, but Lt. Browne knowing that I had received it has been at me all the evening to answer it or as he expressed "go and write to your wife." So here I am writing. It has been a busy day with us. I have been at the store house assisting Lt. B. in his sales and issues. – that is I have looked on while he has done most of the work! he being able to do it about three times as fast as I possibly could. About 3 o'clock I left, and went to attend to the drawing of some Stores, and as I could do that as well as any one, I felt as though I was accomplishing something. This evening I wanted to make out some papers, and tried in vain to do it correctly, for the Lieut. kept interrupting me, by telling me to go and answer my letter, and finely, finding that I was losing time, and neither getting on with my work, or answering my letter bade the Lt. good evening, and followed his advice. The Lt's. Office is in his tent, and I am there all the time at work, except when, as has been the case for the last two days. I assist him in the store-house. There is a chimney in the tent, and I have a colored man to keep fire for me most of the time. When I write letters I come into the Office of the <u>Chief Commissary, Lt. Col. Morgan</u>, for whom my friend Mr. Moller is Chief Clerk. It is very seldom that there is anything being done in Col. M's office in the evening, and its usually a quiet, and good place to write. This evening Mr. Moller and a friend, are playing Chess, a game of which Mr. M. is very fond. and at which he is quite expert. At present I am tenting with Mr Moller, we have a nice fire-place and chimney to our tent, and a fire when we choose. A colored man and builds a fire for us to get up by in the morning, and blacks our shoes, brings water for us to wash with, makes up our beds, and takes care of us, and our tent generally. Yesterday was my first day in the store house! was occupied most of the time in making sales of provisions to officers! this is the principle business of Lt. B. he only issues rations to a few detachments immediately connected with Headquarters. The amount of sales varies from $75 to $125 per day! Yesterday <u>Genl. Grant</u> came into the store house to weigh himself. it was the first good fair sight I have had of him. He is a very unpretending appearing man, has a very slight stoop in his shoulders, and with his back turned, and at a little distance, has a little the appearance of Nelson Brown. I also <u>saw Genls Mead</u>, <u>and Benham</u> yesterday as they rode by. Plenty of great (or would be great) men about here, if one only has the time, and disposition to watch for them. There is much less stir and bustle around there Headquarters, than there used to be at the Headquarters of the 10th Corps. I do not know exactly how to account for it. But its lively enough down on the wharf, and about the storehouses. I am having a very unpleasant time with my teeth, which I supposed I had placed in good "grinding" order, before leaving home. I have one large double tooth, which was still, and was so sensitive while at home that I could not use, but thought that once over the soreness caused by the operation, it would be all right, but intended, it has grown worse. and for the two past days has ached a good portion of the time. I must have it extracted unless it speedily reforms and behaves better. I thank you much for your kind letter of the 16th. I assure you that I have no idea of forgetting my little wife and hope that I may never live to do so. It is good of you to continue to love, and trust me, after having been partially undeceived as to my true worth and merit!– be assured that I appreciate your kindness, and most sincerely hope, and wish that you might not be further disappointed. I am very grateful to Libby for her love, and good opinion of me. I trust that I appreciate her favor and esteem, and earnestly hope that I may never forfeit either. I wish that you might see Sara B. before she leaves. I heartily agree with you that I am mor than blest in having the love of such dear, good sisters. Julia has been very good indeed to me!– the kindness shown me by both her and George can, and never will be forgotten. I have always liked Emma very much. though I have had but little opportunity to become acquainted with her. I do not forget Carrie among the number of those I love. Dexter is more fortunate than I in being able to remain about home. Am very glad indeed that he succeeded in getting the School. – he has been a most excellent friend to me, and knows me better than you do. for he has seen me tried in positions that you never have, but he is naturally very kind hearted, generous, and ready to forgive others faults, and so I doubt not, has thrown the mantle of charity over many of mine. I fear that his testimony in my case would be hardly reliable. I do regret exceedingly that my abilities are not such as would warrent me in accepting a situation in the Bank under Mr Townsend. I could well afford to remain there and work hard six months for nothing until I became acquainted with the business, for I should surely be able to make it up in after years, but it is impossible. I am conscious that three years in the army has ruined more than one of my life prospects. not necessarily perhaps but still most effectively. I should write to Mr. T. as soon as I have time, and thank him for the kind interest that he manifests in my wellfare. It may be that he may yet be able to help me when I need it still more. I do not know of one to whome I would more readily and confidently apply. I think it would be pleasant to live in New Haven, for it is a very pretty place, and I am much pleased with what I have seen of it. I think I shall always find <u>something</u> pleasant anywhere where <u>you</u> are, and would esteem it a great privilege to be premitted to spend even a Sabbeths with you, but as you say, "it may be for the best" that we are thus separated. – at least for the present. We have hope for the future, and mutual love, and trust in God for the present. I must beg your indulgeance for this letter!– it being late when I commenced. I have been obliged to write rapidly, and with little care. You mentioned that you wish some articles of dress, for winter use. – have you sufficient money to purchase them? if not let me know.

With love
Ever Your Own Dear (I was going to say "Ned") Husband.
Edwin J. Barden

No. 6 Rec'd Nov 1st/64

OFFICE, CHIEF COMMISSARY OF SUBSISTENCE
Armies operating against Richmond,
City Point Oct. 16th **1864.**

My Very Dear Wife.

    I am receipt of your three letters of the 18", 20" & 23d. Please accept my thanks for your goodness in writing so frequently. I regret that you will have looked in vain for several nights past for letters from me. I fear that I am inexcusable for the omission: though I have been exceedingly busy. Three nights ago I made two ineffectual attempts to write. but not feeling well on account of a cold, I was rather pettish, and getting vexed at errors that I made, tore up what I had written and went to bed. Tonight I have so much to say that I know not where to begin. I received your last letter this evening while in Lt. Browne's tent. and read it there. Mr. Moller was there, and I gave him your message, with which he was much pleased. Lt. Browne asked if I would deliver you a message from him and I readily agreed to, but am sorry that I did, for I think it entirely inapplicable, and improper, but as he insisted upon my compliance with my agreement, I will do so, but under protest. it was: "Tell her that I hope <u>she</u> will make as good a wife as I am sure <u>you</u> will a husband." As he is more than enough our senior to be our father, he may be excused for sending such an ungallant message. Lt. Browne is an Englishman, who has been in this country about fifteen years: he is I should judge about fifty-five years old, a large tall man with a bald-head, usually very pleasant and jolly: he is married and has one boy about fourteen years old of whom he is of course very proud. His family are in N.Y. City. His time in the army will expire about the first of December, but he will probably be promoted before that time. to a Captain and Assistant Commissary of Subsistence. I dont know as I can succeed very well in giving you an idea of my duties. It is entirely different from any business of which you have any knowledge though it is nothing but what you could easily understand. and readily become acquainted with. Lt. Browne's principle business is selling Subsistence Stores to Officers. and such citizens as are entitled to purchase of the Government. He also issues rations to several small detachments of troops who are on duty about Head Quarters, and to destitute citizens, or refugees. He generally attends to the sales and issues himself. though I have done it several days. and expect to whenever he feels like resting. There are no Books to be kept except one containing the amounts of sales. All the accounts are kept on printed blanks furnished for the purpos. There are I believe about fifteen different sets of papers to be made out each month: all to be in duplicate. and one copy of each to be sent to the Head Quarters of the Commissary Dept. at Washington. I shall be very busy indeede for the remainder of this month and for several of the first days in the next: and then hope to be able to get along more easily having at least less to learn. if not less to do than this month. We are considerably interested tonight in the forward movement that is supposed to be being made. It is understood that <u>Gen. Grant</u> startes tomorrow morning. I hear indirectly that the army is in motion today. All day yesterday the sick men were arriving at hospital here, whither they were being sent from the front:-whole train loads of them continued to arrive during the day. I hear that all the extra teams, not absolutely necessary for the use of the army in its movement, are being sent tonight within the fortifications at this place. I can only hope. and pray that the move may be successfull. A capital band of music is playing in front of the Genl's quarters while I am writing. I wish you could hear them: they do play splendidly. I believe they belong to the <u>10th Infantry (Regulars)</u>. You inquire about my health: it has been excellent thus far, my cold is not severe. and I hope to recover from it in a few days. I have grown fleshy since being here. Yesterday I weighed one hundred-fifty and a half. pounds. which is above par. for me. My weight while at home was about one hundred forty-two. or four. pounds. The weather has been excellent thus far, very moderate, and with out sudden and unpleasant changes. I still wear the same clothing that I did while at home, reserving my woolen shirts, and thick pants for cold weather. There are three of us who occupy one officers, or wall tent, as they are called. We have a nice fire place, and a good fire whenever we choose to have it built. A negro comes in every morning and builds us a fire. to get up by: blacks our shoes and puts the tent in order. Mr. Moller and I bunk together. We have had a nice. new, wide bunk made for us, and are going to get a tick made and filled with straw, and then we shall have a bed fit for a king. I am going to buy me one government blanket, which with my shawl, and Mr. M's blankets. will make us a plenty of covering. The third occupent of our tent is Mr. McClure, who is clerk for his brother <u>Capt. McClure, who is an Assistent Commissary of Subsistence</u>. They are members of one of the best families, and extensively connected with wealthy and influential men, both North and South: they have many relatives, and a host of former friends and acquaintences in the rebel army. Their family residence is in Carlisle, Penna. but they have lived in Washington a great deal. Both the Captain and his brother are gentlemen by birth and education. and being very affible, and entertaining, I enjoy their society very much. and only wish that I had more leisure time to spend with them. The Captain received orders tonight to report to Cleveland, Ohio (I think) for duty:- he does not wish to go. and is going to make an effort to stay with this army. and I sincerely hope he will be successfull. I hope George has made a good investment in Vineland. What does he propose to do there? He could do well here no doubt, but I do not blame him, for remaining within the pale of civilization, even though he should not make it pay nearly so well. Please ask him if he has done my errand to the jeweller at the Falls. and say to him that I shall write to him as quick as I can possibly get time. I have not heard a word from our regiment since I came here. nor from the boys at the office at the Corps Hd. Qrs. I have only written two short letters besides those to you. I am much rejoiced to hear of <u>Wells' safe arrival home</u>:-he will have a world of adventures to tell, and no doubt his friends will keep him busy this winter. I thank Norton for his message, and have him booked for one of the first letters that I get an opportunity to write. You are having a nice time to yourselves at housekeeping. I wish I could be with you now. I think we could all have a quiet sociable time. I intend to write to your Father soon. and if it be your wish to remain there for the present, settle with him definately in regard to terms. I wish you to be as comfortable as possible. and have whatever conveniences you may wish. I have chafed a little over that unpleasant remark. and cannot easily forget it. It was I think very unjust, and unwarrented by the circumstances. You may remember that I expressed once, an unwillingness to be at your Father's so much because I feared that he would think that I was presuming to much, and abusing his hospitality. I have no doubt but that he feels kindly towards me, and I certainly bear him no ill will, but much to the contrary, I do not intend to ever notice the remark further, unless it is repeated. I desire to be on the most friendly terms with all the members of your family, and no one shall ever know from me but what I am. even though there should be cause for ill feeling, which at present there is not. I cannot tell now whether I can come home at Christmas. or not. I may not remain here, or in the army until then. It will be a bad time to leave on account of its being the last of the month when all the accounts for the month must be made up in readiness to send off to Washington. I shall try to get away, if things remain as they are, if it is only for a day. I am booked for a Christmas dinner at Mr. Randalls, and shall use every proper endeavor to be on hand. I hope we may have the pleasure of dining there on that day. It is uttely impossible my dear, for you to live here. With the exception of three or four ladies about Gen. Grants Hd. Qrs. there is not a lady in the place. I do not think they are allowed to come. And even if it were at all practicable, I should not on the whole wish you to be here, but its no use wasting words. the thing is impracticable and I wonder what people can be thinking of when they ask why I did not bring you down with me. I cannot answer your letters fully tonight, having been interrupted for some time, and it is quite late. Will you please mention again anything that I may have omitted to notice, and of course you may wish to know. I will try to write often, but please excuse me if I fail to do so for the coming week or ten days. Be assured that I do not forget you but that I am thinking of. and loving you all the time. Love to all of <u>ours</u>. and a kiss for Lizzie.

Ever Your Own Devoted Husband,
Edwin J. Barden

Editor's Note: The following letter reports on the <u>Battle of Fair Oaks and Darbytown Road</u> where Union casualties were 1,603, Confederates fewer than 100.

No. 7 Rec'd 5th Nov. 1864

HEADQUARTERS, ARMIES OF THE UNITED STATES
OFFICE OF CHIEF COMMISSARY,
City Point Va Oct 29th 1864.

My Dear Wife

I received your letter of the 24th inst. today, and improve the opportunity which this evening affords me for writing, to answer it. I am very grateful to you for writing when you are so weary and tired as you must have been after your hard days work. I fear that if it had been my case I should have considered that I had a sufficient excuse, for going to bed and letting the letter wait another day. Seems to me you are not resting much! I thought you were not going to do anything this winter except for yourself, but here you are blistering your hands, and getting as round shouldered as a washerwoman. I dont blame you for not wishing to keep house, for I know enough of a housewifes duties to be able to judge that they are not only very arduous but many of them very unpleasant. There is a great deal of drudgery to be done, that I can easily sympathize with any lady in wishing to avoid. You need not fear being soon obliged to assume the duties of the housekeeper, for the very best of reasons! There may be a little advantage in being a poor man's wife after all. But I think it a very questionable one! You do not mention that you had any invitation to attend the surprise party, at Mr Wells' so I conclude that the gentlemen have turned you the cold shoulder already, and that henceforth you will find yourself "counted out" when the parties of young people are being made up! that you will find yourself left in the ranks of the "home guard". You may not mind it much at first, but after the novelty has faded away, I fear it will begin to be rather tedious! I have regretted since my return here that I did not exert myself more to see Charlie Couch. and satisfy myself as to his real feelings toward Lizzie! Sometimes I think I must have appeared indifferent about it to Lizzie! but

I was not!– there seem to be no way for me to do anything in the matter unless I went to see him. which would have taken valuable time that it did not seem that I could spare. It does not seem now as though I have been home at all The recollection of the few days spent there, is like that of a dream! This life is becoming a sort of a second nature to me, though I am sure that on the whole I do not like it, and would gladly exchange it for one more congenial to my tastes, were I able to do so. The movement of the army of which I made mention in my last, did not result in any advantage to us! and to no great loss. I believe it is authentically reported that 500 will cover all our loss. It was found impracticable to extend our lives further to the left with our present force. so the army after a little skirmishing retired to their former positions. A Division of one Corps had a spirited engagement in the night. the particulars of which I have not learned but am creditably informed that they were in no way defeated, and that their loss did not exceed 300. Less than 200 were lost in the skirmishing during the day. I suppose the Copperheads will try to make a great deal of capital out of it for election purposes. but trust that they will not succeed. I begin to feel very anxious as the time for election approaches! The Copperheads seem very confident, but it does not seem possible that men in their sober senses, whatever may be their party feelings could submit to the undoing of all that has been accomplished at such a priceless cost, and plunge the country into a still worse condition that it has ever yet been in. My teeth are behaving well now, though the one that is so sensitive still prevents me from using any on that side of my face. I am better of the cold, that I mentioned in my last, and am feeling very well. I did not mean to convey the idea that I was averse to writing to you on the evening referred to! and trust that I may never be so. But having written two evening before, it did not seem quite necessary. and there was considerable work which I thought required immediate attention. I do think that in many respects I am very pleasantly situated considering the circumstances!– it is vastly different from what it was when I was a soldier. You told me in N.Y. that you should require a Cloak, and hat for the winter as well a dress. You only mention sending for a dress. Can you get the other articles near home? I dread terribly to come down to, and vex you with, the practical, so soon, but do not well see as I can avoid it. Will twenty Dollars be sufficient for your present wants? (not including that required to pay for board). I borrowed a small amount while home, rather than ask Mr Randall for the little sum deposited with him, and I should like to repay it this month, though it is not absolutely necessary. and if it will interfere with your wishes. I can easily defer it, and send you more. Please let me know. and do not give up purchasing anything that you had intended to. Excuse me for troubling you thus, but I did not like to send you so small a sum without explaining why I did so. I fear that it will not be sufficient, and by that you will be sure and tell me if it is not! I had intended to have written a letter to my niece Fanny Hill, after finishing this. I have not written to her yet though I promised to do so very soon. and fear that she will think I do not care for her, or have forgotten her which amounts to the same thing. A party with a banjo and violin have been singing and playing just outside the window, and you know how difficult it is for me to do anything when there is any music within hearing. I have had to stop writing and listen so much that I guess that poor Fanny will have to wait still another night. If you find subjects rather mixed up, and more errors, than common in this letter, you must attribute it to the attempt to write under difficulties. I am making such poor progress that I think it best to stop work for the night and go to sleep – the only thing that I can do well, under the circumstances.

With a Kiss I remain
Ever your loving Husband Edwin J. Barden

No. 8 Rec'd Nov 4th 1864

Hd Qr's Armies of the U.S.
City Point, Oct. 30th 1864

My Dear Wife.

I received yours of the 25" this afternoon: it was the more welcome because unexpected. having received one letter from you yesterday. and feeling a little guilty at my own failure to write for several days, I had not thought to be favored again so soon. I think you for taking so much pains to let me know all that is going on. Wish I might have formed one of the party of "sons and daughters" that met at your tea table. I fear that it is to be my lot, to be a wanderer and an exile from all the enjoyments of home. But I will not complain. All's for the best doubtless, and I ought to be gratefull that I am even thus fortunate. God has given me a dear little wife, and while he gives me health and strength to labor for her happiness, it becomes me not. to murmer, or complain that I cannot choose my own station. or kind of employment. How does Julia like the idea of going to Vineland? I think I must be a little prejudiced against that section of the country, for I am quite at a loss to imagine what inducements there can be for New England people to remove there. Does George think he can do any better at any mechanical employment than he can at home? I have never had a very high opinion of New Jersey. and I hardly think that it is considered quite on a par with its sister States in many respects. Emma told me that she wished you to come and stay a long time with her. I cannot remember my reply. I knew that you had some reason for not wishing to remain there long at a time, but do not know that you ever definately explained it:- as I have no doubt it is a good one, it is of no consequence if you do not. I have no idea that it arrises from a lack of sisterly regard for Emma. Am glad that you are pleased with your dress, and that it is such as you wanted. I wish that I had received your last letter before I wrote my letter last night. I would not have written as I did about the money. I fear that you will not have sufficient for your wants, and think that I had better send you the greater part of my this months salary, the little affair that I mentioned can wait. I regret exceedingly my inability to do more for you. it would give me the greatest pleasure to be able to satisfy all your desires. Please not to deny yourself of any thing that it is in my power to get for you. I am making blundering work, in attempting to handle this subject, but I trust that you will understand, and pardon me. Do not worry yourself about being so expensive. I know that articles of dress cost money and know that you would not purchase articles that you do not need. I can only repeat that I exceedingly regret my inability to gratify your every wish. Capt. McClure and his brother left us this morning. I felt very sorry indeed to part with them. I have enjoyed our short, and slight acquaintance very much. Mr. Moller and I have the tent to ourselves now, and shall not have a third party in it again if we can help it. We have a table, at which I am now seated, camp stools and some other little conveniences, and we are expecting to keep a model bachelor's establishment. I took a horseback ride tonight. Mr. Browne having very kindly placed his pony at my disposal. I am no horseman, and can hardly keep on a horse if it is going at all fast, but I let the pony walk or go at a slow trot, and enjoyed it very much. I send you a rose that I picked from a bush that is trained up over the piazza of the house close to which our tents are pitched, and which is occupied as the offices of the Chief Commissary and Chief Quarter Master of these Armies. The house is a very ordinary one story and a half house with an 'L' part larger than the main building extending to the rear and with dormer windows in the roof, on all sides. It was the property of Dr. Epps, who is reported to own a large tract of land in this vicinity extending across the James nearly over to Malvern Hill. The location is a very beautifull one. Some time I will try to give you a description of it. The tents of Genl. Grant and his Staff are pitched on the lawn in front of the house, ours are a little to the rear and a few rods from the Generals. It has been very quiet for a day or two, but this evening I hear occasionaly the report of cannons. I should enjoy being with you tonight dearest, but it is idle to wish. With love, I remain

Yours Devotedly,
Ned.

CHAPTER 18

EXPEDITION TO NEW YORK

From the book:

On the night of November 2nd General Hawley was put in command of a Provisional Division for duty at New York. This division included his brigade, four other regiments from the First Division, Tenth Corps, and five from the Eighteenth Corps. General Hawley commanded the division, Colonel Abbott one brigade and Colonel Rockwell the other.

    The necessity for this expedition arose from two causes. The time of enlistment of the three years' men had expired or was expiring; the immense loss in the actions of the summer had depleted the army. Large bounties only brought to us bounty jumpers who deserted the first time they were placed on the picket line, or if we contrived to keep them they were worthless. It became necessary to resort to a draft. This, on the eve of a presidential election when party spirit ran high, aroused bitter opposition among the rougher elements of New York City. General Dix, who was in command, feared that the opposition to the draft, added to political excitement, might cause a riot before which the civil authorities would be powerless, and in order to avoid this requested that sufficient troops might be stationed near the city to keep order and secure a quiet election.

    The division embarked November 3rd, changing vessels at Fortress Monroe and reported to General Butler at New York November 6th. The infantry disembarked at Fort Richmond, the artillery at Fort Hamilton. On the 8th in the morning (election day), all the troops re-embarked, Colonel Abbott with the Seventh Connecticut on the armed transport "Augusta," took post off Catherine Street Ferry, East River. Colonel

Rockwell, with the Sixth Connecticut and others on the "John Round," off the foot of West 26th Street, North River, and Lieutenant Colonel Randlett with the Third New Hampshire on the ferryboat "Westfield," off West 42nd Street, North River. General Hawley commanded the whole. General Butler's headquarters were at the Hoffman House. Everything was in readiness, the ferryboats with artillery were in the slips, the horses hitched, and the infantry lay off the piers in marching order.

The day passed quietly; no troops were seen in the city except the headquarters' guard. The force remained near the city for about a week, then returned to their camp near Laurel Hill, reaching there about the 17th.

No. 9 Rec'd Nov. 8th 1864.

Head Quarters Armies of the U.S.
City Point Va. Nov 4th 1864.

My Dear Wife.

Your letter of the 30th of Oct. was received tonight. I regret that you should have been so long without a letter from me, but hope that ere this you have received one of my last letters explaining the delay. I assure you that I have been extremely busy, and have had no time for writing. Most of our papers are made up now, and I hope to have the evenings more to myself. I have been attending to the sales, and issues for the last four days, and Lieut Browne has done the writing. My evenings have been employed in assisting him. Last night I attempted to write to you, but getting vexed at the errors made, threw it in the fire and went to bed. Yesterday was a dark, rainy, day and I had worked both hard, and late, it was eleven o'clock when I sat down to write. Today I have not worked so hard, though I have been busy until late this evening. I am quite curious to know what special reason you could have for wishing to visit at Emma's just at this time, and why you defer telling me of it until that afar off "sometime." when you are going to tell me everything. I expect that if fortunate enough to live until that time comes I shall be very wise. There is an immense amount of mystery to be revealed to me when that important day arrives. It will be one month tomorrow since we were married. It seems a long time, but figures are stubborn things and will have their own way. I received my pay on the last of Oct. and am hesitating whether to send the money to you my mail, or wait and send by my friend Mr. Moller, who expects to start north on his leave of absence next Tuesday. As he will spend some time in N. Y. before going to Salisbury, I fear you will need it before he gets there: it will doubtless reach you by mail and I think I will send $10. in this letter at any rate. Please acknowledge the receipt of it without delay. I had the honor of dining with Lt. Browne this evening. We had oysters, roast beef, potatoes, rice, cabbage, pickles, coffee, and some kind of pastery, for desert, of which I do not know the name, but it was very nice. Lt. Browne and Mr. Moller mess with <u>Lt. Col. Morgan who is Chief Commissary</u> of all the armies operating against Richmond. The Col. is away today and we three had the meal to ourselves. Their mess bills are quite high, reaching to thirty and thirty five Dollars per month. They breakfast at about half past eight, have a lunch at noon, and dinner, which is their principle meal at half past five. My mess bill for the time I was here last month was a trifle over six Dollars, that of other members of the mess who were present all the month, about nine Dollars. Not very extravegant. It will be higher this month because some of the members found fault because we did not live well enough to suit them, and the caterer for the month will not be quite so economical. Then too, Thanksgiving come in, which will require some little notice. My daily routine is about in this wise. Rise at seven, breakfast at eight, dinner at one, supper at six, work all the time when not at meals and usually two or three hours after supper. I have scarcely looked at a paper since I came here and am as ignorant as a mule of what is going on around the country. Have written two short and necessary letters besides those to you. I have but little time to think of anything besides business, but assure you that you are not forgotten. My health remains good and I enjoy myself as well as could be expected. If I have a touch of the dumps occasionaly it is nothing more than might reasonably be expected under the circumstances, and my time is too much occupied to allow me to indulge in them long at a time. Please pardon me for a short letter. Remember me kindly to all my <u>new</u> relatives, though <u>old</u> friends.

With love, ever your
Devoted Husband
E. J. Barden

No. 10 Rec'd 12th Nov. 1864

Head Quarters Armies of the U.S.
City Point Va Nov 8th 1864

My Dear Wife.

Had you rather have a short, poor letter from me tonight, than none at all? It has been rainy for the past two days! a warm, drizzling rain, which soaks one through imperceptibly, and renders a fire so uncomfortable that one does not like to remain near even so small of one, long enough to get dry. Mud is every where, and abundant. It is of no use to black shoes, or brush cloths. There is a derth of news, and matters of interest here now! for severel days past it has been unusually quiet! A little firing along our front. just enough to remind us that the war has not yet ended. The excitement will commence in a day or two, when the election returns begin to come in. It has seemed but little like election day here, and I have found it difficult to realize that a great, and mighty struggle between loyalty and treason was really in progress. I am fearfull of the result, and await the tidings with much anxiety. I regret exceedingly that I could not have the satisfaction of voting against the vile miscreants who are seeking to destroy our country, and deliver us over, bound hand and foot to the detestable Southerners. I have been attempting to work at the papers today, in the office: but Lt. Browne has kept me on the run most of the time, doing errands, so that I have not accomplished much. Yesterday afternoon I was out in the rain and mud, drawing stores. I think I have a boil coming on the back of my right hand! it is swollen and inflamed, and looks as though it was going to fester, and run! it is quite painfull this evening. Mr Moller expects to start on his pleasure trip in a day or two. I shall miss him much. he says he shall see you, and I think I shall send some money to give you by him. I wish that you would have some photographs taken the first opportunity. that is. the first chance that presents itself of getting <u>good</u> ones How does the new dress come on? I am entirely to stupid to write this evening, and by that you will excuse me. These few lines will serve to show that you are not being forgotten, but that you are thought of, and loved as ever. I dont know where you are tonight, but if any of my letters are present give them a kiss, and get them to give you a dozen for me. With love I remain

Your Devoted Husband
Edwin J. Barden

No. 11 Rec'd Nov. 14th 1864

Head Quarters Armies of the U. S.
City Point Va. Nov. 11th 1864

My Own Dear Wife.

Your dear, good, long letter of Nov. 4th is just received. Mr. Moller received it when the mail arrived this afternoon, and forgot to give it to me until just now. He is more than excusable however for he is quite sick tonight and I very much fear that he will be much worse, and be sick for a long time. He is sadly overworked, and much in need of rest. Last night he worked until very late, and today has hardly looked up from his desk until compelled to, by sickness. A certain piece of work was required to be completed, and then he could have his long looked for leave of absence; he is in hopes to get through so as to start tomorrow, but the prospect tonight is that he will not be well enough to do so. I do hope he will be able to leave, for I think that rest will recruit, and cure him up. He has been out now nearly three years and has never had a leave of absence, and has always worked very hard. His position at present is a very responsible one, and the great trouble has always been that he was to valuable to be spared. <u>Lt. Col. M. R. Morgan is Chief Commissary</u> of all the Armies operating against Richmond, and Mr. Moller is his Chief Clerk; his salary one hundred and twenty-five Dollars per month. In expectation of his starting tomorrow, I had intended to write to Mr. Randall tonight, enclosing him the money referred to in a previous letter to you, but on receipt of your letter, thought I would write you a short letter tonight and let the other rest for the present. Please not to make any excuses for writing me short letters, for I fear you will receive many of that kind from me, and I am always sure to be your debtor on the score of letters. I am writing at a considerable disadvantage tonight, for I have a boil just ready to break, on the back of my right hand; it is quite painfull, and makes it rather difficult to hold a pen, as it is considerably bundled up. I think it will come to a head tomorrow, and I shall then soon be rid of the troublesom thing. I managed to write Fannie a short letter last night, and hope for an opportunity to write again soon, when I will give her your message. We are having remarkably warm weather for the time of year, though I do not think it as healthy, as though it were cooler. It would be plenty warm enough in my tent tonight without a fire though I have a small one on account of its being so damp. Both Lt. Browne and myself have been at the Store house today. I have not been able to write much and so have attended to the sales, while Lt. B. has been issuing rations to soldiers. I am glad that <u>Lt. Moore</u> at last received a leave of absence. he was intitled to it long ago. I hope I may see him on his return. Had our regiment been where it was when I left the service, I should have visited it before this time, but it is now across the James on the extreme right of our line, and not easily accessable. I have a great respect for Lieut. Moore, and sincerely hope he may survive all the dangers of the war and retire to the enjoyments of civil life, crowned with richly merited honors. I suppose that ere this George is on his way to, if not arrived at, his new home. I shall look for his promised letter with a great deal of eagerness. Lizzie may love me the best of any of her brothers, if she can, I am gratefull for her love, and prize it highly, but if she knew George's worth as well as I do, she would think differently. The army is sure to place every man in it in his true position and character. before his fellow soldiers. I have seen George tried , as I will venture to say, even his wife never did, and I know that a kinder heart, and a truer man never existed. I regretted not being able to see Josie Orton when I was home, and can appreciate your disappointment. Her name brings to my mind many pleasing recollections of happy hours in bygone times. I often think that though the number of my lady friends is quite limited, it is more than compensated for by the goodness, and many virtues of those I have. They all seem very dear to us, and I hope I may never forfeit their esteem, and good will. You must not be to sanguine of seeing me at Christmas. If things go smoothly I shall try to visit you then. But I seem to be entirely the creature of chance and to be always ruled by. instead of ever controlling events. I wish if possible to remain somewhere here where I can continue to receive my present salary at least until Spring. Should it be necessary in order to do so, to give up my anticipated visit, I shall consider it best to do so. I hope however that I shall be able both to visit you, and remain here until Spring. You must not be surprised however if am home for good even before Christmas. I hope you remember that if I come home, our Christmas dinner is to be at Mr. Randall's. I cannot answer your long letter fully tonight, but will bear in mind the unnoticed points, and notice them at a future time. Your request shall be considered. I do not wish to promise hastily. If I make you a promise I shall keep it. Be assured my love that I am not angry or displeased at your request, though I trust you will pardon me for saying that I think it quite unnecessary. There is now no neede of your asking me to think of you every day. I do not fail to do so very many times a day. I cannot write more tonight. Please accept much love from,

Your Devoted Husband,
Edwin J. Barden

No. 12 Rec'd Nov. 17th 1864
By Mr. Moller.

City Point Va Nov 11th 1864

My Dear Wife.

  Enclosed please find a Forty Dollars ($40). Fifteen of which is for yourself, and the remaining Twenty five for Mr. Randall. I will write Mr. R. a letter when I get time, which I will send to you and get you to put the money in it, and send it over some day by your Father, or other safe conveyence. Mr. Moller has just come in, and is going to start North this evening, in a hour or so. I have therefore no time to write to Mr. R. Mr M. will call on you during his visit to Leakeville. I know that I neede not ask you to do whatever may be in your power to make his visit pleasant. my hand is very painful today and I have not been able to do much. It is not yet broken, and is very troublesom.

Please excuse haste
With love,
Your Devoted Husband
E. J. Barden

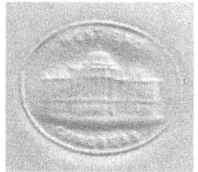

No. 13 Rec'd Nov. 17th 1864

HEADQUARTERS, ARMIES OF THE UNITED STATES
OFFICE OF CHIEF COMMISSARY,
City Point Va. Nov. 13th **1864.**

My Dear Wife.

"It never rains but it pours":- three letters yesterday and one today. Yours of the 8th was received yesterday together with one from Dexter and one from Mr. Cooke. Dexter commenced by rating me soundly for not having written to him. and ended with election news. and a description of the pleasures of Salisbury society this Winter. Mr. Cooke's letter contained a report of the result of the election in Canaan and adjoining towns. I answered Mr. C's letter last night. and think I may possibly write to Dexter tonight. though it will be hardly necessary as Mr. Moller will be there by the time the letter would. and I cannot possibly write anything of interest that Mr. M will not tell him about. Your letter of the 10th came today. I am very grateful to you for writing so frequently. Your letters are a great comfort to me. I am almost sorry that you love me so well. and anticipate so much happiness some day in living with me. That I love you I am certain. that I shall always endeavor to make you happy I have no doubt. but I sadly fear that I can never be your ideal of a husband. that I can never succeede in being what I consider a good husband. I will do my best but my conscience tells me that I have deceived you in regard to my true character. and worth. and that so surely as we are ever permitted to live together long. you will find it out. But enough of this, for you'l not believe me when I talk thus. and I know that only time will convince you. I am quite vexed about the ring, vexed at my stupid carelessness in forgetting it. and sorry that George did not think of it when at the Falls. and pay what Mr. Solmonson asked. I shall enclose the money to Mr. Randall at once and have it settled. I wish the price had been twenty instead of five Dollars. Lieut Browne has just handed me in a parcel of newspapers. they are State papers sent me by Mr Cooke who favors me in this way occasionally. We are experiencing the first real cold weather here at the same time that you describe it as being so unpleasant at home. it was quite warm here though. and the most discomfort we experienced was from the mud. I am "monarch of all I survey" now in our tent. I would feel quite lonesome if I had any time to indulge such thoughts. Mr. Moller has a leave of twenty days. he ought to have had forty. but he is too valuable to be spared. Did I tell you that his salary is one hundred and twenty-five Dollars a month? his pay is better than that of a Captain in the army. I shall keep a bright look out for <u>Lt. Moore</u> at the time specified for I should like very much to see him. I do not know the person named by Mrs. Miles. I have made no acquaintenances here except those with whom I have come in contact in the way of business. There might be fifty persons here by that name and I not know them. I will enquire. and it would help me greatly if Mrs. M. could tell what business Mr. Dewey is in. I am obliged to Auntie for her friendly interest in me. and her kind inquiries and message. please remember me to her when you meet. I suppose Lizzie is enjoying herself with the rest of the young people this Winter at the pleasant surprise parties, and other gatherings which Dexter tells me are making it so pleasant in S--this season. Tell her I often think of her and love both her and Julia very much indeed. I am sure I could not think more of them if we had both had the same mother. Julia has had rather more of a "widows" experience than you. so I suppose she takes her present bereavement quite heroically. She has much the advantage of you, in that her future is not so uncertain. I should feel perfectly independent, and ask for nothing but good health if I were possessed of George's skill as a mechanic. I suppose his health has improved and judge that he would hardly undertake "active operations" in the condition he was in when I left. If there is any new music that you would like please let me know and I will send for it for you. I regret that you could not have visited the music store when in N. York and made your own selections. Any pieces that you may know of and wish for. I can easily get you at any time. I suppose you have seen nothing of my promised Album yet. I have heard no more about it since I came away. So Tom Norton is keeping school. I must say that the Salisbury delegation to Co. "G" has graduated <u>a reasonable</u> number of pedagogues. Should not suppose Tom could muster up dignity enough for a schoolmaster. I believe you told me that Mr Landon made a purchase in the New Jersey "El Dorado." I do not remember that you informed me what Mr. Wolcott and Mr. Butler decided upon. Is there any school kept in the Academy this Winter? What has become of "Abigal?" I don't remember to have heard of her in some time. Maybe she has got married! I wish I might be with you tonight. that we could have the evening to ourselves without interruptions. I am sure I could enjoy your society. as I scarcely had an opportunity to do while at home. There was so much to do in the way of visiting and running about that it kept me so fatigued most of the time that I was scarcely wide awake to any enjoyment. Will you do me the favor to ask your Father if he has any boots, such as he and I talked about. that he thinks will fit me. A large 8. is about my size. If he has any please ascertain the price. It may be cheaper to buy them here. if I was certain of getting a good article. I fear my letter tonight will not prove very interesting. it is with difficulty that I can controll my thoughts at all tonight. Your unanswered question is not forgotte. With love

Your Devoted Husband,
E. J. Barden

No. 14 Rec'd Nov. 21st 1864

Head Quarters, Armies of the United States
City Point Va. November 17th 1864.

My Dear Wife,

It is my night to write to you, and I have seated myself with a full purpos to do so. But I find myself woefully short of material for a letter. If I were with you this evening no doubt I could find enough to say, but there really does not seem to be anything to write about. I often think what a stupid loon I must have appeared much of the time I was with you, when at home. For three long weary years we had been separated, and in our letters had often congratulated ourselves upon the drawing to a close of the period of our separation, and promised ourselves that we would make ample amends for the long pause in our conversation. We should have so much to tell each other. so much to say, that it was not practicable to, on paper. We thought we should never weary of talking. But I think we were both disappointed. I do not remember to have had a single hour of the pleasant chat I had anticipated. Much of the time I was with you, I was so weary from running about that I was unfit for Society at all, and then my proposed return so soon, seemed to drive almost every other thought out of your head. It seems too, as though all the pleasure that I am ever to enjoy, is to be in anticipation only. My visit home, to which I had looked forward for three long weary years, would have been almost a failure, had we not been married. I dont mean to infer that I did not enjoy my short stay very much indeed, for I did most assuredly, but then it fell so far short of what I had pictured to myself, I had thought that I should spend an idle Fall, living on my friends hunting fishing, and visiting when and where I was a mind to!- in short enjoying to the extent of my ability that liberty of which I had so long been deprived. I know that I have every reason to be thankfull, and trust that I am so, for the greatly advantageous turn in my affairs that have taken, but I cannot bring myself to feel kindly towards that property that drives me so remorselessly from home and my little wife and binds me so unremmittingly to toil. for be it known unto you, Dear Wife, that I am idle by nature, and from my earliest recollection have had a mortal aversion to work of all kinds. be it mental, or physical. It is very mild and comfortable tonight after three days of cold blustering weather. I am feeling very well! the boil on my hand has nearly healed up, though I have still to keep it done up. I was in the Store house this forenoon attending to the sales! in the afternoon I was drawing Stores from the Depot Commissary, and attending to other out side business. So that I was out of doors most of the time. I have the promise of a quiet day tomorrow to work at the papers which are getting a little behind, but I know to well how my "quiet" is aft to be intruded upon, to anticipate making any great degree of progress, Lieut. often sayes, "Well I will attend to the Sales today, you need to do nothing but work at the papers," before night he will have had me running all over the Point half a dozen times, on as many different errands, besides interrupting me almost incessantly to look up some little thing for him. I am lucky if I manage to steal a couple of hours out of the twelve, in which to work at the papers. He's a queer "fish" and no mistake, I expect he will go North in the course of a week or so, and leave me to take charge of things alone, which will be to me an unenviable job, with my little experience in the business. I shall do my best, and if I succeede tolerably shall have more courage to ask him for my much desired Christmas holiday. You have doubtless seen Mr Moller before this time, I ought to write to him, but do not know as I shall get time. There is some fireing going on at the front tonight, quite a brisk musketry fire is in progress while I am writing, along the <u>Bermuda line</u> of intrenchments where our Corps lay last Summer, I presume it is nothing more than a picket skirmish. I hope you are well tonight my dear, With love. Good night.

Ever your affectionate husband,
Edwin J. Barden

No. 15 Rec'd Nov. 22nd 1864

Head Quarters Armies of the U.S.
City Point Va. Nov. 18th 1864.

My Dear little Wife.

After finishing my letter to you last evening I espied among the papers on my table. a letter addressed in your hand writing (which by the way. Mr Moller says is very pretty) I caught it up wondering if I had been stupid enough to leave one of your letters to me lying around loose. and found to my surprise that it was unopened and upon further investigation found that it was your letter of the 13" inst. which had been lying at my elbow all the evening while I had been writing to you. My letters are always handied to me by one or another of the clerks, but by some chance that one had been laid on my table. by the orderly I presume while I was at tea. Today I received yours of the 14th. also one from George which I have just finished answering, so I fear your two letters will hardly get justice done them tonight. I am sorry that you must suffer so much with those terrible headaches. I suppose that I cannot sympathize fully with you in your affliction for I am scarcely ever troubled with any difficulty in my head. and never with the sick headaches. I should be very glad to do anything that I could to alleviate your pain though it be nothing more than to hold your head and try my feeble powers at charming away the trouble. So the surprise parties still continue to rage. I presume that each returned soldier will be favored with one. and that my little wife will be left out in the cold each time. and have nothing to do but to reflect upon her folly in marrying a worthless fellow who will not stay and attend her in society himself and after cutting her off from the attendance of other young men leaves her in solitude to repent of her rashness. I think it very sensible in the young people of S- that they are getting over their fear of the violin, and the dance. I have always been an advocate of dancing as an evening amusement. Young people are bound to do some foolish thing to pass away the time when congregated for an evening. and I always held that dancing was the most pleasant, and rational, of all the devices to kill time. that I ever saw tried. We have had no snow down here yet. I am equally fortunate with Mr. Landon in being so long a stranger to the article. We are having a warm rain tonight. with the promise of one of Virginia's most intolerable of muds tomorrow. I dread it. I wonder that Sara B. does not write you just a word at least. but presume she will make it all right some time. My Sabbeths thus far have been spent in hard work. I am sorry:- it is the most unpleasant thing about my situation and gives me more uneasiness than all else. I have foreborne to mention it in my letters to you, because it was an unpleasant subject. but now that you have enquired. I tell you frankly. The Store is only kept open until eleven o'clock, but the Lieut has always had something that he wished done that has kept me employed nearly. if not quite all the rest of the day. I have some hopes of being able to save more of the day by and by but it is uncertain. I think I have received all the letters you have written me, excuse me if I have failed to acknowledge any of them. There was a time when they were an unaccountable long time in coming, but they reach me in better season now. Do not have any fears for my health on the score of confinement to my duties. I assure you that so far as I can see it does not affect me in the least. I am in the enjoyment of excellent health for me. I shall certainly come home at Christmas if I can, and think it best. but as it really is very doubtful. and I found you were looking upon it as a settled fact, when it was by no means one, I thought it best to put you on your guard against disappointment. It was principally because of the uncertainty of Lt. Browne's continuing in his present capacity that I thought probable, or at least likely that I should be home before Spring. I am glad you are agreeable to having the Christmas dinner at Mr Randall's. I should think it all the more appropriate that the day comes on the Sabbeth. A Christmas dinner in New England does not necessitate a brawling, carouse. I sincerely hope that Madam J. has not set her mind upon inflicting a party on us if I should be so fortunate as to be able to run home for a couple of days. I should not want to see anyone but you, my sisters, and a very few dear friends whom I should prefer to see at thier homes. Still I will not be contrary, will even try to be reasonable, but the dinner at Mr. R's. I am resolved on, that over I am resigned to Mrs J's tender mercies for one evening. You have told me about the dress, and I am satisfied, because I suppose you are. Now how about the Cloak and Bonnet? It must be time you were getting them. If the money is wanting, use that I sent you by Mr. Moller for I presume I shall not get about writing to Mr Randall before my this months pay is due, and it does not matter at all if I do not. I wish you to have those articles as well as any others that I may be able to get for you. I have been thinking to hear about them every letter I have received lately, and begin to fear that something is wrong. You must not keep me in the dark my love. for I am a dull hand at thinking of ladies wants, for I never had a wife to look out for before, and I am afraid that if she dont help me look after her wants, or waits for me to anticipate them, I shall long remain in ignorance of them. You have no model husband my dear. but one who loves you. and as it is very late. and he (stupid fellow) very sleepy, wishes you good night and pleasant dreams.

Ever your own devoted husband
E. J. Barden

CIVIL WAR LETTERS 1861-1865

No. 16 Rec'd 24th Nov. 1864

>Headquarters, Armies of the United States,
>Office Chief Commissary,
>City Point Va. November 20th 1864.

My Dear Wife.

Your letter of the 17th was received this afternoon. also one of the same date from Mr. Moller, which I have answered this evening, and as it is getting late I do not expect to only commence a reply to yours. Your letter contains quite a budget of information, for which I am greatly obliged to you. I do not suppose my letter to Mr. Moller will much more than reach him before he leaves Leakeville. I had not expected he would write, for I did think he would wish to use any of his least vacation for that purpos. He tells me that he had another of his terrible attacks of headache while <u>at Millerton</u> where he spent the night. I wonder why Dexter was not there to meet him, for I wrote to him to be sure and be there on both Monday and Tuesday evenings, for Mr. M. would surely be there upon one of them. I do hope he will enjoy himself while in Salisbury. he has been looking forward to his visit there for a long time, and has anticipated much pleasure from it. I hope he will not be disappointed. I told him tonight that if he calls on you again after the receipt of my letter, he must ask you to <u>sing him "Robin Adair"</u> for me!– it used to be a favorite to of his when the Band played it. He had just returned from calling on you when he wrote, and I could tell that he had been filling your little head full of nonsense about me. and of expectation of seeing me next month. even before I read your letter. He is one of the kindest hearted men I ever knew! he never forgets a kindness done to him, no matter how small it be! but magnifies it into a mountain of favor. and seems to think that he can never sufficiently repay it. I will warrent that he talked to you as though I had laid him under the greatest of obligations to me. Whereas the truth is the case is just the reverse. I have never been able to do him any but the most trifing favors. I am almost sorry that he has led you to expect me next month, for I fear that it will not be advisable for me to do so. I would love dearly to see you then, and he knows it, for I have mentioned it to him, and if no new thing occurs to disturb the present state of affairs here. I think with his help I might get away. I hope Lieut Browne will go North, as he some expects to, next week. then if I succeede well during his absence I shall have more confidence to ask him for a few days leave myself. if he should go out of service when his present time expires at the end of Dec. I might not be able to come at the appointed time, but might be home for good in the following month. You see it is as I prophisied!- my little "widow" does not attend any of the parties. I think you are in a fair way to find out who Libby means by "old married folks". You have brought this "terriable fate" upon yourself and though I pity you I am powerless to aid you. Mr. Reid does his work well and strong and I dont think you can find any flaw in it that will enable you to set aside his decree pronounced on the morning of the 7th of October. I can assure you will get no assistance from me in the business. Nov. 21st. Evening. For three days now we have had dark rainy weather. today it has rained almost incessantly. Everything is flooded, and the mud is perfectly awful. My thin shoes have through like a cloth and I have had wet a good part of the day. I am seated before a roaring fire in my tent, and am drying my shoes and heating myself up thoroughly. I do not think I shall take cold, for it has been warm all day, and I have been sitting close by a fire most of the time. I think the "Grain" boot that your Father has, will be the ones I want, but tell him to send the best and most servicable. If I am to remain here during the Winter, I suppose I must be prepared for just such weather as we are now experiencing a good share of the time. If your Father will have the kindness to put them (the boots I mean) in a small strong box, and direct them to me just as you direct your letters, and send them by Express. I think they will reach me safely and I will be very much obliged to him. Lieut Browne goes North tomorrow in charge of about $30,000, Government Funds. His absence will not probably exceede ten days. I wish Mr. Moller were here now, and I should care but little about it, as it is I feel a little nervous principally because I shall be dependent for my assistance upon half a dozen negros whom I shall be obliged to leave in charge of the Stores, a good part of the time. Please excuse me a short letter this time. I had intended to have written quite a long one, but I am tired tonight, and shall have a great deal to do in the morning, and must be up earlier than usual.

With love,
Your Devoted Husband
Edwin J. Barden

No. 17 Rec'd Nov. 29th 1864

Head Quarters Armies of the U. S.
City Point Va. Nov. 25." 1864.

My Dear Wife.

Only a short letter tonight, love, for I am exceedingly tired. Lt. Browne left for N.Y. night before last. I have had two very laborious, and wearisome days of work since, and have the prospect of another tomorrow. I would not mind the work so much if my mind could be at ease when it were done, but there is too much responsibility, to admit of it. I have five negroes under my charge. some of whome I fear are rougish, and the others I <u>know</u> are lazy. The Stores are quite exposed, and very much at their mercy. and it is not very pleasing to think that at the end of the month I may find myself very much short in many articles, without being able to account for the deficiency. However, I guess I can weather it through some way. Yours of the 20th was received last evening and very grateful I was to get it I assure you. I was very weary, and the least bit in the world inclined to the "blues." Should probably have had a good fit of them if I had had time to give way to them, So that your short letter coming just as it did was of more good to me than a much longer one might have been on another occasion. So you are bound to have me home at Christmas. I hope you may succeede in doing so. I regret that Mr Moller in the kindness of his heart should have promised you so definately that I would be home. Now if anything occurs to prevent me from seeing you then, you will be much more disappointed than if you had only half expected me. I do not think I should start at the time Mr. M. mentions, for I do not mean to be absent from here. to exceede ten days in all. I do not think it would be best to be absent longer. Please do not let any one know that you expect me. I saw Mr. Moore on the 22d. but only for a few moments as he did not land here. but went directly on board the boat that was to take him up the river further. and nearer to his destination. I was very sorry indeede for I had made up my mind to keeping him with me over night. and was so disappointed that I could not think of half that I wished to say to him for which our short interview gave me opportunity. He mentioned that he had seen you on the Sabbeth previous to his starting. wished me to give you his compliments, when I wrote. and say to you that he had got thus far on his journey, safely. I think he is looking a little thin. as though he had found visiting very wearing work. He sayes that he has enjoyed himself very highly. I should think he would feel quite lonely to return to the service. particularly as nearly all his old friends and companions have left it. Thanksgiving day has come and past. The ship loads of good things that the people so generously furnished for the soldiers. have come, been distributed. and ere this have gone the way of all "goodies." Somehow a sufficient amount of them lodged in our cooks tent to furnish our table with a bountiful dinner for the day. I was too busy to be hungry. and therefore did not enjoy it much. I was to have closed the Storehouse at eleven o'clock. but found it impossible to get away at that time. and when I did close it, was kept upon the run all the time by other business, and thus passed my fourth Thanksgiving away from home. My hopes are all centered upon Christmas if I lose my promised enjoyment then, I may as well give up reckoning upon any more holidays. I had a letter from Fannie Hill tonight: she sends "love to Aunt Sarah". and says that Mrs. Randall expects me home at Christmas. We have quite cold weather here now. the ground is frozen. and the little puddles of water are covered with ice. The other night it snowed a little but it was so dark that I could not see it. and not enough fell to show upon the ground. There is no news that I am aware of. <u>Gen. Grant is back from his visit North. he came in and weighed himself today: his weight is 140 1/2 lbs</u>. Marrying is getting to be quite the fashion with the Commisssary Clerks. One of our mess who went home on a short leave of absence a few weeks ago. returned the other day having "doubled" himself while away. Another of the mess went home the other day. and we all expect that he has gone to "do likewise."

With love.
Ever your devoted husband.
Edwin J. Barden

No. 18 Rec'd Dec 5th 1864.

Head Quarters Armies of the U.S.
City Point Dec. 2nd 1864.

My Dear Wife.

Your letter of the 29th Nov. was received this afternoon. I have also your letter of the 20th & 21st the receipt of which I believe I have not before acknowledged. You have probably missed a letter from me in the few days past, but presume you will attribute it to the real cause – want of time, and so not feel very much disappointed. Lieut Browne returned today. I shall be exceedingly busy for several days yet, but feel easier now that he is here to assume all the responsibility. I expected Mr. Moller today, but he did not arrive. I received a letter from him dated the 27th, telling me about taking tea with you, and expressing himself highly pleased with his visit, mentioned the music, told me the pieces you sang and played, &c. I think he enjoyed it very much. I have ever so much to say to you tonight but really do not feel like writing, and trust you will excuse for sending you a short letter when you have good reason to expect a long one. having been at work quite late for several nights past I feel quite stupid and in need of sleep. I have occupied the Lieut tent during his absence, and tonight am back in our own! it seems quite homelike but rather lonesom without Mr. Moller. I am thinking of that "Seat by your side" that you mentioned, and how well I should love to occupy it, but should not feel much like talking tonight. It is very fine warm weather here now. a fire is scarcely needed. I get along nicely now without the boots, but hope they will reach me in season for the next wet spell which will doubtless be in a day or two. I did not direct you to pay the Express charges on them as I did not think that they would be required in advance. You need not have paid your Father for them. I could have paid him when I came home as you seem bound to have me do at Christmas. I fear you will be short of money, and will enclose ten Dollars in this letter. Please let me know at once if you receive it all right. My Mess Bill for Nov. was $10 – the value of the ration furnished by Government for the month was $10. So that my board really cost me $21. We do not use but little that the Govt furnishes however, so that in fact my board did not cost me the above amount. Lt. Browne brought me a <u>pair of buckskin Gauntlets</u> in N.Y. costing $2. I think they are cheap, but could have got along well enough without them. I spoke about getting a pair but did not expect he would buy them. Of course I shall keep them, and may have more need of them than I now think. Do you know you little goose that it will cost me at least calculation about $30. to pay you a visit of two or three days? To be sure the pleasure of seeing you is not to be compared with such a paltry sum, but can we poor folks afford such pleasure? I am glad you are comfortably established in your own room, and hope that you will enjoy yourself. It is a great pleasure to me to be able to conclude even in ever so slight a degree to your comfort or happiness. and my only regret is that I cannot do more for you.

        Pardon my short letter, and think of me ever your devoted husband.
                                      Edwin J. Barden

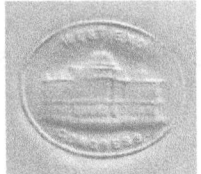

No. 19 Rec'd Dec 15th 1864

HEADQUARTERS, ARMIES OF THE UNITED STATES OFFICE OF CHIEF COMMISSARY,
City Point Va. Dec. 11th **1864.**

My Dear Wife.
I do not know but that you will have begun to think by this time. that you are a "deserted wife." Deserted in a manner, we know you are, but not willingly: and you are by no means forgotten. I think it is over a week since I have written to you: A long time you doubtless think, but I have been so much engaged that I could scarcely note the time as it passed. I have worked very hard during the days, and far into the nights. and feel as though there was a large amount of sleep due me. which I have not had time to get. I have had two letters from you I think since I wrote. In your last you mentioned that I had not acknowledged the receipt of two of your letters. Upon examination I find that they were duly received, please excuse me for neglecting to thank you for them. This is a wretchedly cold, dark cheerless day, muddy under foot, and a leaden sky overhead with a raw wind blowing. the very sound of which sends a shiver through one. at the thought of being obliged to face it. Lieut. Browne has been slightly indisposed for a couple of days, and has left the Storehouse to me. Today being the Sabbeth I closed it at eleven o'clock and finished work about one. I feel decidedly like lying down and taking a nap, but my conscience will not allow me to do so. I hope you are not having so cheerless a day in Lakeville. Night before last. sufficient snow and sleet fell, to cover the earth and trees with a mantle of snow and ice: it had been a long time since I had seen mother earth thus dressed in her northern winter robes. It seemed quite homelike. and strongly reminded me of hand sleds, skates, sleighs, buffalo robes, pretty girls, spirited horses, jingling bells, furs, mufflers, moonlight nights, hot suppers, and a multitude of other sources of pleasure in by gone days. It did not remain cold but a short time. and by noon the beautiful garment of white was exchanged for one of the foulest and most stickey of mud. The boots reached me. but I did not intend to keep them for they were much too large for me. I could readily have sold them, but could not find any others to suit me, and so I was compelled to put them on at last. They will answer to wear

through the worst of the mud this winter. and I shall have me such a pair as I require made by some of our shoemakers at home. I assure you my dear wife. that it <u>will</u> be a <u>great</u> pleasure to me to be permitted to see you, even for a day. I do not think I need add that it would be a still greater pleasure to be permitted to remain with you. I find that I am getting my mind fixed upon my contemplated visit and fear that if anything occurs to prevent it. I shall be as much disappointed as you will. And so you "waded" through that tedious <u>sheet of foolscap</u> that I sent George, I shall have to give him a scolding for showing my letters. You must have been highly entertained, for I dont think you found much in it, but a repetition of what I had already written you. I went over to Bermuda Hundred three days since on business, and was fortunate enough to meet an old friend who was clerk in the A.A.G.O. with me. I was glad enough to see him: he scolded me roundly for not having written to him as I agreed to when I left the office. His time will not be out until next April. I also met a former member of our Company: afterwards Commissary Sergeant of our Regiment. he is now clerk in the Quarter Masters office at Bermuda. We had a lively chat. and through him I learned of many others of our Company from N. Haven. I have just astonished one of our negroes, by telling him that I am married. I have your last letter open beside me. he (the negro) was fixing up things in the tent. and saw the letter. he thought it was very pretty writing. I said yes it is my wife's writing. You should have seen the look of astonishment with which he repeated my words. I have not yet found the time to write the letter to Mr. Randall. and as it is probable that I may be home soon I shall let the matter rest for the present. and in the meantime if you have any neede of the money do not hesitate to use it. The prospect of soon seeing you again operates now very much as it did before my discharge from the army: I am too impatient to write. and cannot confine my mind to it. I am finishing my letter in the evening. having first left off writing to go to <u>supper</u> and was then again interrupted by an evening call from one of the other clerks who spent some time in our tent. Its a dreadful windy night. light as day out of doors. the clouds of the day having all disappeared as night came on leaving the moon shining brightly. We have a bright warm fire burning and are quite comfortable. Mr Moller sitting with coat off. and slippers on. reading. only looking up occasionally as some blast more powerful than usual almost threatens to over turn our tent. to remark "isn"t it terriable" to which I reply "I think it is jolly" as I love to hear the wind blow when I am comfortable. I hope my little wife is well and enjoying herself this evening. and that in a few days more I may be permitted to see her.

With Love.
Ever your devoted husband.
E. J. Barden

No. 20 Rec'd Dec 22nd 1864
City Point, Va.

OFFICE, CHIEF COMMISSARY OF SUBSISTENCE,
Armies operating against Richmond.,
City Point Va. Dec. 18th **1864.**

My Dear Little Wife.

    Do you really begin to think that you are not cared for, and being forgotten by your husband? Your "cry of distress" dated Dec. 14." reached me today. I am sorry you have been so long without letters from me. but I have not had the time to write them:- true I might have sent you just a line, but I delayed hoping for time (to) writ a full letter. I have been very much engaged during the month, working early and late. My eyes begin to trouble me a little making it unpleasant to use them much in the evening. I am using red ink this evening because it is so much easier to be seen. I have a sty coming on one eye which besides being unpleasant. is decidedly unornamental. I expect to start for home on the 21st and you may reasonably expect me on 23d. Please dont tell any one only George. and Sisters, if you can help it. I will not tell you how long I am to stay. so you'll not have to tell wrong stories if questioned about it. I shall probably be on my way before this reaches you. so you need not write again untill you <u>hear</u> from me. I shall be very busy until I start. and shall be obliged to use my eyes more than will be pleasant:- will my dear little wife be pleased to pardon my shortcomings. and short letters and resting assured that I still love her devotedly and am impatient to see her: accept of this, the poorest apology for a letter I have ever sent her. and believe me still her own loving husband.

                                                                     Edwin J. Barden

CHAPTER 19

LAUREL HILL, VIRGINIA

From the book:
January 1st 1865 General Hawley was in command of the First Division and Colonel Abbott in command of the Second (Hawley's) brigade.

As preliminary to our next movement it may be well to turn to an expedition planned by General Grant. This was too close to the enemy the post of Wilmington, and if possible to capture Wilmington itself. This was important for two reasons. First, Wilmington was the principal port for blockade runners, and second, General Sherman was near the coast, and after he reached there and had rested and equipped his troops, it was desirable that we should have Wilmington as a base of supplies and a point where his army could, so soon as the weather should permit, co-operate in the destruction of Lee's army. The time was opportune, as it was known that Bragg had left Wilmington and gone to Georgia.

LETTERS FROM EDWIN JANES BARDEN TO HIS FIANCE
AND LATER HIS WIFE, SARAH MARIA JONES,
WRITTEN DURING THE CIVIL WAR

SECTION 5 – 1865

No. 21 Rec'd Jan 4th 1865
New York.

New York Jany 2d 1865

My Darling Little Wife.

You are doubtless abed. and I hope asleep by this time. as it is nearly 12 o'clock. I say I <u>hope</u> you are asleep because I fear you are not. I fear you are lying awake thinking of me. I love to have you thinking of me. but do not wish that thoughts of me should ever cause you discomfort or wakefull moments. I told you not to expect to hear from me until next week. I did not think then that I should have an opportunity to write to you from N.Y. but while on the cars it occurred to me that I might pencil you a few lines this evening. and that prehaps they would be acceptable. I had to wait a few moments at your father's Store for the Stage. and every moment seemed an hour. so much did I regret that they could not have been spent with you. The Stage was loaded. and all seemed gay but myself. We reached Millerton three quarters of an hour before it was time for the train. I spent the time in the Depot. not caring to go out. When the train arrived I overheard the Conductor tell a friend that all the cars were full except the Smoking car. so I went into that and found a comfortable seat though I was sick and half stupified by the smoke. On the arrival of the train at 26" Street I made my way to the Washington Hotel and asked for a room, was informed that there were no single rooms, could only get a berth in a double bed in a double room, Not relishing the idea of a "picked up bedfellow" especially after having been so well entertained for the past few nights, I concluded to seek lodgings elswhere. and not wishing to waste time running about started immediately for the "International" where I am comfortably quartered for the night in room No 111. I rather like this place to stop at when <u>alone</u>. The beds are clean. the house is quiet. and charges moderate. I met with a touch of ill luck, for while inquireing at the Washington. the omnibuses all left and not one passed me on my way down. so I had to foot it. and lug my own valise. Tomorrow night by this time I expect to be well on my way to Washington. I have had a delightfull visit. and ought not to complain. but I do wish that I need never be separated from <u>you</u>. May the Good God watch over, defend and bless you and make you happy both here and hereafter is the prayer of your loving though most unworthy husband.

Edwin J. Barden

CHAPTER 20

BACK TO CITY POINT

From the book:

We now return to the boys of the Seventh whom we left in camp. They were aroused at three o'clock A. M., January 3rd, by the reception of marching orders. It was known that they were once more to embark on sea going steamers, so they broke camp, not expecting to return. Knapsacks were packed, articles not necessary were destroyed, and old letters burned. At eleven-thirty o'clock in the forenoon of the 4th, Hawley's brigade moved, with orders to go directly to Bermuda Landing and bivouac there. The day was stormy and the marching difficult. It took until 5 P.M. to march the ten miles. No transports being at hand, the brigade bivouacked in a muddy open field without tents or wood for fire, but after a while marched back about a mile to a sheltered position where wood abounded. There they passed an uncomfortable night, and a part of the next day. Snow fell during the night to the depth of some inches, and it is difficult to imagine a more uncomfortable bivouac.

No. 22 Rec'd Jan 10th 1865
City Point

HEADQUARTERS, ARMIES OF THE UNITED STATES,
OFFICE OF THE CHIEF COMMISSARY,
January 5th, 1865

My Darling Little Wife.

Back again to City Point. The comforts of home, and the society of my dear wife exchanged for Camp life and its none too pure and refining apreciations, While well accustomed to and quite at home with the latter. I need not tell you how much I miss, and deplore the loss of the former. The forepart of the day on Tuesday I spent in doing up my errands. I called on Mrs. Browne and also sat for some photographs which I directed to be sent to. if sent as soon as agreed upon they will reach you before this letter. If they are not good ones please do not let any of them go. I would rather you would not give away any of them until I have seen a specimin of them - please send me one or two as soon as you receive them. I enclose you a couple of <u>likenesses of Gen. Grant</u>! they are copies of an engraving and not photographed from life but I think they are as near correct as any pictures of him that I have seen. The afternoon of Tuesday hung rather heavy on my hands as I did not feel inclined to amuse myself with sight-seeing. and my business was done. Most of the time was spent in my room at the hotel. At half past six I took my luggage and started for the train. it is quite a little distance and my valise and bundle grew heavy before reaching the Ferry. I was just in time for the boat, and for the train at Jersey City which started at precisely half past Seven O'clock. We have but just got under headway, probably a quarter of mile from the Depot, and just in the heart of the city, when we were startled by what we supposed was the discharge of a cannon in close proximity to the train I had only time to wonder what could be the occasion of the firing at that time and place, when the train stopped so suddenly as to nearly pitch us out of our seats. A general rush for the doors followed. I kept quiet for a few moments, and was quickly informed by those outside of the cause of our sudden stoppage,- the boiler had burst, making a complete wreck of the engine, and throwing the tender and baggage car off the track, and cross the joining track so as to block off both tracks. I got out in a few moments and went forward to see the wreck. the fragments of it were strewn all about a mere mass of old iron and brass.- a portion of the forward part of the boiler was thrown across the street, striking the upper corner of a three story building, passing through the upper front of the building, and landing on the roof of an adjoining house, leaving a hole through which a <u>hogshead</u> might be passed. Three men on the engine were badly hurt, but I did not learn that any one was killed. A great crowde immediately gathered around, and it was quite difficult to get near. In about half an hour, an engine hitched on to the rear of our train and took us back to the Depot where we remained until a quarter before eleven, when we took a new start. I was very sleepy, and went to sleep immediately upon our arrival in the Depot and did not wake up until between eleven and twelve when the conductor came along for tickets. I slept nearly all night but in such an uncomfortable position that I was about as tired the next day as though I had remained awake. Being off time we were obliged to keep out of the way of other trains, and frequently to wait some time for them to pass us. Our train should have reached Washington at five minutes before six A.M. but did not get in until one P.M. seven hours behind time. However it answered my purpos as well as though we had arrived earlier, for the mail boat for City Point did not leave until three, so that I had ample time to get aboard. Our passage to this place was as pleasant as could have been expected!- the water was still, the boat roomey, and not crowded, and a quiet lot of passengers. I reached my quarters about five o'clock and received a hearty welcom from all my friends. Found everything going on finely, and that I was not expected until today. The fore part of the evening I spent in Lt. Brownes tent with him, and the remainder of it, in our owne tent talking with Mr. Moller and Mr. Widner, and trying to write you a letter. It is now half past eleven, and as tomorrow I have to "get on the harness" and go to work,- think that you will excuse me for not writing more.

Good night dearest, and much love from,
Your Devoted Husband.
Edwin J. Barden

No. 23 Rec'd Jan 11th 1865

HEADQUARTERS, ARMIES OF THE UNITED STATES,
OFFICE OF THE CHIEF OF COMMISSARY,
Jany 8th 1865

My Darling Little Wife.
Your precious letter of the 3d was received this morning. and most welcome indeed it was I can assure you. It should have reached me last night. but the mail boat was

delayed by the fog. and ice and did not get in until three o'clock this morning. The perusel of your letter made me both proud, happy, and sad: proud and happy to know that I am so dearly loved by one so pure and good. and sad to think that I should cause unhappiness and discomfort to one I love so dearly and to whom I earnestly wish to be only a source of joy and pleasure. I cannot help wondering how it is possible that any one can possibly love me so much.--it is strange and unaccountable to me. Honestly Sara though I believe I love you as sincerely. and devotedly as it is possible for my wayward, selfish nature to love. yet I fear that I do not fully appreciate the wealth of your affection for <u>me</u>. That <u>you</u> should love everybody, and your husband in particular above every one else, seems perfectly consistent, provided, your husband were any one else but me. I feel that your love is my greatest earthly blessing and take great pleasure in thinking of my rich possession. You speak of your indebtedness to me for the comforts you enjoy:- please never mention such a thing again. I am entirely powerless to do for you as I wish. and as I <u>ought</u>. at this time of my life. to have been able to do. It is a pleasure to me to do anything for your happiness. and I hope it may always continue to be. One thing I feel assured of, and it is the principle cause of my self distrust as to my ability to always be the careful, devoted husband that it is my duty to be, and which I so much wish to be:- I ought to have been married sooner. and been learning to be self denying. and careful of others happiness, instead of having no one to look after but myself. and thus growing selfish, and careless of others happiness and comfort. I know that there can never possibly be a reason why I should ever be unkind to you in thought, word, or deed, but I sadly fear that if our lives are spared I shall cause you many a heartache, and unhappy moments, and this fear will force itself upon me, and causes me much uneasiness. My dear wife, I do sympathize with you in your loneliness, and feelings of desolation after my departure, the sight of your dear, sad face at the window as I left the gate is ever before me, and I think will haunt me until we meet again. But you are not alone in your feelings of desolation, though surrounded by busy scenes, and actively engaged most of the time, I have many a sad moment, and linger lovingly on the thoughts of the hours we spent together, and which <u>you</u> made so happy. It seems long and uncertain, before we may hope for a renewal of the pleasures we enjoyed during the holiday week, but we must be cheerful, and hopeful, and if we rightly employ the time, we may yet be all the happier for our separation. Do not think my dear that any pinching, or saving, you can do will help the matter. You are not extravagent I know, and it is my wish that you should have every necessary, and comfort that my limited means will allow, and it is my earnest desire that you will at once give up any foolish, <u>Mrs. Ashbel</u>

<u>Landon</u> notions that may have crept into your head. Please oblige me darling and be as comfortable as you can this winter at any rate. A proper regard for the wants of the morrow are most commendable, and I would not wish to forbid it, but there is no occasion for you to deny yourself any of your accustomed expenditures or articles of dress, or convenience: it would not, or rather could not make any difference about my being away from you. Please remember about the wood, coax Eddie to bring it in for you, and pay him well for it. Do not fret yourself about that promise. You did perfectly right in urging me to make it. Honestly I do not think you would have done your duty to yourself as a wife if you had not faithfully endeavored to win it from me. I only regret and am ashamed that there should ever have been a necessity for it. Today is the Sabbeth, I have many times made compairisons between my situation now, and one week ago. It has been compareitively a day of rest to me: We have not opened the Storehouse for business as usual, and I have done no work of any account. This evening I went to a prayer meeting in the Christian Commission tent. It seemed quite like old times and called to mind similar scenes when I was a soldier and good Mr. Wayland used to lead our evening prayer meetings. Since commencing this, your letter of the 4" has been received and I am quite cheered to find by its tone that you are feeling in better spirits. Am glad that I thought to write you those few lines from N.Y. especially as they seem to have done you so much good. My mess bill for December is $6. added to which is $4. my share towards paying for a new Stove for the use of the mess. I mention it merly as "an item". and because I thought you would be interested. and not to terrify you with an account of expenses. I have the requisite amount left. and more too after paying all the expenses of my visit, besides over $30 of my December pay yet undrawn. I can send you more money any time if you wish it. I hope you are well and happy tonight, and that you are getting a little more reconciled to your "lone widowhood." Give my love to all who enquire for me as I am sure none but friends will do so.

Ever and devotedly,
Your loving Husband
Edwin J. Barden

No. 24 missing

No. 25 Rec'd Jan 14th 1865
City Point.

**Office, Chief Commissary of Subsistence,
Armies operating against Richmond.,**
City Point Va. Jny 10" **1864**

My Dear Little Wife.

I have nothing in particular to write about tonight, and it"s only two days since I last wrote. but having the time. thought I would say a few words to you before retireing. as tomorrow evening I might be engaged, and not able to write. This has been one of the most dismal, gloomy, and unpleasant days that Virginia can boast of producing; and that is saying a great deal. It commenced raining last night, and continued to pour down with but few and short intermissions until about sunset. Of course the mud is very plenty, very stickey and very deep; and traveling almost impossible. I have been very busy during the day: it was the day for issuing rations; and I attend to it, besides making sales. everything seemed to go wrong end first, and to drag along most unaccountably slow. Lt. Browne is not very well just now, and did not assist me today. From 10. A.M. until, 5 P.M. I had my hands full. But the days work is done, and with coat and boots off, and slippers on I am seated in my tent before a good fire communing as best I can with my darling wife who is far away. Mr. Moller admires my slippers very much, and they grow more and more in favor with me every day if such a thing is possible. I think if you could realize how much comfort they are to me, you would feel yourself more than repaid for the weary hours of toil you spent on them. I told Mr. Moller how much you liked the toilet glass, and as I told you he would. he scolded me for <u>not</u> leaving it with you. I purchased one in N.Y. but as it does not suit me quite as well as Mr. M's, I am going to exchange with him and send you his: but it will be a present from me, you understand, I shall also send you a bottle of "<u>Sterlings Ambrosia</u>" and if you like it, use it freely and I will send you more when its gone. Shall send the articles as soon as I can find a little box suitable, and time to attend to it. Capt. Adair returned today, you may remember that Moller wrote to me that he left the day after I did. Widner (Capt A's clerk) will leave for home tomorrow, or next day. A very distressing accident occurred to the boat on which Cap. A. was passenger from Baltimore last night. As I hear it explained, the Steamer attempted to pass a schooner and in doing so had to very nearly cross the schooner's track: this she attempted to do <u>ahead of</u>, instead of <u>astern</u> of the schooner. The consequence was a collision in which eight staterooms on board the Steamer were almost completely swept away. and their occupants either killed or wounded. Six are known to be killed. Cap Adair mentions one such incident. A newly married couple were on board, had been married only six days I think. after the crash. Capt saw the lady standing, pale and trembling in her night cloths. with the blood streaming from a great cut that had laid open one side of her face, "Are you hurt" said the Capt. "Oh it doesn't matter" said she "where is my husband." Search was made for him and at length he was found and extricated from the wreck, but in a dying state, being mortally wounded in the spine, I think it very very sad. The boat was able to continue her trip to Fort Monroe. Mr. Moller had a letter from our old friend <u>Capt. McClure</u> the other day in which he gave an account of a railroad accident in which he was a sufferer. Ten were killed, and Seventy wounded--all the killed were in the same car with Capt. McClure who himself received a severe wound in the head which had confined him to his room for a couple of weeks. My little experiences dwindle into insignificance by the side of such accidents, but I cannot say that I thirst for such adventures. I have just received your letter of the 5th, am greatly pleased with such continued evidences of your thoughtfulness of me. I was here in my tent with Mr. Moller that evening and probably while you were writing to me. We did not "partake of the Nutmeg" that evening. On the contrary I wrote you a short letter. I have not yet suffered on account of my promise and do not expect to. Have only been asked to drink once, and then was not urged. The fact is all hands wer pretty merry during the holidays, and will not probably feel like drinking much more for some time yet.

I find myself quite at home on my straw bed, but think I should prefer a feather one, under <u>certain circumstances.</u>
<u>I must bid you good night darling.</u>
With unchanging love I remain your devoted husband.
Edwin J. Barden

No. 26 Rec'd Jan 16th 1865
City Point.

Office, Chief Commissary of Subsistence,
Armies operating against Richmond.,
City Point Va. Jany 13th 1865

My Dear Wife.

    I suppose you are in Lime Rock this evening if you carried out the intention expressed in your last letter,- hope you are having a pleasant visit, and that in Frank's cheerful home you will be able to forget your loneliness, and find your time to pass quickly and pleasantly away. I suppose Frank and his lady are very happy,- they are well situated to enjoy a great deal of this worlds happiness so far as we can judge from outward appearances, and I have no doubt that do. I hope you did not get there in time to attend Mr. Clarkes donation party. he is a rabid Copperhead of the (to me) most disgusting stripe, namely, the "Cursed be Canaan" kind!- those who attempt to prove that slavery is right, by the Bible. Doubtless he was well received, and received a liberal donation at L.R. the present Capital of Salisbury, but enough of him and his kind of religion. We have had three days of beautiful weather – that is overhead! underfoot, mud has had it all its own way. The evenings too, have been fine until tonight which is dark and cloudy giving promise of stormy morrow. Hope it will not storm tomorrow, for I have a great deal to do. No news here, there has been a rumor that Gen. Grant was going to remove his Headquarters to Washington, but it lacks confirmation. I am much pleased to see that Butler has been relieved from all command and sent to rusticate for a while in Lowell. I trust that the Government has got its eye open at last and that they will never again trouble him with any public position. I cannot write a letter tonight, for Carl and Mr Widner have both come in, and I must give up writing. I have had the evenings pretty much to myself since my return, but get cheated out of the greater part of them for some one is sure to be in the greater part of the time. I find I am not as contented as before my visit home.

I feel iritable and cross if things do not go right, and am often wishing that I could find something to do nearer home wher I could have you with me.

It is no use trying to write more.
Believe me darling your devoted husband
E. J. Barden

No. 27 Rec'd Jan 20th 1865
City Point Va.

Head Quarters Armies of the U.S.
City Point Va January 15th 1865

My Dear Wife.

Your letter of the 10th & 11th inst were received last evening, and read with great interest and pleasure. Am glad that you are once more cheerful and hopeful – have recovered as it were from the effects of my visit. A fortnight ago tonight was my last with you, and I have been contrasting the two evenings. it is needless to say that the present suffers sadly by the comparison. So your visit to L.R. was postponed, for good reasons doubtless, though you did not mention them. Bear my kind regards, and best wishes to Frank and his lady when you do visit there, and say that I regret my inability to accompany you and share the pleasure of your visit. Today Mr. Moller and myself have been taking a ride on the military rail-road which runs from this place to the extreme left of our line, a distence of about fifteen and a half miles. We started at ten in the morning and arrived back here at half past four. The weather was fine, though cold and windy. There was not much to see, that to us, was new or very interesting, but we have been talking about the trip for some time, and are satisfied now that we have seen what little there was worthy of note. Doubtless you think it not the best way to spend the Sabbeth, and I am far from contending that it was, but we could not well go on any other day, and it is a question whether I should have been able to have spent the day any more profitably if I had remained in Camp. A rumor has been floating about all day that <u>Stevens, the Vice president of the rebel States</u>, is at Gen. Grants quarters – that he came through the lines yesterday under flag of truth and reached here last night. I have not been able to learn wheather the report be true or not. I have been at work all the early part of the evening, and being dull and sleepy from exposure to the wind, do not feel answering your letters now, but can only acknowledge their receipt, and tell you how much I love you, and how almost constantly I am thinking of you, and hoping that you may never be more unhappy, or have more of care and sorrow, than now. I do not succeede very well with letter writing since my return. I have only written to you, and commenced a letter to Capt Moore which I sadly fear will never get finished. My time evenings is so much intruded upon by callers who drop in to smoke a pipe and have a little chat, that but little is left to improve. You may be sure that I did not forget to do your errand to Carl, and of course he denied that there was any obligations to him, at all, he is as contrary as a mule about such matters. The photographs are pronounced very good, and as there is no opportunity to have any others taken at present, you may as well use them as you like. Please make it a point to recover all those old ones if possible, and destroy them as soon as in your possession. I think I shall send to N.Y. for another dozen and have them sent to me here. So I'll not need any of those you have. I do wish that you (or rather "we") had good photographs of yourself. I presume there is an artist in Barrington. Do you ever visit at your Uncle Jarvis'? And if so would it not be a good plan to unite business with pleasure, and make it convenient to go up there before long? You could stay and see what pictures you are getting, and if necessary try more than once. I only make the suggestion, if it pleases you. the Stage and Cars will take you there any day. I dont mean the Stage either. You can engage a more comfortable method than that of getting to the falls. Tell Libby I appreciate her love, and good opinion, and return them both with all lawful interest. Tell her you are awful jealous and very particular about my loving even sisters too well. Tomorrow, Mr. Widner goes home. I wish that my going would answer his purpos just as well, and that he was willing that I should do it. I have grown very obliging lately and would be happy to do the going home for the whole Department. You dont think to tell me how George is getting along. I should like very much to know. Has he paid that little bill I owed Mr Tupper yet?

Good night my darling,
As Ever Your Devoted Husband,
Edwin J. Barden.

CHAPTER 21

CAPTURE OF FORT FISHER

From the book:
   January 16th General Terry received the following letter from the War Department:
   Steamer S. R. Spaulding.
   Off Fort Fisher, January 16. 1865. Major-General Terry,
   Commanding, etc.:
   The Secretary of War, in the name of the President, congratulates you and the gallant officers and soldiers of your command, and tenders you thanks for the valor and skill displayed in your part of the great achievement in the operations against Fort Fisher and in its assault and capture. The combined operations of the squadron under command of Rear-Admiral Porter and your forces deserve and will receive the thanks of the nation, and will be held in admiration throughout the world as a proof of the naval and military prowess of the United States.

Edwin M. Stanton,
Secretary of War.

Editor's Note: The following letter refers to the Union capture of Fort Fisher.

No. 28 Rec'd Jan 24th 1865
City Point.

OFFICE CHIEF COMMISSARY OF SUBSISTENCE,
Armies operating against Richmond.
City Point Va. Jany 18th **1865**

My Dear Wife.

Your letter of the 15th was received this afternoon. Your favor of the 12th was received a day or two since. I intended to have written last evening, but was obliged to entertain visitors until it was too late. Have just written to N.Y. for some more photographs, not having been able to get to it before: have only sent for a dozen this time, but presume I shall have to send again. for I wish to get the photographs of as many of my fellow clerks as I can, and cannot get them without exchanging. I enclose you one that I obtained last night. Mr. Moller is going to Ft. Monroe tomorrow, to see about having some pictures taken there. I am very desirous of getting a good likeness of him. We are very jubilent over the capture of Fort Fisher, though we do not forget the many noble lives sacrificed in its capture. Today we have another loss of life to deplore caused by the explosion of a magazine which hurled about two hundred men, unwarned, into eternity. It is very sad that so many brave fellows should be deprived of life just as they were enjoying the triumph of their hard earned victory. We may be excused for indulging in a little pride over our Connecticut General who has been able to accomplish what the pettifogging Massachusetts would be General declared impossible. I do rejoice over the downfall of Butler, and trust that ere long the country will be thoroughly convinced of not only his incompetency, but also of his villainy. I hardly think Mr. Wm. J- would defend Butler so stubbornly again as he did last Fall. I would just love to touch him up a little on his "Butler Mania" at this present time. Well my dear I have not yet been sorely tempted to break my pledge and do not expect to be "above what I am able to beare." Really I must confess that I am surprised at your earnestness about the matter. I had not the most remote idea that you were takeing the matter so much to heart. As I have told you before I consider it no hardship or deprivation to make the promise, and think that the only reason why I delayed it all was because I thought you could not be more than half in earnest about it, or that you were "trying your strength" on me, but it seems my good brother in law slandered me in some way and gave you any amount of uneasiness thereby. I supposed you had given up the furs for this Winter, else I should have sent you more money and advised you to get a good set. However if you are satisfied I suppose I have no reason to be otherwise. As I am no judge of furs, they might as well be cats skin, as either mink, or martin for all that I should know. I saw the N.Y. ladies carrying little bits of white muffs with small black spots on them and trimed with white silk cord and tassels. they looked to me precisely like those I used to see little school girls carry before I went to the war, and which I always supposed had at some time been the garment of a defunct "Tabby." So you are not entirely "cut" by all your maiden friends. they call on you occasionly it seems. Suppose they will not entirely give you up as long as I keep out of the way. I should really enjoy seeing Mary An Simons, and also Maria Haley, if as I suppose is the case, she is the same good hearted unassuming girl she used to be. I most heartily wish her all the happiness and joy she can possibly anticipate in the new relation she contemplates assuming. Do you know when she expects to be married? There will doubtless be a grand time. and the Governor's Mansion overflowing with guests and each ones cup of happiness filled to the brim.

    I trust that you will not charge the lack of interest in my letters to lack of love for you: it is far from being so, but since my return I have found it even more difficult than formaly to write a letter: and beg that you will believe that I love you fondly as ever, and that I want to see you very much and that I am your devoted husband.

Edwin J. Barden

No. 29 Rec'd Jan 28th 1865.
City Point

Head Quarters, Armies of the U.S.
City Point Va. January 22nd. 1865.

My Dear Wife.

    Another Sabbeth has passed: its record of blessing and priveliges improved, or misspent, has been added to that great account to which we shall one day be called to answer.   Though separated from you by many a long and wearisome mile, I have, in thought, been with you a great deal today. We have spent the day very differently no dout. Supposing the day to have been fine, and (as I hope is the case) yourself well, you have probably been to church all day, perhaps to Sunday School also;- have spent the day profitably and I hope pleasantly, I have passed the day in my tent, going out only to my meals, except at night on an errand, when I lingered for a few moments along the Docks to watch the scene of activity and bustle, almost always transpiring there. The time has been my own today, have not been called on to do any work. Have sat here reading and thinking; talking a little with Mr. Moller, occasionally interrupted by short visits from Lt. Browne, all day. Have not been to church because I did not feel like it during the day, and this evening it is too dark and muddy to be out unless one is compelled to be. The weather is very unpleasant! All day long yesterday, from early morning until after sun set, the rain poured down uninterruptedly, it ceased then but has not yet cleared up. No rain has fallen today but it has been dark, damp and foggy overhead, and, on account of the mud, almost impassible under foot. A grim unpleasant day, fit to give a man the blues!- a fear of all the horrors that ever terrified poor humanity. Do not imagine my dear, that I have been thus sorely beset; on the contrary I have got along very well. As I have said before I have thought much of you. A subject not by any means likely to distress, or annoy me. I have felt discontented and lonesome i'll admit, but under the circumstances I think that is to be expected, if not even justifiable. I have hard work to convince myself that, but three weeks have passed since I left you, it seems a long time. I cannot afford to wish time away, any more, for in a few days I shall be thirty years old. What little time may be left me will slip away quickly enough heaven knows. but I cannot feel satisfied to live away from you, and so often before aware of it. I find myself wishing and hoping, hoping and wishing that the appointed time of our separation might speedily pass away, and we be permitted to travel the remainder of our portion of life's journey together. I have not been obliged to work near as hard since my return, as I used to. Then, with scarcely time for necessary repose I had none to spend in unavailing regrets and wishes, now I do little else when not actually employed about my business. Mr Moller returned last night after an absence of three days: he had no opportunity to sit for his photograph for which I am exceedingly sorry. I don't see now when he is going to have an opportunity to have any taken. The first five days that comes I mean to coax him into an establishment here and get his Ambratype taken. The mail boat has not arrived today: detained probably by the storm and fog last night, so we are without papers, or letters tonight. I received a bundle of State papers by yesterdays mail, from Mr. Cook which were very acceptable, and have engaged my attention on a good part of the afternoon. Carl and I have been feasting last night and today. When he left Salisbury he accidently left behind a pair of shoes which Newt, has just sent him, taking occasion to enclose in the same box a huge fruit Cake, nicely frosted and in excellent eating condition, together with sundry packages of walnuts, chestnuts &c. Carl opened the box last evening, and we directly commenced a vigorous attack upon the cake, which attack having been frequently repeated with unabated ardor the poor cake now sadly shorn of its fair and Aldermanic proportions, bids fair to soon go the way of all good cakes. There is a little picket firing just going on, the first I have heard since my return. It is very dark tonight, and an excellent time for "scares", Genl. Terry has at last got himself into notice as a great man, and is enjoying his hour of triumph: he is a very good general, but they must not make too much of him, and crowde him beyond his depth. I will enclose you ten Dollars in this letter thinking you may require it after your late expenditur. If you wish more do not hesitate to inform me. I do not like to send much at a time, as Lt. Browne has just lost thirteen Dollars which he sent in a letter to his wife, which letter she never received. With much love darling. I remain ever your devoted husband.

Edwin J. Barden

No. 30 Rec'd Feb 1st 1865
City Point.

**OFFICE CHIEF COMMISSARY OF SUBSISTENCE,
Armies operating against Richmond.**
<u>City Point Va. Jny 25."</u> **1865**

My Dear Wife.

Your letter of the 22d was received today, and the one of the 20th yesterday. I intended to have written to you last night but had the teeth ache so severely that I was obliged to defer it. We are having very cold weather: the coldest of the Winter thus far. I think. It froze hard in our tent last night. and it will be nearly as cold tonight: though there is not such a gale of wind blowing as there was then. I wrote to you on Sunday evening: probably we were both engaged on our letters at the same time. I did not mail mine on Monday because the weather had been so bad that the mail boats had not been able to make their time: and as I had enclosed $10. in the letter thought it better to wait a day until matters were straitened out a little. Hope you are not having such a time with your fire as I am, this evening. Have been laboring faithfully since sun set to get a good fire started, and have not succeeded yet. have nothing but miserable green pine wood, and the draft of my chimney is not sufficient to make it burn. Do not know but I shall have to give up, and go to bed to keep warm. I write a few words and then have to stop and give the fire a punch. If my letter is "cooler" than usual tonight you will know to what to attribute it. We had a little excitement here yesterday morning caused by the attempt of the rebel rams to pay us a visit. They had arraigned to be here at daylight, but were delayed in passing the obstructions, and by getting aground, so that daylight found them in the vicinity of <u>Dutch Gap</u>. They passed our upper batteries with but little or no harm to themselves, as it was dark and no certain aim could be taken, but daylight, and good artillery practice in our batteries about Dutch Gap interferred seriously with their programme. I have no good account of the whole affair, so will refer you to the papers, which doubtless ere this contain full accounts of it. It was thought at one time here that the rams would be successfull in their attempt, and everything was in readiness to remove as many of the Government Stores as possible to places of safety, at a distance from the river. A battery was planted on the Point near Gen. Grant's quarters and within a few yards of our tent, but the guns mounted were to light to offer very much resistence to the approach of an iron clad. The rebels were emboldened to make the attempt which has terminated so disasterously to them, by the knowledge that all our iron clads but one were absent on <u>the Wilmington expedition</u>: the recent heavy rains having swollen the river so that many of the obstructions we had placed in it, were thereby removed, also helped to render their plans feasible. They will hardly attempt the operation again. I think. We had but one Monitor in the river yesterday, and she was without any motive power of her own, having met with some accident to her propeller, which rendered it for the time useless. She was towed into action, but could be of but little service. Am sorry that it is not practicable to visit Barrington at present. I am getting quite anxious to have the photographs taken. You can try the artist at Millerton if you like, but I have but little faith in your getting a decent picture. I learn that at the Hospital, distant about a mile and a half, there is quite a regular church established, and that services are held there every Sabbeth. I should have gone out there last Sunday had it not been so stormy. There is also service held in the chapel of the Christian Commission within a few rods of our mess tent. I know of no good reason why I should not attend worship at one of these places, when not employed on the Sabbeth, and intend to do so in future. I do not like to be away from my quarters much even when not employed, as something might occur to require my presence when least expected. Please ask George some day when he feels like fussing a little, if he will not overhaul my revolver and see if he can put it in shooting trim, and if he has an opportunity, to get some moulds for it also. There is no hurry about it: any time will answer.

    Hoping you are well tonight, and happy. With much love I remain
                     Your Devoted Husband.
                     Edwin J. Barden

No. 31 Rec'd February 3rd 1865.
City Point

>Office Chief Commissary of Subsistence,
>Armies Operating Against Richmond.
>City Point Va. Jany 29th 1865.

My Dear Wife.

 Sunday evening again and I am writing to my dear one far away. What is my darling doing tonight? I hope she is well, though now I think of it perhaps she may not be just at <u>this time</u>. Well, what shall I say to you this evening? Will it do to tell the old love story over again? I have but little else to tell, though four long days have passed since I last wrote. Very cold and windy days they have been too and the frost has each day penetrated deeper, and the ice on the rivers steadily increased, though the sun has shone brightly each day. This forenoon it was very severe out, the wind blew furiously driving across the sky mazes of dark angry clouds, that hid the sun from us, and threatened us with a frozen rain, or a heavy fall of snow, but this afternoon it has been very pleasant. the sun has shown warm and bright, and tried in vain to loosen somewhat the grasp of the ice king, on land and water. Jack Frost has to strong a hold, to be broken by two or three hours of sun shine. The air is as clear as a bell tonight, the stars are very bright and with the faint light of the new moon, make it a comfortable night to be about in, to those who do not object to the cold. The river is full of ice, and fears are entertained that it will impeede navigation seriously. The mail boats are irregular in their trips, and no dependence can be placed upon their time of arrival. The boat due today at three oclock is not in yet. We are feeling quite secure against any further molestations by the rebel rams, as several of our iron clads have arrived in the river. Among them is the New Ironsides which is anchored off Bermuda Hundred, in full view from here, and about a couple of miles distent. Two of their officers were in our Store house yesterday purchasing some stores, they invited me to come on board of her, and I mean to avail myself of the invitation the first opportunity, for I have long desired to gratify my curiosity to see, and examine the interior of the noble vessel. This has been, or rather is, my birthday! A capitol time to review the past, to con over the lessons it cannot help but have taught us, and, being guided by them, to lay plans and resolve upon our future course. But a great portion of my part, is unpleasant to dwell on, and I have formed too many good resolutions only to break them. So the day has passed with me as all other days do. I am not insensable to, and I trust not wholley ungrateful for the many blessings I have received and the goodness of God to me. The faith is my own that they have not been duly prized and improved. One of the richest, and chicest gifts of the Great Giver has been bestowed upon me in the past years, may He teach me to prize it as I ought, and be my remaining years few, or many, help me to devote them to <u>its</u> comfort and happiness, "Until death doth us part." I have forgottten your birthday. I think it is in this month, and one of the two remaining days, but am not certain! Please refresh my memory. I often smile when I think of my awkward predicament in the Town Clerk's Office and the lucky guess I made there. I have not yet sent the little articles I spoke about, having waited in hopes of being able to get one or two other little things which I wished to send you, but as I cannot procure them shall send the others as soon as I can get a suitable box. I think I shall send home my thick woolen shirts. I do not use them, and shall not probably require them at all, and they may get lost if left around here. Some times they may be useful. If you think of it, will you please put two or three pearl shirt buttons and a needle in one of letters. I cannot get them here, and those I have will wash off. Have you used that coat of mine as you thought you could? I hope if you can possibly use it you will, and if it will not make what you wish, let it go and get some cloth that will. Have you sent one of my photographs to Carrie as you thought of doing? I must write to some one of Mr. R's. people soon and if you have sent to Carrie I will write, and send to Mr's R. provided anything, comes to send. Have you had occasion to replenish your wood pile yet? I think such weather as this must make great inroads into it, or would if you had a shivering husband around, to keep warm. I had nearly forgotten to mention that there was an execution here on Friday last. A man belonging to the <u>1st Conn. Heavy Artillery</u> was hung for desertion. I did not go out to witness the scene though it was but a short distence off. I have no taste for such sights, though I cordully approve of the punishment for that crime. Hope you have received the money I sent a few days ago. Will send you more soon. If there is anything that I can get, or do, for you please let me know, and it shall be done. I hope you will not deny yourself any necessary, that is in my power to procure for you. I must repeat what I have told you before, that "I am too stupid to anticipate a ladys "wants," but it will give me pleasure to supply them when made known to me. Now dear wife good night. Be assured of the continued love and devotion of your fortunate husband.

Edwin J. Barden

No. 32 Rec'd Feb 7th 1865
City Point

OFFICE CHIEF COMMISSARY OF SUBSISTENCE,
Armies Operating Against Richmond.
City Point Va Feby 1st **1865.**

My Dear Wife.

Your letter of the 24th was received on Monday, and that of the 27th today! Many thanks for them, and for the spirit of love and devotion they breathe. I see you keep quite as good track of my birthdays, as I do, and much better than I do of yours. You have given me the information respecting yours, for which I asked in my last letter. I will try not to forget it again. You write in your last letter as though you thought I distrusted you in some degree. I cannot conceive why you should think so! I am sure that I have endeavored to prove to you my faith, and confidence in you, and beg to assure you once and for all, that I have the most unbounded faith and perfect confidence in you, in each and every respect. I know your love for me and believe that you are willing to make any sacrifice that you think necessary for my happiness. If you have any reason to think (and very likely in my ignorance I have given you cause to) that I do not confide in you fully, please let me know it at once that I may if possible remove it at once. I do assure you that this absence from you, though voluntary, is most unwilling, and could I see any way open by which I could manage to maintain us, I would cut short at once, but there seems no other way of getting along at present, and God only knows whether there be any better days in store for us. I am ashamed to say it, but I become almost discouraged some times with this miserable tread mill kind of an existence. Could I have any reasonable prospect of its ever coming to a pleasant and profitable end it would be more endurable, but this uncertainty and doubt is most annoying. I have had the fortune, good or bad, or to see three of the leading rebels today. I was going to dinner this noon when I saw them coming from Gen. Grants quarters, they stopped to talk with some officers, and I stopped and looked at them to satisfaction. they wer <u>Stevens</u>, Hunter and <u>Marshall</u> accompanied by a <u>rebel Colonel by the name of Hatch</u>. They are on a peace mission, and at present are waiting to hear from Washington wither word of their being here has been sent.

<u>Feby 3rd.</u> Was obliged to lay aside my letter the other evening on account of a severe attack of tooth ache! Was busy last evening and would not finish it, and have been busy all this evening and will only say a few words before bidding you good night. Hope you are at Emma's when this reaches you. give my love to them all. Tell

little Ellia that I should like to have him down here and show him the big ships. The Steamboats, the Soldiers, and the big guns, and have him hear the nice music we have. Tell him I think he would like it first rate. There is no hurry at all about Mr. Mollers slippers. You can take your time on them as he has an ordinary pair that he uses. When they are done I will give you directions for sending them. I don't know as there is any thing that I wish sent to me. You can do but little for me at present but love me. I know you do that, and am sure that it is no trifling matter to be loved as I am by you. Mr. Moller wishes to be remembered to you! he is lying on the bed reading. My photographs have come, and are nearly all gone. I must send for another dozen at once. I will enclose you three photographs of brother clerks, that I have obtained and I am to have some others as soon as they can be obtained. Mr. Mollers I cannot get to you but I mean to get his ambratype soon. Will you please excuse this very poor letter, and believe me your loving and devoted husband

Edwin J. Barden

No. 33 Rec'd Feb 9th 1865
City Point

**OFFICE CHIEF COMMISSARY OF SUBSISTENCE,
Armies Operating Against Richmond.**

City Point, Va. Feby. 6th **1865.**

My Dear Wife.

Your letter of the 1st inst. was received yesterday. I suppose you wonder why I did not write yesterday. I can hardly give a good reason. I had the day to myself, and spent the most time in the tent. In the evening I took off and attended writing at the Commissary tent. I did not feel like writing, though I assure you it did not arise from any feeling of indifference to you, for I thought of you a great deal yesterday. It was just a month yesterday since my return to this place. I could not realize that it had been no longer time. I cannot see why time should seem so much longer now, than it did previous to my visit home. I think it must be that I learned to love you a great deal more than ever, or else that I have just begun to appreciate the blessing of a good wife. Be the cause what it may I am as discontented and home-sick a fellow as you could wish to see. We are having beautiful weather now. yesterday and today were both very fine, while the evenings are charming. We have no snow, and have had scarcely any since I came back. The streams are free from ice and there is just frost enough in the ground to keep it hard and in good traveling conditions. Mr. Moller was at dinner with us yesterday. We frequently invite him over, and he often accepts and favors us with his company. Yesterday we had a feast of, roast turkey, and Oyster pie, and apple dumplings. Mr. M and myself went over our usual dinner hour, and found the boys much vexed on account of the dinner not being done! The darky cook had got a "negro streak on", and had not done his work. Dinner was postponed until five o"clock and we fell to and made a hearty lunch off the dumplings which had been soaked "their appointed time" in hot water, and wer guiltless of showing the least sign of having guilt, or to be the least bit "high." At the appointed time we returned, and did full justice to the Gobblers, and "bivalves." My Mess Bill for January was $9.15. Allowance was made for the five days I was absent at the first of the Month which I did not expect as such. The Bill for those who were present the whole Month was $10. Not bad I think considering the way in which I live. I had expected it would be much more. I am thinking of sending you some more money, and this inst. shall send it by Express, though I have about as much faith in the safety of the War. On looking over my letters last night, I found yours of the 16th and am at loss to see how I failed to acknowledge the receipt of it! Please excuse the omission. I have all of the "fourteen" letters written by you in January. I have not kept count of the number I have written, but know very well that I have not written as many as you. Mrs. Barnum is very good to still remember you so kindly, though to be sure she has no reason to do another visit. Your bridal presents have not been very numerous, but I think you have yourself to blame for it. if you had married a rich man you might have had a plenty of them. The rebel Peace Commissioners have returned to Richmond, and are this evening to render an account of their success to their men.

Feb – I am told that they found our Colonel very firm and decided Gen. Grant has not allowed the rebels to be idle, but is taking advantage of the fine weather to worry here, and not sending any force to operate against them. Last night Lt. Stone was here. I cannot bear how severe the battle was, or how extensive, but it resulted in the enemies leaving the field, and his dead, in our possession. The capture of Wilmington, or Charleston, or both would be one of the most powerful "terms" that the President could offer to the rebels. I hope you are well and happy tonight, my darling, and that you may ever be thus is the prayer of your devoted husband.

Edwin J. Barden

No. 34 missing

No. 35 Rec'd Feb 16th 1865.

By Mr. Randall

Ans. (Note: Sara would often add "Ans." to indicate that the letter received had been answered.)

OFFICE CHIEF COMMISSARY OF SUBSISTENCE,
Armies Operating Against Richmond.

City Point Va. Feby 6." **1865**

Dear Wife.

I enclose herewith a $100 to you. and send by Express through Mr. Randall.

Having just written you a letter. I have no more to say tonight only that I love you a great deal. and earnestly wish that a merciful providence would place it in my power to be always with you.

Good Night.
Ever your own
Ned.

CHAPTER 22

FORT ANDERSON CAPTURE

From the book:
    From the 1st to the 10th of February we drilled our men and brushed up; mails arrived but were slow.
    On the 9th a general advance was made in the direction of Wilmington. We gained about three miles, had severe skirmishing with slight loss; Abbott's brigade captured a line of Confederate outworks, with sixty prisoners. From the 12th to the 18th bad weather and other things which we did not understand kept us still at Ocean Pond, six miles north of Fort Fisher. On the 19th at 8 A.M. we resumed the advance, and skirmishing all day long, gained six miles, reaching what appeared to be the enemy's main line. Here we intrenched and spent the night.
    When day dawned we discovered that the enemy had left our immediate front, and also evacuated Fort Anderson on the other side of Cape Fear River; this was said to be in consequence of the approach of Schofield's troops.

Editor's Note: this letter is written in faded purple ink

No. 37 Rec'd February 20th 1865.
City Point

                                          Office Chief Commissary of Subsistence,
                                                 Armies Operating Against Richmond.
                                                        City Point Va. Feby 16th 1865.

My Dear Wife.
    Your letter of the 10th was received this afternoon. I assure you I appreciate the effort you made to write when you were sleepy and tired and value your letter accord-

ingly. Mr. Moller sayes, "Mrs Barden can do me a favor," to which I reply, "she will be most happy to do it." Mrs wants a few shirt buttons, small, colored, Agate buttons for white woolen shirts. Will you please send a few in your next letter in the same manner in which you sent them to me. Am glad you are to have Olin's photograph. Wish I could be equally successful in getting those of my other "Companions in arms." By the way, how do yours come on? is it? most time to hear from them? I remember Miss Saunders now, you have mentioned her, but have not thought to enquire of her when home. I did not know she had been or was ill. What is the matter of her, and is she likely to recover? I dread to hear your news which I suppose of course will be "Consumption", and "no hope." I am very sorry for her whatever be her condition for I believe she was always a very good girl. I do not think were I in her condition, and in the state of mind that you describe, that I should care to live, or even to get well. I think she would be a loser by the return of her health. I think there is too much of sorrow and suffering in this life, for one to wish to live who is fully preferred to die. Our severe weather began moderating night before last, and today it has been raining a good part of that time. All is usually quiet about here, and we search the papers for news, and items of interest. Mr. Coak occasionally sends me a bundle of State papers, which seem to me much like the visits of old friends. I received a package from him this week. Whenever you get the Slippers for Mr. Moller ready to send, the best way will be to get George to pack them snugly in a little box and send them by Express. I mention this tonight not because there is any hurry about it, but because I happened to think of it, and to think also that I should like you to send me at the same time a couple of boxes of Seidlitz Powders and may be some other little articles that I may chance to think of by the time you are ready. This epistle can hardley be dignified by the title of a letter, but though I have written very little please remember that I love you a great deal and that I remain as ever,

Your Affectionate Husband.
Edwin J. Barden

No. 39 Rec'd 25th Feb. 1865.
City Point

                                      Head Quarters Armies of the U.S.
                                              City Point Va. Feby 21st 1865

My Dear Wife.

    I hope to be able to write the letter this evening which I promised in my note to you last night. On Sunday Lt. Browne wished me to visit a place called Jane's Landing on business for him. The place is distant from here about six miles in a direct line, but over twice that distance to follow the river around! it is the base of supplies for the <u>Army of the James</u>. I left my tent about ten o"clock and loitered about on the dock until eleven at which hour the boat left for Bermuda Hundred; to which place a few moments ride brought me. there I found two of my old comrades, formerly members of my Co. now clerks in the Quartermasters Department. An hour and a half passed away rapidly enough, during which time we had dinner, at close of which I found that the remaining time was little enough to finish my journey, and my business in. Procuring a horse I set out on my journey through the mud to my destination. After a little skirmishing I found the road through the woods the nearest way, and made the best time I could which, being no rider at all, and the road in a shocking condition was slow enough. However, all things must have an end and my journey was no exception to the rule. So after an hours ride I found myself at the Landing. I finished my business as quickly as possible and started on my return, with the certainty that a few moments extra detention anywhere would make me too late for the last boat over to City Point which would leave the Hundreds at four o"clock. Had got but a little way from the Landing, when, with the horse upon a gallop, I found the saddle was loose and turning around, letting me off on the ground. I landed on my feet however and with a good strong hold on the reins, it took all my strength to hold the horse, which was much frightened by finding the saddle flying about between its legs. I succeeded in quieting him finely, and with the aid of a passerby, put the saddle on again and fastened it firmly! made a hasty examination found myself alright, and unhurt, and then remounted and persued my journey. I tried to make up lost time, but the road was so bad that it was of no use, and I came in sight of the boat just as she was leaving the wharf. I did not fret much for I knew that if I watched my chance, there would surely be an opportunity to get over some time during the evening. I did watch, and was relieved much sooner than I expected, for in about half an hour a tug left for the Point after the mail and I came over in her. I expected of course to write to you in the evening and answer your letter of the 16th which I had received the night before after going to bed, but on my return from supper I found Mr. Moller in bed suffering from one of his terrible head aches! Of course writing was out of the question and I devoted myself to the care of Carl at once! he was worse than I ever saw him before, it seemed as though he would go crazy. Capt Adair came in and we did what we could, and finally after vomiting freely he became easier, and was able to sleep a little but passed a very sick uncomfortable night. As for me, my horse back exercise had completely used me up, and no sooner did Carl get a little easier than I tumbled into bed and lay like a log until morning. Nor have I recovered from the effects of my ride yet, but am so sore and stiff in every joint that I can scarcely get about. I suppose I caught a little cold which would account for my soreness! however the day was beautiful. I enjoyed it very much and think that on the whole the trip did me good. As for Mr. Moller, he got about yesterday and today is feeling quite well again. Yesterday I was quite pleasantly surprised by a visit from a cousin of mine, who is a private in the 2nd Conn. Heavy Artillery. he has been on duty in the Corps Hospital at Winchester for some time and is now on his way to rejoin his regiment which is at the front. He was one of a large party of soldiers from Camp Distribution on their way to their regiments! the Provost Marshal could not finish the distribution of them yesterday, and he remained over night. We of course had a good deal of talking to do. Yesterday your letter of the 16th was received, and I was just about seating myself to write you last evening, when two of my fellow clerks dropped in for an evening call, at the end of which I found myself to sleepy to do more than pen the little note which I started on its way this morning. I have been interrupted frequently this evening. first just as I setting down to write I had to go down to the Store house and get a quart of Whiskey for a thirsty officer, then, contrary to his usual custom Mr. Moller came in from his office quite early, and of course we have had some talking to do, so that it is already quite late and I have only just got ready to answer your letter. So shall be obliged to postpone it once more. I will enclose you the desired photograph, and without stopping to look over and correct my letter, will beg you to accept it "for better or for worse" together with the love of your devoted husband.

E. J. Barden

Editor's Note: The following letter follows the <u>capture of Fort Anderson and city of Wilmington</u>

No. 40 Rec'd 28th Feb. 1865.
City Point

                                      Head Quarters Armies of the U.S.
                                             City Point Va. Feby 24th 1865

My Dear Wife.

    I have not kept my promises you see: the next evening after my last letter was written, Mr. Moller expected company in the tent and I waited for them until it was too late to write and they did not come fineally,- last evening I was busy and could not well write! it is late now but I"ll try to write a few words to my own dear little wife. We have news today of the capture of Wilmington N.C! the rebel stronghold are fast falling into our hands!- the Confederacy is fast approaching "the last ditch." Deserters are fairly crowding into our lines! today a whole company of North Carolinians numbering 45 came in in a body at one point, bringing their arms and equipments with them. It is probable that the rebels are concentrating their forces for one last, mighty effort, and that erelong a great battle may take place, in which if the rebels are successful, they may be able to hold out for a few Months longer, if they are defeated it will be to them their "Waterloo," and put to flight forever all their hopes of success. Your little dear, kind letters of the 14th & 16th give me much pleasure. I am sure you cannot appreciate the pleasure it gives me to be able to do anything for you. I can only hope that I may not be deprived of the privelege of serving you, and so far as lies in my power of supplying your wants. I do not like your photographs <u>at all</u>! Do not have any more of them printed. I shall keep one of them (with your permission) for the present, but dont intend to exhibit it as the picture of my wife. My wits have surely gone a "wool gathering" tonight, for I cannot collect them sufficiently to write. I hardly know whether to have any more shirts sent me or not. At present I am wearing my woolen ones! the ones I have will last some time yet, and on the whole I dont think you need send me any more at present. I do not think of anything that you can send me of which I stand in neede at present. John and Josie have either been very unsure or else are a little unfortunate!- if the latter – its consoling for them to know, that such an <u>accident</u> cannot happen to them again. How do they bear it, and what do the good people of L. say about it? Think you must have had a rare, jollie time of it at your swell Learner's. Wish I could have been with you then - wish I could be with you now -wish I could always be with you. The twenty five cents enclosed to you by Mr. Randall was the balance of a Dollar that I sent to him to pay the Express Charges. The Express Co. will not take pay for parcels forwarded from here, though they require all articles sent here to be pre-paid. My famous big boots are nearly worn out. I have been obliged to get me a pair of thick heavy shoes! They cost me 8 Dollars. think I shall have some new thick soles put on the old boots, and wear them when it is wet and muddy which I suppose it will be for the most of the time during the next two months. Tell George that I hear Capt. Mills is dead, he may know of it, but I was much surprised to hear of it, which I did last Sunday from one of the former members of our old Co. whom I met at Bermuda Hundred. Please pardon me for not writing more tonight. Mr. Moller has just come in and I must go to bed.

With love.
Ever Your Devoted Husband
Edwin J. Barden

No. 41 Rec'd March 6th/65.
City Point.
Ans.

Office Chief Commissary of Subsistence,
Armies Operating Against Richmond.
City Point Va. March 1st. 1865.

My Darling Wife.

Your favors of the 22nd & 24th are received. I think I have received those of the 7th & 14th also, but my letters are in my valise and I have not time to stop and look for them this evening! Will see to it shortly. I am disappointed and vexed this evening. I expected to have had George here spending the night with me, and he has not come. Last Saturday evening I received a note from him saying he was at the Fort & asking me to come down if I could. So on Sunday morning I went down to the Fort, found him at the Hospital, and we spent the night together. In the morning we tried to get a pass for him to come to City Point, but did not succeede! I told him I thought I could get a pass for him at this place, and send it to him! and he agreed to call at the P.O.

for a letter from me last night and if it arrived, and contained the pass he was to come up today. I did get the pass and sent it to him yesterday morning, and calculated on seeing him here this afternoon as much as I did on getting my supper but he has not come. I fear he did not get the letter! though there may be many other good reasons for his not coming! I have but little hopes of seeing him tomorrow, though he may come. I was mighty glad of the opportunity of seeing him at the Fort. I can assure you, though we did not have a very long time to visit together! I had meant to have written you a decent letter tonight, but my usual luck has attended me! and here it is lacking only a few moments of eleven o'clock, and I have but just time to say a few words. I have been hard at work all day and evening, and about two hours ago came to my tent to write, but was interrupted by a call from a fellow clerk. I received a letter from Mr Randall tonight which I must try to answer tomorrow evening, and will also try to write you a few lines then, unless (which I do not much expect) George should be here. Mr Moller says "do not forget to thank Mrs' B. for me, for sending those buttons!" A big thing to be thankfull for isnt it? We are feeling rather anxious tonight about Sherman. The rebels claim to have gained a great victory over him, and cheered for two hours almost constantly all along our line here. We have no intelligence direct from him and of course are in an unpleasant state of excitement in regard to his whereabouts and welfare.

Once more, dear wife, I crave your indulgence for this short letter, and with love, remain your affectionate husband.

Edwin J. Barden

CHAPTER 23

WILMINGTON

From the book:

At daylight March 2nd the brigade returned to Wilmington. General Hawley arrived from the Army of the James and was assigned to the Department of Wilmington, which included all the country in rear of the army operating from the base of Cape Fear River. He was also charged with the duties of Provost Marshall General of the district. General Terry, with the remainder of the force marched toward Weldon. General Schofield, with his army from the West, was assigned to the command of all the forces co-operating with Sherman and moving towards Goldsboro. General Abbott was assigned to the command of the Post of Wilmington.

From the 3rd of March onward, the work of the Seventh, though no less arduous than before, was of a very different nature, consisting of fatigue and guard duty, policing the city and nursing the sick. The schools and churches, the post office and customhouse were reopened; hospitals were improvised in the large warehouses and the great number of sick among the exchanged prisoners were given such attention and care as was possible with the scanty hospital stores at command.

Much confusion and distress prevailed among the poor people of the city and surrounding country. The Sanitary Commission appeared with a shipload of supplies, and to restore order, relieve the needy and nurse the sick, kept everyone hard at work.

No. 42 Rec'd March 10th 1865

                                               Head Quarters Armies of the U.S.
                                                        City Point V<u>a Mch 5th 1865</u>

My Dear Little Wife.

    Your letters of the 26th & 28th of February are before me. I have looked over my letters, and find that I have yours of Feby 5th, but cannot find one of the 12th of the same month. I have one of the 10th, and one of the 14th Feby, but none of the 12th. It has not probably been received. I am behind about writing as usual. The evening after writing my last letter, George was with me and having a great deal of visiting to do with him. while he remained I could not well write. For a full account of his visit here I must refer you to him as he will probably have told you all about it any way before this reaches you. I hope he enjoyed his visit but as it rained about all the time he was here and the mud rendered getting about almost impossible, and we wer disappointed in going over to Bermuda Hundred. I was mightly glad to see him I can assure you, and wanted most awfully to go home with him. He will not have a very pleasant journey home, with a sick man to take care of, but I trust that ere this he has safely accomplished it. It has been very pleasant today for the first time in several days. I undertook a horseback ride this afternoon, for pleasure, but soon gave it up in disgust! and have been wondering ever since how I could possibly have been so stupid as to have thought of such a thing after the recent rains. The last week has been a more busy one than usual with me, and I am still a little behind in correspondence of having let some of the work go, to attend my visiting, I sent you Fifty Dollars by George. My Mess Bill for February was $10 45/100. I spent a few Dollars on my trip to the Fort, and have had occasion to use a little for other purposes! and think it best to keep a little by me. I also sent the Toilet Glass, by George. Guess you had begun to think that I had forgotten it. By direction of Lt. Browne I insert the following sentence, "That Darned old" Browne has just been in and interrupted me in what "I was about to say." He came into our tent to show me some papers, and on leaving dictated the above. He is an odd old Chick, and often annoyes me exceedingly, but he is kind to me and does well by me, and I can well afford to be inconvenienced ocassionally. We have <u>good news from Sheriden</u> tonight, and as I have not heard the report concerning Sherman, (to which I alluded in my last) confirmed! Affairs still wear a cheering aspect. I some times wish that we might be on the move, for I get tired of remaining quiet, and think I should like it if were to be stirring about a little and could see a little of the country! Army life is very dull and tiresom unless when on the move and amid the excitement of active campaigning! But it must still be a

Month or six weeks yet before the sacred soil will be in a condition to admit of the movement of large armies. You have most likely been to church today as it was the Communion Sabbeth. I have not attended any service, having been at work all the fore part of the day. I have not yet answered Mr. Randall's letter! Have just received another paper from Mr. Cook. I think you may send me another of the white shirts, also a little black thread. Hoping you are well tonight my darling, and with much love, I remain your devoted husband.

Edwin J. Barden

No. 43 Rec'd March 13th 1865.
City Point.
Ans.

Office Chief Commissary of Subsistence,
Armies Operating Against Richmond.
City Point Va. Mch 8th. 1865.

My Darling Wife.

Your favor of the 3rd inst. is before me having been received yesterday. I took last evening to answer Mr. Randall's letter and propose to devot this one to you. We have had another rainy day, warm, foggy, and dismal. Yesterday and the day before were bright and clear, and the mud dries up rapidly! It had even become possible getting about, but the rain of today has made the mud as deep and plenty as ever. Army movements must be impossible while this weather lasts. I suppose George has given you his experience of City Point weather and mud before this time. I felt quite lonesome after his departure. I hope he arrived safe home with his sick charge Am sorry you were disappointed that afternoon in not getting a letter, and as I remember that several days passed without my writing I fear you did not meet with better success the next day. <u>My brother is still a soldier</u> I believe, but is not with the regiment. I suppose he is in hospital somewhere I wrote to his wife previous to my visit home for information respecting him, but have never received any answer. You had better make the wrist bands of my shirts to turn back from where they button, as I have no sleeve buttons, having lost those you gave me when on James Island, and not feeling able at present to attempt any rash display. If Maria Holly's intended is a gentleman and a man of good habits, I think the Governor has shown his good sense in giving his assent to the match. I heartily agree with you that Maria is a nice girl! and hope that she may realize all the happiness she anticipates! I trust her husband will be spared the pain of being compelled to be separated from her, for the first six months at least. How does Libbie get on? You dont tell me anything about my pet sister lately. I suppose my sister Sara B. has not returned yet. Mr. Moller has just come in to retire for the night! he is feeling quite bright tonight but is unwell most of the time lately. he works too hard and gets no out door exercise. I have but little to write about this evening, but have plenty of love to send to my little wife, which I beg she will accept from her devoted husband.

Edwin J. Barden

No. 44 Rec'd March 16th 1865
City Point.
Ans.

Hd Quarters Armies of the U.S.
City Point Va Mch 12th 1865

My Darling Wife

Your letter of the 5th was received last night, and I purposed to have answered it in evening, but was unable to on account of the many calls made upon me to go to the Storehouse and draw Whiskey for thirsty officers. Today I received your letter of the 7th. All of your letters dear wife, are eagerly watched for and much prized but it seems to me your letter of the 5th is an uncommonly good one and deserves more than common attention! it has had it at any rate. You ask if I realize all the meaning contained in the words with which it commences "My Own Darling Husband" I tell you frankly (and I think I have told you the same in substance before) that I do not believe I do. I do not think that I am capable of appreciating the extent and wealth of that love which a good and pure woman like you bestows upon the object of her affections. I would that I were able rightly to estimate and value it! I think it my misfortune and not my fault that I cannot do so. My earnest desire is, that he whom you have honored and made happy by the bestowal of your love, may be able to render himself more worthy of it, and never do aught (again) to make you regret your choice. I thought of you on last Sabbeth, and remembered the day. I wish I could say that I spent the day even as profitably as it lay in my power to have done!- but it was a broken day with me, and passed but little like the Sabbeth. I cannot express to you how earnestly I join with you in the wish that the "time may come when we may spend our Sabbeths together."- for long weary years I have wished and waited for it until now it almost seems as though it might never be. I am very very far from being a good Christian, if I can claim any right at all to the name, which I often really doubt. I have great need of your prayers dearest, and it is a comfort to know that I have them! though I am often tempted to have but little confidence in the efficiency of my petitions, yet I firmly believe that "the fervent effectual prayer of the righteous availeth much," and therefore value highly the intercessions of those whom I believe to be God's children. I need say no more about my visit to George and of his visit to me. I will only add what you well know that <u>you</u> were not holly forgotten in our conversations. Am sorry that he should have been detained any longer at the Fort, and dont see how the Furlough came to be delayed!- it should have been at the Hospital before him. I fear you are losing faith in me, how after my repeated assurances of the fact? can you still "wonder if I am glad that I have a wife who loves me so devotedly"? I dont know wherin my wife has been any "trouble" to me yet, but I do know that she is a great comfort, even now and deeply regret that circumstances prevent me from enjoying the "comfort" to a greater extent. Am much gratified at Mrs Reid's kind rememberence of me Please express my thanks to her in whatever words you deem most appropriate whenever you have an opportunity. What a good little woman she is! I hope that Libbie's throat is well before this! These throat difficulties are terrible! though I often think that a person is fortunate if they can dispose of them for a year by even, severe, suffering for a few days <u>with quinsy</u>! At the same time I would not underrate the severity or danger of that disease. So Josie has lost her little one! Well, though the young Mother may for a while miss it, and lament its loss, yet I think none will be disposed to question the wisdom and goodness of the Great Father in the event. I hope that Josie will soon be well again. Some how I cannot bring myself to think of her as anything else but the "prettiest <u>little</u> girl in the village" as she was almost universally conceded to be, when I was in the Store. One would hardly have looked for so much sickness in the Spring, after such steady old fashioned Winter weather as you have had for the past three months. I hope that death may not follow in the footsteps of disease, and oh! I do hope and pray that my dear little wife may be spared a visit from either. There is no hurry about the slippers, as I have often told you, so dont worry about them. Mr. Moller is going to Washington tomorrow, but will not be gone over four days. I think he goes on some public business, though he has several errands of his own that he will attend to. This has been a beautiful Sabbeths day. I went over to the General Hospital Chapel this morning! Mistaking the time, I was quite late, and found the Chapel filled and the preacher just reading his text when I got in. I found a seat near the door and listened to a most excellent sermon. I do not know the preachers name. He preached extempore, and it is a long time that I have heard such a sermon. He had a very poor voice particularly when he attempted (which he seldom did) to elevate it at all, but he used good language, seemed never at loss for the right word in the right place, and handled his subject like a man who understood his work, and whos heart was in it. His text was the 1st Epistle of John 4th Chapter & 10th verse. He dwelt chiefly upon mans utter inability to work out his own salvation without divine aid, of his incompacity to make a satisfactory atonement for sin, of God's goodness and mercy in devising and offering to man the plan of redemption as set forth in the gospel, and ended with an eloquent and earnest appeal to the minds and hearts of his <u>hearers</u> to accept without delay of God's offers of pardon and salvation. I made quite the detour from the direct road, on my way home, to avoid the mud. After dinner Mr. Wyeth (one of our mess and an excellent young man) came over to my tent, and for a while we have had some conversation on religious matters, but Mr. Widner & Mr Moller coming in the conversation changed to other subjects not so reverant. After they left I felt quite sleepy and lay down and after reading a few moments went to sleep and had a two hours nap. This evening has been spent with you as much as it has lain in my power to do so. Mr Moller has been, and still is, in the office, preparing I suppose, to go away in the morning. It is almost ten, and with sending much love to my little wife I will close.

Ever your devoted Husband
Edwin J. Barden

No. 45 Rec'd March 20th 1865
City Point.
Ans.

Hd Quarters Armies of the U.S.
City Point Va Mch 15th 1865

My Darling Wife.

  I have no particular material on hand out of which to make a letter tonight, but as it is the middle of the week and you'l be looking for a letter on Saturday I thought best to write a few words, and see how well I could succeed in making "bricks without straw." I have first returned to my tent after supper; it is a warm evening but as it rained a little just at night and it is therefore a little damp, our colored man thought it necessary to "fire up" a little, which he has done much to my discomfort, for it is so hot I can scarcely remain in the tent. Mr. Browne wished me to play cards with him, and I suppose I ought to have done so, but did not feel like it, and so excused myself on the ground of having a letter to write. Mr. Moller left for Washington on Monday, and I expect him back tomorrow. Am master of the house in his absence. I was most agreeably surprised yesterday by a visit from Capt Deane to the PO in the morning to mail a letter;- was first dropping it into the box and noticed an officer standing near, who at the same instant hailed me. It proved to be our friend James, who was on his way to rejoin his Regt, and had stopped at the hotel here over night and was then waiting for the starting of the morning train to the front. I easily persuaded him to defer his departure until 3.P.M. train and spend the intervening time with me. The forenoon we passed in the Store house, talking when we could get a chance. he took dinner with me, and we passed the remainder of the day until the train time in looking about the Point, and in sitting in my tent, of course finding full employment for our tongues all the while. I had not seen him before since he visited us when in camp at New Haven in /61. I much regretted that I could not have known of his being here the preceding day for we could have spent the night together, and had a nice visit. James is looking quite well, better than I ever saw him before, but it is not certain how active duty at the front may affect him! He appears to me the same as he always did with the slight exceptions that would necessarily occur in any one who had laid aside the minestry for a calling so exactly the opposite. I shall now hope to hear from him ocassionally, and if his Regt remains in its present location long I may go out and see him; which I can easily do any day if I have a desire to. It is thought that the army is about to move; or some portion of it at least; it does not seem as though the weather could be sufficiently trusted to undertake a general movement yet. The few pleasant days just passed have made the roads quite passable, but a half a days rain would effectually blockade them again. City Point has been flooded today by a drove of notables both male and female, a boat load of whom came down from Washington last night. Senators, Congressmen, Cabinet officers with wives daughters and for aught I know sweethearts, they flourished about, poking their noses into every corner until they had mastered the sights here, and then all hands started off on the cars for the front. Well Dear Wife as I did not promise to write a letter, you will be expecting me to stop at any time, which with sending you lots of love, may as well be now. Good night my darling.

Ever Your Devoted Husband
Edwin J. Barden

No. 46 Rec'd March 23rd 1865   Hd Quarters Armies of the U.S.
City Point.   City Point Va Mch 20th 1865
Ans.

My Own Dear Wife.

    Your letter of the 12th & 14th are received, and have made me anxious to know how your getting on with that cold. I hope and trust that it is better-that you are entirely free from it, and its evil effects. I cannot bear to think of your being sick and I so far away from you! I have experienced that misfortune once and do not wish to again. It is not likely that I could make myself of much use at the bedside of a sick lady. I should do more harm than good I know:- but if you must be sick (which God grant may not be) I want to be with you. I received George's letter of the 12th, and have answered it. I commenced this afternoon, thinking that I would write you as long a letter as I could, but had not written half a page when I was interrupted by the entrance of three of my brother clerks, who had come over to witness the Guard mounting at Headquarters, and were an hour too early. I entertained them to the best of my ability, went out to see guard mounting with them, that over with it was too late to commence writing again before supper so I tore up what I had written because it did not quite suite me. I have been interrupted once since commencing this time, and give up trying to write a long letter now. My dear little wife, it was very good of you to write to me when you felt so unwell, but you must not attempt to write much when you feel unwell, just a line or two to tell me how you are. This has been a lovely day. I went over to the Hospital to church in the morning, and heard a good sermon from <u>the 51st Psalm 11 verse</u>, delivered by Mr. Vincent. The speaker is evidently a smart man. his discourse was one of instruction and warning and I think calculated to do good. I consider it a great treat to be permitted to attend well conducted religous services, and hear such good preaching here in the Army. The chapel was filled today: several ladies were present and quite a number of officers. There are now two other places here where there is preaching on the Sabbeth. I anticipate being much more busy hereafter than I have been since my return from my visit home. Lt. Browne is going to issue Rations to some new commands; and as it is likely that the officers of those commands will also purchase most of their Stores from us, that will also add much to our business, besides there will be the additional work on our papers. In spite of your preferring my present location, I cannot help hoping that we shall move before long. If I must be away from you I like a little excitement to keep my spirits up, and prevent me from getting uneasy and discontented. Not that I wish to forget you my darling, or to think of you less, but that I should be less likely to be homesick. It does seem some times as though I could not stay another day. that I must take the next boat for home. Mr. Browne loves to teaze me about going home; offering me opportunities to do so when he knows very well that it is not practicable for me to do so. I more than half wish that something would turn up that rendered it imperative for me to go home. Poverty may be neither a sin or crime, but to me it seems a great curse. I do not mean to complain, the Good God is infinitely better to me than my dearest, but my heart is very rebellious, and its evil passions have too much power over me. I am much afraid our dear sister Sara's will be disappointed about spending the next Winter with us. I wish she might not be, for I love her much and it would give me great pleasure to have her with us, but dearest so long as we cannot live together ourselves, there is little chance of our mutually enjoying the society of a third person dear and much coveted it may be. Give my love to her when you write and tell her that if you and I ever have a shelter for our own heads there will surely be room under it for her. Capt. Adair and Mr. Moller have just come into my tent and commenced talking, and there is no use in trying to write more tonight. Good night my darling.

Love from your devoted husband
Edwin J. Barden

No. 47 Rec'd March 29th /65   Hd Quarters Armies of the U.S.
City Point.   City Point Va Mch 24th 1865
Ans.

My Dear Little Wife

You"ll get no letter from me on Saturday of this week, and for the following reason; the night that it should have been written I was at the Docter's waiting to have a boil dressed. Said boil being a good sized one on the left side of my neck just under the ear. It has broken and commenced discharging today and begins to feel easier. Have kept large poultice's on it for the last two days and nights. It is in rather an uncomfortable place!- in fact I don"t think my "comforters" choose the least troublesom places for their operations. Now I think of it I can"t remember ever to have had one that was entirely out of the way, and in a real good place. On the evening in question when I should have written to you, the Dr. was busy, and as I could not reasonably hurry him, he took his time, and I did not get back to my tent until bed time. Yesterday I received your favor of the 19th. Also a letter from Mrs Randall. She says you have not been over to see her yet!- which of course I knew. Also that she has been playing "Bridget" most of the time this Winter, but is expecting a lady of foreign birth to adorn her household, on the 1st of next month, after which time she hopes to have a little more leisure, and means to call on you. I should be much gratified if it were convenient and <u>pleasant</u> for you to make them a visit, but do not wish to urge you contrary to your inclination. I hope however you will not fail to call there should any chance take you to the village. Very, very good friends they have been, and still are, to me, and very dear they are to me. I cannot expect you to enjoy being there as much as I should, but do think they would be much pleased by a visit from you. So you were having a touch of Spring weather that lovely Sabbeth were you? Having had several days of fine warm weather with considerable wind, we begin to feel quite assured of the approach of Spring. Plum and Cherry trees are in blossom. the grass begins to look green, and for three or four evenings back it has been so warm we have had no fire in our tents The last three days have been very windy, and tonight it is plenty cool enough for fires. And so Lib has been dreaming of me has she? now aint you jealous? I would"ent have thought she would have dared to have told you of it! for fear of having her eyes scratched out. Tell Lib. I dont write my sermons but deliver them extempore, and on the "Mrs. Caudle" plan. My texts are usually, "Missing Buttons." Undarned stockings. "Grievious rents and tares in wearing apparrel" &c &c. for all of which vexations I dont see as a wife is any antedote at all. I wish I could be favored with a good sensible dream of home occasionally, but I cant. I either lie as stupid as a log, or else have some sort of a "Man in the moon" fancy, having no connection with rational or loveable things. I dont think you are beating me so very much as getting up early after all, for during the past week I have been up by seven and most of the mornings as early as half past six. I dont know what you are going to do with so much time redeemed from sleep, but trust that you are plotting no mischief. No stirring news here as yet. Some of Sheriden's men arrived here today; they are in good spirits, and good condition, but show traces of active duty, and rough work. One candle has just gone out and Mr Moller is abed fast asleep. so good night.

With Love
Ever Your Devoted Husband,
Edwin J. Barden

No. 48 Rec'd March 30th /65 Hd Quarters Armies of the U.S.
City Point. City Point Va Mch 21st 1865
Ans.
Dear Wife.

My letter to you today must be written quickly if at all, for its past three o"clock already, and Mr. Moller says that he has invited two or three of our brother clerks to call on us this evening. So there will be writing to my little wife then. Well, I have not been to church today-mores the pity. I was a little misled by a report this morning that we were about to move, and therefore thought it necessary to attend immediately to some work on our papers; and besides I am obliged to have my neck bundled up rather inelegantly and thought I should not appear to very good advantage. To be sure that would not be of much importance, as my "market is made" (and well made too), but then there is no knowing but that the brethren seeing me with such a wealth of white neck tie displayed might take me to be one of the "Cloth", and insist upon my officiating, and you know it would be awkward to have to tell them that I had loaned all my sermons to my little sister Lib, and that I could not preach extempore because I had always accustomed myself to use notes. But jesting aside I am sorry that I did not go, for I find that I might have deferred my work until afternoon, and also that the same man preached, that did last Sunday. So I should have been reasonably sure of hearing a good sermon. I feel so much more human after having been to church on a Sabbeth, besides the spiritual good to be obtained there, it brings to rememberence the pleasant hours of the past, and seems to unite me for the time being, once more with society, and loved friends, and encourages me to hope that I may yet enjoy a better life in this world. In regard to our moving there are many reports and surmises though nothing definite is known. We are in a state of partial preparation to march, and should not be surprised to receive orders to leave any day. On the other hand, there is no known reason why we shall be likely to go for weeks yet. The question is where can we go to? Until the Army has advanced and gained some new and important position at a distance from here, it does not look likely that the Headquarters would be moved, even though the Genl should spend much of his time at the front. The engagements last night and night before, on our left, and the arrival here of Genl Sheriden and his forces, has set every one to thinking that we are to be off at once. Gen. Sheriden was here this morning, I should liked much to have seen him, but was not thus favored. <u>President Lincoln and his wife are here</u>! they went out the road yesterday to visit the left of our lines. I have not been fortunate enough to catch a glimpse of them. Genl. Meade's family are also here. All together there is quite a collection of Washington "Upper ten" here. The clerks on the wharf unanamously vote him to be as homly a lot as is often to be met with. I send you the photographs of two of my fellow clerks. Also a photograph (taken from a drawing) of Genl. Grants Hdqt'l. In the center of the picture are two houses which the artist represents as being alike, though in reality one is considerably smaller than the other! the left one of the two is Genl. Grants. The others are those of his Staff. Of the two photographs, one is that of Mr Widner, Capt. Adair's clerks, he is going to leave us this week, Capt. Adair having been assigned to duty with the Army of the James. Am sorry to lose them as they are both pleasant men and agreeable companions. I fancy the Capt. will not like it much. he is now absent on leave. I had a short call from Rev. Mr. Jarvis this morning. he did not stay but a few moments as we were busy at the time. he has just returned from a visit home. I do not attempt to give you any particulars of the late fights, for two reasons, one is, I have no definite knowledge of them myself, and the other is you will find them all in the N.Y. papers by the time my letter reaches you. I learn that <u>the 6th Corps</u> was engaged, and I feel a little anxious to know if the 2nd Conn. suffered any. It was a cold, raw kind of a morning today, but it is clear and pleasant at this time (just at sunset) I had a splendid dinner today! Roast Beef, baked Shad, potatoes Onions &c. I suppose you dont have Shad yet awhile up home? We have had them for some time here. Do not be too much disappointed if you do not receive another letter from me this week, as I shall be very busy, it being the last of the month and the Papers to be made out. I hope my dear little wife is well today, and happy!- that you may always be thus blessed is the prayer of

Your Devoted Husband
Edwin J. Barden

No. 49 Rec'd April 6th 1865

Hd Quarters Armies of the U.S.
In the Field Va Apl. 2nd 1865

Dear Wife.

I suppose you are wondering why you do not hear from me. I was too busy to write during the middle of the week and on Friday afternoon about 6 o"clock we got orders to leave the Point and join Gen. Grant at his Hd Qr's at the Front. Then I did have a busy job. All our Stores on hand, with the exception of enough for ten days rations, were to be turned into the Depot Commissary, the remainder to be properly packed, and all our troops put up for a move. And we were to be ready to go out on the 6 o'clock train in the morning. I worked until after twelve o"clock, went to bed too tired to sleep soundly, and was up at 5 o clock in the morning and at work again. We could not get off on the 6 o c" train, nor on the 9 o'c train and barely got ready for the one that left at 12 Pm. Just as we wer putting our Stores on the cars, orders came to turn in everything and come on without delay. We reached <u>Humphreys Station the end of the Military R.R.</u> about three o'clock. there we waited until about dark for teams to haul our luggage to camp. It was dark by the time the wagons wer loaded, or would have been but the moon From the Station to camp the distance was about four miles, over bad roads. We arrived in camp about 9. unloaded our stuff, in taking a cold lunch turned in for the night. We occupied the remains of what had been a small house, but the siding being all torn off up to the second floor, we did not want for ventilation. The place was just inside of a rebel entrenchment which had been carried a day or two previous, and about a mile and a half in rear of the then line of battle. Heavy firing was kept up all night but it did not prevent any sleeping like a "top". This morning early the fighting commenced very fiercly! the roar of cannon and musketry was tremendous for a couple of hours, and then after a moments lull, there was a tremendous cheering for five or ten moments. Our boys wer charging the works, which they carried in fine style! then the sound of firing grew more distant, until it ceased altoghether, with the ocassional exception of the report of a heavy gun. At Noon we loaded up and commenced to follow up the advancing army. At present we are within about three miles from Petersburg I think. The rebels are driven into their last line of works in the edge of the city and our forces are I suppose, resting a little before the final struggle for the city. We are within a mile and a half of the line of battle, but not a gun is to be heard. Its a beautifull moonlight night the band is playing,- all is quiet, every one seeming to be resting, and gaining strength for a new struggle. I am very tired, love, so please excuse me for not writing more. I am very well, darling, "as fine as a fish" feel better than I have for six weeks before, have caught a trifling cold, but its of no account. Am writing under difficulties for Mr. Browne is chattering like a magpie, and its impossible to collect my thoughts enough to write. I received the box the evening before we started. Mr. Moller was in ecstasies over the slippers. Says he must acknowledge the receipt of them to you direct. I have received severel letters from you for which I am ever grateful, but I do not remember the dates of them, and I trust you will, excuse me for not answering them in detail. Be assured darling of my continued love and devotion. Please consider yourself kissed and bid good night by your husband.

E. J. Barden

MY DEAR SARA

No. 50 Rec'd April 18th 1865

Hd Quarters Armies of the U.S.
In the Field Va Apl. 6th 1865

Dear Wife.

    Your letter of the 31st has just been received and I take the opportunity offered by the return of the Mail carrier to drop you a line. Shall make no attempt to answer your letter only to send you word of my where abouts, and say that I am well though pretty well tired out. We reached here at twelve o clock last night. two hours were consumed in putting things to rights and getting something to eat (for I had not tasted food since daylight in the morning) then I was obliged to be up before day light this morning in order to be start which should have been made at eight but has been delayed until the present time, though we have been hitched up and in momentary expectation of orders to move. I hear that we are going to the rear, as a fight is expected within three or four miles of us this evening. Gen. Grant went out to Gen. Sheridens last night and has not yet returned. Lee is nearly surrounded by our forces, and it is thought that he will be brought to bay, and prehaps the final struggle of the rebellion will take place near here. This is the junction of the <u>Southside and Danvile R.R.'s</u>. Our line of march since I wrote you last has been along side of the former road. I suppose we are near midway between Petersburgh and Lynchburgh. It is six o"clock, and we have the prospect of another night march. I had rather remain and take my chances with the stray shot and shell, than to try last nights experiences ever again. Will write again the first opportunity. With love my darling I remain as ever your devoted husband.

                                                                Edwin J. Barden

CHAPTER 24

LINCOLN ASSASSINATION

From <u>the book</u>:

On the 6th came the news that General Grant had started a successful movement against Petersburg, which filled us with joy and made us think of home. On the 16th we received New York papers to the 13th with news of Lee's surrender and the progress of Sherman. Our joy at this was chilled a day later by the news of Lincoln's assassination. At first the news came by wire and was only known to a few. In the afternoon of the 14th a vessel arriving from Fortress Monroe confirmed the sad tidings and then a wail went up from the whole city and from the camps. The comrades looked in each other's faces, clasped hands, but could scarcely speak. Each one closed his lips with determination, while grief and a desire for revenge struggled for the mastery. On the 19th an article appeared in the "Wilmington Herald of the Union" signed by J. R. H. In the following extracts the boys will recognize General Hawley's style. After relating the circumstances of the assassination he said in part: "And so this new villainy, legitimate spawn of slavery and rebellion has striken the nation in a sore place. Words are feeble when we think of the deep sorrow that falls upon the people. It will soon prove that never has a ruler so entrenched himself in the affections of a nation. He was clear in his moral and political truth, steady and calm in his purposes, sagacious, patient, long-suffering and filled with love for us all. North and South. None but fanatics as wild and few in number as those who did this most senseless, foul and cowardly murder have ever attacked his purity of motive."

Editor's Note:

Since the close of this 6 April 1865 letter, General Grant's Army has captured Richmond.

9 April 1865 Gen Robert E. Lee surrenders his Army of Northern Virginia to Gen. H. Ulysses Grant.

President Lincoln is assassinated 15 April 1865 and Andrew Jackson takes over as President.

President Jefferson Davis and his staff have fled to Danville, Virginia where he will be captured and imprisoned for two years for treason. He will be released without trial or conviction.

Ned Barden appears to have had a two or three week furlough to home that likely began at or soon after Lee's surrender to General Grant.

His letters resume upon his return to his same clerk job.

No. 51 Rec'd May 15th 1865
City Point
Ans.

City Point Va May 10th 1865

My Darling Little Wife

Back again at my old place of business - I had almost said "old quarters", but that is not exactly so, as the old tent that afforded me shelter for the past six months has been turned over to a colored gentleman, and for the night I am quartered with Mr. Moller in a neat little log hut nicely wainscotted and papered. Tomorrow I expect to take possession of some one of the several big houses still remaining vacant. I reached here without accident or event worthy of notice about two P.M. and was cordially received by Lt. Browne, Mr. Moller and my fellow K.K's. Have not done much but talk, but shall have plenty to do tomorrow as Mr. Browne has not sent of his papers for the Month of April, and wishes me to make the fair copies of them, which go to Washington. I found on my arrival at Millerton, that the time of the evening train, had been changed that day, and that it did not leave there until 3 minutes past Seven o"clock. I thought it doubtful about being able to connect with the 12 AM. train for Washington, and for a while wished myself back to L.- to spend another night with you. I lounged around your Uncle Ben's Store most of time until the cars came:-saw "Aunt Adeline" and Libby Warner who were busy shopping. On being seated in the cars, I became stupid and drowsy as usual, and slept most of the way to N.Y; took a little lunch at Pawlings, and left my silk han"ke"f in the car upon leaving by way of "paying for my lodging." I had it out to lay my head on, and getting up quickly quite forgot it. It was raining hard when I reached N.Y. I immediately took the street cars for City Hall, and from thence started on foot for the ferry City Ferry, determined to do my best to get on that night. It turned out that I was in good time, though with no great deal to spare. I hardly waited for the train to start before I dropped to sleep, and spent the night in a series of "Cat naps" interrupted by frequent "nudges" from the Conductor, and the changes of position rendered necessary by my uncomfortable bed. The train reached Washington about eleven o'clock. I walked down to the Dock a distance of a little over a mile rather than to ride, for I wished to stretch myself after sitting cramped up so long. It was dark and cloudy all day, and rained a great deal, though not all the time. I thought at first that I would see a little of the city having about three hours before the boat would leave; then knowing how often vexations delays will occur when least expected, in military departments concluded that it would be best to get my pass stamped by the Provost Martial and secure transportation, and then if there was still time I would look around a little. I accordingly presented my pass to the P.M. who refused to stamp it unless the Q.M. Dept. said it was all right. in the Q.M. Office they gave me no consolation but rather the reverse. Said that if I had belonged to Gen. Grant"s Hd Qr's I had better get some endorsement on my pass from the same office of his Staff. By the time I began to be rather vexed, and made up my mind that I would go to C.P. on that pass if I had to stay in W. a week to effect it, and though it should cost me twice what it would to pay the regular fare charged. So I started at once to look up the War Department near which I was told Genl. Grants quarters were. I found them without trouble, and Col. Bowers at once gave me the required order, with which I returned in good time for the boat. We had a good, and fast boat, though rather small. there were but half a dozen State rooms, and no births, so we had to pass the night as best we could, in chairs on the cushioned seat that ran around the Saloon, and on the floor. My lot fell upon a section of the seat, and I lay and slept with what comfort I could until the boat reached Ft. Monroe at six. o'c in the morning. The day has been windy and cold until just at night when it came off fine and the evening is splendid. I do not yet know what will be Mr. Browne's fortune. it is possible that he may be assigned to the Small issue here. I do not think he will be sent to Richmond. of course I cannot tell how long I may remain. Col. Morgan will move his office to Washington in a very few days, and then good bye Mr. Moller. It makes me lonesom and home sick to think of trying to stay here after he is gone. I believe I have not much more to tell you tonight except that I am tired and very sleepy-that its past 11 o'c'.-that Mr. Moller is already asleep,-that I love my little wife more than I can tell that miss her very much and long for the better days which we so fondly hope will ere long come.

Ever your devoted husband,
Edwin J. Barden

No. 51 Rec'd May 15th 1865

City Point Va May 11th 1865

Dear Little Wife.

If I am not home sick, lonesom and low spirited tonight, then was never poor mortal afflicted with those diseases. I would give half my <u>fortune</u> to be with you tonight love! I am occupying tonight one of the houses formerly composing Gen. Grants Hd Qr"s. if you look at the picture you will see it; the one next to the Genl's on the left. I took possession today, and may be turned out "neck and crop" tomorrow to make room for some Shoulder Strap. It is quite a good sized hut, and contains two rooms. I took this one because we expect Mr Love every day, and if he comes he will share it with me. While he stays which not probably be but a day or two-I think I told you last night in my letter, that Col. Morgan would doubtless move his office to Washington in the course of a few days. A heavy thunder shower is brewing while I write. I expect to hear the rain pattering on the roof every moment. It has been very hot today. Such a change from the cool bracing air of New England, that I have felt it very much. A couple of hours in the morning were sufficient to divest me of that thick woolen shirt. in a little while I did not miss it, and was only deterred from donning a linen coat by a strong effort of self denial-I feared that it might be rushing matters rather too fast. The country about here is very beautiful now, there is a profusion of flowers where they have not been effectually trodden out of existence by the foot of war, Green Peas, and other early vegetables, are already of a fit growth for table use. Strawberries are profusely exposed for sale, but I think that they come up from Norfolk, but we have ripe cherries on trees here at the Point. I have been at work a little today, that is I have made a beginning;-tomorrow I hope to accomplish something. Mr. Browne has disposed of nearly all his Stores, and as there are none left of our old customers, we are neither selling nor issuing Stores. Probably we may have a few days of leisure, before Mr. B. is assigned to duty elsewhere. He thinks today that he will not leave but if he dont give the whiskey barrel a wider berth. He'll wake up some fine morning and find himself out of business. He got comfortably tight this forenoon-has lain abed reading most of the afternoon, and is complaining woefully of headache this evening, but all of this between you and me. Let it go no further. I am living at my old mess at which there has been no change. I cannot of course give you any definite idea of how long I may remain here. You will doubtless agree with me that as long as I can get a good salary here, and can find nothing to do at home, I had better remain, even though it be unpleasant to do so. I very much dislike this uncertainty in my present circumstances, were I a single man I think I should not care. A roving unconfined life would rather agree with me so long as I could maintain it respectably and not let it partake too much of the vagabond character, but "with a small but very interesting family" upon my hands life assumes a very different aspect.

With love darling, ever your devoted husband.

Edwin J. Barden

No. 52 is missing

No. 53

City Point Va May 12th 1865

My Dear Little Wife.

Do you think that I can afford the time, paper, envelopes, ink, and postage stamps necessary to furnish you with a letter daily? Well, no matter whether I can or not, whenever I feel able, and have the opportunity, and anything to write about, I will <u>write</u>. I received your dear, good, long letter of the 9th this P.M. and gratitude for it, if not (or aside from) the fervent love I bear, for its author, impell me to write tonight. I think I have experienced something of your "hungering and thirsting" after messages from the loved one, for the past two days. I doubt not you know something of the pleasure I enjoyed in the perusal of your letter,-it was like old times when I was a soldier boy, and the letter was more than either food, or rest. Not but what all your letters, are always welcome, and eagerly expected, but you know the things we most highly prize, and that give us the greatest pleasure, are more fully enjoyed some times, than others. What a dear, good, loving, little wife you are!-and what a fortunate dog I am in being your husband. it seems as though I have no business to be otherwise than happy, while in the possession of such a prize. May I ever be able to rightly appreciate my good fortune, and faithfully endeavor to be to you, what I know <u>you</u> are and will be, to me, a loving, devoted, partner, come weal or woe. I saw Mr. M's letter sufficiently to comprehend its meaning which was all I required of it. I am not yet much wiser in regard to the length of my sojourn here, than when at home. Onley it looks now a little as though I might find employment here for a while yet. Mr. Browne told me today that Col. Morgan said that he (Col M) would make arrangements to have Mr. Browne retained on duty as an A.C.S. until his regiment was mustered out of service. So there may be a chance for me to remain some time, and however contrary it may be to <u>our</u> wishes, I think you will agree with me that in the present state of <u>our</u> finances, I had better do so, rather than to leave and take the uncertain chances of finding employment about home. I suppose there is but little comfort for you in the above probabilities, but recollect my dear, that they afford me as little, as they do you, and that I would gladly return to you at once if it were prudent, or even practicable to do so. If we cannot be the arbitrators of our fate, we will do the best we can, still "Hoping on, Hoping ever," and making the most of the few enjoyments vouchsafed us. I enjoyed my brief visit home to the full extent of my capacity for enjoyment, and it adds greatly to my satisfaction, and present recollection of it, to know that my presence gave you so much pleasure, and that you have nought of evil to remember against me. I fear that your love for me causes you to hide a "multitude" of my sins, and that unless you are careful, you will yet do yourself injustice in your anxiety to please me. You must recollect that I have lived a long while, with no one to care for, but myself, and that as a very natural (though by no means necessary) consequence I have grown careless of others welfare, and shall be very apt to come short of my duty, in attending to the thousand and one, little courtesies and attentions, which are a wife's right, and which go far towards making her life happy. You may not observe this failing in me now, but sooner or later I doubt not that you will have just cause of complaint. It was very stupid of me to leave that Diary around in your way, for I knew that it would not be likely to amuse you, to say the least of it, and would be very likely to give you uneasiness, and pain, as in truth it did. I hope I may never more have ocassion to conceal a like distress from you. Yet I do not think it would be wise, or best to trouble you with the recycle of every petty ailment, or fit of low spirits that I have. I cannot thank you enough darling, for your kind, and reiterated assurances of entire devotion, and for your readiness to go any where, or do any thing that may be in your power, for me. It is little that I can do in return for all your favors and I do hope that I shall always be ready and willing to do that little. I had no idea of your being "obstinate" about the work that you wished to do, if you felt able and wished it. I had not the least objection, but I feared that it would be too hard for you. Dont worry your little head about "living in idleness" but enjoy yourself as much as lies in your power, recollecting that while doing thus, you are not idle, but doing <u>my work</u>, and thus giving <u>me</u> pleasure. Cares and trials will come quickly enough, do not anticipate them, or go out to meet them. Secure whatever of pleasure you can now, and then when days of toil & care do come, you will not have to regret that you did not "improve your privileges." I cannot think of one "thoughtless" or "unkind" thing that you did or said while I was home, which I can either forgive or forget. I shall certainly have to visit you again before I can find anything to quarrel with you about. Having never seen aught but good in my little wife I cannot well remember her as otherwise, and when I think of her other than "kindly," may I be bereft of reason, and the power of thought. It was very cool here this morning, after the heavy rain, and thunder of the past night, but it came off clear and fine towards noon, and this afternoon the weather has been most lovely. I have been at work all day, not hard, but just busy. This afternoon I went down on the dock and saw a large quantity of machineary that was captured from the rebels. It belonged to their Armory, and I suppose must comprise nearly all that they had in their armory at Richmond. It was captured on the cars. You may remember reading of their removing it previous to leaving the city. Mr. Moller has been in spending some time with me, and has left for his own house, and bed; and as it is past eleven, I must follow his example.

With love darling, a kind good night.
Ever your devoted husband.
Edwin J. Barden

No. 54 Rec'd 18th May/65
City Point
Ans.

City Point Va May 14th 1865

My Dear Wife.

Last Sabbeth at this time I was by your side in church, how much I wish the like fortune was mine today. I attend service at the chapel of the Christian Commission nearby, this forenoon there is no more service until in the evening. The minister was not a man of much talent, or ability as a speaker, and compared unfavorably with Dr. Reid. I was sorry that I did not start earlier and go out to the hospital, but it is very warm, and I thought I might hear as good a sermon at the large new Chapel the Commission have lately erected at the Point. They evidently put their smartest men on duty at the hospital, and one will seldom fail of hearing a good sermon at their Chapel there. You see I have skipped over one day without writing, and would probably like to know why! and also how I spent the day. I worked a little during the fore part of the day, and at noon Mr. Moller invited me to go out to Petersburg with him and Mr. Stake; they were going on the 3.P.M. train; and back in the evening. I thought I would like to spend two or three hours looking about the place, and concluded to go! Before it was time to start Mr. Love came, and Mr. M.-being detained by the non-arrival of the mail, could not well go. So Mr Stake and myself went without him. Mr. S-was some time in getting a clue to the business upon which he went, and which he had not expected would detain him ten minutes! he finaly had to go quite a little distance from the Depot to attend to it, and did not return upon the same train with me;-which train left P-at half past Six. I rode through the place a little with a friend who is at "K.K." there;-there is but little to see; the place is about the size of Pittsfield I should think, and was probably once very pretty and thrifty. The present business seems to be limited to the "Pea Nut" stands, and huckster shops. A little board with "Eating House," "Restaurent," "Saloon," "Beer" or "Lemonade" is hanging from almost every other house. I do not see how they all find customers to support them! The people must literly "live on each other." On reaching home, I found Mr. Moller and Love, just going to the "Circus" which has been performing here for the last week or two. they asked me to go and I went. The performance was about an average one of the kind. It was near 12, when we got home, and quite that time when we retired for the night. I was very sleepy (as usual) and it was almost eight o'clock when I awoke this morning. Love was still asleep, and I dressed quietly and went away to breakfast without disturbing him. Mr. Browne is quite curious to know what made us sleep so late. I suppose he thinks we were dissipating, but he is much mistaken for we did nothing of the kind. I shall get through with our papers tomorrow, and then shall have nothing to do until Mr Browne is assigned to duty elsewhere. Col. Morgan went to Ft. Monroe yesterday. he is awaiting orders assigning him to some new post. If as I expect, this should prove a week of leisure, I intend to go up to Richmond for a day or two. All the other clerks have been up, and I have only been waiting for a time when I should have nothing to do. I wonder if you got any letter from me yesterday, it is possible but hardly probable. I think I shall get one from you today. I hope and trust that you are well, and are enjoying yourself. I hope you will tell me if you are <u>quite</u> well, and that you will not work too hard. I suppose you will be going either to Emma's or to Sara French's this week; if you are not needed at Emma's. I think you ought to visit Mrs. French. She seemed to desire it so much. You see I have a considerable sympathy for her, for I know if I was in her place I should want you badly. I thought I could write you quite a letter when I sat down, but somehow my ideas seem to have flown-not from pen, but to the regions of the unknown, perhaps they will be back in time for my next letter.

With love my darling.
Ever your devoted husband
Edwin J. Barden

CHAPTER 25

PRESIDENT JEFF DAVIS CAPTURED

From the book:

On the 17th of May news came of the capture of Jeff Davis and this emphasized the fact that the war was over. The people of North Carolina generally seemed to accept the result without murmuring and were glad of an opportunity to collect their scattered families and restore their decayed fortunes.

General Schofield issued orders directing each corps and district commander to send to each county under his jurisdiction a discreet officer with a sufficient force to organize a small company of responsible, loyal citizens to serve as a local police force. They were to be furnished with captured arms and ammunition. They were obliged to take the oath of allegiance to the United States Government, and an oath to preserve the peace, prevent crime and arrest criminals as far as practicable within their counties, and to obey all lawful orders of the military authorities of the United States.

No. 55 Rec'd May 22nd 1865
City Point

City Point Va May 18th/65

Dear Little Wife.

I received your letter of the 11th containing the one from my sister, on last Monday. I had just written to you the day previous, and so postponed answering it until the next evening. On Tuesday I went to Richmond came back on Wednesday, and feeling very hot and tired deferred writing until tonight. You were perfectly right in reading the letter. My sister need not know of it, and if she does I can convince her that her anxiety about it was entirely unnecessary. I should not have visited her had the letter been received in time, for I mean whenever opportunity affords, to visit her in company with you, and do not feel able to incur the expense twice, when once can be made to answer the purpos, besides I could no more aforded to have been absent from you, than you could to have spared me. I have not seen my brother;-he was here inquiring for me the Sunday previous to my return, was much disappointed in not finding me;-was on the point of leaving the place when he called, and I cannot find out from the negro of whom he inquired about me where he was going. I am under the impression that he was on his way to join his regiment, which I suppose was with the rest of 6th Corps;-then at Danville;-now on its way to Washington. I shall write to his wife directly in answer to her letter. Weather during the week has been excessively hot. I have felt it very much, & it dont seem as though I could ever muster up courage to go to work again;-it will come hard to be obliged to, after remaining idle so long. I assure you I am in no hurry for "active operations" to commence. Mr. Browne on the contrary is fretting and fuming about because he has nothing to do. I have tried to dog him off home on a leave for fifteen or twenty days, which he might get just as well as not, but he does not seem to want to go home. I know very well if I was his wife I should not want to have him. Col. Morgan has returned today. he has not yet received any orders for himself and of course none for Mr. Browne. It is thought that Mr. B. will have what is called the "Small Issue." here. Capt Martin is to relieve <u>Maj. Wiley as Depot Commissary</u> here. If Mr. B. is assigned to duty here, he will not probably have anything to do until Capt. M. comes, which will not be under a fortnight or three weeks yet. I should have no compunctions of conscience at remaining idle that length of time and drawing my pay from the Government-I consider that I should only be getting back a little of what I have been giving to it, for the three years past. Well, about my visit to Richmond, I just started at 3.P.M. Tuesday on board the little steamer "Moniter;"-the ride of the river was delightfull, and full of interest. Among the objects seen from the boat, and by name, familiar as "household words" are, the "Dutch Gap" failure, and our batteries on the opposite side of the stream, built to defend its workmen. next our "Crow's Nest," and "Water Batterie's" thrown up for the accomodation of the rebel <u>"Howlett House" battery</u>, which faces them from the opposite end in the river, next to the obstructions placed in the river by our people to prevent the rebel rams from coming down too far, then the "Howlett House" battery in a commanding position on the high bank of the river; it looks desolate and bare, dismantled of its guns and with nothing but the chimney remaining, of the once elegant mansion which stood within a few rods of it, and from whence it derived its name; just below it, a half mile or so, lies the wreck of one of the rebel gun boats that attempted to pay us a visit last Winter. She got aground under fire of our batteries and was blown up by her crew, a portion of her stern is out of the water, and on it is a large gun, still on its carriage and looking "ready for operation." On we go past long lines of rebel batteries on our own left, and those of our owne on the right at, "Chaffin's Bluff" the rebel lines cross the river, and they have heavy works on the "Bluff" on the right bank of the stream;-here too lies another wreck of one of the rebel "Navy," which was coming uninvited to dine off Uncle Sam's bounties, at this place last Winter. She was roughly handled by our batteries and gave up the ghost here on her return to Richmond. About three miles further up, on the left bank, we come to the famous "Fort Darling" quiet and beautiful it looks, crowning the summit of a high bluff that rises abruptly from the waters edge! its nicely blacked guns shining out from between the bright green sides of its embrasures. The slopes and angles of the work are as nicely turned, and grated, as the terraces in a gentlemans garden, and as nicely turfed to. The river at this place makes another of its miriad of turns, and the rebels had sunk boats in the channel so as to obstruct it all but a narrow place near the opposite bank. This would compel any hostile boat attempting to run by to keep out from under the bank, and thus keep within range of their guns. It does not seem possible that even our iron clads could ever have succeeded in passing it. Directly after passing Fort Darling we are in sight of Richmond, and though there are still some rebel works commanding the stream they are of minor importance, and we are so busy with the city that we pay little attention to them. In its better days the city must have presented a fine sight as one approached it up the river. it stands on the right bank, which rises gradually as it recedes from the river thus giving the view of nearly all the city at once. We land at the lower end of the city, first passing the sunken remains of the last of the rebel "James River Fleet," which they blew up on the day of the evacuation. A few rods up the street, and we pass the never to be forgotten "Libby" prison! it is a large brick building four stories high on the street that fronts the river, and three stories, on the opposite street, which is the front. it was formally a warehouse and there are several mor buildings of the same class standing near, and from which at a distance it is difficult to distinguish it. At present there is no glass in any of the windows above the first story! They are all open except being burned with the iron! On the river side it is white washed up to the top of the second story windows. I wanted to go through it, but found that I would have to go up to the Capitol to get an order admitting me and I am so thoroughly sick of dancing attendence on Military Officials for favors that I choose to forego the pleasure, rather than to have any more to do with them than I could possibly avoid.

I cannot finish my letter tonight for Messrs. Moller & Love have been in for some time, and their talking has disturbed me so that I could not write, and it is now late, and I am sleepy. I will send along what I have written and finish when I have an opportunity. With a great deal of love darling, I remain your devoted husband.

Edwin J. Barden

Editors note: The following publication cover shows the starvation of prisoners that Ned references in his letters.

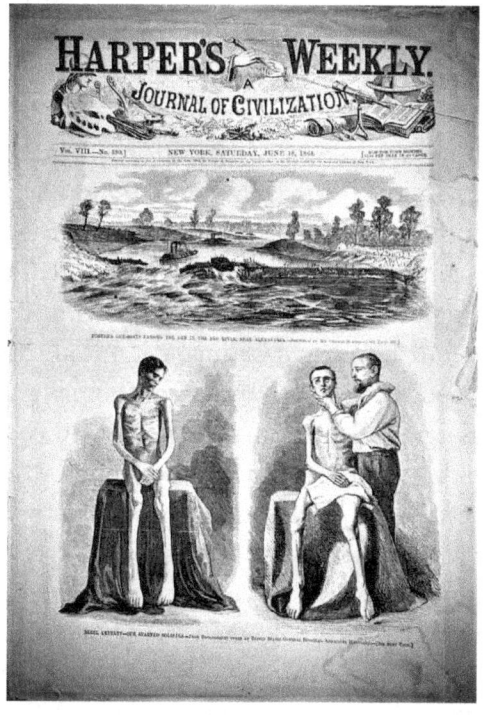

No. 56 Rec'd 23rd May/65
Ans.

City Point Va May 19th. 1865.

Dear Little Wife.

It is a month ago today since I left here on my visit home. The time passed pleasantly, and all too rapidly while there, and yet it seems a long while since the

19th of last Month. I should like to know where my little girl is this morning, and what she is doing. It is another of our hot sultry mornings. Scarcely a breath of air is stirring. Mr. Love has gone to Richmond. I got up a little after 7, dressed leisurely, was very late to breakfast, went to the post office and mailed my last nights letter to you, came back and shaved and have seated myself to finish the account of my Richmond trip. I believe I left off at the Libby. It was with difficulty that I could bring myself to realize that within those quiet harmless looking brick walls, so much of misery and suffering had been endured, and in the bitterness of my heart I could not but doubly curse the inhuman wretches who were the cause of it, and who living within a stone throw of their hellish work, never sought to alleviate the sufferings of their miserable victims, but rather on the contrary added insult to injury whenever opportunity offered. I was shown the cell in which the fiendish wretch "Dick Turner" for a long time the keeper of the prison, was confined after the occupation of the City by our forces, and from which he effected his escape one dark stormy night last week. Why he was not hung or shot at once I do not see. Certainly he deserved no such chance for his life as a trial would have given him. I was told by the guard that several officers who had suffered at his hands while inmates of the prison, had attempted to shoot him through a small hole in the door of his cell. I could not but regret that they had been prevented, but perhaps there is yet a worse punishment than an easy death awaiting him. I also saw both entrences to the hole dug by Col. Straight, and his companions, and through which some three or four hundred I believe got out of the prison. But few of them succeeded in making good their escape. Many of them were shot down by their merciless pursuers, and the greater portion recaptured.-About sixty or eighty I think got away safely; perhaps your memory may serve you better than mine does. In the prison the hole commences in the brick wall of the lower story or basement, about a foot from the floor which is a pavement of small unevenly shaped stones. After removing the bricks they began to burrow in the earth, going straight forward under the street and coming up near the centre of an old shed on the opposite side of the street used for storing tobacco. The distance dug must be forty feet or over. The earth that they removed in the excavation was carefully spread about the floor in such a way as to not attract attention. The basement at that time was only used as a kind of lumber room, and receptical of all kinds of odds and ends. Our men gained access to it by carefully removing a piece of flooring overhead. On the opposite side of the same street, and distant of about twenty rods stands "Castle Thunder," another Military prison. I am at loss to know what gave it such a martial name, for saving the iron barred windows it differs in no preceptible way from a multitude of other surrounding brick buildings, which may have been built for either warehouses or Stores. There are but few prisoners in either Castle Thunder or Libby, at present. they consist of some bad men of our own army, and some rebels. Crossing the street on which the prisons are, we come to Main Street which runs nearly parallel to the river and was the principle business street of the city. it is now one mass of ruins for nearly a mile in length. The loss of property by fire must have been immense. After all I had read, and heard, I was not prepared to see such an extensive destruction of property. besides the burning on main street large Store houses along the bank of the canal, and the large buildings used as the C.S. Armory, Arsenal, and Laboratory, were all destroyed. All this ruin and desolation presents itself as we go up Main Street to the <u>Spotswood House</u>, where I stopped for the night. The Spotswood is a large, fine five story brown stone, front building, much worn and dilapidated inside, and sadly in need of rejuvinating. In the hands of an enterprising Northern man, it might easily be made a first class hotel. in the hands of its present proprietor it is hardly a fourth rate house, where one is charged extortinate prices, for very ordinary accomodations. I had three meals there. my supper and breakfast were no better than we have had at our mess. The dinner was the only meal that made any pretence to respectability. I have sent you the bill of fare for the meal. (Editor: See the bill below) A brother clerk made the trip with me. We had roomed together, it was a good sized room in the fifth story of the building and contained two double beds, which we should have had to have shared with two others had the house been crowded. They were good spring beds and good matresses, the sheets were clean but coarse and warm, the cover lid had been white once, but looked as though it had been used for a horse blanket, and had been a stranger to soap and water for a long time. There was gas, but no running water, the wash bowl was an ordinary cheap white one, and the water pitcher was one of those big gray stone pitchers, such as workmen used out of doors, where they want something that will not break easily, in lieu of a tumbler we had a little coarse white, earthen mug with no handle. The furnature was polished black walnut, and had once been nice, but is now much scratched and soiled. the carpet had been a nice one, but was long since worn out, and threadbare but still struggled to maintain its hold on the floor. For these accomodations for one day we paid $4. each. the price would have been the same, had the room been occupied by four, instead of two, and I am told would have been the same had we procured meals elsewhere. Such were our accomodations at the <u>first</u>, and <u>finest</u> hotel in Richmond. I am told that there are others where guests fare better for less money, but we went to see the city and wanted to see what, at least purported to be, the best of it. I visited the Capitol building, which though once a fine one is much smaller than I should have thought would ever have been erected for the purpos I saw the rooms used as the various houses of the rebel Congress, and took a seat in the Speaker's chair of the house of representatives, the Senate Chamber was locked up, but I had a good view of it through the glass doors. The Capitol stands in the midst of beautiful grounds on an eminence that commands a view of at least half of the city, and a vast extent of beautiful country across the river. Within a few rods of the building stands a fine equestrian statue of Washington, on a lofty pedestal of, I think, grey granite. it was placed in position and inagurated here that <u>Mr. Holley was Governer of our State.</u> You may remember that he visited Richmond and made a speech on the ocassion. I saw other buildings that were occupied by various departments of the rebel government, but the heat was intense, and I did not feel interest enough in them to examine them. I went out and saw Davis' house at present occupied by <u>Genl. Halleck</u>. the picture you have of it is a very good one. I also saw the house occupied by Gen. Lee. at present. it is one of a block of ordinary four story brick houses on Franklin Street and almost within a stones throw of the Spotswood. It has rather a seedy air, the shutters in particular being much dilapidated. I went out near the State penitentiary and from a high knoll hard by, had a good view of the rebel government works, that were destroyed. the famous <u>Tredegar works</u>, and of "<u>Belle Isle</u>." I could overlook the place completely, it was I supposed about a mile distant. The river above the city is divided into two channels by this island. I could not see how long it is, but it is only the lower end nearest the city that is interesting to us, and as the suffering place of so many thousand of our brave boys. Here within sight of the rebel Capitol and not three miles off, on a low narrow neck of land not over two acres in extent, were crowded together at times, as many as <u>thirteen thousand</u> Union soldiers suffering from their heartless persecutions everything but death, which to many of them would have been hailed as a welcom relief from the horrors of their situation. The ground rises to a considerable height directly in rear of the prison pen, and towards the upper end of the island, and here the rebels had pieces of light artillery in position to open a raking fire upon the prisoners, at any sign of an attempt to make their escape. I was sorry that I could not visit the place of burial of our poor boys who breathed their lives out in these fearful prisons, but the heat of the day was so great that I did not feel able to look it up! especially as I heard that it was quite a little distance out. I went to Theatre in the evening, the performances of a negro minstrel troupe, winding up with some sort of a play in which a fierce looking mustachoed, ringlet headed brigand and a simpering, lackidasical young maiden bore the most conspicuous parts, comprised the entertainment. The audience did not exceed a hundred in all, and were mostly soldiers and negroes. I saw about a dozen women in the Dress Circle where we were, but they did not look any too honest, though perhaps I judged them wrongly. We reached the city about half past six in the evening, and I left the hotel at three, on the next day. The rebels on their retreat burned the bridges crossing the James, and as I decided to return by rail, I had to cross the river and take the cars at the village of Manchester on the opposite side; the river is crossed on pontoon bridges! Two omnibuses leave the hotel to carry passengers, the distance being about a mile & a half. As it was very hot and I was quite tired from my tramps around the city I thought I would rather pay a quarter or half Dollar for riding over, rather than to walk in the burning sun. But imagine my surprise when the conductor in collecting the fair coolly asked me for a Dollar! This was the "last feather" and if it did not break the camel's back it completely broke down my little stock of patience and if I had been a swearing man I could have sent the proprietor of the Spotswood, and all his attachee's to the old Nick without remorse! As it was I paid the charge without a word, considering such unparalleled impudence to great for my powers of rhetoric to do justice to. As we waited on the platform for the cars, I found plenty of others who were cursing the hotel and its proprietor roundly, and thus I derived a miserable kind of satisfaction in finding others feeling as badly bitten as I did. I had a pleasant ride home by rail. There is nothing of interest to be seen on the route except the ever present earthworks. I reached home about half past six P.M. Was too tired to write you that evening, and spent it in talking with Moller & Love. And thus ends my account of my visit to the rebel Capitol. it is sadly jumbled up, and I fear you will not be much edified by it, but I have done the best I could. You know that my descriptive powers are very limited, and will not therefore expect great things of me in that line. I have been writing by odd spells all during the day, taking up my pen for a few moments at a time and often interrupted before completing a sentence. Its about five o clock now, and if it were not raining I should go out and see if there was not a letter for me. I "blew up" Mr. Moller soundly yesterday for not getting me one. I begin to think you are going to "diet" me on one a week, "just to see how I would like it." I think you suggested something of the kind when I was home. I can assure you that I do not like even the prospect of such hard fare. I am thinking you may be away from home. I cannot bear to think that you may be sick. that is always my first fear when I fail to hear from you regularly. You must not think that I am out of paper, because I have made such a piece of patch work of this letter. I was to lazy to go up to Mr. Browne's tent after whole sheets, so used up the pieces lying by me, as my letter required. Now darling little wife, I will bid you good by once more. Please remember that I am loving you a great deal, and longing very much for your dear presence with me.

Ever Your Devoted Husband
Edwin J. Barden

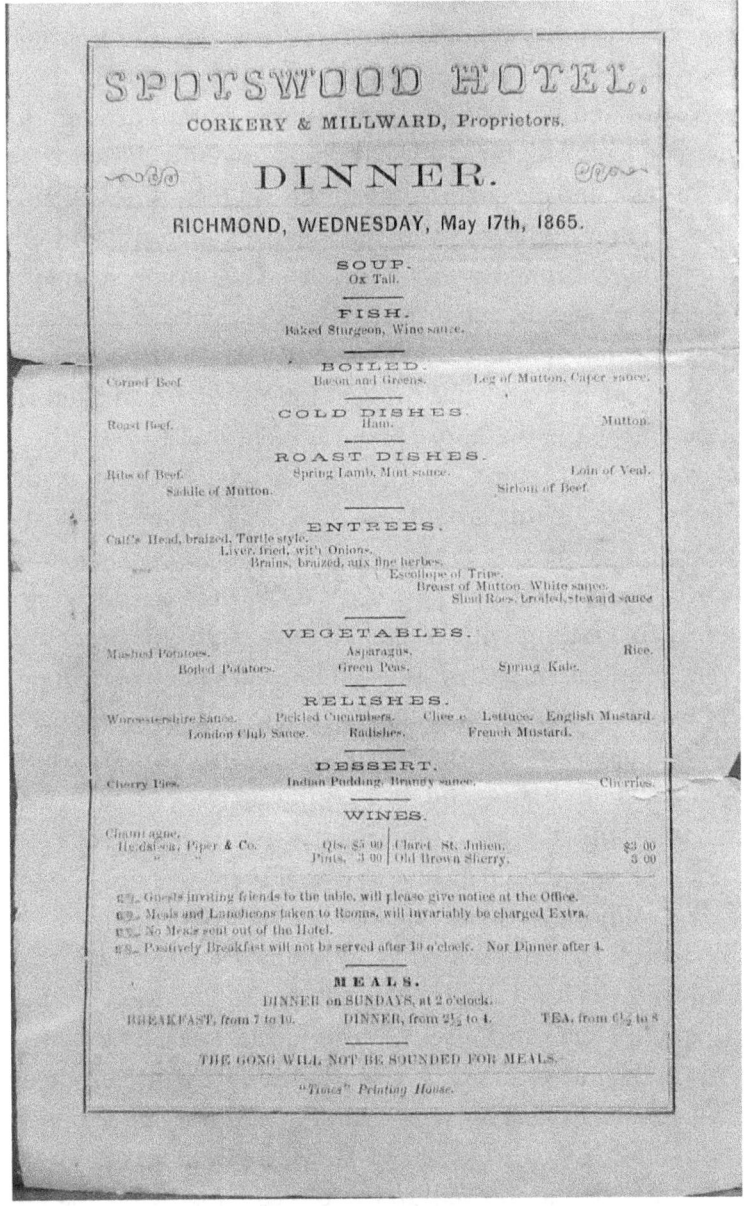

No. 57 Rec'd May 24th 1865
Ans.

City Point Va May 20th/1865.

Dear Little Wife.

Your favor of the 16th was most gladly received this P.M. I was wondering why I did not hear from you, and wishing Oh! so much, for a word from you when it came. It seems there is one back, that I have not received. I cant afford to lose it, for it leaves a blank in our correspondence just where I seem to want it most. I am feeling more than usually lonely this evening. Mr. Love left for home this morning!-he is a very pleasant companion, and I shall miss him very much. He hurried down here at the repeated solicitations of Col. Morgan, who could not wait for him to hardly visit home; and after he arrived, the Col. concluded that he did not want him at present, but should, as soon as he was assigned to duty elsewhere, which assignment he is daily expecting. But Mr. Love was not anxious enough for a job to be shoved off onto some other officer and await Col. M's pleasure! his brother is in N.Y. doing a good business as a commission merchant, and did not want him to come down at all, for he required his services very much himself, and would give him a good chance. So when the Col. returned from Fort Monroe and Mr Love found out how the matter stood, he packed his trunk, and bade us good by, this morning. Mr. Browne is going to do a little work next week. Col. Morgan directed this P.M. that we should make a few issues, for the purpos I suppose of showing that we have not lain entirely idle during the Month. Accordingly I drew a few Stores this afternoon, and am ready to commence operations again, in a small way. Monday morning, Mr. B. is delighted at the prospect of having something to do once more, but I am not "frantic" about it. The weather is too hot for one to worry himself into a fever for the sake of finding work to do. It now seems quite probable that Mr. B.-will be assigned to the "Small Issue" here. he expects it, and others think he will have it. I am preparing myself for it, but am too well acquainted with the sudden changes, and disappointments, incident to military life to make calculations on its being a sure thing. I still consider myself as "living from hand to mouth" and likely any day to be ordered to "Connecticut" for duty, provided I could find any there to do. Since the removal of Gen. Grant's Headquarters and the withdraw of our troops from this region, City Point has been gradually declining in importance, its glory has departed, and from being a busy, bustleing, active place, the base of supplies of one of the greatest Armies of the age, it is fast subsiding into the quiet of a second rate Military Post. There are but few troops stationed here. <u>The 25th Corps. (Colored)</u> is lying here at present but is under orders to embark on transports as soon as possible. Subsistence Stores to the amount of forty days rations for the entire Corps, are now being loaded up, and it will probably leave here next week. its destination is supposed to be, and probably is Texas. As Kirby Smith seems to have some difficulty about surrendering. I suppose it is going to assist him in the operation. A few soldiers will be stationed here, and Petersburg, Burksville, and other interior Military Posts will ship their supplies from here, which will make a little work, and prevent entire stagnation. If I were to remain here through the summer, and could make it comfortable for you, I should want you with me. With the departure of the troops, the principle objection for ladies being here is removed. Still there are no conveniences for their comfort, and no society for them, and in this hot, dreary, and lonely place, you could have but little enjoyment. You could not come alone. I would not go after you, and so that ends the matter, even if the uncertainty of my own stay did not effectually settle it at the outset.

Sunday May 21st. Mr. Moller came in last evening, to smoke a pipe and have a little chat, so I laid my writing aside, and when he left it was too late to resume it. I got up before Seven this morning, and flattered myself that I could take just as much time as I pleased about my toilet, and then be in abundant reason to walk out to the Hospital to church! and I was dressed soon enough but just as I was putting on the "finishing touches" one of our negroes came in and said Mr. Browne was sick, and wanted me to do something for him. I found him suffering with a sever tooth ache-or neuralgia in the head, he thinks!-his face was much swolen and he had evidently had a hard night of it. By the time I had his errands done, it was late and I only got into the Chapel near by in time to hear the end of the sermon. I am spending the afternoon in writing, thinking, and gazing out of the open windows. The river in front of my house is full of vessels of various kinds. Some of them quite large! they are accumilating to take off the 25th Corps. I am thinking you are about sitting down to tea now; you can have the whole end of the table to yourself. I shall not be there to crowd you. How many pleasant Sabbeth afternoons like this we have sat at the tea table together, little thinking how highly we should one day prize such privileges, and how earnestly we should long for them. Tell Lib. when she gets married, to be sure she gets some one that will stay with her, and not to take up with some Wandering Jew, who only pays her flying visits semi-ocassionaly, and then darts off into some obscure place to which she cannot follow him, and then proceeds to make both her and himself as miserable as possible, by telling her how happy they would be, if they were only together &c, &c. I declare I am ashamed of this letter for it is one mess of erasures, and blurrs. I don"t know what has got into me. I only know that I did not commence my letter as I intended to, and have been constant by making mistakes ever since. I would have pitched it into the fire a dozen times only that I was too lazy to rewrite it.

<u>Evening</u>. Your letter of the 18th came to hand by this afternoon's mail. I have been out on an errand for Mr. Browne, which has occupied me so late that I do not feel disposed to write much more tonight, and will defer answering it until another time. I have not received a letter that you must have written on the 15th and in which you probably acknowledged the receipt of my two first letters, of the 10th & 11th inst. Your letters that I have received, are dated respectively, May 9th. 11th. 16th & 18th. The missing one may turn up somewhere yet, but I do not like to miss it.

Good night darling wife. May Heaven bless and protect you is the prayer of.
Your Devoted Husband
Edwin J. Barden

No. 58 is missing

No. 59 Rec'd May 31st 1865.

City Point Va May 27th 1865.
My Dear Wife.

This is one of the most unpleasant days imaginable. A fine sifting rain, driven in scads by a bleak unfriendly wind, a deep mud, caused by heavy rains that have been falling for the last two days, and the most dreary aspect of all nature generally, all conspire to make it one of the most gloomy, dreary days of the year. Night before last, and all day yesterday, with but few in short intermissions, the rain came down in torrents. It still shows no sign of clearing up, though the rain has subsided into the drizzling, blinding mist, before mentioned. I have just returned from my dinner, and left word with my (or our) colored man to come up and make me a fire, which will make my room look more cheerful, and habitable at any rate. I have nothing at all to do today, and if I felt like it might write a number of letters, but dont expect that I should get further than this one, even if I succeede in keeping my courage up long enough to finish it. We are having a very lazy time of it now, on Wednesday we turned over to the Depot Commissary what few stores we had, and literally "shut up shop". a negro family moved in the same afternoon and are going into the "Cake & Beer" business, in my old sanctum. The next two days I was partially busy in getting Invoices & Receipts, and working along our months papers as far as it was feasible to get them. Today my "occupation is gone," and I exhibited my independence by lying in bed until about eight o"clock, and was then persuaded to rise only by the knowledge that I should have to go breakfastless unless I bestired myself at once. The forenoon was spent in my room in conversation with Carl and Mr. Browne, and in reading. It is at last settled that Mr Browne is to have the "Small Issue" here. he has got his orders, and has just informed me that Capt Benedict is to turn over the business to him on the afternoon of the 31st, so if you get this by that time you can think of me as being hard at work. If Mr. B-were a good businessman I should like it I think, for I judge that after the troops get away, I could get along with the work quite comfortably, but he is not. He has not the least idea of systemizing his work so as to do it to the best advantage, though he is forever prating about it, and one not acquainted with him would think him the most systematic man in existence. I feel almost discouraged sometimes, and have often felt tempted to tell him that he must get another clerk. I dont feel sure of staying with him long, as it is. If the work is very hard I shall "resign" at the earliest possible opportunity, for it will be of no use for me to attempt to work hard under him! the vexation and annoyence of mind would be ever much worse to beare than the fatigue of body. Some less nervous person might endure it well enough, and by going about the business their own way and paying no attention to him, keep along without serious difficulty. It may not be as bad as I fear. You see we cant tell how much we will have to do because we cannot tell yet how many troops are going to remain about here. I have dwelt upon this subject longer than I meant to, but you see it is uppermost in my mind at present. Col. Morgan has not received his orders yet; he is at Fort Monroe awaiting them, and Carl is making himself comfortable here. Your letter of the 21st was received two days since. I trust that ere this, you are <u>well</u> again, and hope no serious results will follow from your unwanted visitation. Being unskilled in some particular branches of the healing art, I am not qualified to express an opinion as to the curse of the "phenomenon". I am blessed with another of my good friends! The boils. this time it is on about the centre of my right cheek, a very conspicuous, and not overpleasant place for it, for although yet in its infancy, it already begins to interfere with the operations of my jaws. As usual when I have one of these visitors, I do not feel quite well, but suppose I shall be enough better in the end to pay for it.

With much love Dear wife I remain
Your Devoted Husband.
Edwin J. Barden

No. 60 is missing

No. 61 Rec'd June 8th 1865.
Ans.

City Point Va May 31st/65.

My Dear Little Wife.

 I expect to commence work this afternoon and if I do shall be both too busy and tired to write much this evening, so make a beginning at a letter while I have a few spare moments this morning. Your letter of the 26th was received two days since. Many thanks for it. Your letter of the 15th turned up at last! it seems that one of the clerks in Capt Ames office who messed with us, took it to hand to me at the table; not happening to meet me at that time, the letter lay in his pocket forgotten until he stumbled upon it accidentally one day while rummaging them. I shant attempt to measure "smartness" with you right here, especially for the last Month. I havent pretended to get up before half past six; often not 'till eight o'clock. Oh! I have been lazy I can tell you, and should have enjoyed it tip top but for the annoying thought which would intrude itself, that I would soon have to go to work again. I dont love to work. I hate it Work and I, have always been sworn enemies, and shall always remain so! necessity alone compels me to continue the acquaintance. Now here I am this fine morning, dreading the afternoon and its promised employment, as a dog does a threatened whipping! Whereas I thought to be thankful that I can have something to do! and the ability to do it. I am glad to hear again from Josie Orton, am glad to know that she has much pleasant prospects in view. I hope that her intended husband will prove to be all that she could desire, and that she may never find herself the loser by reason of her early disappointment. Josie's peculiar circumstances may have been good reason for what might seem to some an over anxiety to be married! Whatever her desires may have been upon the subject, it never occurred to me that she manifested them very strongly in public, but of course you being in her confidence know how she really felt about it. I thank her for her kind message! Please convey to her my best wishes for her welfare and happiness. In a very few days now I can tell whether to send for you or not. If I can arrange a satisfactory place to board (get our meals I mean) and find comfortable quarters near at hand, I shall send for you. There can be nothing positive known about the length of my stay. Everything may warrent me in supposing that I shall remain all summer, and some day without the least warning, I may find my occupation gone, and myself out of a job or Mr. Browne may be ordered to a distant post. But having made up my mind on that point I shall have you come if matters arrange themselves satisfactorily in the next few days. The worst of it is, your being compelled to come alone, and you are so unused to travelling. I should feel uneasy and wretched from the time of your starting until I had you safe here. I fear that it would not be practicable for me to meet you at Washington, as it would require an absence of three days from my business, which I fear could not be spared. However I could give you such directions, that you could have no serious inconvenience in getting along; though it would be very uncomfortable for you. Hope you will not be troubled with a visit from the unpleasant disease you mention as prevailing in the village. No, I didn"t pay an "extravegant" price for the music. I purchased eight pieces, for which they asked me fifteen cents apiece, but they fineally took a Dollar for the lot. I gave two of them to Mr. Love, and sent three to Carrie.

 Evening. At 4.P.M. Mr. B. and myself went over to our new post and took account of the Stores, or rather I took the account (which of course was my business) and Mr. B. got tight!- the old fool- I would rather have given five Dollars out of my own pocket. The prospect looks "scaly" enough. The officer whom Mr. B. relieves, does not like it, and seemes determined to leave things so that it may be as disagreeable as possible for us. I dread the morrow. If we can only manage to weather it out for a fortnight or so I think the machine might run pleasantly enough. Its useless for me to say that I am discouraged tonight, for you would know it by my writing. If I <u>only could</u> get a situation elsewhere, that would afford me a living. I am tired tonight though I have not but little today, but I am nervous over the prospect of tomorrow, which I know will be a hard day for me, and I am going to bed early in anticipation of it.

I wish I could kiss you tonight darling.
With love.
Your Devoted Husband.
Edwin J. Barden

No. 62 is missing

No. 63 Rec'd June 9th 1865.
Ans.

City Point Va June 6th/65.

Darling Wife.
 Your letter of June 1st came to hand last night, and was read with great pleasure, but I was to weary to answer it. Am not going to write much tonight-just enough

to assure you of my continued love, and say how much I wish we could be together this evening, and every evening. I have not had to work very hard during the day, which was fortunate, for I have not felt well. The weather during the last four or five days has been intensely hot. I have had to work hard have drank too much water. I shall have to be more careful, for I suffered from the same cause during the heat of last Summer. Well my dear your prospects for a "summer in Virginia" do not brighten much. The order in Mr. B's case has been suspended, and is likely to be revoked, but as an offset to that "favorable symptom" <u>Gen. Carr</u>, who at present commands here, has applied to have a Post Commissary appointed who will if so assigned perform the duties at present devolving on Mr. Browne. True in that event Mr. B might possibly be assigned elsewhere, but I hardly think it probable that he will. I think the old man will be worried into resigning. I am sorry for him on account of his recent bad luck, at the same time I do not actually think him to be a fit man for an officer. I know that he wishes to remain in the service because like myself he has no way of getting a living out of it, but as I have before written he is utterly incompetent to perform duty in his regiment, and would be compelled to resign at once if returned to it. There is hardly a chance of my getting anything to do down here if I should leave Mr. B. or rather if he should leave me. So I think you will probably have me on your hands to support in the course of a few weeks. Col. Morgan is still unassigned, and Mr. Moller is here waiting orders. he is doing all in his power to assist Mr. Browne, but I dont think it will avail much. Ive been looking for him in here all evening and am vexed at him for not coming. I know where he is, he is over playing Chess with one of <u>Maj. Wiley's</u> clerks. I hate Chess, because I dont know anything about it, and because those who do, never know anything else when playing it. I am growing ill natured and out of sorts every moment, and Im going to stop writing and go to bed.

Good night my love.
Your Devoted Husband
Edwin J. Barden

No. 64 is missing

No. 65 Rec'd June 19th 1865.
City Point.
Ans.

                                                                City Point Va June 13th 1865.

Dear Wife.

    I have plenty of spare time at the Store house this afternoon, and it has just occurred to me that I might better improve it in commencing a letter to you, as I intended writing this evening any way. There has been scarcely any thing at all doing during the day. I could have done all the work in two hours easy! It is quite annoying to be compelled to remain here until 5 o'clock waiting for some stray negro to come along after twenty-five Cents worth of flour. I would not care any thing about if I did not have to work evenings to make up the papers. Your letters of the 8th & 9th were received last evening. It was very good of you my dear to write me an account of the Wedding so soon. I felt some curiosity to hear about it. Though I did not expect that it would differ materially from other grand affairs of the kind, that have at times "agitated" our quiet community. I think I should enjoy hearing your account of it better than I should to have been there and seen for myself. What a figure I would cut in Gov Holleys parlors amid such a company! I think if caught there I should "O for a lodge in some vast wilderness" very speedily! Would have liked to have seen that room of yours on the morning in question. I might perhaps have witnessed some "startling developments." Wouldn"t I like to have caught Miss Lib asleep in my bed though? Guess she would be careful where she took up her lodgings another night. I suppose Lakeville will now subside into its accustomed quiet. it will probably be some time before another event of the kind and like magnitude occurs to disturb its repose. I dont know of any other young lady in the village who can lay claim to quite so extensive a display in the event of her changing her name. After all I think it is the best plan to have just as good a time, and make just as much display as the ability and inclination of the parties may warrent on such occassions. Getting married is a part of life, and like all the other good parts, should be made the most of. I have come to consider myself as a sort of an outsider-who is denied the principle pleasures and enjoyments of life! but I do not affect to despise the pleasure of social intercourse or to feel bitter towards those more fortunate who can enjoy them.   Evening. Well darling wife I have been at work until my lamp gave out for want of oil, and now I'll add a few words to this, and bid you good night. Last evening I wrote to Carl, and expected surely that I would hear from him tonight, but he probably did not have the chance to write that he expected. If you could have had the rain which has fallen here today I think the Strawberries would be safe from drought for at least one week to come. I have got along very well today, and kept my temper pretty well, but then there has been nothing doing and Mr. Browne has been at his quarters most of the afternoon. I dont seem to have much to say to you this evening though I think I could talk some if I had a chance, or at least could prove a tolerable good listener.

With love dearest.
Ever Your Devoted Husband
Edwin J. Barden

No. 66 is missing

No. 67 Rec'd June 26th 1865.
Ans.

City Point Va June 22nd/65

Dear Little Wife.

Your letters of the 16th (2) and 18th are received. Once more allow me to thank you for your goodness in writing so often. I know that you must have felt badly-almost abused in fact, at not hearing from me more frequently. I have felt and do feel uncomfortable almost guilty in fact, at my remissess, but it has seemed nearly impossible for me to write for several days past. And it is not because I have been so hard at work either, for our work has been very light. Neither have I been downright unwell, but I have been troubled with an unaccountable, stupid, sleepy kind of feeling, which at times during the day would nearly master me, so that I could scarcely tell what I was about. The sensation usually increases towards evening, making it almost impossible for me to keep wide enough awake and my thoughts sufficiently collected to do anything. Such is my condition this evening, and nothing but the most obstinate determination to the contrary keeps me from tumbling onto the bed, where I would fall asleep almost instantly. I have been troubled this same way before. I do not know what causes it, but it is much more disagreeable than serious. The weather still continues intensely hot. there is no escape from the heat. Go where you will, you are subjected to the same stifling, sweltering atmosphere, unrelieved by a breath of cool air, until it really seems as though the act of melting away entirely and leaving no other trace of yourself than a "grease spot," were not only not improbable, but very likely to occur. Well, things "go on" about as usual. there seems no prospect of it being any more pleasant here for me, while I remain with Mr. B. Not that there has been any "rupture" since I wrote last-on the contrary I have managed by keeping my tongue still to "keep peace in the family," but as I have told you before crosswords have passed between us! I have taken a settled dislike to him, and no matter what he does I am bound to dislike him. I cannot but blame myself for my conduct to him, and for allowing such feelings to master me. I feel sorry for him, for he is an elderly man, with a wife and child whom (so far as I know) he has no other means of supporting than what his pay affords. then too he has been most unfortunate during the month, and is likely to be more so. He sold his horse to an officer who pleading the absence the Pay Master only paid him $80 of the $150 purchase money down, the officer has left the place and has been, or is about to be mustered out of the service. The horse is still here in charge of another officer who has given Mr. B. some ambiguous kind of a promise that all shall be made right, but it looks to me as though it would turn out to be about $70 wrong. No doubt therefore Mr. B. has much to make him irritable and cross!-which feelings by the way he does not vent on me by any means!-but I contend he has no excuse for getting "tight" almost every afternoon and making a fool of himself before every one who happens to see him. The "Point" is almost deserted now!-the <u>25th A.C.</u> have all left except two or three little detachments and a General. ther are only two full regiments here now, and one of them is about to be mustered out of the service!-they drew rations of us today as they fondly hope for the last time. Regiments mustered out of the service and on their way home, are continually passing through here, and we have them to supply with three, five, or eight days Rations, as they may require. We have fed two such regiments today. There are also daily several hundred <u>ex-rebels who have taken the oath</u> and are on their way home, who have to be supplied with a days ration each. If I could only manage the business myself, and in my own way, and have you with me. I would like to remain here and get a $100. per Month. I am confident that I could arrange the business so that I could do it alone in half the time now spent over it, and still have it more satisfactorially done. I may be very egotistical, but I consider Mr. Browne just about as fit for a Commissary, as I am to fill Dr. Reids position.

    Dear darling little wife-you cannot want me with you than I wish to be there. It seems at times as though I could not, and would not stand it any longer! this in particular has been a long dreary month. A Month of perplexity and discomfort. More than twenty times I have nearly made up my mind to tell Mr. Browne that he must get some one in my place, and then would conclude to let it pass a few days longer, hoping all the time that something would turn up that would decide the matter for me.
                      Good night dear wife. I cannot write more now.
                                   Ever Your Devoted Husband
                                       Edwin J. Barden

No. 68 Rec'd June 28th 1865.
Ans.

City Point Va June 24th 1865

Darling Little Wife.

This is Saturday night and I am strongly tempted to let alone writing until tomorrow, but I more than half expected to be busy then and think it better to make sure of a few words to you tonight. Your favor of the 19th was received last evening. I have not answered three or four of your last letters, but believe I have acknowledged the receipt of them all. With regard to writing when not busy, at the Store-house;-it is up hill work on account of the frequent interruptions I am subject to. As sure as I get interested at all in writing - just in the middle of a sentence, some darkey is sure to come along and want a half Dollars worth of Sugar! and away flows the "bright idea" that was just about being put on paper to "electrify and charm the reader". I have tried reading, but it is of little use. it spoils the pleasure of reading any book to be continually broken off just when in the enjoyment of the finest passages. I find no fault at not being kept hard at work all day, by any means. I only wish that I could have the business come all at once, so that I could do it up and have it out of the way leaving me free to employ the remainder of the day as I like. So I am an "Outsider" am I? Very well! If you prefer Miss Lib for a bedfellow, all I can say is that you are having plenty of opportunities to gratify your choice. I do not know how it will be about your seeing me next Month. I am quite undecided what to do, and of course entirely ignorant of what may occur to decide the question for me. It is mighty unpleasant as it is and I fear it would be still more so to find myself out of employment in the middle of the Summer. Mr. Browne talks a little of resigning, and I do not think it right to dissuade him!- in fact have no inclination to do so, because I think that if he can find any thing at all to do at home he will be more comfortable there, and because I do not consider him fit for the service. We have gotten along very quietly together during the last week!- in fact I can usually get along pretty well when there is but little to do. It is when we are very busy that the old fellow "crosses my track" the most. Prehaps I dont miss those Strawberries any, but its my opinion that I do. This is the fourth year that I have been denied them, and I am of the opinion that it is decidely to bad. It is not likely Mr. Moller will come here, for if the Col is not at Washington, he is surely at Ft. Monroe. So there could be no reason for his coming to City Point. I heard from him indirectly tonight, and shall probably have a letter from him tomorrow. I have received one from him giving an account of his taking tea with you, and of you being a "little out of sorts" because you did not get a letter from me. I received a short letter from Mr. Randall a few days since, in which he informed me of the marrage of several of my old acquaintences in Canaan, and also of the contemplated 4th of July celebration at the Falls. I am not partial to such occassions as a general thing, but feel as though I would like to be there on the coming Fourth. A day of rest and quiet with a pleasant dinner with a few friends, plenty of Strawberries and ice cream, and a ride in the cool of the evening with my wife, is my idea of a comfortable, if not patriotic 4th of July celebration. Speaking of rides! Do you remember one we took on a 4th of July evening years ago? I dont suppose you thought then that you would ever be my wife. What a long time ago it seems! and <u>eight</u> years <u>is</u> a good portion of ones brief existence too! It seems to me at times as though our married life really commenced nearly five years ago. I have been with you so little since I have had the happiness to call you wife, that there seems but little difference between our married life of nearly nine months, and our single life for four years previous. I want to be with you very much darling wife, and look as earnestly for fortunate circumstance to re-unite us, as I used to for my deliverence from service.

Good night my darling.
As Ever Your devoted Husband
Edwin J. Barden

No. 69 Rec'd June 30th 1865.
City Point.
Ans.

City Point Va June 27th 1865

Dear Little Wife.

Your letter of the 23rd was received Sunday afternoon. I intended to have answered it last evening, but was prevented by being obliged to wait all the evening over at Maj. Martin's office for the arrival of a regiment, to which I expected to have to issue rations! It came in on the 9 o'clock train, but had drawn its rations in the morning, so I lost my job, of which I was very glad. Returning to my house, I had an hours work to do, and retired a little before 11". very sleepy.

Mr. Browne had a letter from Mr. Moller yesterday, he is in Washington, and not likely to come to City Point! He had succeeded in getting Mr. B's two months pay, and wanted to know how he should forward it. Mr. B. has also got $50. from the man who purchased his horse! this makes $130. he has received which is all he will get. Mr. B. has been in very good spirits today. I got a pitcher of ale for our dinner this noon, and I guess he did not have to take more than one drink of Whiskey on top of it to "settle" him. He retired in "good order" about 3 o"clock, and when I went to his quarters an hour ago, he was fast asleep. When I first came with him, he never drank anything until in the evening when he was pretty sure of not being disturbed! then he would "worry down" his half pint and turn in before the night and come out all right the next morning. But lately he has taken to drinking in the afternoon, and gets fuddled about three or four o clock. Nothing has yet transpired to decide my return home. Sometimes I think I will speak to Mr. B. about it without delay, then again I think perhaps I had better wait, for it cant be long before some change will be made here, which will settle the question without any action on my part. And I cant help thinking too, that it is my duty to remain as long as I possibly can on account of the difficulty of procuring suitable employment elsewhere. It seems a little like tempting fate to relinquish a good paying situation, and trust to luck for obtaining another. But I am dreadful discontented and uncomfortable here, and I want to be with you so much. Tell Lib. she will have to look up another beau for the 4th for I cant get there in time!- besides I am a little jealous of her, for I hear she has been making herself very much at home, in my wife's room. How about the bonnet! Have you gotten it yet? If not had"nt you better be about it? Has your Father been to N.Y. yet? Am glad to hear that Mrs Deane is so much better. I could never think, when writing, to enquire after her. Should liked to have seen Sadie! the dear little chick. You can tell Sara, that I know just as little about "keeping boarders" as she does. It rather puzzles me to know what they would find to eat. I guess if they found anything, they would do better than I could. Perhaps Sara could muster up a little "French Gingerbread" for them. By the way, how is Mrs. French? Your promised visit did not take place after all. Well you must settle that with her yourself. I tried all I could to have you go. I shall tell Mrs F. that I have no control over you. You will do as you are a mind to. I suppose Mrs. Barnum never cried her eyes out because she was not at home when we called! I would not have done so if I had been in her place. Dont think that I would speak ill of Mrs. Barnum I think to much of her to do so. Still I cant help thinking that if I were only rich people would not be in so much of a hurry to drop your acquaintence. It is all right of course. I have no reason to find fault, but spiteful, rebellious thoughts will arise. Never mind! in the army we say of such things "it is the fortune of war." in your case it is the fortune of getting married. I am feeling very well now, the weather has been more endureable for a couple of days past though still very hot. With love darling.

Ever Your devoted Husband.
Edwin J. Barden

No. 70 Rec'd July 6th 1865.
City Point
     Ans.

                                                 City Point Va July 3rd 1865

Dear Little Wife.

    Your letters of June. 28th & 30th were received this evening. very good and thoughtful, is my little darling, to write so often. Pretty good evidence (were any needed) that your love is not growing cold. You are a dear good little wife, and I regret that you have not a husband worthy of you. It is not you that should have any fears of "fitness" for a life companion!- When I reflect upon my own ill temper and disposition I almost fear to think of living with you constantly for I fear that I shall not only cause you hours of discomfort and sorrow, but that you will also cease to love me as you do now. This has been one of the most unpleasant days in my calendar!- petty annoyences and perplexities commenced early in the morning, and did not cease during the day. I am completely discouraged! there is no use in my attempting to stay. I cannot do the work under the existing circumstances. If I could leave tomorrow, I would gladly do so. I have had no outbreake with Mr. Browne today, or for many days past, but his unfitness for his position any way, and the condition in to which he now almost daily gets himself, makes the position of a clerk for him unbearable to me. I did not write to you yesterday because I was busy at work all day and evening. I intended to have done a lot of work this evening, but I have been interrupted, and hindered so that I shall not accomplish anything. I am very dull and sleepy in need of rest, and I know that my dear little wife will excuse me for sending her such a short letter.

                                                               With love darling!
                                                       Ever your devoted Husband
                                                                   E. J. Barden

No. 71 Rec'd July 10th 1865.
Ans.

                                               City Point Va July 6th 1865.

Dear Little Wife.

    Your letter of the 2nd was received last evening, as was also one from Carl, in which he requested to be remembered to you. his letter required an immediate answer because of reference in it to business of Mr. Browne's. So I was compelled to devote last evening to it. It is almost 9 o'clock. I have been trying to work a little, but only succeeded in spoiling a paper upon which I had previously bestowed an hours work. The fact is I ought not to be expected to make up papers in the evening which require thoughts, and care, after spending one of the long hot days at the Store house!- even though I do nothing there but to sit and watch the clouds and bite my nails. I have hardly touched the June papers yet, and dont know when they will ever get made up! They are quite numerous, and must work on them this Month. I could not be with you to attend the celebration of the 4th at the Falls. am not sure but that I should have preferred remaining quietly at home with you, had I been there. I spent the forenoon of the day at the Store house, and intended to have remained at my quarters at work the remainder of the day, but upon going to dinner I met Maj. Martins Clerks, who insisted upon my spending the afternoon with them! to which I consented. We got the loan of one of the Tug boats, and started on an expedition up the river. After debating awhile whether to visit Dutch Gap or Malvarn Hills, we finally decided upon the latter place. We landed at a little wharf at a place called Haxall's Landing,- distant about ten miles from C.P. and hid from view of it by a bend in the river. There is a moderate sized cottage standing a few rods from the dock. to this we directed our steps to enquire the direction, and distence. In vain we "pumped" the darkies, and the lady of the house - a fleshy, motherly looking woman of probably forty or fifty-five, for the desired information. We could only learn from them the direction (which we already well knew) the whereabouts of a path which led to a main road which the lady "reckoned" would lead us to it. At this stage of the proceedings a young lady appeared upon the scene. She was much more intelligent than the old lady, and conversed freely, and correctly, with only an occassional "I reckon" thrown in. She is suffering from the chills and fever she said, and the dull, yellowish-white color of her skin bore testimony to the fact. She directed us to the right path, and as it turned out, gave a shrewed guess at the distance, which she thought was about three miles. It looked like a great undertaking to attempt such a distence in the afternoon of one of the hottest of July days, but as the road for a good part of the way lay through the woods, we concluded to try it. We took our time for it and walked slow, stopping often to rest and pick blackberries, of which there was a great abundance along the path. After a while we struck the main road, which still shows abundent evidence of the passage of large armys along it. A walk of about two and a half miles brought us to the hill, at the place where McClellen had his headquarters during a portion of the first days fight - the second day you may remember, he was safe and sound on board of the gun boat Galena, with the French Princess! while his troops under subordinate Generals were unsuccessfully repelling the assaults of the rebels and ended the day by whipping them soundly, so that it only needed the leadership of a competent Commander in Chief, to have retrieved the disasters of the preceeding six days, and nearly destroyed the rebel army. The heaviest part of the fighting was further on. Where we stood was under the fire of rebel guns, evidences of which were not wanting. One of our party was there during the fight, and from him, we got a genereal idea of the disposition of the contending forces. I would have liked to have gone about a mile further on, which would have taken me over the most hotly contested part of the field, but the boys were too much exhausted to be coaxed along any further and were content to view the field from where we were. The place at present gives no sign of strife that once disturbed its quiet, the field is overgrown with a luxuriant crop of grass, in to which the foot sinks at each step, causing no little fatigue in walking. After a half hours rest, we started on our return. We stopped again at the house, rested ourselves, and made way with a large bottle of cool, sweet milk, which "Priss" as the old lady called the younger one, served out to us with her own hands. The young lady had as fine a set of teeth as I ever saw, and she was not backward in showing them! but as an assett to them she had a pair of eyes with which she could watch both sides of a house at once. We stopped at Bermuda Hundred a few moments on our way back, and finally reached home about 8 oclock, weary enough I can assure you, but feeling that we had not spent the afternoon unprofitably. Am glad that the music is received. I saw them advertised and thought perhaps they might be good. You complain of not being able to sleep because of the flies! Why not get some mosquitoe netting and have it arranged over the bed? Get a large piece enough to completely curtain the whole bed. George I have no doubt will put it up for you. I wish if it is not too much trouble you would do it, and I think you will find it quite a comfort. I suppose George is well yet, and at work for Mr. Merwin.

With love darling.
Ever your devoted Husband
Edwin J. Barden

No. 72 Rec'd July 12th 1865.
City Point.
Ans.

City Point Va <u>July 9th 1865</u>

Dear Little Wife.

Your letter of the 4th inst. was received two days since, but it has not been practicable to answer it until now. I try to appreciate your goodness, and the sacrifices you make in writing to me so punctually without regard to your own comfort. I think I am a mean, selfish fellow to make you go without letters so long, just because I am a little tired or sleepy. I think you made out to celebrate the 4th quite satisfactoryly, though perhaps, it would have been just as well to have left out the headache! Which I wish with all my heart I could have had the curing of. You see that I am not the only friend you have, by any means. Perhaps you fared better on the 4th than you would had I been at home! I think you get more rides when I am absent, than when I am with you. Some of your friends are better able to please you in that respect than I am, but certainly none are more willing or desirous. In that as in many other respects my ability is not equal to my inclination. After a week of the most intense hot weather, we are at last being blest with a cool, refreshing shower! Vegitation had begun to experience the effects of the heat, and to wither and turn brown, while the roads were exceedingly dusty and uncomfortable for travel. I went up to Petersburg yesterday afternoon, and returned in the evening. I went on an errand for Mr. Moller. There seemed to be more people about than when I visited the place before. There are a large number of Stores open, but there are scarcely any full assortments of goods. trade did not seem lively though it was Saturday afternoon, and market day. I strolled through the market, which was plentifully supplied with, meats, Fish and Fresh Vegetables. I say plentifully, but after all I think it was only so because of the few fires. It was nearly night when I left it, and judge that most of the dealers had fully one half of their mornings stock on hand. Most of the articles were of a poor quality, and they were prepared, and exposed for sale in that slovenly, slip-shod way in which everything is done at the South, but little regard being paid to making them look fresh and interesting. There is but little money in circulation in the City, and most of the people have got to live pretty close until they can raise new crops, and commence manufacturing again. I have less and less sympathy with the Southern people every day, and feel indignant to think how cheaply they are being let off after all their crimes. It is nonsense to talk about what they have suffered! All that they have suffered heretofore they have suffered willingly, and gloried in it! now give them a taste of a different kind of suffering, which shall do a little towards punishing them for their atrocious crimes. I purchased some music for you, but dont think much of the selection. the Store could not boast of a very large assortment, and most of their stock was old. Met one of my old friends formally clerk for Maj. Wiley and a member of our mess, on the side walk, received a pressing invitation to come up and spend a day with him, which I should like to do very much. I dont know whether or not, I have told you that Maj. Martin's lady and daughter are here! they came about a week ago.

The Major occupies with his family a house close by our mess house. Their cooking is done in our cook room, but they take their meals in the house by themselves so that we are not favored with the society of the ladies at all. I have not met either of them though I have seen the young one out doors frequently. She is fifteen I believe, and very plain looking. From my quarters to our place of business it is nearly a quarter of a mile, and our mess room is about half way between. It makes an uncomfortable walk this hot weather. I go to breakfast about half past seven, and from there to the Store house, and do not return to my quarters until we close for the day at five oclock. Mr. Browne says that we shall close up earlier this week, and I hope he will, for there is nothing doing in the afternoon, and we have a sight of work to do on the papers for June, and it will never get done unless he gives me more time. I have been very sorry that I did not have you come down, but still I am satisfied that I acted for the best. It would not be well to send for you now, for it is quite probable that there may be a change any day. I think I could have made you pretty comfortable here, and know that your presence would have been a great <u>help</u> to me, aside from the great pleasure it would have been to have had you with me. I know you must be much annoyed at being asked so frequently when I am coming home and probably nine tenths of the questioners do not care a fig whether I ever come back or not, but only ask because they must say something. I would come home at once if I could find anything to do. But there is now a great scarcity of employment, and there will be still less as time passes and men are discharged from the army and seek for work among the various branches of industry. Am obliged for the information about the bonnet! I would have liked you to have had a nice hat <u>this</u> Summer, and hope that you have not allowed our poverty to prevent your getting it. The subject is one about which I should say very little any way, for the styles are (to me) so ludicrous that there is but little probability of my falling in love with any of them. Whatever pleased you would be agreeable to me. I hope you are suited with the one you have, and that you have not been too economical about it. You are a dear, good, thoughtful, little wife, and I love you more and more, every day, and if I dont continue to do so as long as I live, I shall be a brute, and unfit to have a wife. I have been hard at work all day! it's a wearisom life, when there is no rest for mind or body, day after day, and Month after Month. No wonder that men in the Army become demoralized, skeptical and infidel in character and religion. I think there is something more that I wish to say to you, but I cant think of it readily. I have to write a note to Carl this evening yet, so must bid darling wife good night.

Ever Your Devoted Husband.
Edwin J. Barden

No. 74 Rec'd July 22nd
    City Point. Va.   City Point Va July 19th 1865.
    Ans.
My Dear Wife.

    You are doubtless wondering why you do not hear from me. It is because I have not been able to sit up long enough during the last three days and a half to write. I was taken last Sunday forenoon with a severe pain in the small of my back which increased until it made me nearly frantic. I had the Docter in the evening, and towards midnight succeeded in getting relieved of the pain. I am well every way only quite weak particularly in my back. Have not sat up half an hour at a time during the last three days. Cannot sit up to write this evening, but thought I must send you a word of explanation for my long silence. It will yet be several days before I shall be able to write you a letter. In the mean time be patient, and remember that I am still your loving husband.

                                                Ned

July 20th 1865

    I am some better this morning, but get along most provokingly slow. It will be several days before I can write you a letter, for all the time I can sit up must be employed in assisting Mr. Browne who with no one to help him has his hands full. Have three letters I think from you that have not been answered. By the time you get this I shall probably be well and about my work.

                              With love, ever your own
                                                Ned

No. 75 Rec'd July 26th 1865.
City Point.
Ans.

                                        City Point Va July 23rd 1865
Dear Darling Little Wife.

    Your letters of the 12th, 14th, 16th & 18th are all received, and I believe none of them have been answered! they each contain much of interest as well as abundant evidences of your love for unworthy me. And each one is worthy of and deserves a separate answer, but dear wife I shall have to be your debtor for them all, and can do but little towards answering them. I sent you a short note a few days since to explain the reason of my delay in writing, for I feared that you would begin to think that I was growing very lazy, when really, I could not write. Last Sunday I got up feeling about as well as usual, went to breakfast but ate very little, not feeling hungry. After breakfast I went over to the Store house, having some work to do there which took me until nine o clock! then I came back to my quarters and went to work writing! In about an hour I commenced feeling a pain in my back and lower limbs, which continued to increase steadily. I laid down for a few moments two or three times and was surprised to find that I obtained no relief by resting. I went over to dinner thinking that a little exercise might help it, but it was with some difficulty that I walked back to my quarters. I threw off my clothes at once and laid down, but not to rest, or find relief. The pain increased until it seemed as though I should go crazy. In no position which I could assume, either lying sitting or standing, could I obtain one instants relief. I never knew what pain was before. All that long bright summer afternoon I passed in perfect agony. There was no one to do anything for me! at length as night came on, I could stand it no longer, and so going to the door, called out to a young man, a clerk in the Q.M.Dept. who lived in one of the little houses near by, and with whom I was slightly acquainted,- and requested him to go after the Docter. The Docter came without delay, and after a few questions said he thought he could soon relieve me of my pain which he thought must be an attack of rheumatism. Mr Terry (the name of the clerk) went back with the Dr. and soon returned with the remedies, one of which was ether! one other kind of medicine to be taken internally, and a bottle of liniment with which to "bathe the part affected". Mr. Terry very kindly remained with me, and assisted me to apply the remedies. They afforded a partial relief for about an hour, and then the pain seemed worse than ever. Mr. T. made another trip to the Docters and came back with two more vials of medicine,- one of which contained Morphine, and some Musterd for a poultice! he spread a plaster just half the size of the one of my linen h-d-k-f's, and applied it to my back, where it was soon drawing, and smarting in a manner that promised to reach, and remove the cause of my troubles. Meanwhile I kept up regular half hour attacks upon the Morphine bottle, and in a few hours was rejoiced to find that I was becoming oblivious, to both pain, and plaster. Mr. Terry remained with me until after midnight, and then as I became easier, went to his own quarters. I did not sleep during the night but remained about half conscious, and half stupid. In the morning I found myself free from pain, but sick from the effects of the Morphine, which I expected as a matter of course. I hardly left my bed that day! in the afternoon I had a spell of vomiting, and having thus disposed of the opium, began to consider myself in a fair way to recover. On Tuesday I felt much better, and sat up some, and on Wednesday my back was worse, and I lay abed all day to pay for it! on Thursday I was better and sat up about half the time,- wrote the little note to you.

    Friday I determined to get out, for I began to be real sick from close confinement, and so went up to the Store house, and staid there until three o clock having my dinner sent up to me. It was pretty hard work, but I stuck it out, though I made for the bed immediately upon arriving at my quarters. Yesterday I was able to resume my duties but felt no disposition to over work myself. Today I have been bending over the table a good deal, ruling and writing, and my back aches some! it is quite weak yet, and I have to be very careful how I move, or do anything to strain it. I am at loss to know to what I am indebted for such an attack! it came without any warning, and excepting it, I was usually well, until I began to feel the effects of the medicine. You see my love that I am sometimes sick on Sundays as well as you. I can assure you that I most earnestly hope my attacks will not be frequent. I suppose that your cure is passed hope. I am sorry my dear that you must be so sick and sorrier still that I cannot be with you to do the little that may lie in my power to assist, and comfort you. I regret to hear that you are growing poor and thin, and wish that if you have an opportunity to visit the sea side you would embrace. Wish that you had accepted Mrs Barnum's invitation - that is if it would have been agreeable to you. About coming home!- I know nothing definite about it as yet, but I think business is going to wound up here very soon. There is less than a regiment of troops here now. The employes of the Q.M. & Comms. Depts, have been discharged, all but a very few. Maj. Martin has discharged all his clerks but two. The Rail Road between here and Petersburg, and the South Side R.R. are to be turned over to their original owners on Tuesday of this week. Everything looks like the speedy dissolution of everything "Government" at this place very soon. I think that the Month of August will witness the abandonment of this place by all but the negroes. It is currently reported that there is to be a Freedmans home established here, but if there is it will probably not affect our stopping here any. I have but little that Mr. Browne will be assigned to duty at any other Post, when he is relieved from here!- he will probably be ordered to his regiment, and if so, will at once resign. I consider that it is my duty to remain with Mr. B. as long as possible however unpleasant it may be to do so, because of the difficulty of procuring employment elsewhere. The country is flooded with the discharged soldiers, for whom at present there are not situations for one quarter! Men are every day sending to their friends and acquaintences now with the army, to procure them places. While not a day hardly passes, but that Government discharges more or less of its employees. I have but little hope of finding employment any where about home. I dont see what there is about there that I can do! Something may "turn up," but the prospect is bleak enough at present. I have not heard from Carl in over a fortnight, and do not know what to make of it!- True I have not written to him but then, I am not sure that I owe him a letter. I intend to write him a few words tonight. Mr. Browne and I get along very comfortably together now. he had to work pretty hard while I was sick, but did not grumble. I had one of our negroes to wait on me! had all my meals brought to me! and with the exception of Sunday, was very comfortably sick. I am wanting to see you dear wife, and I trust that you do not do me the injustice to suppose that this separation from you is at all agreeable to my feelings. I am doing what I think to be the best for us, in remaining here, though it is sadly against my inclination.

With love darling
Ever your devoted husband.
Edwin J. Barden

No. 76 Rec'd July 29th 1865.
City Point.
Ans.

City Point Va July 26th 1865

My Dear Little Wife.

    Your letter of the 19th came to hand this evening, and was read with great pleasure. You are a dear, good, loving little wife, and too good for a dozen such <u>scape graces</u> as I am. You must not let yourself get into such a fever every time I am a little unwell! people will be sick you know, and but few are exempt entirely from painful, or unpleasant attacks of some kind. I did wish for you many times while I was confined to my quarters. I can assure you it was dull enough! I have completely recovered from the attack I think, and am feeling as well as usual. The weather is so terrible hot and debilitating that one can scarcely muster up courage to move. While I am writing, the prespiration is fairly streaming from every pore. There is no doubt but that the Military Depots here will soon be broken up and removed, but the notice in the papers was a little premature. its quite as near the truth as nineteen twentieths of the newspaper reports. It may be necessary to have a Commissary here, to do the same duty that we are doing now, even though the Depots are broken up. I make no calculations on staying here long, but still think it best to remain as long as I can. Much as I would love to have you with me darling I do not think it best for you to come at this late day. We must still wait and hope. I wrote you a letter last Sunday which has probably reached you ere this, and put you at ease in regard to my welfare. I wrote you a while ago that we had commenced closing up the Store house at 3.P.M. and that I hoped to have more spare time. But now my evenings are completely broken up, and I have as little leisure as before. Regiments mustered out and on their way home are daily passing through, and to many of them we have to issue Rations!- there is one train that should arrive here at 8.P.M. but is frequently an hour or two behind time. Almost every night a regiment and sometimes more, comes in by this train. About half of them want rations, and I have to go up to the Store house and issue them in the night, and am frequently until 12 o clock and later. Every evening I have to be on hand when the train arrives, to see if my services are required. Our old Regt. passed through this morning! they are mustered out and on their way to N.H. to be disbanded, and that is <u>the end of the old "7th"</u>. As good a Regt. as ever left the State. The 6th C.V. was along with it! they went on board the boats which are to take them direct to N.H. immediately upon leaving the cars. Our Regt. had the "Oriental" and the 6th "Eastern State." I knew that they were coming this morning, and was on hand when the train came in, to see the boys, and staid with them until they left the dock. There were thirteen of the original company with whom I left the State nearly four years ago! the rest of the Company are drafted men and substitutes. I saw Capt. Moore and had a few moments chat with him. he was with Genl Hawley! they are stationed at Richmond now, and I am going up in two or three days to spend the night with him. He has been at R. all the Months but I did not know it. Havent heard from Mr. Moller yet, and dont know what to make of it. Government has turned over the R.R. between here and Petersburg, to its original owners, and they have also stopped running a Mail Boat between here and Washington. So I expect I shall have to pay for my rides after this.

    I can write no more tonight darling. With love ever your devoted husband
                                                        Edwin J. Barden

No. 77 Rec'd August 5th 1865.
Ans.

                                        City Point Va July 30th 1865
My Dear Little Wife.

    Your letter of the 26th was received last evening, together with the Paper, for which I am much obliged to you and George. Your letter of the 25th was recieved a day or two since. I was sorry to hear that you were worrying so much about me. I thought that I had explained my situation sufficiently in the note sent you to put you at ease in regard to my situation, but it seems that Aunty came in with her "typhoid" bug, and almost frightened you out of your wits. If I am unfortunate enough to have another such attack, I"ll either keep it to myself, or not tell you of it until I am quite well and about my work. You must not be so much worried about me. I am not so "frail and delicate a thing," as to succumb to every evil breath that stirs without a struggle. There is no danger of my over working myself. I am far to lazy for anything of the kind, and certainly the hot and debilitating weather we have here is a poor incentive to overexertion. This is the first comfortable day we have had in a long while! it rained nearly all last night, and remains cloudy with an occassional sprinkle of rain still. the air is quite refreshing. I had almost forgotten that there could be such a luxury as a day like this in Virginia. Well, one day more and July will be gone, and two thirds of the Summer have passed. We little thought of being separated so long when I left you last, but I dont know but that this promises to be the longest of my absences from you since our marriage. It is very unpleasant, but the case might be much worse, and though my heart cries out against it,- I am constrained to believe that it is the best we can do at present. I am at loss to know what I want to do, supposing I could do as I like - that is could get employment at whatever I chose. I am pleased and gratified to know that you are willing to go with me where ever I wish to. But I dont know where I want to go myself. I like New England better than any other part of the Country I have been permitted to see, but no where that I can turn, do I see any prospect of success for me. I dont want or mean, to write despondingly, for no doubt there is work enough some where for me to do, at which I can support us both, but I dread hard work, and would like to avoid it if possible. At the same time the phantom of my ragged cousin (which at times "haunts me still") will I think - should other, and wortherier motives fail - proves sufficient incentive to so much of labor, be it what it may, as will maintain us decently, so long as God grants me health and strength. There is little danger of my interfering to prevent your visit to Emma's where it is quite probable this letter may find you. Give my love to them all and tell Frank & Ellen that maybe I will be up to help them pick up apples, or go a fishing with them this Fall. I dont know anything about how beneficial these seaside trips are, as they are generally undertaken and carried out! but I do think that a few weeks spent anywhere on our New England coast during the heat of the Summer, with a proper amount of exercise and bathing, cannot help but be of service to one whom the heat of the season debilitates, and causes to get "poor and thin." As to our being able to afford it - we can afford all that we have or can get to restore and secure the health of our darling little wife. For what I concieve to be good reasons, I should not like to attempt clerking at Millerton. Your Grandfathers humors I might endur, as being an old mans right but fear that I could not bear your uncle Ben's fetishness so well. I fear that he and I are too much alike on that point to agree well. But <u>this</u> between <u>us</u>. I may yet be glad to try my hand there. I think Mr. Miner's would have been the place, if I were to stop in a Store any where about home. Say to your Grandfather whatever you like for me, and tell him that I am pleased to be remembered by him. It never does any harm to amuse old people in any reasonable fancy. I began to think there would be but little to do this Month (coming), but Mr. Browne told me this morning that we have got to issue to Contrabands, and other destitute citizens. if it be so I think we shall be tolerably busy, but there is little fear of being so hard pushed as we were in the Month of June. There is quite a number of transports lying in the river in view from my quarters, and I conclude that more troops are expected through here. There has been none through since our regiment passed. I have heard that most of <u>Kilpatrick's Cavalry</u> were coming through this way. Our Store house was entered night before last, and two gallons and a half of Whiskey stolen. It was done no doubt by the guards themselves. They belong to the 39th Ills. and are as wicked and undisciplined a lot of Scamps as I have seen in the army! they are anxious to be mustered out, and dont care what they do! their officers are young, careless, and unreliable, and seem to have little care for, or control of the men. I wish they might be cleared out, and a good colored regiment sent here to do duty. I shall hardly expect a letter from you today, having received one yesterday, but should be pleased to find myself disappointed. I have some work to do and cannot stop to write more now. Tomorrow will be one of our busy days and I must prepare for it a little. I hope to be more punctual in writing to you the coming Months than during the one nearly past. With love darling.

Ever your devoted husband.
Edwin J. Barden

No. 78 Rec'd August 7th 1865.

City Point Va August 3rd 1865

My Dear Little Wife.

Your welcom letter of the 30th was received yesterday, and very greatly am I indebted to you for it. You take a great del of trouble to write such good long letters to me. I should think you would tire of doing it, when you are so poorly repaid. You must practice a deal of self denial, if it be as unpleasant a task for you to write as it be for me. This is one of the hottest of hot days, with scarcely a breath of air stirring. I am at the Store house; there is nothing doing, and I thought but to commence a letter instead of remaining entirely idle. It is very quiet, with the execption of the guards in front of the building who are "fighting their battles over again" and speculating on the chances of being discharged soon. Lt. Browne has been sick all the week; he was better yesterday and came up to the Storehouse, but the exercise was to severe for him and last night he was worse. He is an obstinant old cur about doctoring when anything ails him, and wont do as any one else would, similarly affected. Am quite well myself, or would be if the heat did not so completely use me up. Our regiment of thieves (the 39th Ill's) has left, but not until they had entered the Storehouse the second time and stolen five gallons more of Whiskey. We have a great deal of trouble with our Stores by reason of their being damaged! it keeps my men constantly employed in assorting, and cleaning them. I drew ten barrels of potatoes from the Depot today, and nearly one third of them are entirely decayed! the stench from them is such that I can hardly stay in the house. I received a good long letter from Carl last Sunday. he made no excuse for not writing sooner,- had not felt like it I suppose. His letter was very interesting to me, but would probably not be particularly so to you, as it relates principally to matters of business. He writes that he is well and that he has not been troubled with the headache since he has been in W. - which he attributes to the amount of exercise that he has taken. I think I told you that he walks about two miles to each meal. He wished to be remembered to my "kind wife". I suppose that you are at Emma's today. I hope it is more comfortable there, than here. I do not think it very judicious for you to attempt walking this hot weather when you are not accustomed to it. I should much prefer to have you hire a conveyance whenever you wish to make a visit. But it may be that you are out of money, and obliged to go on foot. I must send you some at once. I was all last Month doing it, and fineally failed. I hope you did not suffer for the want of it. Poor Lib is sorely affected. I am afraid that with her, poor throat, sore fingers, and intemperate habits (witness the Celler way full of empty bottles) I shall have hard work to get her married. Give her my love, and tell her I am real sorry for her poor fingers for the felons are not to be trifled with, and that she is a real good sister in spite of the "bitters"

 Aug. 4th Morning. Upon returning to my quarters yesterday afternoon I found Mr. Browne very ill indeed, in a raging fever, and quite delirious. I went for the Dr at once and have remained with since doing what I could. The remedies applied broke the fever in the night, but he is very week and ill this morning, and there are strong symptoms of the return of the fever. He ought not to be left to the care of a negro, but I can do no better. I must go to the Store house. It promises to be another fearful hot day, enough to throw a well man into a fever. I enclosed 10 Dollars fearing you may be in immediate want, and will send more as soon as I can attend to it. With abundance of love I remain

Your devoted husband
Edwin J. Barden

No. 79 Rec'd August 10th 1865.
Ans.

City Point Va August 6th 1865

My Dear Wife.

Your letter of the 1st was received yesterday, full of love and affection for unworthy me, and interesting and entertaining (as are all your favors) as usual notwithstanding you complained of being short of material for letter. Well, I have but little to say tonight. I feel much more like going to bed. I have been broken of my rest more or less for the last four nights, and I am feeling very stupid and heavy in consequence. I have slept the greater part of the time each night, but have been aroused frequently to attend on Mr. Browne. Have not been in bed during the time but made a comfortable couch out of some camp stools. I am thinking of going to my house tonight, and trying for a good nights rest. I think it will be safe to leave Mr. B. in care of the negro who is in attendance upon him during the day. He has been very comfortable during the day. Sleeping most of the time, and I think he will rest well through the night. Yesterday he passed a dismal day being very sick most of the time, and was very uneasy the most of the night. On the whole I think he is in a fair way to recover now, if he will only do right, and be careful. I have been at work most of the time today, but have not accomplished much, feeling too listless and inactive to apply myself very steadily, or energeticly. I am in good health now which is quite fortunate. I should have liked to have assisted you in the capture of those lilies, or joined with you in a rumble about the farm. I dont think either of us would have been of much assistance to Bert, in his farming but we could have shown our good will. I suppose you are at home today, and if well, have been to church. We have no official intimation yet of the abandonment of this Post. Although unofficially the day is set for it! but there is no truth in or foundation for the rumors, so far as I can learn. The place is pretty much abandoned now by all but negroes. One after another the Sutter shops and Restaurents have shut up, until there are now but two or three left. At one time we were pretty well supplied with. Vegetables, fresh eggs, poultry, and berries, by negroes who came in with them from Petersburg, but since the Government has turned the rail road over to its original owners, they have raised the fare so that they cannot afford to bring their stores here to market. The fare from here to Petersburg by rail, is one Dollar, the distence is only nine miles. A pretty exorbitent charge I think. I am too sleepy to write tonight. With love dear wife. Ever your own

Ned.

No. 80 Rec'd August 17th 1865.
Ans.

City Point Va Aug. 13th 1865

Dear Little Wife.

Your letters of the 6th, 8th & 10th, are all before – were duly received and although unanswered, hearty thanks to the dear writer were silently rendered upon the receipt of each of them. I have no doubt but that you are worrying your little head into a perfect furor of anxiety because you have not heard from me as often as usual, thinking me sick and afflicted with at least half of ills that flesh is heir to. But such is not the case. I am, and have been, quite well, but so busy that I have hardly had time to think. Mr. Browne does not get about yet, he sets up part of the time, and has been at work writing a little to day! I think he will be able to get around a little during this week. At present he is kept down by a perfect swarm of boils! he has them nearly all over, and gets but little rest in consequence of them. The bad Whiskey is coming out I suppose!- hope he will know enough to let it alone when he gets well. He could'nt stand another such an attack - at least - down here. Our business does not seem to diminish much. The Govt Bakery at this place has been turned over to Mr. Browne, and we now have that to attend to. Luckily, there is little business doing in it, but there is a large quantity of property to be looked after, and kept sight of. Our July accounts are not made up yet. I have been hard at work on them all day and am as tired as a dog. I hope to get them done the coming week. The Sub's Depot here is to be broken up the last of this Month. A large portion of the Stores now here, will be sent off, and what remains will be turned over to Lt. Browne. Major Martin the Depot Commissary,- goes to Richmond and City Point becomes simply a "Post", probably to be abandoned in a month or two. The weather is getting more endurable now and I feel much more like taking hold of work,- which is lucky for I have a plenty to handle. I begin to think I may not be home to spoil your proposed visit to Hartford in Sept,- if I am not, you must go - and I hope you will have a pleasant trip. I am too sleepy to write more tonight, if I have time will say a few words to you in the morning. Good night my love.

Ned.

Monday Morning.

I spent a restless uncomfortable night, and feel rather dull this morning, but guess I can soon work it off. I have but just dressed, and its breakfast time; past seven o clock, and cannot stop to write. Be assured of my continued love, and I think of you none less though my letters be few.

Ever your devoted husband
Edwin J. Barden

No. 81 Rec'd August 23rd 1865.
Ans.

City Point Va Aug. 18th 1865

My Dear Little Wife.

You are fareing poorly for letters now days. I suppose you consider yourself badly used but I cant help it. I am very busy, and have no time to call my own! Its past ten o clock now and have been working all the evening. Mr. Browne does not get along at all. He did very well for a few days, then seemed to hold on for two or three days, and yesterday he had a decided fall back, again. He has had quite a number of large boils which ought to have relieved his system of a great deal of impure matter. I think he has been injudicious about his diet - has eaten too much, and of things which he aught to have avoided. I have no time to see about his food, and it would make no difference if I had,- for he is an obstinate old coote, and as he has a negro to wait on him, he would keep him on the go, after something for him all the time. I feel sometimes like getting discouraged, but know that I have no right to be, so long as I am well, and there is nothing but work to trouble me. It is not the work that I care so much for either, but it is the uncertainty of affairs, and the consequent anxiety to know how matters will turn out. If Mr. Browne were well we should be having it quite easy. Maj. Martin the Depot Commissary, is waiting very impatiently for his recovery, so that he can turn over to him all the Sub's Stores in his possession here, and enable him (the Major) to get away the last of this Month. I see no probability of Mr. B's. getting about under a fortnight yet at least, and it is quite as likely that he does not get out at all. Meanwhile he has already a large quantity of Property on hand for which he is responsible, and which it is my duty to look after. On the whole, I will say that it is not very pleasant here just now. I am pretty well now and as long as I can keep so, guess I can stand the pressure. Your letter of the 13th was received a day or two since. Many thanks for it. I received a short note from Capt. Moore the forepart of the week, telling me that he was going home, and asking me to meet him at the dock when the boat arrived here. Unfortunately, the letter did not reach me until the Capt. must have been nearly to Salisbury, so that the meeting was out of the question. He asked me to write to him at S - and if you see him please tell him that I will try to do so. Heard from Carl the forepart of the week, he is well. I enclose Ten Dollars in this, have been trying to send you a larger amount by Express, but can't get at it. I see that they have had <u>a severe accident on the Housatonic R. road</u>. Also that my "one time" "superior officer" Jerry, has got back to New Haven "under guard." Wonder how he likes it?- he has achieved notariety, but its a bad time to show off good on account of there being so many Wall St. men in the same "search for fame," who divide the "honors" with him. Cant write more tonight love.

Ever Your devoted husband
Edwin J. Barden

No. 82 Rec'd August 26th 1865.
City Point.
Ans.

City Point Va <u>Aug. 23rd 1865</u>

Dear Darling Little Wife

Its past Ten o clock, and "yours devotedly" is awful sleepy, but must say a few words to his dear, patient little girl way off up Conn. before he sleeps. Your letters of the 16th & 18th are both received, and gave me much pleasure. I regret to hear of Harvy's illness, and fear that he will not recover for I remember how poorly he was looking when I was home in the Spring. "Poor Minnie", indeed. She is being sorely tried. I earnestly hope she may be spared the threatened affliction. Well darling I am busy as ever. not hard at work all the time, but with something to see to, and to be chasing about after nearly all the time. One of Maj. Martins clerks is helping us make up the July Papers, which Mr. Brownes illness has prevented my doing. Mr. B. does not get well yet, and gives little promise of doing so at present. He has grown very thin and weak, and sets up but little. Still the Dr. says that he is getting along, and will recover without having to go North. I should be heartily glad to see him about it. If he remains here and relieves Major Martin, he will have another clerk, in which event I should hope to have a chance to draw a free breath occassionally. Mr. Moller is going to <u>Fort Leavenworth Kansas</u>! expects to go any day now. Am sorry to have him go so far away. He goes with Col. Morgan who is to be Depot Commissary there. It is said to be one of the most important Posts we have now. How would you like to go out there? I suppose it is foolish to ask the question when I well know you do not wish to leave New England, but if we must go some where in search of a living it may be that you have some choice as to which way we may start.

Please excuse this hurried note tonight and believe me,

Ever and truly your devoted husband
Edwin J. Barden

No. 83 Rec'd August 31st 1865.

Office of Small Issue
City Point Va Aug. 27th 1865

Dear Little Wife

Your letters of the 18th & 23rd, inst. are both before me, as usual full of love for my unworthy self. I know darling that it is very trying to you to be compelled to hear from me so seldom, and hope soon to be able to write more regularly. Mr. Browne is still very unwell, and I am aprehensive that his recovery out here is very doubtful. This morning upon going to his quarters I found him quite out of his head - and in a state of nervousness that was really quite alarming. The Dr. had been giving him some medicine yesterday and last evening intended to quiet him, but it evidently had had exactly the contrary effect, and had made him quite deranged. I got the Dr to see him without delay, and he gave his some medicine which has worked well thus far, and this evening Mr. B. is quite in his right mind, though his nerves are in such a state that it is quite impossible for him to be quiet, and in attempting to do the least thing - to take a drink of water for instance - he has to seize the glass with both hands, and then it will rattle against his teeth like a pair of Castinets. Notwithstanding his condition, he is if possible, more anxious than ever to remain here, and take charge of all the business that Maj. Martin has soon to turn over to some one. I was to have spent the day in taking an account of the property, and stores which the Maj. has to turn over, but Mr. B.'s condition was such that I have spent most of the day with him. In company with Maj. M's. clerk I rode out to the Cattle Herd this P.M. and took account of a quantity of property there which it would not be convenient for me to do on a week day. Major Martin says that if Lt. B. is not better tomorrow, and also gives promise of a speedy recovery, he shall be compelled to report his condition to the Chief Comm's. of the Department, and doubtless another officer will be sent at once to relieve him. It is my duty to Mr. Browne to do all in my power to enable him to remain here, but I really have but slight hopes of his recovering here, and think that prudence would dictate his immediate removal north. The state of perplexity and suspense in which his condition has kept me during the Month has been very unpleasant, and I have more than once caught myself wishing that it might be ended in any way, not much caring how. If we do remain here, Lt. B. will have another clerk next Month, suppose that it will be quite easy for us. Maj. Martins present clerk is to remain with us. He is a nice young fellow and a first rate clerk. I made his acquaintence last Summer at Hatcher's farm; he was a soldier then as well as myself, and was at that time clerking it for Lt. Browne. Major Martins wife appeared to him quite unexpectedly yesterday, having come on from her home in Groton. Ct. without notifying the Maj. of her intention. I regret to hear that Harvy and his little one, still continue in danger, and await your next letter with much anxiety. Harvy looked very unwell when I was home in the Spring, and little able to bear the shock of a severe attack of any disease. I suppose you must have told me when Julia was taken sick, and what is the matter with her, but I have forgotten about it. I find you in your last two or three letters talking about her condition, and of your being with her. Am sorry that Emma must be deprived of your society, for I know by sad experience, what she loses thereby, but what is her loss, is Julia's gain and as it is "all in the family" she must make the most of it. And so Libby Bostwick has captured the Captain? I think she might have gone much further and fared worse. I believe <u>Captain Moore</u> is in every way worthy of her, and will make the prince of husbands. If I could have seen him here, he might possibly have enlightened me a little as to his intentions, though it is not certain for he is usuly very close mouthed upon all such matters. No doubt you are right, and that my letter will not be missed, but still if you have an opportunity, I should be pleased if you would mention to him the reason why I have not written. I was informed by Carl of his intended departure West. Am very sorry that he is going so off from us. I will take care to inform him of your willingness to go to K. with me, and perhaps he will be able to get me a situation out there. I hardly think that it is best for you to come down here darling wife, much as I want to see you. I could hardly make it pleasant for you I fear, and things are in such an uncertain state, that really it does not seem hardly advisable. I do not think that my own stay can exceed a Month more at the longest, and as that Month is not very healthy here for people not acclimated, I do not think it would be best for my frail little wife to jepordize her health and may be life for the sake of being with me. I am very sleep tonight darling, and with much love, will bid you good night.

Ever and devotedly your husband
Edwin J. Barden

No. 84 Rec'd Sept 4th 1865.

City Point Va Aug. 30th 1865

My Dear Little Wife.

I am in receipt of your letters of the 25th & 27th,- the latter just read. Thank you my darling for your good, loving letters. I dont know what I should do without them. I havent any time to write now, but as I do not expect to be more at liberty for several days, I may as well take the needed time now as ever. Lt. Browne is going to take all the Stores in Maj. Martins possession, and do all of the Commissary business of the Post. The Major has been in much doubt about the propriety of turning over the Stores to him, for the last few days on account of his feeble condition, but he has appeared so much better yesterday and today, that he has concluded that it would be safe to do so. I have no relish for the increse of Lt. B.'s business, for I consider him incompetent to manage properly, what he has now on hand, but as there is no escape from it I can make the best of it. I do not expect to have to work any harder. Maybe not so hard, but I know that the unpleasantness of my service will increase, with the amount of Lt. B's business. I more than once caught myself heartily wishing that the Major would conclude that it was not prudent to turn over the business to him, in which event you would have been favored with the presence of "yours devotedly" in a very short time. I have no right to wish to be thrown out of employment and a good salary when both are so hard to be found, but I get so vexed and out of sorts at times, and with all, so home sick that I seem to have little power to reason about the matter. I suppose Carl is well on his way to Kansas by this time, if not quite there. I received a note from him yesterday bidding me good bye, and assuring me that his first business upon settleing himself in his new home, would be to look out for a place for me. I am greatly obliged to you dear wife for your willingness to go anywhere with me, where it may seem expedient to go in search of a living. I most heartily wish darling that it lay in my power to provide for ourselves without inflicting on you the pain of the separation from your family and friends. For myself it matters little, but for you, I feel sure that I should quickly become a useless piece of drift wood, living with but little aim or purpos, and it mattering little where I wandered to. Still I am not without a strong love for my native land, and boyhoods home, and would greatly prefer to remain quietly there, if it would only afford me a decent living, and its climate agreed with my health. I would prefer going to Florida with George for many reasons. I know that it is healthy there, and that the climate agrees with me. it would be much pleasanter for you, because you would still have one dear sister left you in your isolation from the rest of your kindred and friends. We should also always be sure of having a friend and a brother in George, whom I know by experience can be relied on, and trusted in. But if I leave Salisbury with "my family" I must be sure of a situation before I start! Were I master of a good trade, or what is of more importance good health, I might start off and take my chance of making my way with strong hands, and good will, but as it is I do not think it would be prudent. You mentioned in one of your letters that James Deane was home; has he entered upon any business yet, or do you know what he thinks of engaging in? Do you see Sara B. often, and what is she about? Please give my love to her, and ask her how she would like to come down here and keep house for me? tell her I dar"ent have you come cause its unhealthy, but thought it possible that she might feel anxious to make a myrtar of herself, and that this would, or might be, a good chance. Tell her I am sadly in need of reclaiming, before I can be fit for civilized life, or to be a good husband, and that I think its a capitol chance for her to play the missionary on a small scale. If she succeeds well with me, I"ll recommend her for a place in the Feefe Mission, or among some other gay and festive people. Has <u>the 2nd Conn Heavy Artillery</u>, been mustered out of service yet? I have seen no account of it. Have you girls given up your Hartford trip yet? I hope not for I think you would enjoy it better than you think for, and there is little chance of my being home to spoil your fun with my untimely presence, and I fear there is still less prospect of my every being able to attend you on a visit with your friends. I have a sister living in Norwich Ct Whom I have not seen in a great many years. I should like very much to visit her, and see a little of that part of our State. I am told that the country is very beautiful there. I was talking with Mrs. Martin at the tea table tonight, about it. She thinks Norwich is a beautiful place, and that if she had money, she would like to live there. I suppose there is little chance of my ever seeing it though! and as for my brothers and sisters they are so scattered about, that if we meet in heaven, it is all I can hope for. I wish it were pleasant for you to visit at Mr. Randall's. I should very much like to have you. You seem to be a very domestic little thing, and only to be coaxed abroad by some errand of mercy. Well its exceedingly fortunate for me that it is so. Have you money enough for the Hartford trip if you conclude to make it? I guess this is a queer kind of letter, but it can"t be corrected and revised by me, and if you find it full of errors, and poorly written you mus"nt read it over many times, and it wont seem so bad.

With love darling
Ever your devoted husband
Edwin J. Barden

No. 85 Rec'd Sept 7th 1865.
Ans.

City Point Va Sept 3rd 1865

Darling Wife.

Your letter of the 29th is this evening received and read with great pleasure. It's very close and hot tonight,- and I am sleepy and dull. So you"ll not get much of a letter. Always some excuse you see, for not doing my duty in the way of writing. Why dont you get angry at me for them, and say that you dont believe me, that I dont try to write to you &c, &c. Cause you are so good and confiding, I suppose. Cause you do just as near right as you can yourself, and are so charitable as to suppose every one else does. Are you really always going to be such a "burdensom" wife, and aint you never going to be a big stout, home wife, able to use the broom stick upon me with effect, or to pitch me out of doors if occasion should require it? And if you aint agoing to be such a giantess in strength and endurance, what did you marry a poor fellow like me for? Dont you know that what I want for a wife is some big, stout, stupid buxom woman, who should be such a prodigy of strength and endurance, as would make her the terror of every form of disease – and I might add – of her husband? Oh Dear! what a worthless little piece of household furniture I am tied up to poor life. Why didnt I stay in the army and get decently killed, instead of coming home to wed myself to such a fate? "Woe is me, for I am undone." You dear darling little chicken, I only wish that in all our future lifes I might never do a single thing to cause you unhappiness or pain, and that I might always be able to maintain you free from care and labor! It is a great gratification to me to be loved as you are loving me, and to have my presence longed for, as it is by you! but dearest I cannot premit your love and your longings for me to interfere with your present happiness. It would spoil much of the satisfaction I have experienced (selfishly no doubt) in being able to place you in a position of comparetive freedom from labor and care, for a short time,- to know that you were so much, and so constantly wishing for me, that it spoiled all of your enjoyments. I must earnestly desire your happiness darling wife, though it may be that I have lived my selfish bachelor life so long and so carelessly, that I cannot, or do not, fully appreciate the wealth of your women's love. I regret to hear that the sick ones, get no better, poor Maria, she is being sorely tried. I think Mr Browne is gaining slowly, but his nerves are so unstrung that he is still unable to write his name. My own health still remains good. So you dont think you will remain a "devoted wife" but little longer? And don't know how you"ll avoid it.- Well I suppose you will be absconding with some nice young man. Dont know how you"ll "manage it" any other way. Please come this way, and bid me good by will you? Oh Dear! I cant write, and dont want to if I could. I just want to take you in my arms and have a good chat, and I want to have that time come when we shall not be compelled to separate any more. Good night, and good by if I do not get a chance to say a word in the morning.

Ned

No. 86 Rec'd Sept 14th 1865.
Ans.

City Point Va Sept 10th 1865

Darling Little Wife.

Your letters of the 3rd, 6th, & 8th were each duly received and for them you have my hearty thanks for they were most welcom. I have been very hard at work all the week. We have some papers which it was desirable to have Maj. Martin sign before he left, and I worked on them what time I could get, night and day until they were done. Major M. left yesterday, and Lt Browne is sole representative of the Comm's Dep't here now. There are but few stores here, and the issues are very small, still there is a good deal to look after and attend. There is a herd of over four hundred cattle, horses, and mules, wagons and harnesses, and a large amount of Commissary Property about which you would know nothing if I should tell you,- all to be watched over and kept from loss. The clerk that I have mentioned once before, remains with us at present, and if he will only stay a Month or so, or until we wind up, all right, but he hates Lt. Browne the worst way, and another officer has spoken to him about coming with him, and I am afraid he will be off in a week or so, and then it will be sharp work for me to "run the machine." If he will only stay we will have it quite easy. Lt. Browne is nearly well again, having improved wonderfully during the past week, though the weather has been intensely hot and almost insupportable. The papers you sent were received today and I have spent a couple of hours in reading them! they are very interesting to me, and bring me all the State news I have at present, for Mr. Cook seems to have forgotten me lately. But I am afraid you are trespassing on George's good nature in using so many of them. I am sorry that you were disappointed in your proposed visit to the Fall's, & to S.C. and still hope that you will be able to make it. I think it is very kind of you to take the trouble to visit my friends for I do not suppose that you would really enjoy it very much. You seem to meet with many disappointments in your plans. I can not but wish, and hope however, that you may never be subjected to any of a more serious nature. Still I suppose it is no trifling matter to get all the "muslins" and "fixings" in order. I am sure I should not think it was if I had it to do. By the way, I am about out of "Muslins." I shall have to come soon or I shall be a veritable "Flora McFlimsey." I would make a Ragman's sign now, and dont know what I"ll do a month from now. My shirts are worn out, stockings used up and lost, handkerchiefs gone to look up the stockings, pants full of holes, and past repairing, pockets all given out, and on the whole I am decidely shabby. I have refrained from buying or sending for anything because I thought each Month would be my last here, and now it is more uncertain than ever, and I want to get along without expending anything for cloths until I get home if it is possible. I am as much surprised as you could possibly have been, to learn of the state of affairs between Lib. and George Olin. I tell you plainly I am sorry, but of course it is of no business of mine and I shall certainly not presume to interfer or offer advice gratis. Olin is a good fellow in his way, and there is really nothing bad about him, that he would make some girl a good husband I have not the least doubt, but it does not seem to me that he is the man for Lizzie, and I hope if the affair has not progressed to far she will refuse his offer – but not a word of this to her – as you love me. But I do not mind you telling her from me if she should ask you what I thought of it – nothing else, that I do hope she will not listen to Auntie's solicitations, but lay them entirely aside, and consult her own heart. I consider her life's happiness of more importance than Aunties house and fields, and I should hate to have her barter the one for the other. The plea that it is necessary to have some one there to attend to the place is in this case simply ridiculous. And if Lib marries George it is not at all certain that he will marry Auntie too, or that he will always remain there and take care of the place. George is very obstinate, and if he took it into his head to go off and take Lib with him, a ten acre lot full of "Aunties" would not prevent him. I fail to see any immediate necessity for Lib's getting married. It does not seem to me that George is at all such a man as she wants and without saying a word to the predjudice of my old friend and comrad. I hope that Lizzy will not marry him!- unless she is satisfied that she really loves him, and feels that he is indispensible to her happiness – and not to the <u>care of the farm</u>. Do you mean to say that Libbie is afraid "she will never have any one else"? I cannot believe you do, for I am loth to think that she considers marrying absolutely indespinsible. And if it is Auntie who is so much worried about it, I should advise Lib. to be fifty old maids, rather than marry just to please a third person. I am afraid that, that rascally little whiffet of Docter has done Lizzy some real injury than we have been aware of. I hope he has got a shrew for a wife who will worry his worthless life out of him. I have not yet written to Mr. Moller but intend to do so immediately now. I do not wish to drag you off so far from home, for it is more than probable that if we go west or South we shall bid good by to New England for ever when we leave. Perhaps if Carl succeeds in finding a situation for me out there, I had better go out first alone, and see for myself what I can do and whether it is best for us to commence our battle of life there. It seems to me that there is no prospect of my being able to get a decent living about home, and if I must dig or grub for it, for you sake I would rather do it somewhere else than in Salisbury. I do not feel like writing more tonight darling, though I have hardly commenced to answer your dear letters. Don't laugh or be vexed at the immense sum of money I inclose. I send it because the currency is new, is more pleasant for a lady to carry and use than the dirty and mutilated in constant use. Dont think you must keep it in your pocket to look at. I suppose you have to buy little "fixens" occasionally.

With love. Ever your devoted husband
Edwin J. Barden

No. 87 Rec'd Sept 21st/65.
Ans.

City Point Va Sept 15th 1865.

Dear Wife.
Your letter of the 10th was recd yesterday, and today that of the 13th and also

the letter from Mr. Moller. Why did you not open Carl's letter? You were perfectly welcom to have done so, and could have re-enclosed it to me yourself. He sent it to you by my direction, for I thought it probable when he left that I might be home by this time. I have not written him a word since he left. Isnt it wicked? I am going to write to him just as soon as I possibly can. His letter was a short one written while stopping over night at a hotel in St. Louis. He had had a comfortable journey thus far, and had there met with an old friend Capt McClure who is going to Ft. Union in Texas, and was going on by way of Ft. Leavenworth, which was very pleasant for both. Am glad that you enjoyed your visit among my friends so well. I feel as though you could not but feel it rather a task to go over to S. Canaan, but I am very glad you went! the more especially as I have only written them one letter since I came away in the Spring. I would not like them to think that I am neglecting them. Did you see, or hear of Fannie Hill? poor girl, I promised faithfully to write to her, and have never done so. Did you see Mr Peck who has been, or rather is, visiting Carrie? Mr. Cook told me about him once – nothing to his disparagement or credit, but Mr C was not favorably impressed with him. I do want her to do well when she marries, for I cant bear to think of her throwing herself away on some worthless city fop. I rather thought when I was home last that she might have George Hall. It would be a safe – if not a brilliant match. I think she would have told me, once, but dont suppose she would hardly trust me with her love secrets now. No matter for that however, I only hope she will do well. I have written to Mr. Randall since receiving your letter, but of course did not refer to the prohibited "Grave" subject. You ask for information as to the probable length of my stay. My dear, I cannot enlighten you much any way. Mr. Browne's regiment was mustered out and went home a fortnight ago, but he has never been notified of it officially, or relieved from duty in the C.S. Dept, or received any intimation that he would be likely to be discharged soon. Matters are progressing smoothly and quietly here now and for aught I can see he may remain a couple of Months yet, and perhaps all Winter. At the same time he may be ousted in a week. It is evident that some Commissary will be kept here probably all Winter, and Browne will be mighty glad to remain as long as they will let him. I feel it is my duty darling, much as I wish to be with you, to remain as long as I can get my present salary. So my dear wife your patience may yet be tried for another Month. I know it is hard but I think it is best. It is almost a year since we were married, and we have been together but very little! we yet hardly know one another as husband and wife. I know it is very trying to you, but I cannot see how I can do better. You must have patience with me a little longer dearest – it may not be as long as we fear. It was a year yesterday since I was mustered out of the service but the longer I stay the less fit I become for any other kind of life. Well my dear I hope you will enjoy your visit at Emma's. I dont expect to "surprise" you there this time, but almost wish I might. I have made a poor answer to your excellent letters, but I am not feeling very bright this evening and trust you will excuse me.

With love darling.
Ever Your devoted husband
Edwin J. Barden

CHAPTER 26

HOME SWEET HOME

From the book:

GOLDSBORO, N.C. AND HOME. JUNE 23 TO JULY 29, 1865.

The stay of the Seventh at Goldsboro from June 7th to July 20th was uneventful. At last on the 20th of July they were mustered out of the service of the United States and returned by rail to City Point and thence by steamer to New Haven. When Generals Terry and Hawley, Maj. Adrian Terry and Capt. E. Lewis Moore learned that the old regiment was on its way home they took a steamer and went to City Point to meet them. There learning that they had gone to Petersburg, they procured a carriage and joined them there, spending the night of the 25th in jovial companionship. On the morning of the 26th the regiment took cars to City Point, and on reaching there at once embarked and steamed for home. General Hawley and Captain Moore accompanied them a few miles on the "Blackbird" and took their final leave.

The regiment proceeded to New Haven under the command of Col. S. S. Atwell where it arrived on the evening of July 29th, and was received by a committee at the wharf, and escorted with the usual triumphal display through illuminated streets to a supper at the State House. Mayor Scranton welcomed the soldiers to the hospitalities of home and Colonel Atwell briefly responded. They then went into camp at Grapevine Point. The work of making out muster and pay rolls and final statements occupied them until August 11th, when they were formally discharged and made their way speedily to their homes.

This history cannot be more appropriately closed than by copying the first and last words of General Grant's final report to the Secretary of War:

Headquarters Armies of the United States.
Washington, D. C. July 22, 1865.
Sir:

I have the honor to submit the following report of the operations of the armies of the United States from the date of my appointment to command the same.

From an early period in the rebellion I had been impressed with the idea that active and continuous operations of all the troops that could be brought into the field, regardless of season and weather, were necessary to a speedy termination of the war. The resources of the enemy and his numerical strength were far inferior to ours, but as an offset to this we had a vast territory, with a population hostile to the government, to garrison, and long lines of river and railroad communications to protect, to enable us to supply the operating armies.

The armies in the East and West acted independently and without concert like a balky team, no two ever pulling together, enabling the enemy to use to great advantage his interior lines of communication for transporting troops from east to west, re-enforcing the army most vigorously pressed, and to furlough large numbers, during seasons of inactivity on our part, to go to their homes and do the work of producing for the support of their armies. It was a question whether our numerical strength and resources were not more than balanced by these disadvantages and the enemy's superior position.

From the first I was firm in the conviction that no peace could be had that would be stable and conducive to the happiness of the people, both North and South, until the military power of the rebellion was entirely broken. I therefore determined, first, to use the greatest number of troops practicable against the armed force of the enemy, preventing him from using the same force at different seasons against first one and then another of our armies, and the possibility of repose for refitting and producing necessary supplies for carrying on resistance; second, to hammer continuously against the armed force of the enemy and his resources until, by mere attrition, if in no other way, there should be nothing left to him but an equal submission with the loyal section of our common country to the constitution and laws of the land. These views have been kept constantly in mind, and orders given and campaigns made to carry them out. Whether they might have been better in conception and execution is for the people, who mourn the loss of friends fallen, and who have to pay the pecuniary cost, to say. All I can say is. that what I have done has been done conscientiously, to the best of my ability, and in what I conceived to be for the best interests of the whole country.

It has been my fortune to see the armies of both the West and the East fight battles, and from what I have seen I know there is no difference in their righting qualities. All that it was possible for men to do in battle they have done. The Western armies commenced their battles in the Mississippi Valley, and received the final surrender of the remnant of the principal army opposed to them in North Carolina. The armies of the East commenced their battles on the river from which the Army of the Potomac derived its name, and received the final surrender of their old antagonist at Appomattox Courthouse, Va. The splendid achievements of each have nationalized our victories, removed all sectional jealousies (of which we have unfortunately experienced too much), and the cause of crimination and recrimination that might have followed had either section failed in its duty. All have a proud record, and all sections can well congratulate themselves and each other for having done their full share in restoring the supremacy of law over every foot of territory belonging to the United States. Let them hope for perpetual peace and harmony with that enemy whose manhood, however mistaken the cause, drew forth such herculean deeds of valor.

I have the honor to be. very respectfully, your obedient servant,

U. S. Grant
Lieutenant-General

A Parting Word to My Brothers in Arms. (from the 1905 book by Stephen Walkley)

Whatever its defects, I think this volume will recall to your minds many things which you had forgotten, for we can forget a great deal in forty years. I hope also that it will tell you some things which you never knew, for the private soldier who does his duty well, especially in the hour of battle, knows only what passes in his immediate presence. I believe the events here narrated will be like hooks on which you have hung past memories, which, when brought to light, will enable you to live over again the stirring years from 1861 to 1865, so that when your little grandchild climbs upon your knee and says. "Grandpa, tell me a story," you will have a story to tell.

In this utilitarian age we like to know what good we have accomplished — what we have to show for those four years of suffering and death. As in a game of chess it will sometimes happen that a single pawn interposed at the right time will save the game, so in the game of war, it may be that a single regiment, standing in the right place at the right time and doing its duty heroically will save a brigade, if not the whole command. Instances are not lacking in which you were privileged to be that lucky pawn.

When at Olustee you stood for three hours and fought superior numbers behind intrenchments without flinching, when at length the charge was made and your unerring fire melted gaps in the charging columns and finally sent their scattered ranks back to their intrenchments, when at last after nearly twenty-four sleepless hours of skirmishing and fighting you safely guarded the rear, you surely kept what was a disastrous attack from becoming a disastrous rout.

The devoted 150 men who on that foggy morning at Drewry's Bluff stood with full magazines in the light intrenchment which you had dug with your knives and plates and kept back many times their number until the rest of our force reached the cover of the woods, did as brave and effective work as Leonidas and his three hundred Spartans at the pass of Thermopylae. It was a great honor to have been killed or captured in that trench.

At Newmarket Road October 7, 1864, it was the deadly fire of your rifles which stopped the rush of Hoke's division flushed with the hope of success, and turned what began as a rout into a victory, establishing a line near Richmond which was never after given up until the city was evacuated.

It is more than possible that between Fort Fisher and Wilmington these same trusty rifles saved a section of the <u>grand old First Heavy Artillery</u> from <u>Libby Prison</u>.

Not to multiply exceptional cases you have the honor shared by more than two thousand other brave regiments; namely, that of being a part of the great whole which saved the Union. You did not fight for war, but for peace. The South tried to separate the states by force of arms. The North had only the choice either to tamely submit, or to resist by the same force. When we remember the bitterness which prevailed on the opposite sides of Mason and Dixon's line in 1860, when we reflect that if our nation had become two, that bitterness would probably have increased, when we compare the strife of 1860 with the harmony of 1905, who is there among us who does not thank God that he was permitted to bear even the slightest part in the war which led to that result? Who does not pray that our grandchildren will be brave enough and good enough to make the Union which we helped to save a benign mother to a hundred million people and a kind friend among the nations of the earth. So, in the words of "Tiny Tim" we say for the South as for the North "God bless us everyone."

Stephen Walkley
PRIVATE, COMPANY A, SEVENTH CONNECTICUT VOLUNTEERS
From his book:
History Of The Seventh Connecticut Volunteer Infantry
Hawley's Brigade, Terry's Division Tenth Army Corps
1861-1865

EPILOG – THE FAMILY
(LINKED TO 'NED' BARDEN)

Readers of this historic account have asked where Edwin and Sarah settled after the war. Based on letters from their son while at West Point in 1894 it appears that they lived in what is now a three-story brick townhouse at 914 P Street N.W., Washington, D.C. while Ned worked for the Treasury Department, Second Auditor's Office. The office name changed in 1894 to the Office of the Auditor for the War Department until abolished in 1921. Ned died 16 June 1902 and was buried across the nearby Potomac river in Arlington National Cemetery. Sarah Barden remained in Washington, D.C. until her death 13 June 1917 and was buried with her husband at Arlington.

Edwin(Ned)Janes Barden (1835-1902) Corporal, Co G, 7th Connecticut Volunteer Infantry Regiment
    Married Sarah Maria Jones (Barden) (1837-1917)

    Their son and only child – Edwin Jones Barden (1870-1942) West Point Class of 1894, Colonel, US Army Corps of Engineers
        Married Claudia Rhett Stuart (1871-1942)

        Their daughter Claudia Stuart Barden (1906-1991) Married William Renwick Smedberg III, Naval Academy Class of 1926, Vice Admiral, US Navy

        Their daughter Claudia Smedberg Adams-Estes (1928-2016)- (book editor) Married Robert Stark Adams (1920-1982) Naval Academy Class of 1943, Captain, US Navy

        Their son (book editor) Robert Stark Adams, Jr. (1951-alive) Naval Academy Class of 1973 COL(SEAL) US Army Medical Corps
            Married Jeri Ann Dull (Adams) (1954-alive)

            Their son Robert Stark Adams III (1983-alive)
                Married Maggie Fitzmaurice (Adams) (1988-alive)
                Their son Robert Stark Adams IV (2019-alive)
                Their daughter Nelson Fitzmaurice Adams (2000-alive)

OF RELATED FAMILY INTEREST (The authors other great-great grandfather in the Union Army on the Smedberg side of the family).

William Renwick Smedberg (1839-1911) Lieutenant Colonel(brevet), 14th US New York Infantry, Union Army Civil War and Colonel California National Guard
    Married Fannie M. Smedberg (1839-1930)

Their son William Renwick Smedberg Jr. (1871-1942) West Point Class of 1894, BGEN US Army – played football with his West Point classmate - Edwin Jones Barden (noted above).

> Edwin Jones Barden married Louise Chaffin Smedberg (1879-1977)
>> Their daughter was Claudia Stuart Barden (noted above) who married William R. Smedberg III (noted above).

AND
Claudia Adams-Estes (editor) had a younger brother Edwin Barden Smedberg, U. S. Naval Academy 1958, CAPT (Navy)
> Edwin Barden Smedberg had a son Edwin Barden Smedberg, Jr.
>> Edwin Barden Smedberg, Jr. had a son Edwin Barden Smedberg III

So, the son and daughter of 1894 West Point Classmates (Edwin Jones Barden and William R. Smedberg, Jr.) married and became this author's grandparents.

And these two 1894 West Point classmates were also both sons of Civil War Union Army soldiers (Edwin Janes Barden and William R. Smedberg) who played together on the first West Point team to beat the Naval Academy in 1891 (a source of great pride to them passed down through the generations).

AND
Claudia Adams-Estes (editor) had another younger brother
> William R. Smedberg IV, RADM (Navy) U.S. Naval Academy 1951 who had a son William R. Smedberg V

AND
Claudia Adams-Estes had two other children
> Claudia Stuart (Adams) Monteith (daughter)and
> William Renwick Smedberg Adams (son)

INDEX

1st Connecticut Heavy Artillery, 535, 622
1st Massachusetts Cavalry Regiment, 157
1st US Colored Infantry Regiment North Carolina, 324–325
2nd Connecticut Heavy Artillery Regiment, 1, 543, 612
3rd New Hampshire Infantry Regiment, 1, 160, 161, 193
3rd Rhode Island Regiment, 138, 196
4th New Hampshire Infantry Regiment, 208
4th New Jersey Light Artillery Battery, 428
5th Virginia Infantry Regiment (Confederate), 633
6th Cavalry Regiment, 17
6rh Connecticut Infantry Regiment, 1, 137, 138, 315, 339, 378, 468, 558
6th Corps, 558, 571
7th Connecticut Volunteer Infantry Regiment, 1, 51, 52, 59, 126, 127, 138, 192, 250, 309, 348, 430, 468, 601
7th New Hampshire Infantry Regiment, 1, 264, 315
8th Maine Infantry Regiment, 179, 258
8th Michigan Infantry Regiment, 123, 126, 127, 131, 203, 245
9th Maine Infantry Regiment, 227, 228, 231, 339, 392
10th Army Corps, 222, 424, 439
10th Connecticut Volunteer Infantry Regiment, 449
10th Infantry Regiment (Regulars), 487
17th Connecticut Volunteer Infantry Regiment, 339
18th Corps, 466
21st South Carolina Infantry Regiment, 339
24th Massachusetts Infantry Regiment, 337, 348
25th Army Corps (Union—Colored), 579, 588
28th Connecticut Volunteer Infantry, 214
28th Massachusetts Infantry Regiment, 126, 127, 131, 137
46th New York Infantry Regiment, 107, 124, 127, 218
47th New York Infantry Regiment, 467
47th Pennsylvania Infantry Regiment, 183, 193, 214
48th New York Infantry Regiment, 112–113, 315, 348
48th Pennsylvania Regiment, 179
54th Massachusetts (colored regiment), 312, 315
55th Pennsylvania Regiment., 138, 216
61st North Carolina Infantry Regiment, 338
76th New York Regiment, 42
76th Pennsylvania Regiment, 170
79th New York Regiment, 126–127
97th Pennsylvania Infantry Regiment, 197, 203, 348
100th Pennsylvania Regiment, 127
112th New York Infantry Regiment, 389
117th New York Volunteers, 356
169th New York Volunteers, 389
176th Pennsylvania Regiment, 328

A

Abbott, John C. (BGEN) (brevet), 493, 513, 540, 547
Adams Express, 203, 266
Alabama, CSS (ship), 451
Ambratype (ambrotype) definition, 103, 137, 231, 531, 537
Annapolis, Maryland, 1, 23, 25, 60, 158, 191
"Annie Lisle" song, 83
Arago, SS (steamer), 186, 262, 306, 308, 310, 313, 317, 318, 324, 328, 359, 371, 384, 385, 405, 408, 417, 414, 428
Army of the James, 542, 547, 558
Army of the Potomac, 360, 372, 410, 621
Augusta, USS (armed transport), 493

B

Banks, Nathaniel P. (MGEN), 157, 184, 201, 211, 212
Barden, Jesse (a brother), 1
Barrington, New Hampshire, 527, 533
Barton, William Brainerd (COL), 315

Battery Halleck, 78
Battle of Chantilly, 130
Battle of Drewery's Bluff, 429
Battle of Olustee, 411
Battle of Pocotaligo, S.C., 193, 208, 256
Battle of Secessionville, 127
Baynard Plantation, South Carolina, 162, 164
Beaufort, South Carolina, 39, 54, 136, 147, 159, 169, 181, 199, 201, 206, 212, 215, 217, 222, 223, 225, 229, 230, 237, 256, 262, 268, 278, 309, 315, 324, 345, 348, 360, 362, 364, 368, 370, 378, 383, 387, 397, 419, 448
Beauregard, P.G.T. (GEN, Confederate), 93, 330
Belle Isle, Virginia, 576
Ben D. Ford (steamer), 91, 138, 143, 179, 196, 234, 377, 378, 397, 401
Benham, Henry W. (MGEN, brevet), 41, 109, 418, 484
Bermuda Hundred Campaign (Virginia), 427, 446, 511, 532, 545, 548
Bermuda line, 502
Bienville, USS, 212
bilious intermittent fever (yellow fever), 196
Birney, David B. (MGEN), 455, 462, 467, 469, 474
Bissell, Lucius W. (DR), 351
Bluffton, South Carolina, 49, 172
Boat Howitzers, 184
Boston, USS, 179, 184, 185, 193, 239, 256, 257, 281, 288, 304, 312
Braddock Point (South Carolina), 34, 35, 57
Bragg, Edward S. (BGEN), 372–373, 513
Brannan, John Milton (BGEN), 169, 192, 193, 205, 222
Buckingham, William Alfred (Connecticut Senator), 2, 5, 271, 279, 292, 410, 474
buckskin gauntlets, 509
Buell, Don Carlos (MGEN), 93
Bull Run, First Battle of, 5, 21, 145
bullets, poisoned, 325
Burdick, James H. (CPT), 318
Burnside, USS, 261, 264, 265, 266, 439
Burrall, George, 422, 480
Butler, Benjamin F. (MGEN), 169, 439, 449, 455, 493, 494, 52, 529

C

Camp Distribution, 453
Camp Finnegan, 413
Camp Kettle newsletter, 39
Camp Misery, 118
Camp of the 7th Reg. C.V., 51, 52, 59, 71, 74, 77, 78, 81, 85
Camp Palmer, 198, 200
Camp Starr, Fernandina, Florida, 248
canon
    Columbiad, 40, 44, 61, 71, 89, 138, 183
    James rifled, 71, 124
Canonicus, USS, 365
Carr, Eugene Asa (MGEN), 585, 591
Carroll, Samuel S. (BGEN, brevet), 140
Castle Thunder prison, 574, 575
Chaffin's Bluff, Henrico County, Virginia, 443, 462, 572
Chamberlain, Valentine B. (CAPT), 318
Chatfield, John Lyman (COL, 6, 312, 315
City Point, Virginia, 20, 435, 449, 457, 477
Collins line ships, 398
Columbiad cannon, 8 inch and 10 inch, 40, 44, 61, 71, 89, 138, 183
Commissary of Subsistence, Office of, 305, 440
commodious definition, 141, 268
confinement at the Tortugas, 246
Copperheadism definition, 319
Cosmopolitan (steamer), 116, 117, 137, 138, 143, 179, 185, 240, 309, 315

D

Dahlgren, John A. (ADM), 285, 311, 335
Dawfuskie Island, Georgia, 13, 60, 61, 62, 75, 76, 102, 108, 162, 165
Dead march music, 34, 66, 202, 253
Deane, James (REV MAJ), 10, 173
Deane, John M. (MOH), 10, 153, 156, 173, 422, 553, 612
Deep Bottom Virginia, Second Battle of, 460, 461, 462, 464
Delaware, USS, 137, 228, 236, 237, 256
Dempsey, Robert (LT), 304, 412
Department of the South (Union), 1, 41, 285, 344, 348
Depot Hospital, Point of Rocks, Virginia, 450

Dovers powder, 366
Drewery's Bluff, Battle of, 429
DuPont, Samuel Commodore, 262, 330
Dutch Gap, Virginia, 462, 533, 535, 593

E
Eaton, Amos B. (MGEN, brevet), 315, 481
Edistoe Island, South Carolina, 113, 117, 122, 125, 131, 136, 137, 141, 143, 145, 148, 158, 245, 398
Ellsworth, Elmer (LT), song tribute, 83
encomiums definition, 139
Errickson (steamer), 179

F
Fair Oaks and Darbytown Road, Battle of, 489
Falls Village, Canaan (Connecticut), 61, 468
Farris (Farrars) Island, 443
felon definition, 604
Fenton, William M. (COL), 123, 149, 245
Fernandina, Florida, 208, 227, 230, 232, 236, 241, 248, 255, 274, 279, 290
Ferry, Orris S. (MGEN, brevet), 262, 462, 467
Fingal, SS, 97
Finnegan, Joseph (BGEN; Confederate), 230
Folly Island, South Carolina, 317, 323, 329, 348, 355, 357, 360, 369, 371, 374, 401
foolscap definition—sheet of foolscap, 519
Fort Anderson and Wilmington, 540, 544
Fort at Bay Point, 197, 306
Fort Beauregard, 2, 29
Fort Clinch, 230, 280
Fort Darling, 429, 572, 578
Fort Fisher, 528–529, 622
Fort Gregg, 339, 345
Fort Johnson, 330, 337, 392
Fort Leavenworth, 609
Fort Marion, 285, 287
Fort Monroe, 25, 26, 29, 100, 155, 166, 427, 448, 450, 476, 524, 579, 582
Fort Moultrie, 116, 341, 352
Fort Pulaski, 40, 42, 43, 44, 45, 50, 56, 61, 76, 87, 94, 97, 100, 102, 104, 106, 108, 110, 112, 162, 166, 209, 210, 296
Fort Royal, Battle of, 29
Fort Sumter, 116, 136, 263, 315, 330, 335, 337, 342, 378
Fort Union, 550
Fort Wagner, 309, 313, 316, 332, 333, 335, 337, 338, 339, 342
Fort Walker, 2, 29, 33
Fort Wells, 37
Foster, Robert Sanford (BGEN), 291, 364, 373, 386, 459
Fredricksburg occupation, 214
Fulton, USS, 309, 317, 321, 327, 366–367

G
Galena, USS (gunboat), 593, 594
General Hospitals at Fort Monroe, 450
George McClellan (steamship), 91
George Peabody, USS (transport ship), 207
George Washington Parke Custis, USS, 263
Gideonite definition, 262
Gillotts' Number 3 pen, 170
Gilmore, Quincy Adams (MGEN), 40, 41, 310, 313, 317, 337, 397, 420, 446, 449
Gordon, George Henry (MGEN), 386
Gould, Charles G. (CPT, 5th VT Vol), 413
Grahams' plantation, South Carolina, 161
Grain Boot, 506
Grant, H. Ulysses (GEN), 410, 436, 449, 484, 508, 519
Grant's headquarters, 449, 477
Gregory, John (MD), 339

H
Halleck, Henry W. (MGEN), 576
Hamilton (CAPT), 386
Hamilton's Light Battery, 136, 386
Hatchers, Virginia, 431
Havelock, Henry (MGEN), 224
Havelock—definition, 74
Hawley, Joseph Roswell (MGEN, brevet, and 42nd Gov. of Connecticut), 5, 6, 35, 150
Hilton Head, South Carolina, 30, 37, 39, 44, 45, 49, 57, 67, 74, 75, 76, 91, 98, 131, 133, 136, 139, 141, 143, 147, 149, 151, 154, 155, 157, 159, 161, 164, 165, 167, 171, 175, 176, 178, 182, 187, 189, 190, 192, 193, 196, 208, 225,

228, 240, 250, 262, 264, 265, 267, 279, 296, 298, 304, 305
Hine, Peter E. (MD), 221
Hitchcock, Edwin S. (CPT), 19, 22, 178
Hogshead definition, 308, 519
Holley, Alexander H. (Gov. of Connecticut), 576
Holley, Maria (Gov. Holley's daughter), 152
Hospital Morris Island 7th Connecticut Volunteers, 344, 347
Housatonic, USS (gunboat), 412
Housatonic railroad, 608
Howlett Battery, Virginia, 463, 571
Humphreys Station, the end of the Military R.R., 59
Hunter, David R. "Black Dave" (MGEN), 41, 148, 154, 245, 262
Hunter's Regiment—former slaves, 154

I
Illinois, SS (steamship)
Irish waiters in hospitals, 175
Ironsides, USS, 321

J
Jackson, James S. (BGEN), 157
James Island, South Carolina, 128, 132, 135, 158
James Island campaign, 378
James rifled canon, 71, 124
James River, 572
John Adams, USS (gunboat), 258
Johnston, Albert Sidney (MGEN, Confederate), 456

K
Keokuk, USS (steamship), 265, 330
Kilpatrick, Hugh Judson (MGEN, brevet), 603
Kilpatrick's Cavalry, 603

L
Lakeville, Massachusetts, 293, 320, 339, 340
Lamar, Thomas J. (COL, Confederate), 126
Landon, Ashbel, 520–521
Legarsville, South Carolina, 377
Libby prison, 572, 622
Light House inlet, 354, 378

Loco-focos party, 253, 277

M
malaria, 221
Marion, SS (steamship), 42–43
Marshall, Humphrey (BGEN, CSA), 464
Mary Benton, USS, 315
May Port Mills, Florida, 186
McClellan, George C. (MGEN), 110, 141, 142, 145, 146, 208, 261
McClure, Charles (COL), 487, 524
Mead, George G. (MGEN), 484
measles, 99
Meigs, Montgomery C. (MGEN), 481
Meridian Hill, Washington, DC, 20
Milburns, William H. (REV), 142
Military railroad, 559
Mills, Charles C. (CPT), 432, 434, 435
Mitchell, Ormsby (MGEN), 175, 197
mixing quinine, 449
Mohawk, USS (gunboat), 248
Monitor, USS, 533
Monohunsett, SS (steamer), 378
Montauk (ironclad), 265
Moore, E. Lewis (CPT), 500, 527, 610
Morgan, Michael Ryan (COL), 440, 484, 495, 497
Morris Island (South Carolina), 317, 380
mortar shells, 41, 121
Moultreville, South Carolina, 341
Mount Prospect, Maryland, 166
muskets, smooth bore, 202

N
nabobs, definition, 115
Nantasket, USS, 368
Neptune, USS, 271
New Haven, Connecticut, 14
"nutmeg" State, 208

O
oath of loyalty for Confederate troops, 588
Olin, George, 200
Olmstead (COL), 41
Olustee, Battle of, 411
Ossabaw Island, 257

P

Palmetto Herald (newspaper), 422
Parrot guns, 124, 193, 368, 463
Paul Jones, USS, 138
Pawnee, USS, 67, 138
Pawne Landing (Folly Landing), 401
Petersburgh and Richmond railroad, 428, 469, 470
Petersons Magazine, 213
pettifogging, definition, 529
Pierpont, William H. (CPT), 3
Pinckney Island, South Carolina, 160, 161
Planter, USS (gun boat), 139, 225
Pocotaligo, S.C., Battle of, 193, 208, 256, 339
Point of Rocks, Virginia, 465
Pope, John (MGEN), 108
Port Royal, South Carolina, battle of, 2
Port Royal Island, 215
Port Royal Ferry, 211, 215
Porter, George L. (MD), 449
Portuguese hymn, 253
Price (ship), 368
Puttnam, Haldimand S. (COL, brevet), 315

Q

quinine, 449
quinsy, 551

R

Reid, Adam (REV), 472
Rhode Island light artillery battery, 157
Robbers Row, South Carolina, 262, 367, 404
"Robin Adair" song, 505
Rockwell, Alfred P. (BGEN, brevet), 126, 493
Rodman, Daniel C. (BGEN, brevet), 230, 309, 318, 363
Rosecrans, William S. (MGEN Confederate), 350
Ruby, SS (blockade runner), 330

S

Sabine, Thomas T. (MD), 65
Salisbury, North Carolina, 469
Sammons, Simeon (COL), 412
Sanford, Oliver S. (MAJ), 41, 364
Sanitary and Christian Commissions, 448
scapegrace, definition, 600
Seabrook plantation, 141
Sealy, Israel R. (CPT), 324
secessh, definition, 91, 103
Secessionville (Ft. Lamar), 126, 330
Secessionville, Battle of, 126, 127, 330
Second Battle of Deep Bottom, Virginia, 460, 464
Sewells Point, Virginia
Seymour, Horatio (NY Gov), 201, 314, 319, 423
Seymour, Truman (MGEN, brevet), 222, 315, 368
Shaw, Robert Gould (COL), 315
Sheridan, Philip H. (GEN), 548
Sherman, William Tecumseh (GEN), 1
Skinner, Benjamin F. (CPT), 265
Smith, Kirby (GEN), 580
Smith, William F. (MGEN), 439
smoking cap, 24
South Side railroad, 560, 599
Spencers breech loading rifle, 365
Spotswood House, 575
Sprague, William (RI SENATOR), 451
S.R. Spaulding, SS (steamer transport), 424
St. Augustine, Florida, 285, 310, 410
St. Helena Island, 405, 413
St. Johns Bluff, 186
St. Johns island, 213
St. Johns river, 184
Stanton, Edwin (Secretary of War), 481
Stephens, Alexander H. (V.P. Confederacy), 526
Sterlings Ambrosia, 524
Stevens, Clement H. (BGEN, CSA), 115
Stevens, Isaac (MGEN), 130
Stono River, 235, 377, 398
Strong, George Crockett (BGEN), 315
Suttlers stores, 367, 384
Swamp Angel battery, 350

T

Tattoo, military, 227
Terry, Alfred H. (MGEN), 1, 5, 6, 108, 150, 177, 192, 208, 386, 428
Three months men, 1, 21, 99, 149
Tompkins, Charles H. (BGEN, brevet), 222
Tortugas confinement, 246, 250

Tourtellotte, John Eaton (BGEN, brevet), 318
Townsend, James Matthew, 36, 384
Treasury Department, Second Auditor's Office, 624
Tredegar works, 576
Turner, John Wesley (BGEN), 337, 386, 462
Turner, Richard "Dick," 574
Tybee Island, Georgia, 40, 42, 44, 52–53, 58, 71, 74, 77, 78, 81, 85, 108, 158, 240, 468

U
Uncle Gideon, 451

V
Vandalia (sailing ship), 67
Vodges, Israel (BGEN), 386

W
Wabash, USS, 31
Walker, William S. (COL, CSA), 192
Walkley, Stephen (PVT, 1905 book author: 7th Connecticut history), 3, 4, 5
Wallack's Theatre, 478
Weehawken, USS, 340, 387
Weldon Railroad, 470
Welles, Gideon (Sec. Navy), 451
Wells, George L. (PVT), 436, 488
Wiley, Daniel Day (BGEN), 571
Wilmington Expedition, 533
Wilson, Henry (SENATOR), 451
Wilson, James H. (MGEN), 41, 318
Winfield Scott, SS (steamer), 39
Wolcott, Samuel W., 24, 95, 340, 357, 441, 464
Wood, Charles A. (2LIEUT), 256, 431
Woods, Charles R. (MGEN, brevet), 367
Wright, Horatio B. (BGEN), 1, 139
Wyandotte (sloop of war), 67

Y
yellow fever (bilious intermittent fever), 196
Yulee, David L. (FLA Senator), 230

ABOUT THE EDITORS

Claudia Adams-Estes and her son Robert Adams, Jr. carefully transcribed each letter as written. This labor of family and history took years to complete due to the death of Claudia and passage of the letters to the next generation.

Claudia was married to CAPT (Navy)Robert Adams, U.S. Naval Academy Class of 1943. Following his death she married MGEN (Air Force) Howard Estes, U. S. Naval Academy class of 1950.

She is the daughter of VADM (Navy) William R. Smedberg III. VADM Smedberg, U.S. Naval Academy Class of 1926, and he also served as Superintendent of the Naval Academy.

Claudia's brother CAPT (Navy) Edwin Barden Smedberg, Naval Academy Class of 1954, had a son Edwin Barden Smedberg, Jr. who had a son Edwin Barden Smedberg, III. All were named after Edwin Barden the writer of these letters.

COL (Army) Robert Adams, MD, U.S. Naval Academy Class of 1973, served 18 years in the Navy including 12 years in the Navy SEAL Teams. He then changed uniforms to go to medical school on an Army scholarship and served 18 years in the Army retiring as a Medical Corps Colonel. He then founded an award-winning family practice clinic in Knightdale, North Carolina.

He is married to the love of his life Jeri D. Adams who is a proud descendant of Charles Stoneburner who served with the Confederate Army in Company C, <u>5th Virginia Regiment</u>, and was Post Commander of the Staunton, Virginia Stonewall Jackson Camp #25.

www.ingramcontent.com/pod-product-compliance
Lightning Source LLC
Chambersburg PA
CBHW060357010526
44109CB00051B/2480